T0336674

Information Systems Research and Exploring Social Artifacts:

Approaches and Methodologies

Pedro Isaias
Universidade Aberta (Portuguese Open University), Portugal

Miguel Baptista Nunes
University of Sheffield, UK

Managing Director: Lindsay Johnston
Editorial Director: Joel Gamon
Book Production Manager: Jennifer Romanchak
Publishing Systems Analyst: Adrienne Freeland
Development Editor: Christine Smith
Assistant Acquisitions Editor: Kayla Wolfe
Typesetter: Travis Gundrum
Cover Design: Nick Newcomer

Published in the United States of America by
Information Science Reference (an imprint of IGI Global)
701 E. Chocolate Avenue
Hershey PA 17033
Tel: 717-533-8845
Fax: 717-533-8661
E-mail: cust@igi-global.com
Web site: http://www.igi-global.com

Copyright © 2013 by IGI Global. All rights reserved. No part of this publication may be reproduced, stored or distributed in any form or by any means, electronic or mechanical, including photocopying, without written permission from the publisher. Product or company names used in this set are for identification purposes only. Inclusion of the names of the products or companies does not indicate a claim of ownership by IGI Global of the trademark or registered trademark.

Library of Congress Cataloging-in-Publication Data

Information systems research and exploring social artifacts: approaches and methodologies / Pedro Isaias and Miguel Baptista Nunes, editors.
p. cm.
Includes bibliographical references and index.
Summary: "This book discusses the approaches and methodologies currently being used in the field of information systems viewed through socio-technical aspects of the design of IS artifacts as well as the study of their use"--Provided by publisher.
ISBN 978-1-4666-2491-7 (hardcover) -- ISBN 978-1-4666-2492-4 (ebook) -- ISBN 978-1-4666-2493-1 (print & perpetual access) 1. Information storage and retrieval systems--Research. 2. Information storage and retrieval systems--Social aspects. I. Isaias, Pedro. II. Nunes, Miguel Baptista.
Z699.5.R47I54 2013
025.04072--dc23
2012023140

British Cataloguing in Publication Data
A Cataloguing in Publication record for this book is available from the British Library.

All work contributed to this book is new, previously-unpublished material. The views expressed in this book are those of the authors, but not necessarily of the publisher.

Editorial Advisory Board

Regina Connolly, *Dublin City University, Ireland*
Gabriel J. Costello, *Galway-Mayo Institute of Technology, Ireland*
Rachel Harrison, *Oxford Brookes University, UK*
Janet Holland, *Emporia State University, USA*
Hannakaisa Isomäki, *University of Jyväskylä, Finland*
Tomayess Issa, *Curtin University, Australia*
Clare Martin, *Oxford Brookes University, UK*
Guo Chao Peng, *University of Sheffield, UK*
Junko Shirogane, *Tokyo Woman's Christian University, Japan*

List of Reviewers

Fenio Annansingh, *University of Plymouth, UK*
Roger Blake, *University of Massachusetts, USA*
Mike Brayshaw, *University of Hull, UK*
Cheryl Brown, *University of Cape Town, South Africa*
Si Chen, *University of Sheffield, UK*
Mario Covarrubias, *Politecnico di Milano, Italy*
Bruno Delgado, *University ORT Uruguay, Uruguay*
Jan Derboven, *Centre for User Experience Research (CUO), Belgium*
Helena Garbarino, *University ORT Uruguay, Uruguay*
Roelien Goede, *North-West University, South Africa*
Steven Gordon, *Babson College, USA*
Jorge Tiago Martins, *University of Sheffield, UK*
Elspeth McKay, *RMIT University, Australia*
Reza Mojtahed, *University of Sheffield, UK*
Sergio Lujan Mora, *University of Alicante, Spain*
Julius T. Nganji, *University of Hull, UK*
Sara Pifano, *Universidade Aberta, Portugal*
G. Shankaranarayanan, *Babson College, USA*
Johanna Silvennoinen, *University of Jyväskylä, Finland*

Nicky Sulmon, *Centre for User Experience Research (CUO), Belgium*
Estelle Taylor, *North-West University, South Africa*
Alessandro Trivilini, *University of Applied Science of Southern Switzerland (SUPSI), Switzerland*
Elena Zaitseva, *University of Zilina, Slovakia*
Lihong Zhou, *Wuhan University, China*

Table of Contents

Chapter 6
Taxonomy of IT Intangible Assets for Public Administration Based on the Electronic Government Maturity Model in Uruguay 103
Helena Garbarino, Universidad ORT Uruguay, Uruguay
Bruno Delgado, Universidad ORT Uruguay, Uruguay
José Carrillo, Universidad Politécnica de Madrid, Spain

Section 2
Interpretivist and Social Constructivist Approaches

Chapter 7
Online Ethnographies........ 124
Hannakaisa Isomäki, University of Jyväskylä, Finland
Johanna Silvennoinen, University of Jyväskylä, Finland

Chapter 8
Grounded Theory in Practice: A Discussion of Cases in Information Systems Research........ 142
Jorge Tiago Martins, The University of Sheffield, UK
Miguel Baptista Nunes, The University of Sheffield, UK
Maram Alajamy, The University of Sheffield, UK
Lihong Zhou, Wuhan University, China

Chapter 9
Creating Interpretive Space for Engaged Scholarship 161
Gabriel J. Costello, Galway-Mayo Institute of Technology, Ireland
Brian Donnellan, National University of Ireland – Maynooth, Ireland

Chapter 10
Exploring Higher Education Students' Technological Identities using Critical Discourse Analysis 181
Cheryl Brown, University of Cape Town, South Africa
Mike Hart, University of Cape Town, South Africa

Section 3
Case Study Approaches

Chapter 11
The Role of Case-Based Research in Information Technology and Systems........ 200
Roger Blake, University of Massachusetts Boston, USA
Steven Gordon, Babson College, USA
G. Shankaranarayanan, Babson College, USA

Chapter 12
Mapping Participatory Design Methods to the Cognitive Process of Creativity to Facilitate Requirements Engineering 221
Nicky Sulmon, Centre for User Experience Research (CUO), Belgium
Jan Derboven, Centre for User Experience Research (CUO), Belgium
Maribel Montero Perez, ITEC – Interdisciplinary Research on Technology, Education, and Communication, Belgium
Bieke Zaman, Centre for User Experience Research (CUO), Belgium

Section 4
Mixed-Methods Approaches

Chapter 13
Combining Research Paradigms to Improve Poor Student Performance 243
Roelien Goede, North-West University – Vaal Campus, South Africa
Estelle Taylor, North-West University – Potchefstroom Campus, South Africa
Christoffel van Aardt, Vaal University of Technology, South Africa

Chapter 14
Experiences in Applying Mixed-Methods Approach in Information Systems Research 266
Guo Chao Peng, University of Sheffield, UK
Fenio Annansingh, University of Plymouth, UK

Chapter 15
Examining Web 2.0 E-Learning Tools: Mixed Method Classroom Pilot 294
Janet Holland, Emporia State University, USA
Dusti Howell, Emporia State University, USA

Chapter 16
Evaluation of Web Accessibility: A Combined Method 314
Sergio Luján-Mora, University of Alicante, Spain
Firas Masri, University of Alicante, Spain

Section 5
Evaluation Research Approaches

Chapter 17
Information Systems Evaluation: Methodologies and Practical Case Studies 333
Si Chen, University of Sheffield, UK
Nor Mardziah Osman, University of Sheffield, UK
Guo Chao Peng, University of Sheffield, UK

Chapter 18
Supporting Unskilled People in Manual Tasks through Haptic-Based Guidance 355
Mario Covarrubias, Politecnico di Milano, Italy
Monica Bordegoni, Politecnico di Milano, Italy
Umberto Cugini, Politecnico di Milano, Italy
Elia Gatti, Politecnico di Milano, Italy

Chapter 19
Integrated Methods for a User Adapted Usability Evaluation 379
Junko Shirogane, Tokyo Woman's Christian University, Japan
Yuichiro Yashita, Waseda University, Japan
Hajime Iwata, Kanagawa Institute of Technology, Japan
Yoshiaki Fukazawa, Waseda University, Japan

Chapter 20
A Protocol for Evaluating Mobile Applications.. 398
Clare Martin, Oxford Brookes University, UK
Derek Flood, Dundalk Institute of Technology, Ireland
Rachel Harrison, Oxford Brookes University, UK

Compilation of References 415

About the Contributors 458

Index 468

Detailed Table of Contents

This chapter aims to demonstrate how the online survey tool was used during the PhD and postdoctoral research stages to evaluate and assess the New Participative Methodology for Developing Websites from the Marketing Perspective. After reviewing techniques from numerous disciplines for methodology integration, a new methodology was produced and an online survey to assess the methodology was completed by website industry participants and information systems professionals in Western Australia. A new online survey was developed on the basis of the PhD results and 210 Web developers and information technology professionals from Australia assessed the new methodology. This chapter will discuss three factors: the advantages and disadvantages of using an online survey in helping to facilitate high quality research and an overview of the functionality of the online survey tool(s) from a researcher's point of view. Finally, a practical set of recommendations and endorsements is presented regarding the adoption of an online survey from the researcher's standpoint.

Many studies have raised awareness of the importance of trust in the online commercial environment, and it is widely acknowledged. Due to the international nature of eCommerce, it is likely that the influence of culture may extend to online consumers' trust responses. In this chapter, a trust measurement instrument that had been previously validated in Hong Kong was applied in both the United States and in Ireland—countries that differ in terms of individualism, uncertainty avoidance, and power-distance. Survey methodology was used to collect data. The results provide a refined understanding as to the influence of national culture on the generation of online consumers' trust beliefs. In doing so, they advance the understanding of information systems and diffusion researchers as well as contributing to the understanding of online vendors who seek to gain insight into the factors that can engender consumer trust in their websites.

Pedro Isaías, Universidade Aberta, Portugal
Sara Pífano, Universidade Aberta, Portugal
Paula Miranda, Polytechnic Institute of Setúbal, Portugal

In a research project, the selection of the sample method is crucial, since it has repercussions throughout the entirety of the study. It determines how the population under scrutiny will be represented and with what accuracy. Hence, it has an important impact in terms of the reliability and validity of the research in general, and consequently, its conclusions. This chapter aims to explore snowball sampling as a chain-referral sampling method. An introductory review of the relevant literature highlights its main characteristics, benefits, and shortcomings, and provides a broader insight to circumstances where it can be successfully applied. This theoretical prologue is followed by the analysis of its employment in an online questionnaire and the presentation of the lessons learned from this sampling decision.

Reza Mojtahed, The University of Sheffield, UK
Guo Chao Peng, The University of Sheffield, UK

Explaining the factors that lead to use and acceptance of Information Technology (IT), both at individual and organizational levels, has been the focus of Information Systems (IS) researchers since the 1970s. The Technology Acceptance Model (TAM) is known as such an explanatory model and has increasingly gained recognition due to its focus on theories of human behaviour. Although this model has faced some criticism in terms of not being able to fully explain the social-technical acceptance of technology, TAM is still known as one of the best IS methodologies that contribute greatly to explain IT/IS acceptance. It has been widely used in different areas of IS studies, such as e-commerce, e-business, multimedia, and mobile commerce. This chapter discusses, describes, and explains TAM as one of the well-known information system research models and attempts to demonstrate how this model can be customised and extended when applyied in practice in IS research projects. In order to illustrate this, the chapter presents and discussed two case studies, respectively, applying TAM in the areas of mobile banking and mobile campus in the UK. It is also proposed that comparing with the traditional questionnaire approach, mixed-methods designs (that contain both a quantitative and a qualitative component) can generate more meaningful and significant findings in IS studies that apply the TAM model. The practical guidance provided in this chapter is particularly useful and valuable to researchers, especially junior researchers and PhD students, who intend to apply TAM in their research.

Julius T. Nganji, University of Hull, UK
Mike Brayshaw, University of Hull, UK

The exploration of social artifacts for the disabled is an important and timely issue. The affordances of new technologies like the Semantic Web allow more intelligent handling of educational learning resources that open up the potential of personalisation of services to individuals. Contemporary legislation calls for "reasonable adjustments" and "reasonable accommodation" to be made to services in order to accommodate the needs of disabled people. Here, the authors examine, from a design perspective, how this might be done in the context of higher education. Specifically, they advocate a design based upon

Chapter 8
Grounded Theory in Practice: A Discussion of Cases in Information Systems Research 142
Jorge Tiago Martins, The University of Sheffield, UK
Miguel Baptista Nunes, The University of Sheffield, UK
Maram Alajamy, The University of Sheffield, UK
Lihong Zhou, Wuhan University, China

A growing number of Information Systems (IS) research is drawing upon Grounded Theory (GT), as an inductive research methodology particularly suited to the development of theory that is grounded in rich socio-technical contexts that are understood through the analysis of data collected directly from those specific environments. Studies in IS are eminently applied research projects; therefore, GT seems the ideal approach to enable rich understandings and inductive explanatory theories on the socio-technical human activity systems studied by the discipline. Nevertheless, GT is still an underutilised methodology in IS and several scholars critique the scarcity of theories developed specifically to account for IS phenomena. This chapter makes a contribution to IS research through providing a discussion of GT's application in the field by presenting an outline of the method's modus operandi and a case-based overview of its use in three different IS research projects in the fields of patient knowledge sharing in healthcare environments, IS strategic planning in academic libraries, and E-Learning adoption. As the potential of GT for IS research remains to be maximized, the objective of this chapter is to counter the lack of explicit and well-informed views of GT use in this discipline, whilst sharing lessons learned from the use of the method in context-dependent interpretive studies.

Chapter 9
Creating Interpretive Space for Engaged Scholarship 161
Gabriel J. Costello, Galway-Mayo Institute of Technology, Ireland
Brian Donnellan, National University of Ireland – Maynooth, Ireland

The purpose of this chapter is to argue that the approach of engaged scholarship provides interpretive space for practitioners who are introducing change in their organization. In this case, the change involved implementation of process innovations, which continue to be an important challenge for business and public sector bodies. The research domain was a subsidiary of APC by Schneider Electric located in Ireland and involved a two-year study where the principal researcher had the status of a temporary employee. A new form of Action Research (AR) called dialogical AR was tested in this study. Key finding from an analysis of the interviews showed that the approach was both helpful and stimulating for the practitioner.

Chapter 10
Exploring Higher Education Students' Technological Identities using Critical Discourse Analysis 181
Cheryl Brown, University of Cape Town, South Africa
Mike Hart, University of Cape Town, South Africa

This chapter applies a critical theory lens to understanding how South African university students construct meaning about the role of ICTs in their lives. Critical Discourse Analysis (CDA) has been used as a theoretical and analytical device drawing on theorists Fairclough and Gee to examine the key concepts of meaning, identity, context, and power. The specific concepts that inform this study are Fairclough's three-level framework that enables the situating of texts within the socio-historical conditions and context that govern their process, and Gee's notion of D(d)iscourses and conceptualization of grand societal "Big C" Conversations. This approach provides insights into students' educational and social identities and the position of globalisation and the information society in both facilitating and constraining students' participation and future opportunities. The research confirms that the majority of students regard ICTs as necessary, important, and valuable to life. However, it reveals that some students perceive themselves as not being able to participate in the opportunities technology could offer them. In contrast to government rhetoric, ICTs are not the answer but should be viewed as part of the problem. Drawing on Foucault's understanding of power as a choice under constraint, this methodological approach also enables examination of how students are empowered or disempowered through their Discourses about ICTs.

Section 3
Case Study Approaches

Chapter 11
The Role of Case-Based Research in Information Technology and Systems 200
Roger Blake, University of Massachusetts Boston, USA
Steven Gordon, Babson College, USA
G. Shankaranarayanan, Babson College, USA

Increasingly, academic research in Information Technologies and Systems (ITS) is emphasizing the application of research models and theories to practice. In this chapter, the authors posit that case-based research has a significant role to play in the future of research in ITS because of its ability to generate knowledge from practice and to study a problem in context. Understanding context—social, organizational, political, and cultural—is mandatory to learning and effectively adopting best practices. The authors describe some examples of case-based research to highlight this point of view. They further identify key topics and themes based on examining the abstracts from prominent case-based research over the past decade, analyzing their trends, and hypothesizing what the role of case-based research will be in the coming decades.

Chapter 12
Mapping Participatory Design Methods to the Cognitive Process of Creativity to Facilitate Requirements Engineering 221
Nicky Sulmon, Centre for User Experience Research (CUO), Belgium
Jan Derboven, Centre for User Experience Research (CUO), Belgium
Maribel Montero Perez, ITEC – Interdisciplinary Research on Technology, Education, and Communication, Belgium
Bieke Zaman, Centre for User Experience Research (CUO), Belgium

This chapter describes the User-Driven Creativity Framework: a framework that links several Participatory Design (PD) activities into one combined method. This framework, designed to be accordant with the mental process model of creativity, aims to integrate user involvement and creativity in the early stages of application requirements, gathering, and concept development. This chapter aims to contribute to recent discussions on how user-centered or participatory design methods can contribute to information systems development methodologies. The authors describe a mobile language learning case study that demonstrates how an application of the framework resulted in system (paper) prototypes and unveiled perceptions of learners and teachers, effectively yielding the necessary in-depth user knowledge and involvement to establish a strong foundation for further agile development activities. This chapter provides engineers or end-user representatives with a hands-on guide to elicit user requirements and envision possible future application information architectures.

Section 4
Mixed-Methods Approaches

Chapter 13
Combining Research Paradigms to Improve Poor Student Performance 243
Roelien Goede, North-West University – Vaal Campus, South Africa
Estelle Taylor, North-West University – Potchefstroom Campus, South Africa
Christoffel van Aardt, Vaal University of Technology, South Africa

The aim of this chapter is to demonstrate the advantages of combining methods from different research paradigms. Positivism, interpretivism, and critical social theory are presented as major paradigms in Information Systems research. The chapter demonstrates the use of methods representative of these three research paradigms in a single research setting. The main problem in the research setting is the poor performance of students in a specific module of their academic programme. This problem is addressed by initiating an action research project using methods representing different research paradigms in the different phases of the project. The argument for using mixed methods is presented by providing information on research paradigms, discussing the problem environment, describing the research process, and finally, reflecting on research paradigms and their application in this environment.

Chapter 14
Experiences in Applying Mixed-Methods Approach in Information Systems Research 266
Guo Chao Peng, University of Sheffield, UK
Fenio Annansingh, University of Plymouth, UK

Mixed-methods research, which comprises both quantitative and qualitative components, is widely perceived as a means to resolve the inherent limitations of traditional single method designs and is thus expected to yield richer and more holistic findings. Despite such distinctive benefits and continuous advocacy from Information Systems (IS) researchers, the use of mixed-methods approaches in the IS field has not been high. This chapter discusses some of the key reasons that led to this low application rate of mixed-methods design in the IS field, ranging from misunderstanding the term with multiple-methods research to practical difficulties for design and implementation. Two previous IS studies are used as examples to illustrate the discussion. The chapter concludes by recommending that in order to apply mixed-methods design successfully, IS researchers need to plan and consider thoroughly how the quantitative and qualitative components (i.e. from data collection to data analysis to reporting of findings) can be genuinely integrated together and supplement one another, in relation to the predefined research questions and the specific research contexts.

Chapter 15
Examining Web 2.0 E-Learning Tools: Mixed Method Classroom Pilot 294
Janet Holland, Emporia State University, USA
Dusti Howell, Emporia State University, USA

With so many fields using new technologies in e-learning, we are all challenged with selecting and effectively implementing new Web 2.0 tools. This chapter provides a mixed method research approach to quickly evaluate available Web 2.0 tools and instructional implementation. Class observations and pilot study surveys were used to determine students' levels of satisfaction after using various numbers of Web 2.0 tools and varying student work group sizes. The pilot studies were designed to model initial classroom examinations when integrating emerging Web 2.0 technologies. Use of this type of pilot study approach is necessitated as many individual class sizes are too small for a full research study, and the

time needed to conduct a full study using multiple classes could cause the results to quickly be out of date, thus not providing the needed immediate classroom data for just in time learning. Fast emerging technologies pose a unique challenge to traditional research methodology. Where immediate specific classroom data is needed, a needs analysis with a pilot study is the best option. Note, with emerging technologies, it is difficult to find appropriate literature to determine its effectiveness in the classroom. If desired, compiling the results from many small pilot studies offers an additional benefit of fleshing out key issues to be examined later in greater detail using a full research study for extending theory or scientific practices.

The Web is present in all fields of our life, from information and service Web pages to electronic public administration (e-government). Users of the Web are a heterogeneous and multicultural public, with different abilities and disabilities (visual, hearing, cognitive, and motor impairments). Web accessibility is about making websites accessible to all Internet users (both disabled and non-disabled). To assure and certify the fulfillment of Web accessibility guidelines, various accessibility evaluation methods have been proposed, and are classified in two types: qualitative methods (analytical and empirical) and quantitative methods (metric-based methods). As no method by itself is enough to guarantee full accessibility, many studies combine these qualitative and quantitative methods in order to guarantee better results. Some recent studies have presented combined evaluation methods between qualitative methods only, thus leaving behind the great power of metrics that guarantee objective results. In this chapter, a combined accessibility evaluation method based both on qualitative and quantitative evaluation methods is proposed. This proposal presents an evaluation method combining essential analytical evaluation methods and empirical test methods.

Due to the prevalent use of Information Systems (IS) in modern organisations, evaluation research in this field is becoming more and more important. In light of this, a set of rigorous methodologies were developed and used by IS researchers and practitioners to evaluate the increasingly complex IS implementation used. Moreover, different types of IS and different focusing perspectives of the evaluation require the selection and use of different evaluation approaches and methodologies. This chapter aims to identify, explore, investigate, and discuss the various key methodologies that can be used in IS evaluation from different perspectives, namely in nature (e.g. summative vs. formative evaluation) and in strategy (e.g. goal-based, goal-free, and criteria-based evaluation). Six case studies are also presented and discussed in this chapter to illustrate how the different IS evaluation methodologies can be applied in practices. The chapter concludes that evaluation methodologies should be selected depending on the nature of the IS and the specific goals and objectives of the evaluation. Nonetheless, it is also proposed

because their work allows them to recognize the more serious problems. Consequently, the software can be modified to resolve the usability problems and better meet the end users' requirements. In the future, the authors strive to integrate more diverse usability evaluations, including heuristic evaluations, to refine integration capabilities, to identify problems in more detail, and to improve the effectiveness of the usability evaluations.

The number of applications available for mobile phones is growing at a rate that makes it difficult for new application developers to establish the current state-of-the-art before embarking on new product development. This chapter is targeted towards such developers (who may not be familiar with traditional techniques for evaluating interaction design) and outlines a protocol for capturing a snapshot of the present state of the applications in existence for a given field in terms of both usability and functionality. The proposed methodology is versatile in the sense that it can be implemented for any domain across all mobile platforms, which is illustrated here by its application to two dissimilar domains on three platforms. The chapter concludes with a critical evaluation of the process that was undertaken and suggests a number of avenues for future research including further development of the keystroke level model for the current generation of smart phones.

Foreword

It is a pleasure to write a foreword to this important, new text on Information Systems (IS) research. I have to confess that I was one of the originators to the UKAIS definition of information systems referred to at the beginning of the Introduction. It was the product of a number of meetings of a subset of the (then) board of the UK Academy for IS (UKAIS). The intention was to be inclusive, since we recognised, as the editors describe in their introduction, that IS is a broad church. UKAIS members worked in departments ranging from psychology to creative arts, and from engineering to mathematics. We sought not to exclude researchers who used very different methods to address very different problem types, but the upshot is a definition that, in trying to be all things to all people, perhaps did not help to establish IS as a discipline in its own right. Nearly a decade later, much the same group, this time under the auspices of the Council of IS Professors suggested a new definition that sought to reflect the multitude of changes that had taken place in the intervening period. However, this definition does not seem to have supplanted the earlier version.

Pedro Isaias and Miguel Baptista Nunes have valiantly waded into the mire that is information systems research, but as they identify, this is not necessarily a mire in a bad way. It is perhaps more a garden that has run wild with some flowerbeds obscured by others, some that are misplaced, and some glorious blooms not recognised for what they are. I think those of us who have grown up with the development of the information systems discipline—and it is amazing that nearly all the IS researchers I met when starting out on my own research journey in the 1980s are still active—forget how difficult it can be for the outside world, for funders, and for new doctoral students to understand what it is that we do, and more importantly, why we do it. This is why this text is so important. By gathering together the work of 54 researchers from 28 institutions in 13 different countries around the globe, Isaias and Nunes present us with a vital snapshot of the current state of information systems research approaches.

Multi-disciplinarity and inter-disciplinarity are now well established and are advocated by funding bodies. Information systems research has always exhibited characteristics of multi- and inter-disciplinarity, so might be seen as being the future, though a future that arrived a little too early for its own good. There are many ways to categorise IS research approaches. In their valedictory editorial in the *Information Systems Journal*, Guy Fitzgerald and David Avison analyse the changing nature of research approaches used in papers published in the journal. They identify the rise in interpretive-empirical and critical methods and the decline of descriptive/conceptual/theoretical methods and the increasing use of mixed methods. Fifteen categories of method are identified with surveys and case methods dominating. Fitzgerald and Avison note that the largest number of articles continue to investigate research issues related to the IS development category, but big "gainers" are papers on IS usage, and the big "loser" is the IS management category. They conjecture that the greater emphasis on IS usage may be a reflection of the greater maturity of the discipline as it seeks to investigate best practice in IT use

This book helps us to make sense of approaches and methods. It identifies five main approaches: positivist and deductive, interpretivist and social constructivist, case study, mixed-methods, and evaluation. The editors argue that this diversity of approaches has led to a flourishing and adaptive IS research community and that this has enabled a natural co-evolution of research topics and technology. Hence, this book is rich, and it is diverse. It needs to be read by all new doctoral students and by masters and other students seeking a concise, coherent analysis of the state of IS research approaches. The editors acknowledge that this work is ephemeral as the socio-technical contexts being addressed change. However, unless we are to "capture reality in flight" now, we are never to understand the present nor prepare well for the future.

Philip Powell
University of London, UK

Philip Powell *is Executive Dean and Professor of Management at the School of Business, Economics, and Informatics at Birkbeck, University of London. Formerly, he was Deputy Dean at the University of Bath and Director of the IS Research Unit at Warwick Business School, having worked in Australia, Africa, USA, and Europe. He is author of 12 books, and his work has appeared in over 100 journals and at over 100 conferences. He is Editor-in-Chief of the Information Systems Journal and on many other editorial boards. He is a Fellow of the British Computer Society and an Academician of the Academy of Social Sciences. He holds an Honorary Chair in IS Economics at the University of Groningen.*

Preface

According to the UK Academy for Information Systems (UKAIS), Information Systems (IS) are the "means by which people and organisations, utilising technologies, gather, process, store, use, and disseminate information" (http://www.ukais.org.uk/about/DefinitionIS.aspx).

This type of definition is very common and has contributed to great confusion on what really constitutes IS as an academic discipline and has led to different fields of science appropriating the term in very different and not necessarily convergent ways. Computer science views "technologies," in the definition above, as software. Therefore, the term IS is often used interchangeably with software in computer science literature. Moreover, a myriad of academic and training programs bear the name of IS when, in fact, these are either programming or software engineering courses. On the other hand, information science focuses on the human processes that "gather, process, store, use, and disseminate information," subsuming IS into information management, knowledge management, and even librarianship. Finally, business and management studies use the type of definition exemplified above to focus on both organizational behavior, business processes, and IS as a management strategy.

However, although IS as a field of research lies in the intersection of these three main fields of computer science, information science, and business studies, its focus is clearly different. IS researchers focus on studying the impact of Information and Communication Technology (ICT) on socio-technical environments that support human activity systems. This focus distinguishes IS research from computer science, that concentrates on the design and development of ICT artifacts, information science that puts emphasis on information and business studies that centers its research around the organizational behaviors.

Nonetheless, IS draws on theoretical propositions, research frameworks, and experiences from all these three disciplines. This inherent multidisciplinary results in both strengths and weaknesses. Firstly, and probably foremost, there are very few pure IS academic departments, which means that IS academics and researchers can be found in a number of different and more main stream departments such as business schools, computer science departments, information schools, psychology departments, or even in more creative academic units, such as schools of arts. This puts IS conferences among the more interesting ones to attend due to rich environment for interchange of ideas, clashing of conceptualizations, and challenging of established theories and perceived truisms from the individual disciplines. Secondly, this diversity brings with it different conceptualizations on what research is, which ontological and epistemological paradigms should be used and which research approaches, methods, and designs are appropriate for the field. Finally, there is little consensus in the field on what exactly are typical research questions in IS.

This scenario may seem at a first glance negative and discouraging. However, quite in the contrary, it has led to a thriving, flourishing, and very adaptive IS research community and enabled a natural co-evolution of research topics and technology. In fact, IS has become probably the most flexible and adaptive of all research communities. The apparent weakness of not having well-established research questions means that these can continuously evolve to face the challenges of ever changing ICT, ever evolving human activity systems, and even faster occurring societal evolutions. In this sense, IS research, contrary to many other disciplines, will never become obsolete, as it never settles on established theoretical frameworks or dogmatic and epistemologically rigid perceptions research.

However, this evolutive nature of IS as a discipline poses serious challenges to anyone intending to characterize IS research in a book with the nature of ours. First, we need to give voice to different epistemological positions in the field and allow for the traditional dichotomous dialogues such as positivism versus interpretivism, deductive versus inductive, or qualitative versus quantitative. Second, we need to also include those who reject the incompatibility of these traditional opposing positions and propose to integrate them as complementary contributions to the understanding of IS phenomena, such as the proponents of mixed-methods approaches or design research. Finally, we need to create a coherent and understandable structure, so that the reader is able to make sense of these different research voices, positions, and proposals. After much discussion, consultation, and some heated controversy, and with the full awareness that such a structure would never get universal agreement, we finally decided to divide this book into five sections: Positivist and Deductive Approaches; Interpretivist and Social Constructivist Approaches; Case Study Approaches; Mixed-Methods Approaches; Evaluation Research Approaches.

The Positivist and Deductive Approaches section comprises six very different chapters on aspects such as experimental research, survey research, theories of reasoned action, design research, and desktop research.

Chapter 1, titled "Online Survey: Best Practice" and authored by Issa, provides an overview discussion and an example of application of online survey research. This chapter explores advantages, disadvantages, as well as the application of online survey and the functionality of online tools to support it. Finally, the author provides practical recommendations on the adoption of an online survey from the researchers' point of view.

Connolly's Chapter 2, titled "eCommerce Trust Beliefs: Examining the Role of National Culture," presents positivist and deductive research design, using a trust measurement instrument that was beforehand validated in previous research. The study reported applies that survey instrument to two different countries (United States and Ireland) and tries to explore whether and to what extent national culture can manipulate online consumers' trust in e-commerce services.

Chapter 3, dedicated to snowball sampling and written by Isaías, Pífano, and Miranda, bears the title of "Subject Recommended Samples: Snowball Sampling." This chapter focuses on sampling strategies for survey data collection, focusing on snowball sampling as a "chain-referral sampling method." The authors argue that snowball sampling supports the access to hidden subjects that would otherwise not be found and questioned. Isaías *et al.* contextualize their discussion and insights through the experiences gained when conducting a research project on online behavior of Web 2.0 users.

Chapter 4, "Practically Applying the Technology Acceptance Model in Information Systems Research" by Mojtahed and Peng, explores the Technology Acceptance Model (TAM), which is known as one of the best Information Systems approaches to study ICT acceptance and usage. These authors argue the applicability and advantages of this model by presenting and discussing two different UK research projects, one on mobile banking acceptance and another on mobile campus.

Chapter 5, "Designing Personalised Learning Resources for Disabled Students Using an Ontology-Driven Community of Agents" by Nganji and Brayshaw, focuses on the exploration of social artifacts for the disabled, making use of the affordances of new technologies based on the Semantic Web. The chapter uses a very traditional scientific approach for the design of IS predicated on a very prevalent but increasingly controversial computer science principle of reductionism, i.e. the notion that an IS can be seen as discrete components and the design and evaluation processes focusing interactively and incrementally on subsets of such components. Specifically, the chapter advocates a design process using an ontology-based personalization of learning resources for disabled students.

The last chapter in this section, Chapter 6, titled "Taxonomy of IT Intangibles Assets for Public Administration Based on the Electronic Government Maturity Model in Uruguay," is authored by Garbarino, Delgado, and Carrillo and presents an example of desktop research using a taxonomy of Information Technology intangible assets for public administration based on the Electronic Government Maturity Model (EGMM). The authors illustrate the application of desktop research through a project aiming at setting up a taxonomy of assets indicators for e-government in order to enable a better quality of public services online.

Section 2 of this book focuses on "Interpretivist and Social Constructivist Approaches" and comprises four chapters in the areas of ethnographic studies, grounded theory, action research, and discourse analysis.

In Chapter 7, "Online Ethnographies," Isomäki and Silvennoinen offer a comparison and analysis of ethnography, Webnography, network ethnography, cyber-ethnography, and digital ethnography. This very interesting chapter argues that the exponential growth of ICT usage associated to very rapid technological development has fostered the emergence of new research procedures and methods. Therefore, these authors argue that the creation of new applications of ethnography (in this case online ethnography) can be seen as a natural evolution of more traditional strands. The authors use a case comparison method to illustrate their discussion and propositions.

Chapter 8 discusses grounded theory as one of the better established inductive methodologies in use in IS. This chapter, titled "Grounded Theory in Practice: A Discussion of Cases in Information Systems Research," is authored by Martins, Nunes, Alajamy, and Zhou. The purpose of this chapter is to share insights based on the use of this method in the three case study examples and defend this type of inductive research as an effective methodology to create solid theory based on real contexts of practice and developed from the field.

Chapter 9 presents an example of an action research study. This chapter by Costello and Donnellan is titled "Creating Interpretive Space for Engaged Scholarship." In this chapter, the authors propose a new form of action research called Dialogical Action Research, which provides an interpretive framework for practitioners who are introducing change in their own organizations. This approach is based on the core concepts of interpretive space and engaged scholarship and is illustrated through the use of an applied research process.

Chapter 10 on "Exploring Higher Education Students' Technological Identities using Critical Discourse Analysis" by Brown and Hart discusses and reports on the implementation of Critical Discourse Analysis (CDA). The project reported explores the role of ICTs using South African scenarios involving higher education students and their perceptions of the impact of ICTs in their lives. The authors use the method CDA with the objective of gathering more insights and data regarding the social relationships and social identities. The authors propose the use of categorization of discursive types in the text in order to identify explicit, socially negotiated, and hidden meanings.

Section 3 on "Case Study Approaches" focuses on a well establish research practice in IS and comprises two chapters.

Chapter 11 by Gordon, Shankaranarayanan, and Blake, titled "The Role of Case-Based Research in Information Technology and Systems," focuses on case study research as a method with a significant part to play in the future of research in IS due to its capacity to create knowledge from practice and to study phenomena in context. The authors present some very interesting examples of case study research in order to emphasize the role of case-based research. Additionally, the authors provide an analysis of abstracts from more than 200 articles that use case-based research by using a latent semantic analysis to identify their trends and anticipate what the role of case-based IS research will be in the coming decades.

Chapter 12 on "Mapping Participatory Design Methods to the Cognitive Process of Creativity to Facilitate Requirements Engineering" by Sulmon, Derboven, Montero, and Zaman presents a user-driven creativity framework that unites various participatory design activities into one combined method. This framework, designed to be accordant with the mental processes of creativity, aims to integrate user involvement, innovation, and vision in the early stages of application requirements, gathering, and concept development. The research project presented in this chapter discusses a mobile language learning case study that demonstrates how an application of the framework resulted in system prototypes and unveiled perceptions of learners and teachers, effectively yielding the necessary in-depth user knowledge.

Section 4 addresses the controversial use of triangulation of methods and mixed-methods approaches. This section reflects an increasing awareness of the complementary use of different data collection methods that resulted in growing trend in IS Research to adopt this type of research design. This section includes four chapters.

The purpose of chapter 13, "Combining Research Paradigms to Improve Poor Student Performance" by Goede, Taylor, and van Aardt, is to present an action research approach using methods representing different research paradigms in the different phases of the project. The authors illustrate their propositions through the use of a project aiming at developing a blended learning environment for computer science students at a South African university. At different stages of the project, the research used informal discussions, questionnaires, and Web-log analysis. The argument for using mixed methods is developed by discussing established research paradigms and their utilisation in the problem environment and reflecting how these were integrated in the action research process implemented.

Chapter 14 discusses "Experiences in Applying Mixed-Methods Approach in Information Systems Research" and is authored by Peng and Annansingh. These authors use two different research projects that used mixed-methods approaches to illustrate and justify how quantitative and qualitative research methods can be combined in order to study one specific phenomenon. This chapter also provides useful guidelines and proposals to aid IS researchers when selecting and designing a mixed-methods research as well as providing more precise and meaningful insights regarding this type research design.

Chapter 15 provides a mixed-method research approach to evaluate Web 2.0 tools and their application in educational environments. The chapter by Holland and Howell is titled "Examining Web 2.0 e-Learning Tools: Mixed Method Classroom Pilot" and uses a combination of observation and surveys to determine adequacy of emergent Web technologies, levels of student satisfaction in using them, and finally, an assessment of the levels of usage. It represents a good example of mixed-methods making use of observation as a data collection method, which is particularly useful in IS research.

Chapter 16 offers yet another significant combination of research methods that is both innovative and very different from the other research designs presented in this section. This chapter presents a combination of analytical evaluation methods and empirical evaluation methods. The chapter authored by Luján-Mora and Masri focuses on aspects of Web accessibility and is titled "Evaluation of Web Accessibility: A Combined Method." Accessibility has been both a topical and a very politicized issue in IS research for the last decade, but the authors take a very technical and neutral view of the problem, that makes this chapter a very interesting one to read.

Section 5, "Evaluation Research Approaches," comprises four chapters.

Chapter 17, "Information Systems Evaluation: Methodologies and Practical Case Studies" by Chen, Osman, and Peng, seeks to identify, examine, and discuss the most common and established methods currently in use in Information Systems (IS) evaluation. It looks at the process of evaluation from different angles, specifically in terms of its nature and the corresponding strategies that can be adopted. The authors presented four case studies in order to demonstrate how different IS evaluation procedures can be put into practice.

In Chapter 18 by Covarrubias, Bordegoni, Cugini, and Gatti presents an evaluation research method of a haptic guidance device to support people with disabilities in sketching, hatching, and cutting shapes. This haptic guidance device has been used as an input system for tracking the sketching movements. The chapter, titled "Supporting Unskilled People in Manual Tasks through Haptic-Based Guidance," is an excellent example of a very specific purpose criteria-based evaluation that is very common in IS.

Chapter 19, "Integrated Methods for a User Adapted Usability Evaluation" by Shirogane, Yashita, Iwata, and Fukazawa, proposes a usability evaluation process by integrating two well-established methods and one that is original and proposed by the authors. The authors propose this combined methodology to trace operation histories of the target software. Specifically, the research focuses on employing automatic usability evaluations to identify problems ranging from common errors to more complex aspects of efficiency and learnability.

Chapter 20 by Martin, Flood, and Harrison is titled "A Protocol for Evaluating Mobile Applications" and presents a process for the evaluating of interaction design and outlines a protocol for capturing a snapshot of the present state of the applications in existence for a given field in terms of both usability and functionality. In addition, the protocol was designed to be able to create a list of features offered by existing applications and allow new application (such as mobile technologies) developers to establish the current state-of-the-art before embarking on new product development.

To conclude this already very long introduction, we would like to make one final remark. This book aims at illustrating the current diversity and richness of research in the area of IS. It works well as a window into the world of IS, and it aims at shedding light on the theoretical and methodological discussions in the field today. However, due to the evolutive nature of IS research, it should not be taken as a definitive contribution, but one that reflects current trends and opinions. We are sure that in 10 years' time, a similar book could include a different structure and contain very different chapters, both in terms of the nature of the research designs used and, certainly, in terms of the socio-technical contexts being addressed.

Finally, we would like to thank all the authors in the book, apologize for the many reviews and the very strict demands made on their contributions, and congratulate all for the group effort that resulted in this book.

Pedro Isaías
Universidade Aberta (Portuguese Open University), Portugal

Miguel Baptista Nunes
University of Sheffield, UK

Section 1
Positivist and Deductive Approaches

Chapter 1
Online Survey:
Best Practice

Tomayess Issa
Curtin University, Australia

ABSTRACT

This chapter aims to demonstrate how the online survey tool was used during the PhD and postdoctoral research stages to evaluate and assess the New Participative Methodology for Developing Websites from the Marketing Perspective. After reviewing techniques from numerous disciplines for methodology integration, a new methodology was produced and an online survey to assess the methodology was completed by website industry participants and information systems professionals in Western Australia. A new online survey was developed on the basis of the PhD results and 210 Web developers and information technology professionals from Australia assessed the new methodology. This chapter will discuss three factors: the advantages and disadvantages of using an online survey in helping to facilitate high quality research and an overview of the functionality of the online survey tool(s) from a researcher's point of view. Finally, a practical set of recommendations and endorsements is presented regarding the adoption of an online survey from the researcher's standpoint.

INTRODUCTION

The Internet, or the Global Internet, as some people call it, is the name given to a certain network of computers around the world. The Internet is also known by other names such as Cyberspace or the Information Superhighway. The Internet itself is a network of thousands of computer networks utilising a common set of technical protocols to create a worldwide communications medium. Users reach the Internet through their computers and terminals at home or at educational institutions,

DOI: 10.4018/978-1-4666-2491-7.ch001

Copyright © 2013, IGI Global. Copying or distributing in print or electronic forms without written permission of IGI Global is prohibited.

or through commercial Internet access providers and other organisations (Mitchell, Lebow, Uribe, Grathouse, & Shoger, 2011; Subrahmanyam & Smahel, 2011; Sun, 2011).

The contents of the Internet range from high technology research papers to low technology childcare. The total amount of data on the Internet has not been measured in recent years and an estimate of several hundred terabytes probably falls far short. Its development is not slowing at all. The Internet provides common services such as electronic mail, online shopping, electronic news, and information access, via Gopher and other info-bases. These services are accessed through various application programs available for a variety of computer operating systems. All these services work over the common network structure of the Internet. The Internet itself is just a massive communication medium (Kung, Picard, & Towse, 2008; Tsai, et al., 2009; Van Deursen & Van Dijk, 2009).

The Internet is not limited to e-commerce and information. Now includes an online survey approach which was created on the Internet to work simultaneously with telephone interviewing and mail surveys, since the rate of response is decreasing as the majority of people decline to answer these surveys (Boyer, Olson, Calantone, & Jackson, 2002; Couper, Traugott, & Lamias, 2001; Dillman, 2007; Porter, 2004).

Several studies (Couper, et al., 2001; Dillman, et al., 2009; Fleming & Bowden, 2009a, 2009b; Umbach, 2004) indicate that there has been a shift and change in survey research in recent years, as a majority of researchers have started to use the online survey facility to collect survey data quickly and inexpensively via the Web. However, it is important that, before conducting the online survey in any research, researchers should understand the positives and negatives behind online survey usage. This chapter will assess and investigate the current literature review with respect to the online survey, provide examples from the researcher's own experience of using online surveys, and based on the researcher's perspective a set of practical recommendations regarding online survey adoption is presented.

This chapter is organized into three sections: first, the researcher will highlight the current literature with respect to the new website development methodology, then expand the literature review on the research methodology, which was adopted in her PhD and postdoctoral stages, and discuss the advantages and disadvantages of mixed methods. Secondly, the researcher will present her experience of using online surveys during the PhD and postdoctoral stages. Finally, a set of recommendations regarding online survey tool adoption will be discussed on the basis of the researcher's experience.

Finally, this chapter will examine the advantages and disadvantages of online survey usage in information systems research.

BACKGROUND

This section will highlight the current literature with respect to the new website development methodology and emphasize the research methodology, which was used in the PhD and postdoctoral research.

New Website Development Methodology: New Participative Methodology for Marketing Websites

In order for systems (or websites) to be widely accepted and used effectively, they need to be well designed. To achieve this, designers and users need to use a specific methodology to produce the "system" (or website). A methodology "should tell us what steps to take, in what order and how to perform those steps, but, most importantly, the reasons why those steps should be taken, in that particular order" (Jayaratna, 1994, p. 242).

The term "methodology" is used significantly in information systems development, as each methodology should have a set of stages and steps, which need to be followed in sequence if the work is to be done successfully. 'Stage' is a "convenient breakdown of the totality of the information systems life cycle activity" (Olle, et al., 1988, p. 21), while 'step' is "the smallest part of a design process" (Olle, et al., 1988, p. 21). Each stage consists of a set of steps. The sequence of the stages may not always be fixed. In some projects, iteration between stages will occur and this may have a different impact on the methodology as an iteration may "take different forms and thus impact differently on what one can do with a methodology" (Olle, et al., 1988, p. 30).

According to Avison and Fitzgerald (1993, p. 264), adopting a methodology in developing a website or an interface can lead to improvements in these three categories: "A better end product; A better development process; and A standardized process." Therefore, a designer needs to understand users' requirements for the project before choosing the methodology, in order to complete the work successfully and to accomplish profitable results.

In the research study, various types of models and methodologies were analyzed, including: lifecycle models; IS development methodologies; methodologies with explicit human factors aspects; website methodologies; marketing methodologies; and additional detailed techniques such as task analysis and detailed website design and implementation. There are numerous similarities with respect to the stages between methodologies for developing information systems, websites, or marketing strategies. Integrating stages from information systems methodologies into website and marketing methodologies is very beneficial in the development of websites that are more effective and efficient. Human factors experts should be involved in these methodologies to make sure that transaction processes, tracking, maintenance, and updating of the website meet the users' requirements and needs (Issa, 2008; Issa & Turk, 2012).

Firstly, the researcher reviewed the methodologies to identify two aspects: 1) the stages needed for the system development process, and 2) the four key principles (user participation, usability, iteration, real interaction), in order to check the availability of these four key principles in IS development, website, and marketing methodologies. Secondly, the stages of information systems development methodologies were checked to assess how effectively they match the four key principles at each stage and to identify the strongest stage in each methodology. Thirdly, for the website and marketing methodologies, the researcher checked the availability of techniques covering the four key principles in these methodologies; listed the extra stages to be added to the new methodology; and identified the strongest stage in each methodology. Finally, additional detailed techniques of task analysis and website design and implementation (navigation design, promotion, and staff training) were added. Such additional detailed techniques will play a key role in the new methodology, as most of the existing methodologies have neglected these (Issa, Turk, & West, 2010).

Research Methodology

Both quantitative and qualitative approaches were used in this research: i.e., interviews and online surveys were used to allow the researcher to collect a wide range of information from the industry in order to provide a more complete picture of the major and minor research questions of the PhD and postdoctoral research (Maudsley, 2011; Teddlie & Tashakkori, 2009; Wiggins, 2011).

Throughout the research, an inductive (social science) approach was used as the main research method. This approach "begin[s] with [a] detailed observation of the world and move[s] toward more abstract generalizations and ideas" (Neuman, 2000, p. 49). The researcher observes and refines the concepts in order to "develop empirical generalizations and identify preliminary relationships," to "build the theory from the ground up"

(Neuman, 2000, p. 49). During the first step, the researcher identifies the purpose and the aims, and this was achieved by using an explanatory method, seeking to discover why these events and factors are occurring in this way. This approach is aimed at focusing on, and looking for, causes and reasons, as indicated in Figure 1.

The explanatory step was performed in the research by exploring various methodologies for website design to identify the reasons why many users are frustrated and confused when working with websites. The researcher outlined the basic concepts behind methodologies, including life-cycle models, IS development methodologies, methodologies with explicit human factors aspects, website methodologies, marketing methodologies, and additional detailed techniques such as task analysis and detailed website design and implementation. From this information, the New Participative Methodology for Developing Websites from the Marketing Perspective was developed. This methodology will assist the designers and users to fill the gaps in the current methodologies and to avoid the frustration currently experienced by website users.

It was indicated previously that mixed methods were used in this research to examine the Web industry's perspective toward the new methodology. The qualitative method is centered mainly on an ethnographic approach (Creswell, 2003; Hinton, Kurinczuk, & Ziebland, 2010; Sullivan, 2011; Teddlie & Tashakkori, 2009), in that data are collected during interviews, focus groups or observations. The qualitative method was used in this study to examine people's reactions and perspectives toward the new methodology, and to study their direct interaction and non-verbal communication during the interview session (Tashakkori & Teddlie, 1998; Weinreich, 1996). On the other hand, the quantitative method was used to investigate and measure attributes and answers to "what" and "how many" questions in relation to the new methodology, as the data collected by this approach were used to analyze the similarities, differences and relationships between the respondents' answers (O'Neill, 2006; Tashakkori & Teddlie, 1998; Teddlie & Tashakkori, 2009; Weinreich, 1996).

Qualitative data sources include "observation and participant observation (Fieldwork), interview

Figure 1. Research methodology

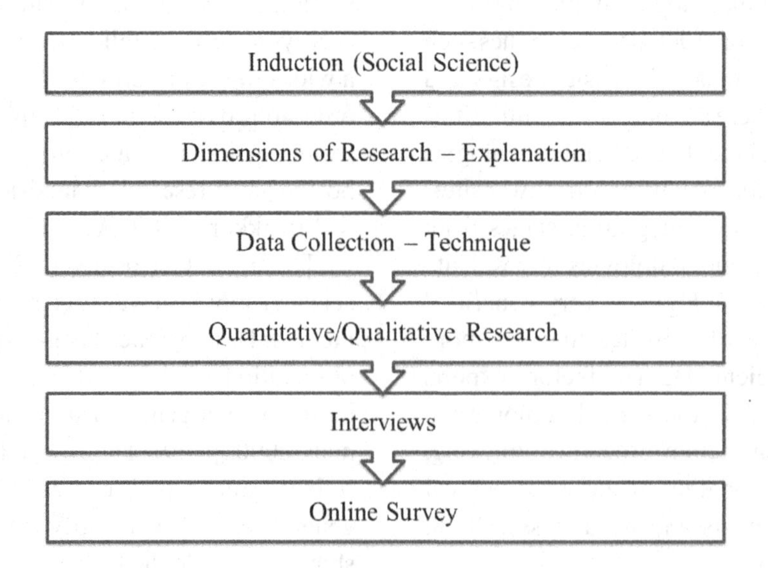

and online surveys, documents and text, and the researcher's impressions and reactions" (Myers, 2003). The advantage of using qualitative methods is that they generate "rich detailed data that leave the participants' perspectives intact and provide a context for health behavior" (Weinreich, 1996).

In addition, according to Marcus and Robey (cited in Kaplan and Duchon [1988, p. 573]), qualitative methods can "yield data from which process theories and richer explanations of how and why processes and outcomes occur can be developed." However, a key weakness of this method is that it is time-consuming.

As mentioned above, this research is mainly focused on interviews leading to the use of an online survey. Both stages are necessary in order to gather the wide range of data needed to support the research objectives and to minimize the shortcomings of each approach. It is useful to examine the weaknesses and strengths of these qualitative methods.

An interview is "obviously and exclusively an interaction between the interviewer and interview subject in which both participants create and construct narrative versions of the social world" (Miller & Glassner, 2006, p. 125). In addition, Mahoney (1997) states that "the use of interviews as a data collection method begins with the assumption that the participants' perspectives are meaningful, knowable, and able to be made explicit, and that their perspectives affect the success of the project." An interview provides "access to the meanings people attribute to their experiences and social works" (Miller & Glassner, 2006, p.126).

Interviews are utilized in this research to explore the type of methodology, tools, and techniques that are adopted by the website development industry in Western Australia and to learn more about their technical experiences and knowledge of how to develop a website. The interviews also identify existing problems and provide an indication of the likely usefulness of the principles behind the proposed new methodology.

The purpose of the interview in this research is: 1) to identify the type of methodology that is carried out by the industry in Western Australia to develop a website, 2) "to inquire about possible measures and focus of the study" (Kaplan & Duchon, 1988, p. 577), and 3) to generate the online survey for the second phase of this research.

The strengths of the interview include the following: it is useful when participants cannot be observed directly. participants can provide historical information; it allows the researcher "control" over the line of questioning (Creswell, 2003, p. 186); it enables the researcher to experience the affective as well as cognitive aspects of responses and it allows the interviewer to explain or help clarify questions, increasing the likelihood of useful responses (Mahoney, 1997).

The weaknesses of the interview include the following: it is expensive and time-consuming; the interviewee may distort information through recall error, selective perceptions, or the desire to please the interviewer; flexibility can result in inconsistencies across interviews; the volume of information may be too large; it may be difficult to transcribe and reduce data (Mahoney, 1997).

The researcher interviewed representatives from website companies in Western Australia. The main objective of the questions used in the interview was to learn about the methodology, tools, and techniques that are used to develop a website. The researcher also discussed the four key principles (i.e. usability, user participation, real interaction, and iteration) behind this research. The researcher was interested to know whether and to what extent these key principles are reflected in the companies' methodologies. During each interview, the researcher discussed with the interviewees her new participative methodology for developing websites from the marketing perspective. Very positive responses were received from the industry representatives, which encouraged the researcher to continue development of the methodology by adding extra aspects suggested by the interviewees. After examining the data gained from all the interviews, the researcher identified the new information about methodologies provided by the industry, which was very useful as an addition to the New Participative Methodology for developing websites from the marketing perspective. This will allow the new methodology to become more practical.

On the other hand, it is ideal to use the quantitative approach in this research, as participants will provide rich and historical information to enable the researcher to learn more about their methodology for developing a website. Therefore, this research approach has been selected as the fundamental research method to be utilized in this study.

The second approach used in this research was an online survey, which was generated and developed from the qualitative research after "analyzing the interviews and observations to derive categories for questions that focused on the primary expectations expressed by interviewees" (Kaplan & Duchon, 1988, p. 578). An online survey is a "pre-formulated written set of questions to which respondents record their answers, usually within rather closely defined alternatives." In addition "online surveys are an efficient data collection mechanism when the researcher knows exactly what is required and how to measure the variables of interest" (Sekaran, 2003, p. 236).

The purposes of using this approach in this research are: 1) to evaluate the "practicality" and "benefits" of adopting the proposed new methodology in the website industry in Western Australia; 2) to consider the various requirements for developing a website; and 3) to evaluate whether it is possible to achieve effective user participation in website design.

The strengths of online surveys include the following: an online survey offers greater anonymity; it is less expensive; the respondent can take more time to respond at their convenience; an online survey can be administered electronically, if desired (Cavana, Delahaye, & Sekaran, 2001, p. 245). Furthermore, an online survey takes less project time than the paper and pencil survey, since once the survey has been created online it is ready for sending, so there are no printing times to consider, and, as a result, online surveys are more sustainable than the traditional methods (Porter, 2004; Prahalad & Rangaswami, 2009; Prasad, Saha, Misra, Hooli, & Murakami, 2010).

The weaknesses of online surveys include the following: the response rate is almost always low; a 30 percent rate is quite acceptable; follow-up procedures for non-responses are necessary (Cavana, et al., 2001, p. 245). Furthermore, Fleming and Bowden (2009a, p. 285) indicated that the most commonly cited disadvantages of Web-based surveys are "sample frame and non-response bias." Other disadvantages of online surveys are technical failures, computer viruses, Internet crimes, and hacking into the Web-based survey, these factors can lead to a decrease in the response rate, according to Fan and Yan (2010) and Umbach (2004). Finally, online survey-based research raises several ethical questions, from privacy to informed consent. The researcher should protect participants' privacy and confidentiality, as any information provided by them through the survey must be held as strictly confidential, and the most important aspect is that survey outcomes should not be disclosed to any parties besides the researchers, unless required to do so by law (Cho & LaRose, 1999; Cho, Park, & Han, 2011; Eysenback & Till, 2001). This approach will provide data to answer the minor research questions supporting the major research question.

Finally, the researcher adopted mixed-methods approaches to recognize and identify the complete picture behind the respondents' perspective toward the new methodology, and, most importantly, to minimize discrepancies in the findings and provide a substantial amount of data to identify the positive and negative aspects of the new methodology (Creswell, 2003; Crump & Logan, 2008; Gilbert, 2006; Hesse-Biber, 2010; Maudsley, 2011; Teddlie & Tashakkori, 2009).

For this chapter, the researcher will emphasize and discuss the advantages and disadvantages of online survey usage in research, especially from the Information Systems perspective.

ONLINE SURVEY DESIGN

With the arrival of modern information technology and information systems in organizations, education and research, new opportunities have emerged to shift from using paper and pencil administration to using Internet facilities. This shift has brought several advantages and disadvantages to researchers who are using online surveys via the Internet. Several studies indicate that online surveys are very powerful tools for maintaining respondent interest in the survey and encouraging users to complete it on time (Couper, et al., 2001; Gordon & McNew, 2008). The online survey is self-administered and involves computer-to-computer communication over the Internet, asking respondents to respond to the survey by clicking on radio buttons and to add further comments regarding the survey questions in a specific area within the survey. The literature review indicates that the majority of online surveys can offer plenty of advantages compared with the traditional methods, including saving money, speed, and a high response rate (Carlbring, et al., 2007; McBurney & White, 2007). Furthermore, the online survey design can provide more dynamic interaction between the respondent and the survey than can be achieved in email or paper surveys (Dillman, 2007; Dillman, et al., 2009). Finally, data downloads from the online survey are ready for immediate data analysis in SPSS or other statistical programs, and the online survey presents the outcomes in various formats including tables and charts, with the mean and standard deviation or cross-tabulation calculated according to the researcher's requirements (Cho, et al., 2011; Fan & Yan, 2010; Fink, 2010; O'Brien & Toms, 2010). Based on the online survey strengths the researcher conducted her research by using online surveys for both PhD and postdoctoral stages.

A five-point Likert scale was used in each part of the online survey to "examine how strongly subjects agree or disagree with statements" (Sekaran, 2003, p. 197; Likert, 1932). Cavana et al. (2001, p. 205) stated that the midpoint in the Likert scale (e.g. the third point in a five-point scale) "is either neutral ('neither agree nor disagree') or a passing level (e.g. 'satisfactory')." The five points on the scale are: 'Strongly Disagree,' 'Disagree,' 'Neutral,' 'Agree,' and 'Strongly Agree.' Besides using the Likert five-point scale for this online survey, the researcher provided a section for participants to write down other comments regarding each part. All the pages of the survey contained instructions at the top of the page and a progress bar along the bottom to provide feedback to users about their proximity to completion. The formal letter and information sheet were emailed to the respondents with the survey link. Pages 1-10 presented the survey items with three questions per page to minimize scrolling, and the concluding page thanked respondents for their participation. A description of each part was provided to the participants to explain its purpose.

To accomplish the second and final phase, the researcher prepared two letters concerning the online survey, one of which was to be sent with the online survey, while the second letter was sent via e-mail. The former letter provided the participants with information about the online survey, the time frame, and the time required to complete the online survey. The second letter contained instructions on how to complete the online survey, and a request to nominate other staff from their organization who would also be willing to complete the online survey. In addition, each participant received a PDF file containing information about the new methodology to help the participant to assess the new methodology (in part five of the online survey).

The online survey was divided into seven parts, each of which discussed one key principle for this research. A description of each part was provided to the participants to explain its purpose. The seven parts were as follows:

Part 1. User Participation: This thesis distinguishes between two types of users: end-users (internal to the client organization) and client-customer users (external). End-users (internal) are the real users in the client organization who test and evaluate the website and use it to respond to the client/customer's queries. The client/customer users (external) are those who interact with this website to accomplish their goals.

Part 2. Real Interaction: Website use statistics or click tracking: The designer will track users' behaviour to help understand what attracts or repels users. This can be achieved by adding two options to the website: 1) a feedback form to elicit users' opinions; or 2) a counter on a webpage, which will provide detailed statistics (log file) to the designer.

Part 3. Human-Computer Interaction and Usability: Human-Computer Interaction (HCI) "is a discipline concerned with the design, evaluation, and implementation of interactive computing systems for human use and with the study of major phenomena surrounding them" (Preece, Rogers, Benyon, Holland, & Carey, 1994, p. 7).

Usability evaluation is used to confirm that the website design is efficient, effective, and safe; has utility; is easy to learn, easy to remember, and practical; provides job satisfaction; and defines performance measures that effectively assess the users' requirements and requests.

Part 4. Iteration: Use of prototypes to allow for evaluation of effectiveness: This approach will assist the designers to build up the new website and make sure that the project will be tested repeatedly until it meets users' requirements. Steps within the methodology may be repeated if necessary.

Part 5. New Participative Methodology for developing websites from the marketing perspective (integrated and contingent): This integrated methodology was created from basic concepts derived from lifecycle models, IS development methodologies, methodologies with explicit human factors aspects, website methodologies, marketing methodologies, and additional detailed techniques (e.g. task analysis and detailed website design and implementation). The main focus has been on defining users' requirements and needs, planning, analysis, design, testing, implementation, evaluation and maintenance. These stages are very useful in any methodology as they enable the designer to ensure that the system is running according to the needs of users and the client organization. The new integrated methodology needs to be "contingent" with both analyst and client choosing the particular techniques, which suit the problem situation.

Part 6. General Questions: The key results of the interview stage of the research project are summarized in this section, and the researcher requests comments on these aspects of the new methodology.

Part 7. Background Information: In this section, the participants provide some details about their level of formal education and main field(s) of study.

DESIGN OF ONLINE SURVEY DURING THE PHD RESEARCH STUDY: STRENGTHS AND WEAKNESSES

Interviews were carried out with personnel from the website design industry in Western Australia, and analysis of this data resulted in the design of the online survey, which is the second and last phase in this research. Designing the online survey involved interpretation of the interview data and analysis together with consideration of the major and minor research questions for this research.

The online survey had to receive the approval of the Ethics Committee at the University before being sent to the industry participants and IS professionals.

The method for phase two of this research (see Figure 2) consisted of the following steps:

1. Designing the online survey (hard copy version).
2. Receiving the approval from the Ethics Committee at the University.
3. Designing the online version of the survey.
4. E-mails to the nine website companies (from interviews) with information about the online survey.
5. E-mails to ten IS professionals to obtain the IS perspective regarding the new methodology.
6. E-mail reminder letters to companies and IS professionals who did not complete the online survey.
7. Receiving all the responses.
8. Reviewing and analyzing the results.
9. Executing changes to the new methodology.
10. Releasing the new revised methodology.

After careful consultation with the research supervisors, the researcher decided to use the online survey tool from her school as the method to make the online survey available to the participants. The researcher was the first person to try the online survey tool externally. The reasons for adopting this tool were:

1. The tool can be accessed from anywhere and anytime.
2. The tool is easy to manage.
3. The tool is inexpensive and practical.
4. This tool can have a "high response rate" (McBurney & White, 2007, p. 245).
5. This tool can "provide a more dynamic interaction between respondent and online survey than can be achieved in e-mail or paper surveys" (Dillman, 2007, p. 354).
6. The tool is quick to deliver the results.

Figure 2. Summary of online survey phase: PhD stage

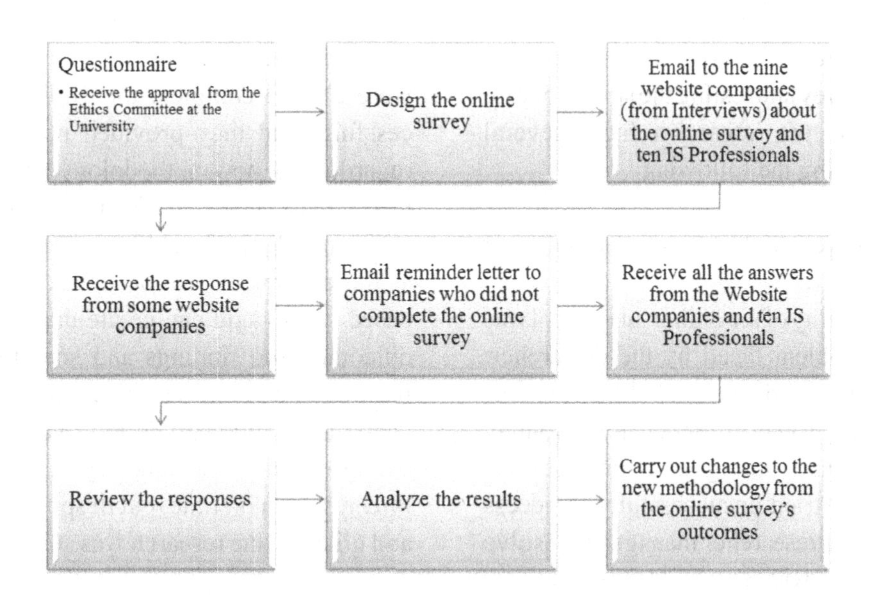

7. The tool will reduce errors in the results since the results are captured as an Excel spreadsheet, which makes it easier for the researcher to analyze the results.
8. The online survey tool from her school will allow the researcher to ensure that any information provided by respondents through the online survey will be held as strictly confidential. Information will not be disclosed to any parties beside the researcher and her supervisors.

To design and test the online surveys for this research, the IT Department at her school provided the researcher with some instructions to be followed. First of all a username and password were assigned to the researcher by the IT Department to allow her to access the main page of the online survey. After accessing this, the researcher created nine other pages. Seven pages were dedicated to the online survey, while the other two pages were the welcome and concluding pages.

For each page, the researcher added to the template the survey description; survey questions for each part; scale; size; question number and instructions to the participants on how to complete the online survey. At the beginning, the researcher believed that this tool would be very usable. However, after spending more than sixty hours entering and updating all the information for the online survey and testing the survey before putting it online, the researcher faced several problems, including the following:

1. Online survey tool instructions were not easy to follow.
2. The availability of the online survey tool was another problem faced by the researcher: during the second weekend, the server was offline from Friday afternoon until Monday afternoon, and during that time, several participants tried unsuccessfully to access the tool. The researcher managed to resolve the problem with the IT Department and contacted the participants to inform them about the server availability.
3. The IT Department started to make some changes to the online survey tool template without informing the researcher. At this stage, a problem occurred, which meant that the last two participants completed the online surveys but the server did not collect their answers. The researcher solved this problem by emailing the online surveys (as a Word document) to the last two participants again to complete and e-mail it back. The participants completed the online surveys within one day and the researcher highly appreciated their efforts.
4. The last problem faced by the researcher was the downloading of results from the system. Most of the participants used a comma in the comment section. The system considered the words after the comma as a new column. Therefore, when the researcher downloaded the results from the system, an IS professional assistant was needed to resolve the problem. The researcher highly appreciated his efforts in this matter.

Dealing with the online survey tool in her school was very challenging and presented exciting new opportunities for the researcher, and the positive aspect was that most of the participants were able to complete the online surveys successfully and they provided positive feedback regarding the new methodology. Finally, from the researcher's perspective it is recommended that the above problems should be considered seriously, as these problems can occur in any online Web-based survey and can create gaps in the survey outcomes and findings and sometimes lead to frustration and exasperation for the researchers. Using the online survey tool was a challenging exercise for the researcher trying to complete her PhD research on time, but despite the challenges and glitches, the research was submitted on time.

DESIGN OF ONLINE SURVEY DURING THE POSTDOCTORAL RESEARCH STAGE - STRENGTHS AND WEAKNESSES

A new online survey was developed for the postdoctoral research stage on the basis of the researcher's PhD results. The new survey received the approval of the ethics committee at the university as the researcher carried out slight changes only in part five of the survey (New Participative Methodology for developing websites from the marketing perspective [integrated and contingent]); later the new online survey employed the Qualtrics online Web-based survey tool (http://www.qualtrics.com/) (See Figure 3).

Currently, there are different online surveys tools on the Internet, including SurveyMonkey, PeoplePulse, Swiftdigital, SurveyMethods, Zoomerang, WebSurveyCreator, and Qualtrics. For the postdoctoral research stage, the research selected the Qualtrics Web-based survey tool, as the functionality, navigation, presentation, design and editing are outstanding, and the maintenance and assistance under Qualtrics is exceptional, helpful, and friendly. On the basis of a literature review (Chen & Macredie, 2010; Cho, et al., 2011; Couper, et al., 2001; Fink, 2010; Gordon & McNew, 2008; O'Brien & Toms, 2010; Rego, Moreira, & Grarcia-Penalvo, 2010) and the researcher's perspective, this section will discuss the advantages and disadvantages of online survey usage in the postdoctoral research stage.

1. This survey offers researchers and college administrators a low-cost option for data collection compared with the traditional methods of paper and pencil or email.
2. This survey offers various facilities for sending reminders and thank you messages, and downloading data in various formats.
3. This survey reduces errors resulting from coding and retyping.
4. **Flexibility in Design:** The researcher has full control in designing, editing, deleting, and viewing the survey in line with his/her requirements; for example:

Figure 3. Summary of online survey phase: postdoctoral stage

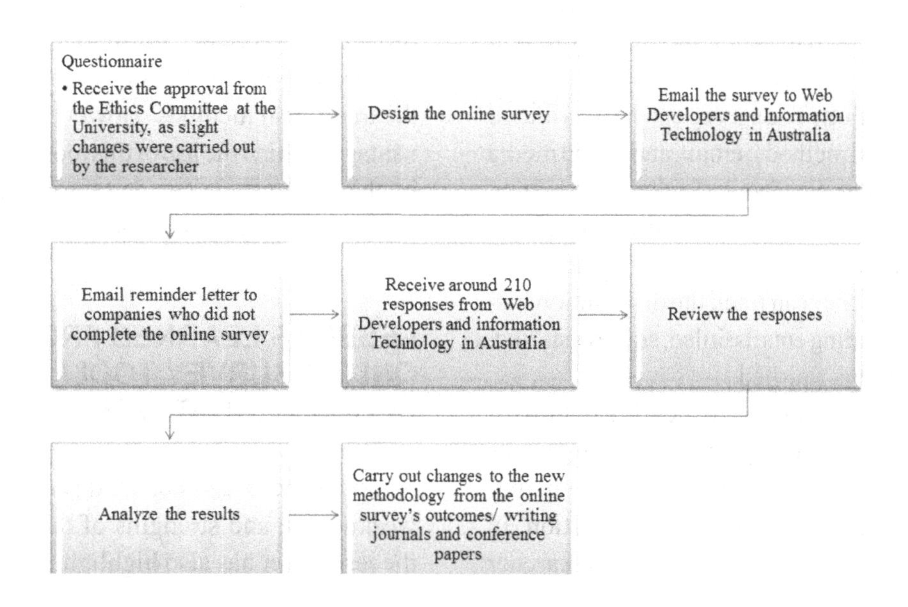

a. This survey contains several question types, including multiple choice, matrix table, text entry, text/graphic, constant sum, slider, rank order, side-by-side, pick, group and rank, drill down, hot spot, heat map, timing, and meta-info questions. Question types can range from single to multiple choices and the position can be horizontal or vertical.

b. This survey allows the researcher to change the look and feel of the survey, including color, font, and adding headers and footers for the survey. For example, for my survey I selected light blue or green for page borders, as these colors are associated with cold, calm, trust, and relaxation (Bonnardel, Piolat, & Le Bigot, 2011; Cyr, Head, & Larios, 2010). Furthermore, within this survey there is a survey option, which contains items, i.e. survey experience (i.e. 'back,' 'save,' and 'continue'), survey protection ('open/close access'), survey termination ('send additional thank you email to the respondents'), inactive survey, partial completion (setting a time and day for completing the survey), and finally, response set, which allows the researcher to view results of the same survey for different collection periods.

5. **Distribution:** The survey can be distributed via several methods: email, and social media (i.e. Twitter, Facebook, LinkedIn, Blogger, Digg, StumbleUpon, Reddit, Delicious, MySpace, Orkut, QR Code). Furthermore, the researcher can track the distribution history, including emails failed, surveys started, and surveys finished.
6. **Data Analysis:** The statistical breakdown of the surveys is well presented, since the results can be presented in various formats, including graphs, tables, and cross-tabulation. In addition, the survey will allow the researcher to export the survey outcomes in various formats to Word, PowerPoint slides, Excel, or PDF files.

Dealing with Qualtrics in my postdoctoral stage was challenging and exciting, as the majority of respondents completed the survey on time and were pleased with the structure and layout of the online survey (see Figure 4). It was an outstanding and relaxing exercise from the researcher's perspective, since designing and editing the online survey took less time than the PhD research stage. From the researcher's perspective, so far, there have been no glitches with Qualtrics, since the support and assistance under it are outstanding, helpful, and friendly. Finally, from the researcher's perspective, an online survey must be selected on the basis of the functionality, service, building and distributing, analysis and reporting, as these aspects are essential for conducting an online survey, in order to ensure your research journey is running effectively without any glitches.

The literature review and researcher's perspective confirmed and corroborated the research question in relation to online surveys. It was noted that the online survey becomes an essential tool for conducting quantitative research, as this tool will provide rich and historical information to enable the researcher to learn more about their research objectives. Furthermore, this tool is handier, less expensive, and more sustainable than the traditional methods. However, several disadvantages behind the use of this tool were confirmed by the literature review and the researcher's perspective in relation to privacy, security, and ethics.

LESSONS LEARNED FROM ONLINE SURVEY TOOL/ RECOMMENDATIONS

Although the researcher has discussed the design, weaknesses, and strengths of the online survey, the researcher has also highlighted a practical set

Figure 4. Screen shot of my online survey: postdoctoral stage

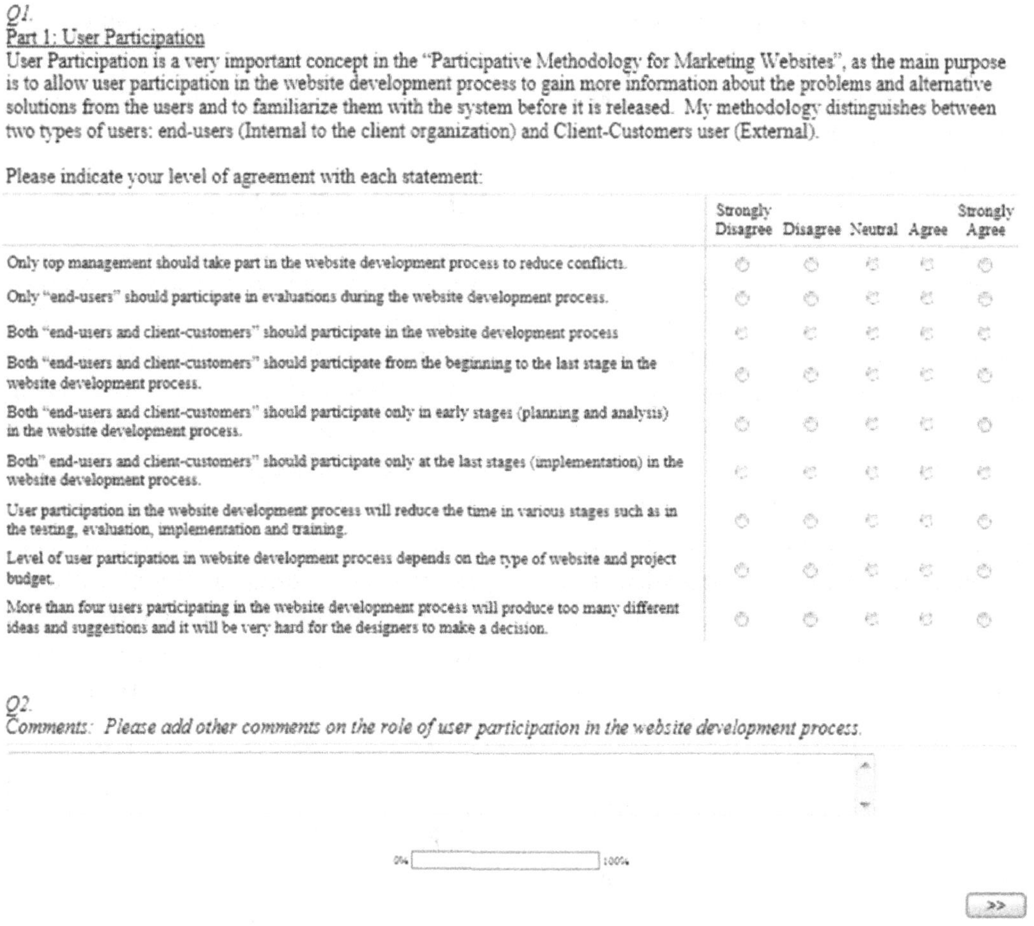

Q1.
Part 1: User Participation
User Participation is a very important concept in the "Participative Methodology for Marketing Websites", as the main purpose is to allow user participation in the website development process to gain more information about the problems and alternative solutions from the users and to familiarize them with the system before it is released. My methodology distinguishes between two types of users: end-users (Internal to the client organization) and Client-Customers user (External).

Please indicate your level of agreement with each statement:

	Strongly Disagree	Disagree	Neutral	Agree	Strongly Agree
Only top management should take part in the website development process to reduce conflicts.					
Only "end-users" should participate in evaluations during the website development process.					
Both "end-users and client-customers" should participate in the website development process					
Both "end-users and client-customers" should participate from the beginning to the last stage in the website development process.					
Both "end-users and client-customers" should participate only in early stages (planning and analysis) in the website development process.					
Both" end-users and client-customers" should participate only at the last stages (implementation) in the website development process.					
User participation in the website development process will reduce the time in various stages such as in the testing, evaluation, implementation and training.					
Level of user participation in website development process depends on the type of website and project budget.					
More than four users participating in the website development process will produce too many different ideas and suggestions and it will be very hard for the designers to make a decision.					

Q2.
Comments: Please add other comments on the role of user participation in the website development process.

0% 100%

>>

of recommendations and endorsements, which researchers must consider before conducting an online survey. These ratifications were driven from the researcher's perspective and the literature review (Cho, et al., 2011; Cornicelli, et al., 2011; Daniels, et al., 2011; Dillman, 2007; Dillman, et al., 2010; Fan & Yan, 2010; Gordon & McNew, 2008; Graefe, et al., 2011; Heiervang & Goodman, 2011; Kaufman, 2006; Lesser, Yang, & Newton, 2011; Rego, et al., 2010; Robinson & Martin, 2010; Sexton, et al., 2011; Smyth, et al., 2010; Tsai, et al., 2009; Umbach, 2004; Vaske, 2011). Before conducting an online survey, please consider the important issues:

1. Before the online survey is released to participants, a welcome page must be available to explain the following aspects: the survey's purpose, the time required to complete the survey, the number of survey questions, survey link, and deadline to complete the survey, and the researcher's privacy and confidentiality obligations and ethical approval. Finally, researchers should include in their welcome page the point that participation in this research is completely voluntary. Participants may withdraw at any time without prejudice or negative consequences.
2. Seek approval from your University before releasing the survey.

3. To increase the response rate, try to personalize the email letters. Furthermore, an online survey is requiring researchers to provide follow-up reminders; therefore, researchers must email the reminders ten days after releasing the survey.
4. Designing a survey is a challenging exercise for researchers; therefore, to attract responses and increase the response rate, the first question should be motivating, interesting, and easy to answer. Moreover, in order to increase the response rate, make sure the survey design is simple, and easy to follow. Furthermore, limit your survey page to two or three questions.
5. The survey should be concise, brief, succinct, and short: no longer than twenty minutes.
6. To reduce confusion among the participants, try to use a similar format, style, and structure to a paper-based survey.
7. To provide flexibility in your survey make sure that you add 'back' and 'previous' buttons to your survey form.
8. All the pages of the survey must contain instructions at the top of the page and a progress bar along the bottom to provide feedback to users about their proximity to completion.
9. All the pages of the survey must contain header and footer (e.g. survey name/researcher's name).
10. Color usage must be limited to one or two colors, to ensure that the survey is readable and easy to navigate.
11. Prevent participants scrolling from one side to another in the survey by limiting the line length.
12. Test your survey on different operation systems and Web browsers to test the style, structure, layout, appearance, resolution, and screen configurations.
13. Test your survey with your colleagues, fellow professionals, and friends to test the style, structure, question layout, and survey appearance.
14. During the testing stage, examine the survey results and ensure whether the conclusions match the research aims and objectives.

FUTURE RESEARCH DIRECTIONS

A new survey will be carried out by the researcher to assess the global perspective, especially in Europe, toward her New Participative Methodology for marketing websites. The Qualtrics Web-based survey will be employed for this project aligned to gain new knowledge and acquaintance, as new facilities and resources will be available soon. To ensure the success of the new survey, the researcher's colleague in Europe will administer the online survey; on the basis of this challenging exercise, the researcher and her colleague will present their perspective to local and international conferences and journals.

CONCLUSION

Currently the Internet provides an excellent platform for researchers to collect data. Adopting this technology in research will allow the researchers to collect more data inexpensively over a large geographic area. This chapter investigates the online survey tools, which were used during the PhD and postdoctoral research stages to evaluate the New Participative Methodology for developing websites from the marketing perspective. The researcher discussed the literature concerning the new methodology, the online survey; later she added her perspective toward online survey tools by listing their main weaknesses and strengths, and some recommendations to be considered before an online survey tool for a research study is selected. These aspects will assist researchers in preventing and reducing errors. Online survey tools are becoming very popular these days, since they are inexpensive and quick, reduce errors in interpretation and analysis of the data, and are more professional than the traditional methods. However, technical failures, computer viruses, Internet crimes, and hacking into the Web-based survey; can lead to a decrease in the response rate. Finally, the researcher will carry out a new survey to examine and assess the new methodology in other countries globally beside Australia via the Qualtrics online survey tool. This exercise aims to gather global perspectives toward the new methodology and to absorb new attitudes toward the online survey tool, as new facilities and resources will be available soon i.e. new reporting facility. In conclusion, collecting and analyzing data via traditional methods is messy and time-consuming, especially with massive volumes of materials and findings, as "finding themes and extracting meaning can be a daunting task" (QSR International, 2011). Therefore, this chapter has highlighted the importance of the online survey in conducting qualitative research, as this tool is inexpensive, easy, handy, and more sustainable than the traditional methods. Nevertheless, researchers should consider the drawbacks behind using this tool regarding privacy, security, and ethics.

REFERENCES

Avison, D., & Fitzgerald, G. (1993). *Information systems development: Methodologies, techniques and tools*. New York, NY: Alfred Waller Ltd, Publishers.

Bonnardel, N., Piolat, A., & Le Bigot, L. (2011). The impact of colour on website appeal and users' cognitive processes. *Displays*, *32*(2), 69–80. doi:10.1016/j.displa.2010.12.002

Boyer, K., Olson, J., Calantone, R., & Jackson, E. (2002). Print verus electronic surveys: A comparison of two data collection methodologies. *Journal of Operations Management*, *20*, 357–373. doi:10.1016/S0272-6963(02)00004-9

Carlbring, P., Brunt, S., Bohman, S., Austin, D., Richards, J., & Ost, L. (2007). Internet vs. paper and pencil administration of questionnaires commonly used in panic/agoraphobia research. *Computers in Human Behavior*, *23*, 1421–1434. doi:10.1016/j.chb.2005.05.002

Cavana, R. Y., Delahaye, B. L., & Sekaran, U. (2001). *Applied business research: Qualitative and quantitative methods*. Canberra, Australia: John Wiley & Sons.

Chen, S. Y., & Macredie, R. (2010). Web-based interaction: A review of three important human factors. *International Journal of Information Management*, *30*(5), 379–387. doi:10.1016/j.ijinfomgt.2010.02.009

Cho, H., & LaRose, R. (1999). Privacy issues in internet surveys. *Social Science Computer Review*, *17*(4), 1–19. doi:10.1177/089443939901700402

Cho, Y., Park, J., & Han, S. K. S. (2011). Development of a web-based survey system for evaluating affective satisfaction. *International Journal of Industrial Ergonomics*, *41*, 247–254. doi:10.1016/j.ergon.2011.01.009

Cornicelli, L., & Grund, M. (2011). Assessing deer hunter attitudes toward regulatory change using self-selected respondents. *Human Dimensions of Wildlife*, *16*(3), 174–182. doi:10.1080/10871209.2011.559529

Couper, M. P., Traugott, M. W., & Lamias, M. J. (2001). Web survey design and administration. *Public Opinion Quarterly*, *65*, 230–253. doi:10.1086/322199

Creswell, J. W. (2003). *Research design qualitative, quantitative, and mixed methods approaches* (2nd ed.). Thousand Oaks, CA: SAGE Publications.

Crump, B., & Logan, K. (2008). A framework for mixed stakeholders and mixed methods. *The Electronic Journal of Business Research Methods*, *6*(1), 21–28.

Cyr, D., Head, M., & Larios, H. (2010). Colour appeal in website design within and across cultures: A multi-method evaluation. *International Journal of Human-Computer Studies*, *68*, 1–21. doi:10.1016/j.ijhcs.2009.08.005

Daniels, A., Rosenberg, R., Anderson, C., Law, J., Marvin, A., & Law, P. (2011). Verification of parent-report of child autism spectrum disorder diagnosis to a web-based autism registry. *Journal of Autism and Developmental Disorders*, *42*(2), 257–265. doi:10.1007/s10803-011-1236-7

Dillman, D. (2007). *Mail and internet surveys "the tailored design method"* (2nd ed.). New York, NY: John Wiley & Sons, Inc.

Dillman, D., Glenn, P., Tortora, R., Swift, K., Kohrell, J., & Berck, J. (2009). Response rate and measurement differences in mixed-mode surveys using mail, telephone, interactive voice reponse (IVR) and the internet. *Social Science Research*, *38*, 1–18. doi:10.1016/j.ssresearch.2008.03.007

Dillman, D., Reipus, U., & Matzat, U. (2010). Advice in surveying the general public over the internet. *International Journal of Internet Science*, *5*(1), 1–4.

Eysenback, G., & Till, J. E. (2001). Ethical issues in qualitative research on internet communities. *British Medical Journal*, *323*, 1103–1105. doi:10.1136/bmj.323.7321.1103

Fan, W., & Yan, Z. (2010). Factors affecting response rates of the web survey: A systematic review. *Computers in Human Behavior*, *26*, 132–139. doi:10.1016/j.chb.2009.10.015

Fink, A. (2010). Survey research methods. *Education Research Methodology: Quantitative Methods and Research*, 152 – 160.

Fleming, C., & Bowden, M. (2009a). The most commonly cited disadvantages of web-based surveys are sample frame and non-response bias. *Journal of Environmental Management*, *90*, 284–292. doi:10.1016/j.jenvman.2007.09.011

Fleming, C., & Bowden, M. (2009b). Web-based surveys as an alternative to traditional mail methods. *Journal of Environmental Management*, *90*(1), 284–292. doi:10.1016/j.jenvman.2007.09.011

Gilbert, T. (2006). Mixed methods and mixed methodologies: The practical, the technical and the political. *Journal of Research in Nursing*, *11*(3), 205–217. doi:10.1177/1744987106064634

Gordon, J., & McNew, R. (2008). Developing the online survey. *The Nursing Clinics of North America*, *43*, 605–619. doi:10.1016/j.cnur.2008.06.011

Graefe, A., Mowen, A., Covelli, E., & Trauntvein, N. (2011). Recreation participation and conservation attitudes: Differences between mail and online respondents in a mixed mode survey. *Human Dimensions of Wildlife*, *16*(3), 183–199. doi:10.1080/10871209.2011.571750

Heiervang, E., & Goodman, R. (2011). Advantages and limitations of web-based surveys: Evidence from a child mental health survey. *Social Psychiatry and Psychiatric Epidemiology*, *46*(1), 69–76. doi:10.1007/s00127-009-0171-9

Hesse-Biber, S. (2010). Emerging methodologies and methods practices in the field of mixed methods research. *Qualitative Inquiry*, *16*(6), 415–418. doi:10.1177/1077800410364607

Hinton, L., Kurinczuk, J., & Ziebland, S. (2010). Infertility, isolation and the internet: A qualitative interview study. *Patient Education and Counseling*, *81*, 436–441. doi:10.1016/j.pec.2010.09.023

International, Q. S. R. (2011). *What is qualitative research.* Retrieved from http://www.qsrinternational.com/what-is-qualitative-research.aspx

Issa, T. (2008). *Development and evaluation of a methodology for developing websites.* (PhD Thesis). Curtin University. Perth, Australia. Retrieved from http://espace.library.curtin.edu.au:1802/view/action/nmets.do?DOCCHOICE=17908.xml&dvs=1235702350272~864&locale=en_US&search_terms=17908&usePid1=true&usePid2=true

Issa, T., & Turk, A. (2012). Applying usability and HCI principles in developing marketing websites. *International Journal of Computer Information Systems and Industrial Management Applications, 4*, 76–82.

Issa, T., Turk, A., & West, M. (2010). Development and evaluation of a methodology for developing marketing websites. In Martako, D., Kouroupetroglou, G., & Papadopoulou, P. (Eds.), *Integrating Usability Engineering for Designing the Web Experience: Methodologies and Principles*. Hershey, PA: IGI Global. doi:10.4018/978-1-60566-896-3.ch006

Jayaratna, N. (1994). *Understanding and evaluating methodologies -NIMSAD- A systemic framework.* London, UK: McGraw-Hill International.

Kaplan, B., & Duchon, D. (1988). Combining qualitative and quantitative methods in information systems research: A case study. *Management Information Systems Quarterly, 12*(4), 571–586. doi:10.2307/249133

Kaufman, D. (2006). Tools keep web content flowing. *Television Week, 25*(30), 60.

Kung, L., Picard, R., & Towse, R. (2008). *The internet and the mass media.* Thousand Oaks, CA: SAGE Publications Ltd.

Lesser, V. M., Yang, D. K., & Newton, L. D. (2011). Assessing hunters' opinions based on a mail and a mixed-mode survey. *Human Dimensions of Wildlife, 16*(3), 164–173. doi:10.1080/10871209.2011.542554

Likert, R. (1932). A technique for the measurement of attitudes. *Archives de Psychologie, 22*(140), 55.

Mahoney, C. (1997). Overview of qualitative methods and analytic techniques. In J. Frechtling, L. Sharp, & Westat (Eds.), *User-Friendly Handbook for Mixed Method Evaluations.* Washington, DC: NSF Program Officer Conrad Katzenmeyer.

Maudsley, G. (2011). Mixing it but not mixed-up: Mixed methods research in medical education: A critical narrative review. *Medical Teacher, 33*, 92–104. doi:10.3109/0142159X.2011.542523

McBurney, D. H., & White, T. L. (2007). *Research methods* (7th ed.). New York, NY: Thomson Learning.

Miller, J., & Glassner, B. (2006). The "inside" and the "outside": Finding realities in interviews. In Silverman, D. (Ed.), *Qualitative Research Theory, Method and Practice* (2nd ed., pp. 125–139). London, UK: SAGE Publications.

Mitchell, M., Lebow, J., Uribe, R., Grathouse, H., & Shoger, W. (2011). Internet use, happiness, social support and introversion: A more fine grained analysis of person variables and internet activity. *Computers in Human Behavior, 27*(6), 1857–1861. doi:10.1016/j.chb.2011.04.008

Myers, M. (2003). *Qualitative research in information systems.* Retrieved from http://www.qual.auckland.ac.nz

Neuman, W. L. (2000). *Social research methods: "Qualitative and quantitative approaches* (4th ed.). Reading, MA: Allyn & Bacon.

O'Brien, H., & Toms, E. (2010). The development and evaluation of a survey to measure user engagement. *Journal of the American Society for Information Science and Technology, 61*(1), 50–69. doi:10.1002/asi.21229

O'Neill, R. (2006). *The advantages and disadvantages of qualitative and quantitative research methods.* Retrieved from http://www.roboneill.co.uk/papers/research_methods.htm

Olle, T. W., Hagelstein, J., Macdonald, I. G., Rolland, C., Sol, H. G., & Assche, F. J. M. V. (1988). *Information systems methodologies: "A framework for understanding.* Reading, MA: Addison-Wesley Publishing Company.

Porter, S. (2004). Pros and cons of paper and electronic surveys. *New Directions for Institutional Research, 121*, 91–97. doi:10.1002/ir.103

Prahalad, C. K., & Rangaswami, M. R. (2009, September). Why sustainability now the key driver of innovation. *Harvard Business Review*, 56–64.

Prasad, A., Saha, S., Misra, P., Hooli, B., & Murakami, M. (2010). Back to green. *Journal of Green Engineering, 1*(1), 89–110.

Preece, J., Rogers, Y., Benyon, D., Holland, S., & Carey, T. (1994). *Human computer interaction.* Reading, MA: Addison-Wesley.

Rego, H., Moreira, T., & Grarcia-Penalvo, F. (2010). Web-based learning information system for web 3.0. [Berlin, Germany: Springer-Verlag.]. *Proceedings of WSKS, 2010*, 196–201.

Robinson, J., & Martin, S. (2010). IT use and declining social capital? More cold water from the general social survey (GSS) and the American time-use survey (ATUS). *Social Science Computer Review, 28*(1), 45–63. doi:10.1177/0894439309335230

Sekaran, U. (2003). *Research methods for business "a skill building approach"* (4th ed.). New York, NY: John Wiley & Sons.

Sexton, N., Miller, H., & Dietsch, A. (2011). Appropriate uses and considerations for online surveying in human dimensions research. *Human Dimensions of Wildlife, 16*(3), 154–163. doi:10.1080/10871209.2011.572142

Smyth, J., Dillman, D., Christian, L., & O'Neill, A. (2010). Using the internet to survey small towns and communities: Limitations and possibilities in the early 21st century. *The American Behavioral Scientist, 53*, 1423–1448. doi:10.1177/0002764210361695

Subrahmanyam, K., & Smahel, D. (2011). Internet use and well-being: Physical and psychological effects. *Advancing Responsible Adolescent Development*, 123-142.

Sullivan, O. (2011). An end to gender display through the performance of housework? A review and reassessment of the quantitative literature using insights from the qualitative literature. *Journal of Family Theory & Review, 3*(1), 1–13. doi:10.1111/j.1756-2589.2010.00074.x

Sun, S. (2011). The internet effects on students communication at Zhengzhou Institute of Aeronautical Industry Management. *Advances in Computer Science, Environment. Ecoinformatics and Education, 218*, 418–422. doi:10.1007/978-3-642-23357-9_74

Tashakkori, A., & Teddlie, C. (1998). *Mixed methdology: Combining qualitative and quantitative approaches.* Thousand Oaks, CA: SAGE.

Teddlie, C., & Tashakkori, A. (2009). *Foundations of mixed methods research - Integrating quantitative and qualitative approaches in the social and behavioral sciences.* Thousand Oaks, CA: SAGE Publisher.

Tsai, H. F., Cheng, S. H., Yeh, T. L., Shih, C.-C., Chen, K. C., & Yang, Y. C. (2009). The risk factors of internet addiction--A survey of university freshmen. *Psychiatry Research, 167*(3), 294–299. doi:10.1016/j.psychres.2008.01.015

Umbach, P. (2004). Web surveys: Best practices. *New Directions for Institutional Research, 121*, 23–38. doi:10.1002/ir.98

Van Deursen, A. J. A. M., & Van Dijk, J. A. G. M. (2009). Using the internet: Skill related problems in users' online behavior. *Interacting with Computers, 21*(5-6), 393–402. doi:10.1016/j.intcom.2009.06.005

Vaske, J. (2011). Advantages and disadvantages of internet surveys: Introduction to the special issue. *Human Dimensions of Wildlife: An International Journal, 16*, 149–153. doi:10.1080/10871209.2011.572143

Weinreich, N. K. (1996). *Integrating quantitative and qualitative methods in social marketing research.* Retrieved from http://www.social-marketing.com/research.html

Wiggins, B. (2011). Confronting the dilemma of mixed methods. *Journal of Theoretical and Philosophical Psychology, 31*(1), 44–60. doi:10.1037/a0022612

Chapter 2
eCommerce Trust Beliefs:
Examining the Role of National Culture

Regina Connolly
Dublin City University, Ireland

ABSTRACT

Many studies have raised awareness of the importance of trust in the online commercial environment, and it is widely acknowledged. Due to the international nature of eCommerce, it is likely that the influence of culture may extend to online consumers' trust responses. In this chapter, a trust measurement instrument that had been previously validated in Hong Kong was applied in both the United States and in Ireland—countries that differ in terms of individualism, uncertainty avoidance, and power-distance. Survey methodology was used to collect data. The results provide a refined understanding as to the influence of national culture on the generation of online consumers' trust beliefs. In doing so, they advance the understanding of information systems and diffusion researchers as well as contributing to the understanding of online vendors who seek to gain insight into the factors that can engender consumer trust in their websites.

INTRODUCTION

Trust is frequently posited as a key factor influencing the success of eCommerce environments (Rofiq & Mula, 2010; Guenther & Möllering, 2010; Golan, 2010). Understanding the predictors and dynamics of consumer trust is an issue of enduring interest for both researchers and practitioners. That interest derives from the understanding that trust is integral to transactions and its absence creates a chain reaction that manifests in lost sales, damage to reputation and market share gains to competitors. In a commercial world that is becoming increasingly commoditised, consumer trust is

DOI: 10.4018/978-1-4666-2491-7.ch002

Copyright © 2013, IGI Global. Copying or distributing in print or electronic forms without written permission of IGI Global is prohibited.

the defining factor that characterizes winners, and in its absence, losers. This has never been truer than in the online exchange context, a context that is characterized by perceived risk, lack of control and increased consumer vulnerability. Trust is consequently viewed as crucial for the success of eCommerce (Kumar & Sareen, 2009; Koufaris, et al., 2002), and it confers rich rewards. As Rajiv Dutta, eBay's chief financial officer notes, *"[At eBay] we do $2.25 billion worth of gross sales a quarter entirely on trust"* (Anders, 2001).

Perceived risk is frequently cited as one of the key factors inhibiting online transactions (Saprikis, et al., 2010; Xu, 2010; Chen & Barnes, 2007). However, added to this, a new factor, the global economic crisis, has entered the equation, resulting in a consumer population that has far less disposable income than previously, a fact that is manifested in sharply reduced online sales (ComScore, 2009). This has implications for Web vendors who are now competing for a vastly reduced pool of available consumers. For Web vendors seeking to gain and retain loyal market share, the imperative for them to ensure that consumers trust their brands and their transaction environments has never been greater.

Gaining that trust is not a simple task and studies have shown that between sixty to seventy-five percent of customers terminate their online transactions when asked to provide personal and financial credit card information as they do not trust the website (Rajamma, et al., 2009; Meziane & Kasiran, 2008). On a practical level, each incidence of shopping cart abandonment represents lost sales to the online retailer (Mullins, 2000), translating into a loss of more than $6.5 billion per year for the total online retailing industry (McGlaughlin, 2001). Clearly, a strong financial incentive exists for understanding how to successfully engender trust in online consumers.

Research on trust in online environments is diverse and includes research on trust in global virtual teams (Mitchell & Zigurs, 2009), trust in virtual organisations (Young, 2008), trust in virtual communities (Johnson & Kaye, 2009), trust in eGovernment (Mutulu, 2010), trust in IT artefacts (Wang & Benbasat, 2008) as well as trust in eCommerce (Dutton, et al., 2009; Connolly & Bannister, 2007). This chapter focuses on the latter. Despite its importance, the extant literature on trust in an online environment is comparatively sparse in contrast to the large body of work that exists on trust in a traditional bricks-and-mortar context. However, whilst it is true that the vast proportion of trust studies pertain to the offline environment, the findings of some of these studies transcend the distinction of traditional versus electronic commerce (Kracher, et al., 2005) and consequently the extant literature should be viewed as a source of deep and valuable insights, particularly for those seeking to understand the predictors and inhibitors of consumer trust in Internet commerce.

CHAPTER FOCUS

This chapter seeks to advance our understanding of trust in an online context by examining whether and to what degree national culture can influence online consumers' trust in Web vendors. Firstly, the importance of trust in an online context is outlined. The conceptual confusion that surrounds the construct is discussed, as are the antecedents, psychological and contextual determinants that facilitate the production and maintenance of trust beliefs. While these sections of the chapter draw from the established offline trust literature, the findings are germane to the online context as traditional and online environments share many commonalities in relation to the factors that can influence consumer trust responses and are therefore worthy of consideration by online trust researchers. In the subsequent section, the methodology used to achieve the research objective is outlined. This involved the application a trust measurement instrument (that had been previously validated in Hong Kong) to samples in the

United States and in Ireland—counties that differ in terms of individualism, uncertainty avoidance, and power-distance. The results are then outlined and this is followed by a detailed discussion of the findings that provides a refined understanding as to the influence of national culture on the generation on online consumers' trust beliefs.

TRUST

Trust is an interpersonal and a collective phenomenon that facilitates human interactions and consequently is considered essential for psychological health (Young, 2006). It has been described as a social glue that improves cooperation between people and a means of reducing perceived risk in situations of uncertainty and complexity (Grabner-Kraüter & Kaluscha, 2003; Mayer, et al., 1995), such as is the case in an electronic commerce context. In fact, it has been suggested that man's ability to trust is elemental to his social orientation (Van Den Berg & Van Lieshout, 2001). Sociologists (Gambetta, 1988), psychologists (Deutsch, 1973), organisational behaviour scientists (Kramer, 1999; Kramer & Tyler, 1996), as well as economists (Bradach & Eccles, 1989), anthropologists (Ekeh, 1974), and political scientists (Barber, 1983) have contributed to the wide body of work that exists on this topic. In fact, Golembiewski and McConkie (1975, p. 131) remark that there is *"no single variable which so thoroughly influences interpersonal and group behaviour as does trust,"* whilst Fukuyama (1995, p. 7) posits that "*a nation's ability to compete is conditioned by a single, pervasive cultural characteristic: the level of trust inherent in a society.*"

Defining Trust

The concept of trust can have many different shades of meaning. This stems from the fact that there is no unique, universally accepted definition of trust. The conceptual diversity that surrounds this construct is a consequence of the varying disciplines of the researchers and the foci of their research, as differing academic emphases, research objectives and insights have resulted in multiple conceptualisations of the construct. Moreover, there are many types of trust relationship ranging from trust between individuals through trust between organizations, and even trust between machines. Trust researchers have noted this considerable conceptual diversity (e.g. Hosmer, 1995) and have remarked that the consequent lack of conceptual clarity has fragmented insight into this construct (Bluhm, 1987). Hosmer (1995) summarises the situation by stating that there appears to be *"widespread agreement on the importance of trust in human conduct, but unfortunately there also appears to be equally widespread lack of agreement on a suitable definition of the construct"* (1995, p. 380).

In an attempt to reduce the conceptual confusion, Brenkert (1998) points to three views of trust in the literature, which he identifies as *Attitudinal, Predictability,* and *Voluntarist* views of trust. The attitudinal perspective of trust stresses dispositional or attitudinal characteristics. Therefore, it considers trusting behaviour to be a function of attitude and inclination, based on a specific set of beliefs, rather than a cognitive state of mind. Thus, Sabel (1993, p. 1133) states that *"trust is the mutual confidence that no party to an exchange will exploit another's vulnerabilities."* Likewise, Mayer *et al.* (1995, p. 712) view trust as "*a willingness to be vulnerable to the actions of another party based on the expectation that the other will perform a particular action important to the trustor.*" The predictability view of trust stresses the importance of expectations about the behavioural predictability of the other party, e.g. Dasgupta (1988, p. 51) emphasises "*expectations about the actions of other people.*" The voluntarist view of trust is that individuals voluntarily take the action of placing themselves in a position of vulnerability, believing that the other party has intentions of goodwill and not of harm. This is evident in Thomas' (1989, p. 181) statement that "*To trust another is to voluntarily make oneself vulnerable with respect to some good, having been led to believe by the other's actions toward one that no loss or harm will come to one as a result.*" Although attempts to categorise trust in terms of perspective are useful in imposing structure on the research field and on the terms of analysis, they are limited in that they predominantly evaluate trust in terms of the dispositional and attitudinal attributes of the individual and do not pay adequate attention to the influence of context on the individual's behaviour. For example, the role of context is particularly significant when that context is electronically mediated and communication is via a website. A more comprehensive approach is required such as that proposed by Sitkin and Roth (1993) who subdivide trust research into the four distinct categories of trust as an individual attribute, trust as a behaviour, trust as a situational feature and trust as an institutional arrangement. This approach reflects a more realistic understanding of the complex nature of the trust construct.

Despite the diverse definitions and categorisations, some points of commonality are evident. For example, the view of trust as positive expectation, and trust as confidence, frequently emerge in the literature. Both views are predicated on the predictability of behaviour and goodwill/ positive intention of the trustee. Thus, researchers such as McAllister (1995) stress the use of positive trusting expectations as a determinant of subsequent behaviour, whilst Good (1988) considers trust to be grounded in specific expectations relating to the other party's behaviour based on claims of that party and Deutsch (1973) stresses optimistic expectancy about another's behavioural motives. Barber's (1983) also defines trust in terms of optimistic expectations, which are assumed by the trustor in relation to a trusted party, and Fukuyama (1995) focuses on trust as a positive expectation grounded in shared norms and values systems. Researchers from an economics base also regard trust as expectation, rather than rational decision (Gambetta, 1988). However, they are more specific as to the source of that expectation. They consider that the trustor's expectation of positive behaviour is a function of the governance mechanisms and incentives which society uses to structure economic transactions in the marketplace. Consequently, trust is viewed (Bradach & Eccles, 1989) as the expectation that an exchange partner will not behave in an opportunistic way because of awareness of these governance mechanisms. In an online context, the threat of opportunistic behaviour is magnified in the mind of the consumer and consequently these governance mechanisms acquire a greater significance. This is discussed in more detail later in this chapter.

Moving beyond the concept of trust as a set of optimistic expectations, Golembiewski and McConkie (1975) instead perceive trust in terms of confidence. They posit that trust indicates confidence in some event, process, or person, based upon personal perceptions and experiences. Furthermore, they consider trust to be a dynamic rather than static phenomenon, which is strongly connected to overall optimism regarding the behaviour of the other party (1975, p. 134). This indicates that trust evolves over time and can be influenced by positive experience. However, Brenkert (1998) makes the important distinction between the confidence that enables trusting behaviour, and trustworthiness. He contends that trust is an attitude or disposition to put oneself into a situation of vulnerability that is dependent on the goodwill and good behaviour of the other party, whilst trustworthiness is an evaluative appraisal of other party in terms of whether they are worthy of trust. This distinction is valuable as it separates out the twin issues central to any examination of trust when defined in terms of confidence, i.e. the dispositional issues unique to the trustor and the perception that the target (e.g. the Web vendor) as trustworthy. While some of these arguments seem quite theoretical in nature, their value lies in that they focus attention on the issues that are essential in any examination of trust issues, i.e. that trustors are made to feel confident that their expectations will be satisfied because measures are in place to protect them against opportunistic behaviour, and that the goodwill of the trusted party is effectively communicated to the trustor. In this study, trust is viewed from an interpersonal perspective and is defined as an attitude of confidence directed towards the online vendor that may be influenced by the personality of the trustor and the attributes of the trustee (Mayer, et al., 1995).

The Antecedents of Trust

Researchers who consider trust to be a dependent variable suggest that a perception of trustworthiness results from the perception of a number of characteristics (e.g. Lee & Turban, 2001). Many attempts have been made to identify these characteristics (e.g. Mayer, et al., 1995). Whilst there may not be complete agreement, common themes such as the characteristics of ability, benevolence, and integrity consistently surface—albeit under differing titles and with diverse emphases.

Ability is perceived to relate to technical competence within a specific domain (Barber, 1983) and is widely considered (e.g. Mishra, 1996) to be an important determinant of trust. Mayer *et al.* (1995, p. 718) define ability as *"that group of skills, competencies, and characteristics that enable a party to have influence within some specific domain."* In a computer-mediated marketplace environment, the Web vendor's ability is likely to be evaluated in terms of the technical characteristics of the website such as website design, ease of use, the presence of security features, the reliability of the website, the speed of the transaction, and the delivery of the correct product within the agreed time frame (i.e., fulfillment of the transaction).

Researchers such as Barber (1983) and Larzelere and Huston (1980) emphasise benevolence as a characteristic that indicates trustworthiness. Benevolence implies a perception of positive intent and good motives which Mayer *et al.* (1995, p. 718) define as *"the extent to which a trustee is believed to want to do good to the trustor, aside from an egocentric profit motive."* In a computer-mediated marketplace context, the presence of privacy policies on websites (i.e. a guarantee that the vendor will not pass on information regarding customers to third parties) can reassure the consumer regarding the Web vendor's benevolence and trustworthiness (Bandyopadhyay, 2009).

Integrity has been proposed as an antecedent to trust (e.g. Butler, 1991). Early research (Gabarro, 1978) indicates one of the determinants of trust to be 'character,' of which integrity is a key component. Later research by Butler (1991) also suggests multiple characteristics associated with trustworthiness: - consistency, discreteness, fairness, and integrity and promise fulfillment. The proposed characteristics correlate closely with 'integrity.' Integrity is a derivative of the trustor's perception that the trustee behaves in a manner that indicates consistent and positive values. Factors, which may influence the perception of trustee integrity, include: the consistency of the party's past actions, credible communications about the trustee from other parties, belief that the trustee has a strong sense of justice, and the extent to which the party's actions are congruent with his or her words (McFall, 1987). In an online context, the consumer's evaluation of a Web vendor's integrity is influenced by communications from other parties (i.e. the experiences of the consumer's peers). In the absence of such communications, the vendor's integrity is likely to be evaluated in terms of whether the vendor provides product guarantees, no-quibble refunds and exchanges, does not overcharge, allows the customer to access their account history on-line, and most importantly of all, that in the event of a problem they stand by their guarantees.

Researchers such as Hardy, Phillips, and Lawrence (1998) and Lewis and Weigert (1985,

p. 456) contend that trust can be viewed as communication, i.e. as a sense-making process that generates bonds between groups. The shared meaning which develops through a successful communication process acts as a signal or cue of trust, and thus as a foundation for positive behaviour. Several studies (Tamimi, et al., 2000; Cho & Park, 2002) provide evidence that contact information and additional information services influence customers' evaluation of websites. Similarly, Van de Iwaarden and Van der Wiele's (2002) study shows that responsiveness and empathy—both of which are related to communication—are important influences on trustors' evaluation of the computer-mediated transaction environment. Research by Lee *et al.* (2002) suggests that vendor provision of online customer service is a measure that can reduce customer concerns. However, there is evidence (Tamimi, et al., 2003) to suggest that Web vendors do not understand the importance that on-line consumers attribute to website service and support features. Those services and support features contribute to the consumers' sense of control, which has been shown to influence both attitude and purchase intention in an online context (Koufaris & Hampton-Sosa, 2002).

Propensity to Trust

In the literature, the effect of the propensity to trust characteristic on the individual's trust response is a matter of dispute. Personality-based psychologists (e.g. Hofstede, 1980) contend that each person has a unique propensity to trust that is influenced by personality type, culture, and developmental experiences. They suggest that the individual's dispositional propensity to trust determines the amount and level of trust that a person has for another party in the absence of available or experiential information on which to base a judgment. On the other hand, organisational psychologists consider that situational factors exert a greater influence on the trust response than does the individual's tendency to trust (Kramer, 1999). This raises the question as to whether examinations of trust in an online context should include a measure of dispositional trust as a control variable. While there is sufficient evidence to suggest that individuals differ greatly in their tendency to trust others (Lee & Turban, 2001), a recent (Connolly & Bannister, 2007) examination of the antecedents of trust in eCommerce found no evidence to support the influence of this characteristic on the individual's trust response. However, the researchers suggested that the influence of this factor might be culture dependent as earlier studies in Asia (Cheung & Lee, 2000) have found this factor to exert significant influence on trust beliefs. More research is required to determine if this is in fact the case.

Trust and Perceived Risk

Trust is always discussed in relation to risk and uncertainty. As Mayer, *et al.* (1995, p. 711) note *"the need for trust only arises in a risky situation."* For researchers such as Nooteboom *et al.* (1997, p. 316) that relational risk has two dimensions, which they define as size of loss and probability of loss. They contend that risk of opportunism can be restrained by measures such as direct supervision or by means of a legal contract. While the dimensions proposed by Nooteboom *et al.* (1997) are relevant, equally important dimensions of risk such as the nature of the loss and the consequences of the loss should also be considered and are likely to be just as important to consumers as size and probability of loss. For example, purchasing online involves a leap of faith with the consumer being required to provide both personal and financial information and trusting that it will not be misused. Whilst much attention has been paid to consumers' online security concerns, it has been shown that the consequences of online privacy concerns include a lack of willingness to provide personal information online, rejection of eCommerce, or even unwillingness to use the Internet (Bandyopadhyay, 2009).

National Culture

There are definite indications in the literature that national culture may influence the generation of consumer trust beliefs (Golan, 2010; Gefen & Heart, 2006; Gefen, et al., 2004; Doney, et al., 1998). However, Gefen, and Heart (2006) note that despite repeated theorisations of trust and national culture as intricately related constructs, eCommerce trust researchers have for the main part ignored the potential effects of national culture. The majority of eCommerce trust research has been conducted in the United States, a country that exhibits high levels of individualism and uncertainty avoidance (Hofstede, 1980) and thus Gefen and Heart assert that conclusions based on studies conducted in one country cannot and should not be automatically applied to other cultures.

The cross-comparisons that exist speak to the fact that the antecedents of trust clearly change across cultures. For example, a recent study by Golan (2010) found control over anonymity (amongst other factors) to be a critically important online trust building mechanism for Israeli youth. Transference-based antecedents of trust have been found to be more significant in Korea than in the US and privacy concerns have been found to exert a stronger influence on the generation of trust beliefs in the United States (Kim, 2008). Other studies (e.g. Cyr, 2008) have shown that the relative importance attributed to website design trust antecedents differs across Canada, Germany, and China. Studies such as these confirm that trust antecedents are likely to vary significantly across culture and indicate that dimensions of national culture such as collectivism and uncertainty avoidance are key elements of culture that influence trust antecedents (Gefen & Heart, 2006).

The four dimensions of national culture as identified by Hofstede (1980) are individualism-collectivism, power distance, uncertainty avoidance, and masculinity. The first of these, individualism (IDV), is described by Hofstede as the degree to which individuals are integrated into groups. On the individualist side are societies in which the ties between individuals are loose: everyone is expected to look after him/herself and his/her immediate family. The opposite to this is collectivism which comprises societies in which people from birth onwards are integrated into strong, cohesive groups often extended families (with uncles, aunts and grandparents) which continue protecting them in exchange for unquestioning loyalty. In a collectivist society individuals tend not to trust strangers (Fukuyama, 1995). On the other hand, in an individualist culture trust of strangers tends to be higher. The second dimension is the Power Distance Index (PDI), which is the extent to which the less powerful members of organizations and institutions (like the family) accept and expect that power is distributed unequally. This represents inequality (more versus less), but defined from below, not from above. It suggests that a society's level of inequality is endorsed by the followers as much as by the leaders. Research (Shaffer & O' Hara, 1995) has shown that individuals from countries that with high PDI scores tend to have less trust for service providers than do individuals.

The third dimension of national culture that Hofstede discusses is the Uncertainty Avoidance Index (UAI). This dimension deals with a society's tolerance for uncertainty and ambiguity and indicates to what extent a culture programs its members to feel either uncomfortable or comfortable in unstructured situations. Uncertainty avoiding cultures try to minimize the possibility of such situations by strict laws and rules, safety and security measures. The opposite type, uncertainty accepting cultures, are more tolerant of opinions different from what they are used to; they try to have as few rules as possible, and on the philosophical and religious level they are relativist and allow many currents to flow side by side. The fourth of Hofstede's cultural dimensions is masculinity. Masculinity (MAS) refers to the distribution of roles between the genders, which is another fundamental issue for any society to which a range of solutions are found.

As the previous sections have outlined, the literature provides considerable evidence that a number of factors have strong predictive importance and are therefore deserving of consideration in any examination of trust. These factors include the characteristics of the online vendor (Chen & Dhillon, 2003; Bhattacharjee, 2002; McKnight, et al., 2002), third party certification (McKnight & Chervany, 2001; Hoffmann, et al., 1999; Jarvenpaa & Grazioli, 1999), the individual's propensity to trust (Kim & Prabhakar, 2004; Lee & Turban, 2001), a personality characteristic (Gurtman, 1992) and the influence of perceived risk (Verhagen, et al., 2006, 2004; Pavlou & Gefen, 2004; Van der Heijden, et al., 2003). In choosing a model to conduct the study it was therefore of particular importance that each of these constructs were adequately represented.

RESEARCH PHILOSOPHY

In terms of the social sciences, the field of Information Systems (IS) research is a relative newcomer. Consequently, in particular in earlier research, the methodologies used in IS research frequently borrowed theoretical approaches from research in other, more established, fields. Sometimes this has been done without reflection on the underlying assumptions of these approaches. This has resulted, on occasion, in a lack of clarity regarding the ontological and philosophical basis of the research approach chosen by the researcher. As the information systems field matures, researchers have emphasised the need to clarify the underlying philosophy and assumptions of IS research (e.g. Hirschhein & Lyytinen, 1996; Orlikowski & Baroudi, 1991). According to Williamson *et al.* (1982, pp. 31-32), the purpose of social research is *"to add to knowledge through exploration, description and explanation of social reality."* However, how that social reality is explored, described, and explained will be influenced by the belief systems of the researcher. In looking at any IS research strategy, it is therefore important to understand the epistemological and underlying philosophical (or metaphysical) assumptions on which it is based.

As is the case for other disciplines, the choice of IS research method will be strongly influenced by the theoretical lens that is used to conduct the investigation. Because there is no one best approach, the choice of research methodology should be based on pragmatic considerations. As Robey (1996, p. 406) states: *"theory and method are justified on pragmatic grounds as appropriate tools for accomplishing research aims."* The nature of the research problem should therefore decide the choice of methodology and the issue becomes one of methodological appropriateness i.e. selecting a method that is considered suitable to address the particular research questions under review (Downey & Ireland, 1983).

In approaching the research question, a positivist approach was deemed suitable, as a number of important factual issues need to be determined as these may influence the on-line consumer's perceptions and behaviours. For example, the on-line consumer may have been defrauded by an Internet vendor, may have received the wrong product, may have been overcharged or received the product after the agreed the delivery date. These facts are not subject to social construction. Hence, there needs to be some consideration of factual realities regardless of the individual's construction of these issues. Consideration of those specific issues is best achieved by use of a positivist research method.

RESEARCH DESIGN

The research was undertaken in a number of major stages, each made up of a number of steps. The first main stage was to design and execute a survey of two groups of potential on-line consumers - one group with a technical background and the other group with a business background. The purpose of the survey was to establish a base of factual information regarding the perceptions and external factors that predict trust beliefs, and to establish the extent to which the propensity to trust characteristic influences the respondents' trust response. The survey also examined a series of variables relating to the profile of the respondents.

According to Pinsonneault and Kraemer (1993) survey research is most effective when the central questions of interest about the phenomena are '*what is happening?*' and '*how and why is it happening?*' Survey research is especially well suited for answering questions about '*what,*' '*how much,*' and '*how many.*' The purpose of the survey in this research was to produce a quantification of trust issues that related to the study population. It is primarily concerned with relationships between variables, and with projecting findings descriptively to a predefined population. Four of the variables are concerned with perceptions that can predict or inhibit trust. Two of the variables are concerned with contextual aspects of the on-line environment, e.g. third party seals of approval and the legal environment. How the individual's propensity to trust moderates the relationship between each of these variables and their trust response was also examined. The information was collected by asking people structured and predefined questions and their answers became the unit of analysis. Finally, information was collected from the sample in such as way as to be able to generalise it to the population or specific groupings within the population. In this survey, there were clearly defined independent and dependent variables and a specific model of the expected relationships, which was tested against observations of the phenomenon. The purpose of the survey analysis is explanation. Survey research aimed at explanation asks about the relationships between variables and does so from theoretically grounded expectations about how and why the variables ought to be related.

MEASUREMENT INSTRUMENT

Comprehensive reviews of all trust studies and trust models were undertaken and a number of models were found to address many of the key issues of concern in this study (e.g. Kini & Choobineh, 1998; Tan & Thoen, 2000, 2001; Egger, 2000). However, the model deemed to be the most suitable for the purposes of this study was that proposed by Cheung and Lee (2000) as it captures the most significant set of trust antecedents, derived from different lines of previous research, and presents them as an integrated entity that can provide direction for empirical testing. For example, the measurement instrument contains 30 items measuring trust antecedents such as perceived security controls, perceived privacy controls, the vendor's perceived integrity, the vendor's perceived competence, personality, cultural environment, experience, third party recognition, legal framework, and perceived risk.

In their model (Figure 1) Cheung and Lee show that consumer trust in on-line shopping is predicted by two sets of antecedents—factors that create a sense of vendor trustworthiness and factors related to the external environment. The former relate to the vendor's perceived integrity and competence and the vendor's security and privacy controls. The latter (external environment) encompass third party recognition (e.g. seals of approval) and the legal framework. The model shows that the effect of both sets of factors on the consumer's trust beliefs is moderated by the consumer's propensity to trust. It also acknowledges the relationship between perceived risk and the online consumer's trust response.

The model was developed and validated in Asia (Hong Kong), a country that exhibits very low levels of individualism, has a high rank on the power distance index and low levels of uncertainty avoidance. These scores are in marked contrast to those of the United States and Ireland. For example, the US shows a very high individuality index with a score of 91. Ireland, on the

Figure 1. A conceptual model of trust in internet shopping (Cheung & Lee, 2000)

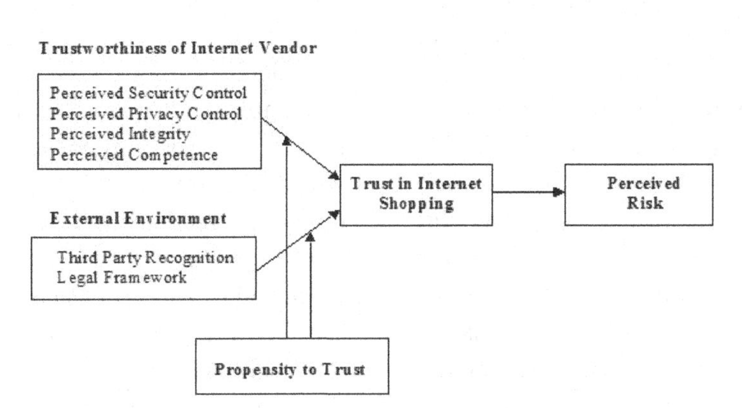

other hand, scores about 70 on the individualism index while Hong Kong scores markedly lower with a score of only 25. The PDI score for Hong Kong is high at 68 while the same score for the US is 40 (in an 11-104 scale) but Ireland scores only 28 on this ranking. In terms of the UAI index, the U.S ranks low at 46 (in an 8-112 scale) but Ireland ranks far lower at 35 with Hong Kong lowest at a score of only 29. This indicates that Hong Kong consumers and Irish consumers tend to be more tolerant of uncertain situations than are Americans. As all three countries are similar in the masculinity index with Hong Kong at 57, the US at 62 and Ireland at 68, masculinity is not hypothesised as an aspect of national culture that could provide an explanation for differences in trust beliefs between these countries. In addition, it should be noted that while the US and Ireland are by no means polar opposites on each of the cultural indices, differences between them are evident in terms of three of the four indices, in particular the individualism index and to a lesser degree the power-distance and uncertainty avoidance indices.

The hypothesized effects of national culture on online consumer's trust beliefs will be examined by comparing the same model developed and validated in Hong Kong with data collected in the US and Ireland. If, regardless of national culture, the trust antecedents are similar, this points to the culture independence of the model. However, if the results differ markedly, this confirms the concerns of researchers (e.g. Fukuyama, 1995) regarding the generalisation of US trust studies.

The measurement instrument proposed by Cheung and Lee was initially developed and tested in Hong Kong. Therefore, several measures were employed to ensure the reliability and validity of the survey instrument for use in an Irish context. Instrument validity was ensured as follows:

Face Validity: In order to assure face validity the measurement instrument was pilot tested on a sample of Irish people. No problems were detected with the language used in the measurement instrument (in either the Cheung and Lee 30-item section or the section added by the researcher). This pilot testing was repeated with the same sample to ensure results were reliable. The results were consistent and again no problems with the language were observed.

Content Validity: The measurement instrument encompasses all constructs essential to this study

DATA COLLECTION

Data collection describes how the sample was chosen and how the information was collected from participants. The unit of analysis in this survey is an individual. This is determined by the research purpose and research questions. The ideal sampling frame for this study would be all Internet users in Ireland and the US who have either purchased or contemplated purchasing over the Internet. However, this frame presents a number of difficulties. The first of these is identification of frame members. As information on this was not available from Internet Service Providers, the only way to survey such users would be a general population sample. However, a general population sample would have almost certainly resulted in an extremely low response rate for three reasons. Firstly, many of those who are users are not adults and it is possible that many of those who are adults would neither have purchased using the Internet nor have contemplated such a purchase. Secondly, the response rate from this type of survey is normally low. Thus, the sampling frame for this research was necessarily limited because to elicit the required information necessitated the involvement of participants with the capability (financial means and adequate technical skills) to purchase on-line—whether or not they did so. Further sampling frame limitations related to the comparative analysis aspect of the research. For example, it was necessary for one sample to have respondents with highly technical backgrounds and for the other sample to have respondents with non-technical backgrounds. For this reason, it was decided to focus on two groups that were highly likely to be Internet users, where a reasonable response rate could be expected, but which had quite different characteristics as potential purchasers. Given these limitations, the approach adopted in this research was to use a variation of cluster sampling.

As the objectives for the sample frames were groups who were likely to purchase on-line; one group with a strong background in technology; one group with a strong background in business; permission to access the respondents and likelihood of an acceptable response rate, the samples that comprised the Irish section of the research was derived from the MBA Graduates Association and from The Irish Computer Society. Both of these groups met the frame criteria and were willing to provide full access to their membership database for sampling purposes. The sample obtained from the MBA association consisted of 620 individuals who had completed an MBA degree. Their undergraduate disciplines and work backgrounds were predominantly non-technical. Their participation in the study was requested via post. They also received the questionnaire and accompanying documentation via post. The following week, follow-up letters were sent to all subjects who had not yet replied to remind them to complete the questionnaire. The sample obtained from the Irish Computer Society (ICS) consisted of 220 consumers with highly technical backgrounds. To become a professional member of the Irish Computer Society necessitates having a degree in a technical discipline along with a minimum of 3 years work experience in a technical or technical-related position. These requirements improved the likelihood that the sample would have adequate disposable income and technical competency to engage in on-line shopping. However, it was reasoned that an individual could have recently joined the ICS but that their technical knowledge could be considerably outdated, (which would reduce their knowledge and experience of on-line shopping). To overcome this potential limitation, an age limit of 45 years of age was imposed on the participants selected. The participation of this second sample in the study was requested via email. They also received and returned the questionnaire via email.

For the United States section of the research, the first sample chosen for this study was obtained from two postgraduate classes at Northeastern University Boston and comprised 75 individuals who were completing a master's level programme in Business Studies. The second sample consisted of employees of the Computer Services department at a major east coast university in the United States. These employees by nature of their occupation are highly technical in both their educational background and level of technical skill. This latter sample comprised 35 individuals. The participation of both of these samples was requested via email and they received and returned the questionnaire via email.

In selecting these groups, an element of judgement was involved. Of judgement sampling generally, Remenyi *et al.* (1998, p. 194) note, *"the composition of a judgement sample is not made with the aim of it being statistically representative of the population. Such samples comprise individuals considered to have the knowledge and information to provide useful ideas and insights."* Hence, this survey is considered representative of a section of the Internet population that are highly informed about on-line shopping and were capable of providing the insights necessary to further our understanding of the trust construct.

DATA ANALYSIS

The survey findings were analyzed using the following protocol:

1. Data from both the Irish and US samples (business and technical) used in the survey research was input into an SPSS[1] package.
2. A frequency analysis was conducted on data from both samples (descriptive statistics).
3. Checks for the internal reliability of each construct were made using Cronbach's Alpha values and factor analysis.
4. The factor analysis results indicated a strong association between certain items and constructs other than had been indicated in the original measurement instrument. Checks for the internal reliability of these newly formed constructs were made using Cronbach's Alpha values.
5. Correlation, partial correlation, and regression analysis techniques were used to test the thirteen proposed hypotheses from Cheung's work.

RESULTS

Table 1 shows the Cronbach's alpha values for each of the constructs. The three constructs of 'Personality,' 'Experience,' and the 'Legal Framework' worked particularly well with both samples. However, checks for internal reliability showed the 'Perceived Security,' 'Perceived Privacy,' and 'Cultural Environment' constructs to have scores of less than 0.65 for both samples. As the 'Perceived Security' construct is composed of only 2 items, this is likely to contribute to the low reliability measure for that construct.

The measure for the 'Cultural Environment' construct (which consists of only 2 items) was low at 0.263 for the Irish sample and at 0.252 for the US sample. In a previous test of this measurement instrument, Borchers (2001) also reported a low measure (0.5307) for the 'Cultural Environment' construct. He concluded that his use of a sample consisting of multiple ethnicities might have confounded his results. However, the Irish section of the research was composed of predominantly Irish respondents. Therefore, conflicting interpretation of items due to the respondents' multi-ethnicity does not explain the lack of a good measure for this construct. It may be that the high alpha result that Cheung and Lee (2000) obtained for the 'Cultural Environment' construct is particular to the Chinese environment[2] and that the respondents in their study were influenced by elements unique to that environment.

Nonetheless, as both samples has three constructs showing scores of less than 0.65, it was decided that the nature of the variables i.e. their relationships with each other and their relationship with the response variable (trust) should be examined more closely. In order to test construct validity, principal components analyses were conducted for each sample.

The principal components analysis of the data set obtained from the Irish sample resulted in a number of significant changes to item associations. The original eleven factors were reduced to eight.

Table 1. Reliability analysis: scale (alpha)

Construct	Number of items	Cronbach's Alpha Ireland	Cronbach's Alpha USA	Cronbach's Alpha Cheung and Lee (2000)
Perceived Security Control	2	0.625	0.546	0.794
Perceived Privacy Control	3	0.624	0.611	0.810
Perceived Integrity	2	0.687	0.663	0.764
Perceived Competence	3	0.685	0.680	0.846
Personality	4	0.886	0.872	0.881
Cultural Environment	2	0.263	0.252	0.833
Experience	3	0.887	0.879	0.880
Third party recognition	3	0.700	0.695	0.795
Legal Framework	2	0.797	0.762	0.882
Trust in Internet Shopping	3	0.6538	0.666	0.860
Perceived Risk	3	0.7262	0.711	0.864

Five factors remained unchanged: 'Perceived Security, Perceived Competence, Experience, Third Party Regulation, and Legal Framework. The three constructs containing new item associations were subjected to a new reliability test. The reliability analysis showed that the three constructs containing new item associations, i.e. 'Perceived Integrity,' 'Personality,' and 'Trust in Internet Shopping,' worked particularly well with this sample. When a principal components analysis was applied to the data set obtained from the US sample a number of significant changes to item associations also resulted with only three factors remaining unchanged: 'Perceived Competence, Third Party Regulation, and Experience. As was the case for the Irish sample, the items that comprised the Personality and Culture constructs loaded strongly together as did the items that comprised the trust and risk constructs. However, unique to this sample, the items that comprised the legal framework construct merged with those that comprised the Perceived Security construct (The implication of this is that it would not be possible to measure the influence of the Legal Framework variable on Trust for this sample). The constructs containing new item associations were then subjected to a new reliability test and showed far stronger results. Having secured stronger reliability measures for the variables that had previously caused some concern, the relationship between the Trust construct and the variables that are considered to influence its formation were measured using Pearson correlation techniques. The correlation coefficient results for the Irish and US samples are outlined in Tables 2 and 3, respectively.

The coefficient results show that a positive relationship exists between trust and the independent variables. The strongest results for the Irish sample were provided by the Perceived Integrity and Perceived Competence variables, thus confirming the importance of the perceived characteristics of the vendor (as mediated by the technology) in generating trust beliefs. The correlation coefficient result for the relationship between 'Experience' and Trust was notably stronger than was the result for the relationship between 'personality' (i.e. tendency to trust) and Trust indicating that the Irish consumer's experience of the on-line purchase environment exerts a stronger influence on the formation of their trust beliefs than does their tendency to trust. The strongest result for the US sample was provided by the Experience variable. While the relationship be-

Table 2. Correlation coefficient results: Irish sample

Correlations Irish Sample	1	2	3	4	5	6	7	8
(1) Trust	1.00	0.39	0.49	0.45	0.41	0.19	0.34	0.27
(2) Perceived Security		1.00	0.44	0.34	0.28	0.25	0.31	0.10
(3) Perceived Integrity			1.00	0.43	0.28	0.22	0.53	0.26
(4) Perceived Competence				1.00	0.35	0.23	0.27	0.24
(5) Experience					1.00	0.20	0.11	0.24
(6) Third Party Recognition						1.00	0.26	0.20
(7) Legal Framework							1.00	0.27
(8) Personality								1.00

tween trust and the characteristics of the vendor (i.e. Perceived Security, Perceived Integrity, and Perceived Competence) was stronger than that shown by other variables, the relationship between trust and experience was remarkably stronger.

In the model proposed by Cheung and Lee (2000) the moderating variable 'Propensity to Trust' is a composite of two sets of items. These are (1) Personality: items relating to the individual's tendency to trust and (2) Experience: items relating to the individual's previous experience. In this study, these sets of items have been treated separately in order to examine their effects more closely. Two tests were therefore conducted, the first controlling for personality items (i.e. the respondent's tendency to trust) and the second test controlling for experience. The results of these tests are shown in Table 4. The variable providing the strongest result is the perceived integrity variable. The coefficient result for this variable is stronger when controlling for personality than when controlling for experience, indicating that the individual's on-line experience has a greater influence on their perceptions of Web vendor integrity than does their tendency to trust.

Multiple regression techniques were then used to establish whether the set of independent variables could explain a proportion of the variation in the dependent variable at a significant level, and to establish the relative predictive importance of the independent variables. The results, outlined in Table 4, showed that the independent variables explain 40% of the variation in trust for the Irish sample and 36% of the variation in trust for the US sample.

The coefficient results obtained for the Irish sample (outlined in Table 5) indicate that two of the independent variables—perceived integrity

Table 3. Correlation coefficient results: US sample

Correlations (US sample)	1	2	3	4	5	6	7
(1) Trust	1.00	0.47	0.47	0.41	0.66	0.20	0.31
(2) Perceived Security		1.00	0.56	0.33	0.07	0.30	0.32
(3) Perceived Integrity			1.00	0.47	0.04	0.21	0.29
(4) Perceived Competence				1.00	0.35	0.23	0.20
(5) Experience					1.00	0.22	0.01
(6) Third Party Recognition						1.00	0.21
(7) Personality							1.00

Table 4. Regression analysis

IRISH SAMPLE		US SAMPLE	
R Square .413	**Adjusted R Square** .394	**R Square** .378	**Adjusted R Square** .361
Predictive Variables: Perceived Competence, Personality, Third Party Regulation, Legal Framework, Experience, Perceived Security Control, Perceived Integrity.			

(coefficient beta weight 0.239) and experience (coefficient beta weight 0.227)—exert the strongest effect on the dependent variable. Perceived security and perceived competence are significant independent variables—but to a lesser degree. Each of these four variables is positively related to the dependent variable. Age also has significant predictive importance but is a negatively related variable.

We failed to reject the null hypothesis that the true coefficient value of third party recognition, legal framework, and personality are zero at the 0.05 level. In the case of all other variables, it is possible to be confident at a 0.05 level that they do have significant explanatory powers.

When the coefficient results obtained for the US sample (outlined in Table 6) are examined it is clear that perceived security (coefficient beta weight 0.257) and experience (coefficient beta weight 0.269) exert the strongest effect on the dependent variable. Perceived integrity has predictive value—but to a lesser degree. Each of these variables is positively related to the dependent variable.

We failed to reject the null hypothesis that the true coefficient value of third party regulation, personality and perceived competence are zero at the 0.05 level. In the case of all other variables, it is possible to be confident at a 0.05 level that they do have significant explanatory powers.

In order to ascertain whether there was any evidence of multicollinearity among the predictor variables in this study a number of studies were conducted. As the highest correlation coef-

Table 5. Predictive importance of independent variables: Irish sample

Coefficients[a]

Model		Unstandardized Coefficients		Standardized Coefficients		
		B	Std. Error	Beta	t	Sig.
1	(Constant)	.257	.311		.826	.410
	Perceived Security Control	.150	.054	.154	2.765	.006
	Perceived Integrity	.293	.079	.239	3.700	.000
	Perceived Competence	.189	.059	.181	3.211	.001
	Experience	.235	.055	.227	4.238	.000
	Third Party Regulation	-8.77E-02	.058	-.078	-1.516	.131
	Legal Framework	8.120E-02	.060	.079	1.350	.178
	Age	-.136	.050	-.134	-2.704	.007
	Personality	7.841E-02	.050	.081	1.562	.120

a. Dependent Variable: Trust

Table 6. Predictive importance of independent variables: US sample

Coefficients[a]

Model		Unstandardize Coefficient		Standardize Coefficient		
		B	Std.	Beta	t	Sig.
1	(Constant	.055	.315		.175	.862
	Perceived Security	.314	.079	.257	3.975	.000
	Perceived Integrity	.246	.083	.203	2.972	.003
	Perceived Competence	.137	.067	.130	2.052	.041
	Personality	.045	.055	.046	.820	.413
	Third Party Regulation	-.037	.066	-.032	-.561	.576
	Experience	.253	.060	.249	4.230	.000

Dependent Variable: Trust

ficient results was 0.53 for the Irish sample and 0.66 for the US sample this is the first indication that the results are not influenced by multicollinearity. As reliance on correlation coefficients alone may miss linear combinations among the independent variables that could also give rise to multicollinearity, a Variance Inflation Factor (VIF) was also computed. However, the highest VIF for the Irish sample was 1.775 (associated with Perceived Integrity) and 1.635 for the US sample again indicating that the regression results are not influenced by multicollinearity. Finally, the condition number for the data used in this regression analysis was found to be only 32 for the Irish sample and 26 for the US sample. In summary, there is no evidence that multicollinearity among the predictor variables used in this study has influenced the regression results. In fact, there is strong evidence to the contrary. Consequently, the regression results can be interpreted with confidence.

DISCUSSION

The factor that is the strongest predictor of consumers' perceptions of vendor trustworthiness differed across the samples. However, one point of commonality that emerged relates to the importance of experience as a predictor of trust beliefs. For both samples, experience was the variable with the second strongest predictive influence on trust beliefs. For the Irish sample, it had greater predictive effect on the generation of trust beliefs than did perceptions of vendor competence. This means that not only is experience a strong predictor of Irish consumers' trust beliefs, but it is more influential than perceived competence and stronger than security controls. For the US sample, experience had greater predictive effect on the generation of trust beliefs than did perceptions of vendor integrity or competence.

The fact that experience exerts greater influence on customers' trust beliefs is not surprising. The on-line purchase environment is fraught with risk and a negative on-line transaction experience would undoubtedly influence subsequent beliefs regarding that environment. However, the fact that experience is a stronger predictor of trust beliefs than perceived security controls for the Irish sample and a stronger predictor of trust beliefs than perceived integrity or perceived competence for the US sample is highly significant and has two main implications. Firstly, it indicates the need to ensure that each customer's transaction experience is satisfactory and that they view their experience of the website interaction positively. However, as it is not possible to anticipate the concerns of every customer, it is essential to have some measures in place that will overcome this problem and increase the perception of the on-line transaction as a positive experience. Enabling the consumer to communicate with the vendor's customer service representatives is one means of addressing the consumer's concerns, reducing perceived risk and ensuring a positive outcome. Secondly, it points to the value of peer referrals. Web vendors should use incentive schemes to encourage referrals from satisfied customers. These schemes reward customers when they refer other customers who subsequently purchase from the on-line vendor. Vendors should also provide review sections on their websites and provide a peer ranking system. This allows potential customers to see how the vendor is rated by customers who have previously purchased from that vendor. These features provide the prospective on-line shopper with evidence that other consumers who purchased from that vendor experienced a positive outcome. Consequently, it encourages them to believe that should they purchase from that vendor, their experience will be equally positive.

The results obtained from the Irish sample shows some support for the hypotheses that certain characteristics convey a perception of vendor trustworthiness and thus are significant antecedents of trust beliefs. The factor that showed the strongest predictive relationship with trust beliefs is perceived integrity (combining privacy). While perceived vendor competence and perceived security showed some association with trust beliefs, it was to a far lesser degree than either of these variables. This result, indicating the influence perceived integrity in the generation of a trust response is consistent with the general trust literature (e.g. Butler, 1991) and shows that such a perception also extends to the online environment.

Experience was the variable with the second strongest influence on the trust response. Relationships between other factors and trust were insignificant or negative. For example, the relationship between the external factors (third party recognition and legal framework) and trust was remarkably weak. Similarly, the results obtained for the US sample provide limited support for the hypotheses that trust-related vendor characteristics predict the online consumer's trust response. While the results for this sample show that perceived security does influence consumer trust in online shopping, it must be borne in mind that that variable is an equal composite of security items and legal framework items. Moreover, this sample showed a very strong relationship between experience and the online consumer's trust response. However, the finding the emphasis placed by the US sample on security and legal framework confirms that view in the literature that mechanisms which reduce perceived risk in an online environment can generate positive trust outcomes (e.g., Grabner-Kraüter & Kaluscha, 2003; Bandyopadhyay, 2009).

In marked contrast to Cheung and Lee's model, the personality construct did not appear to moderate the relationships between the variables and trust for either sample. A previous test of this model (Borchers, 2001) also found that the propensity to trust construct did not have any impact on the relationships between the independent variables and trust. This research supports Borchers' original conclusion. However, when the items comprising the propensity to trust (tendency to trust and experience) are treated separately, experience was shown to have very strong predictive power across both samples—in fact it was the variable with the second strongest predictive influence on trust beliefs for both samples.

The implication of these findings is that Web vendors seeking to be successful in the Irish marketplace should focus on ensuring that they convey a perception of integrity. This can be achieved by providing clearly defined terms and conditions regarding the transaction and by providing customers with guarantees that their rights are protected. US consumers value evidence of security mechanisms and therefore Web vendors should ensure that security of transactions on their websites is emphasised to the online consumer. The management of both of these factors is within the control of the vendor.

As the results show only partial support for Cheung and Lee's model, one question that remains is whether the results obtained in this study indicate that Cheung and Lee's model is intrinsically wrong or whether Irish and US on-line consumers differ from Hong Kong/Chinese consumers in terms of the factors that predict their trust response in an on-line purchase context. The present research found that many of the relationships proposed by the Cheung and Lee model were weak or not significant. For example, the moderating influence of the propensity to trust construct did not have a significant influence on Irish or US on-line consumers' trust response. Similarly, third party regulation did not influence Irish or US on-line consumers' trust response to any significant degree. Therefore, it appears that the model proposed by Cheung and Lee contains significant weaknesses, lending weight to the argument that conclusions based on studies conducted in one country cannot and should not be automatically applied to other cultures.

CONCLUSION

This study examined the factors that influence consumer trust in online shopping in Ireland and the United States. It sought increased insight into the nature of the trust construct as observed in the behaviour of users and potential users of online shopping. The findings show that while differences between the samples exist, the importance of previous experience as a predictor of trust beliefs is consistent across both samples and in both cases, it had even greater predictive effect on the generation of trust beliefs than did perceptions of vendor competence. For Web vendors, this speaks to the need to ensure that customer service representatives are available to address consumers concerns, thereby reducing perceptions of risk and increasing positive outcomes for the consumer. It also speaks to the need to focus on peer referrals from satisfied customers.

The results indicate that Irish online consumers place great emphasis on a perception of vendor integrity, thus pointing to the need for Web vendors to provide clearly defined terms and conditions that convey to consumers that their rights are protected. US consumers place great emphasis on evidence of security mechanisms, pointing to the need for Web vendors to convey the security of transactions on their websites. This study is the first large-scale empirical study of its kind comparing the antecedents of consumer trust in online shopping in Ireland and the United States, thereby providing insight into the influence of national culture on online consumer's trust responses. Whilst confirming the role of national culture and the need for researchers to consider same in future online trust studies, it also provides practitioners with the insights necessary to help them improve consumer trust in their websites. In doing so, it provides a valuable contribution to information systems research and to the overall body of marketing, trust and diffusion research.

Despite researchers' increasing interest in the online trust construct, it is clear that the literature on online trust is very much at an early stage of development when compared to the rich body of research that exists on trust in an offline environment. Many gaps in our knowledge remain. For example, further research on the importance of peer ratings on trust formation and how online trust is affected in the presence of multiple outlets and channels is needed to progress our understanding of the construct. In addition, there is a need for more precise research on the trust formation process, which Urban *et al.,* (2009) contend could be achieved through an examination of a longitudinal database of site visits and trust levels. As overall trust cannot be measured via a single scale, an attitude bank of questions that can be used to measure trust across multiple environments would be particularly useful and transcend the problem of the differing disciplines and foci of trust researchers. Such an attitude bank would provide stronger, more reliable trust measures thereby facilitating more systematic examinations of online trust. Much remains to be done.

REFERENCES

Anders, G. (2001). Business fights back: eBay learns to trust again. *Fast Company*. Retrieved 10th September, 2011 from http://www.fastcompany.com/magazine/53/ebay.html

Barber, B. (1983). *The logic and limits of trust*. New Brunswick, NJ: Rutgers University Press.

Belanger, F., Hiller, J. S., & Smith, W. J. (2002). Trustworthiness in electronic commerce: The role of privacy, security and site attributes. *The Journal of Strategic Information Systems*, *11*, 245–270. doi:10.1016/S0963-8687(02)00018-5

Bhattacherjee, A. (2002). Individual trust in online firms: Scale development and initial test. *Journal of Management Information Systems*, *19*, 211–241.

Briggs, P., Simpson, B., & De Angeli, A. (2004). Trust and personalisation: A reciprocal relationship? In Karat, C.-M., Blom, J., & Karat, J. (Eds.), *Designing Personalized User Experiences for eCommerce*. Dordrecht, The Netherlands: Kluwer Academic Publishers. doi:10.1007/1-4020-2148-8_4

Chen, S. C., & Dhillon, G. S. (2003). Interpreting dimensions of consumer trust in e-commerce. *Information Technology Management*, *4*, 303–313. doi:10.1023/A:1022962631249

Cheung, C., & Lee, M. (2000). Trust in internet shopping: A proposed model and measurement instrument. In *Proceedings of the 2000 Americas Conference on Information Systems (AMCIS)*, (pp. 681-689). Los Angeles, CA: AMCIS.

Chopra, K., & Wallace, W. A. (2003). Trust in electronic environments. In *Proceedings of the 36th Hawaii International Conference on System Sciences*. IEEE.

Cyr, D., Bonanni, C., Bowes, J., & Ilsever, J. (2005). Beyond trust: Website design preferences across cultures. *Journal of Global Information Management*, *13*(4), 24–52. doi:10.4018/jgim.2005100102

Davis, F. D., Bagozzi, R. P., & Warshaw, P. R. (1989). User acceptance of computer technology: A comparison of two theoretical models. *Management Science*, *35*(8), 982–1003. doi:10.1287/mnsc.35.8.982

Deutsch, M. (1962). Cooperation and trust: Some theoretical notes. In M. R. Jones (Ed.), *Nebraska Symposium on Motivation*, (pp. 275-319). Lincoln, NE: University of Nebraska Press.

Doney, P. M., Cannon, J. P., & Mullen, M. R. (1998). Understanding the influence of national culture on the development of trust. *Academy of Management Review*, *23*(3), 601–620.

Egger, F. N., & de Groot, B. (2000). *Developing a model of trust for electronic commerce: An application to a permissive marketing web site*. Paper presented at 9th International World Wide Web Conference. Amsterdam, The Netherlands.

Ekeh, P. P. (1974). *Social exchange theory: The two traditions*. London, UK: Heinemann Educational.

Erickson, G. S., Komaromi, K., & Unsal, F. (2010). Social networks and trust in e-commerce. *International Journal of Dependable and Trustworthy Information Systems*, *1*(1), 45–59. doi:10.4018/jdtis.2010010103

Fukuyama, F. (1995). *Trust: The social virtues and the creation of prosperity*. New York, NY: Free Press.

Gambetta, D. G. (Ed.). (1988). *Trust: Making and breaking cooperative relations*. New York, NY: Basil Blackwell.

Gefen, D., & Heart, T. (2006). On the need to include national culture as a central issue in e-commerce trust beliefs. *Journal of Global Information Management, 14*(4), 1–30. doi:10.4018/jgim.2006100101

Gefen, D., Rose, M., Warkentin, M., & Pavlou, P. (2004). Cultural diversity and trust in IT adoption: A comparison of USA and South African e-voters. *Journal of Global Information Systems, 13*(1), 54–78. doi:10.4018/jgim.2005010103

Golan, O. (2010). Trust over the net: The case of Israeli youth. *International Journal of Dependable and Trustworthy Information Systems, 1*(2), 70–85. doi:10.4018/jdtis.2010040104

Grabner-Krauter, S., & Kaluscha, E. A. (2003). Empirical research in on-line trust: A review and critical assessment. *International Journal of Human-Computer Studies, 58*, 783–812. doi:10.1016/S1071-5819(03)00043-0

Guenther, T., & Möllering, G. (2010). A framework for studying the problem of trust in online settings. *International Journal of Dependable and Trustworthy Information Systems, 1*(3), 14–31. doi:10.4018/jdtis.2010070102

Gurtman, M. B. (1992). Trust, distrust, and interpersonal problems: A circumplex analysis. *Journal of Personality and Social Psychology, 62*, 989–1002. doi:10.1037/0022-3514.62.6.989

Hacker, S. K., Willard, M. A., & Couturier, L. (2002). *The trust imperative*. New York, NY: American Society of Quality.

Hirscheim, R., & Lyytinen, K. (1996). Exploring the intellectual foundations of information systems. *Accounting. Management and Information Technology, 2*(1-2), 1–64.

Hoffman, D. L., Novak, T. P., & Peralta, M. (1999). Building consumer trust on-line. *Communications of the ACM, 42*(4), 80–85. doi:10.1145/299157.299175

Hofstede, G. (1980). Motivation, leadership, and organization: Do American theories apply abroad? *Organizational Dynamics, 9*(1), 42–63. doi:10.1016/0090-2616(80)90013-3

Jarvenpaa, S. L., & Grazioli, S. (1999). Surfing among sharks: How to gain trust in cyberspace. *Financial Times, Mastering Information Management, 15,* 2-3.

Kim, K. K., & Prabhakar, B. (2004). Initial trust and the adoption of B2C e-commerce: The case of internet banking. *The Data Base for Advances in Information Systems, 35*(2), 50–65. doi:10.1145/1007965.1007970

Kini, A., & Choobineh, J. (1998). *Trust in electronic commerce: Definition and theoretical considerations*. Paper presented at the 31st Hawaii International Conference on System Sciences. Hawaii, HI.

Koufaris, M., & Hampton-Sosa, W. (2004). The development of initial trust in an online company by new customers. *Information & Management, 41*(3), 377–397. doi:10.1016/j.im.2003.08.004

Kumar, M., & Sareen, M. (2009). Impact of technology-related environment issues on trust in B2B e-commerce. *International Journal of Information Communication Technologies and Human Development, 3*(1), 21–40. doi:10.4018/jicthd.2011010102

Lee, M., & Turban, E. (2001). A trust model for consumer internet shopping. *International Journal of Electronic Commerce, 6*(1), 75–91.

Liu, C. (2010). Human-machine trust interaction: A technical overview. *International Journal of Dependable and Trustworthy Information Systems, 1*(4), 61–74. doi:10.4018/jdtis.2010100104

Mayer, R. C., Davis, J. D., & Schoorman, F. D. (1995). An integrative model of organisational trust. *Academy of Management Review, 20*(3), 709–734.

McKnight, D. H., & Chervany, N. L. (2001). What trust means in e-commerce customer relationships: An interdisciplinary conceptual typology. *International Journal of Electronic Commerce, 6*(2), 35–59.

McKnight, D. H., Choudhury, V., & Kacmar, C. (2002). Developing and validating trust measures for e-commerce: An integrative typology. *Information Systems Research, 13*(3), 334–359. doi:10.1287/isre.13.3.334.81

Mulpuru, S. (2006). US ecommerce outlook for Q4 2006: A recap of Q3 2006 online retail sales and overview of the upcoming holiday. *Forrester Research Business View Research Document.* Retrieved December 14th 2011 from http://www.forrester.com/rb/Research/us_ecommerce_outlook_for_q4_2006/q/id/40586/t/2

Orlikowski, W., & Baroudi, J. J. (1991). Studying information systems in organisations: Research approaches and assumptions. *Information Systems Research, 2*(1), 1–28. doi:10.1287/isre.2.1.1

Pavlou, P. (2003). Consumer acceptance of electronic commerce – Integrating trust and risk in the technology acceptance model. *International Journal of Electronic Commerce, 7*(3), 69–103.

Pinsonneault, A., & Kraemer, K. L. (1993). Survey research methods in management information systems: An assessment. *Journal of Management Information Systems, 10*(2), 75–105.

Reichheld, F. F., & Schefter, P. (2000). E-loyalty: Your secret weapon on the web. *Harvard Business Review, 78*, 105–113.

Reitsma, R. (2006). Europe's 2006 online shopping landscape. *Forrester Research, Business View Research Document.* Retrieved Dec 14th 2011 from http://www.forrester.com/rb/Research/europes_2006_online_shopping_landscape/q/id/40479/t/2

Remenyi, D., Williams, B., Money, A., & Swartz, E. (1998). *Doing research in business management: An introduction to process and method.* London, UK: Sage Publications.

Robey, D. R. (1996). Research commentary: Diversity in information systems research. *Information Systems Research, 7*(4), 400–408. doi:10.1287/isre.7.4.400

Rofiq, A., & Mula, J. M. (2010). *Impact of cyber fraud and trust on e-commerce use: A proposed model by adopting theory of planned behaviour.* Paper presented at 21st Australasian Conference on Information Systems. Brisbane, Australia.

Saprikis, V., Chouliara, A., & Vlachopoulou, M. (2010). Perceptions towards online shopping: Analyzing the Greek university students' attitude. In *Proceedings of the Communications of the IBIMA*. IBIMA.

Shaffer, T. R., & O'Hara, B. S. (1995). The effects of country of origin on trust and ethical perceptions. *The Service Industries Journal, 15*, 162–179. doi:10.1080/02642069500000019

Sillence, E., Briggs, P., & Fishwick, L. (2004). Trust and mistrust of online health sites. In *Proceedings of Computer Human Interaction CHI 2004*. Vienna, Austria: ACM. doi:10.1145/985692.985776

Tan, Y.-H., & Thoen, W. (2000). Toward a generic model of trust for electronic commerce. *International Journal of Electronic Commerce, 5*(2), 61–74.

Urban, G. L., Amyx, C., & Lorenzon, A. (2009). Online trust: State of the art, new frontiers, and research potential. *Journal of Interactive Marketing, 23*, 179–190. doi:10.1016/j.intmar.2009.03.001

Van der Heijden, H., Verhagen, T., & Creemers, M. (2003). Understanding online purchase intentions: Contributions from technology and trust perspectives. *European Journal of Information Systems, 12*(1), 41–48. doi:10.1057/palgrave.ejis.3000445

Verhagen, T., Meents, S., & Tan, Y. (2006). *Perceived risk and trust associated with purchasing at electronic marketplaces*. Retrieved 28th October, 2011 from http://ideas.repec.org/s/dgr/vuarem.html

Verhagen, T., Tan, Y., & Meents, S. (2004). *An empirical exploration of trust and risk associated with purchasing at electronic marketplaces*. Paper presented at 17th Bled eCommerce Conference. Bled, Slovenia.

Williamson, J. B., Karp, D. A., & Dalphin, J. R. (1982). *The research craft*. Boston, MA: Little Brown.

Xu, Y. (2010). *The study of online shopping perceived risk under the China's e-business circumstance*. Paper presented at the Information Management, Innovation Management and Industrial Engineering (ICIII), 2010 International Conference. Kunming, China.

Zucker, L. G. (1986). Production of trust: Institutional sources of economic structure, 1840 – 1920. *Research in Organizational Behavior*, *8*, 53–111.

ADDITIONAL READING

Bart, Y., Shankar, V., Sultan, F., & Urban, G. L. (2005). Are the drivers and role of online trust the same for all web sites and consumers? A large-scale exploratory empirical study. *Journal of Marketing*, *69*(4), 133–152. doi:10.1509/jmkg.2005.69.4.133

Buttner, O. B., & Goritz, A. S. (2008). Perceived trustworthiness of online shops. *Journal of Consumer Behaviour*, *7*(1), 35–50. doi:10.1002/cb.235

Cazier, J. A., Benjamin, B., Shao, M., & Robert, D. (2003). Addressing e-business privacy concerns: The roles of trust and value compatibility. Paper presented at Symposium of Applied Computing. Melbourne, FL.

Chellappa, R. K., & Sin, R. G. (2005). Personalization versus privacy: An empirical examination of the online consumer's dilemma. *Journal of International Management*, *6*(2–3), 181–202.

Chong, B., Yang, Z., & Wong, M. (2003). *Asymmetrical impact of trustworthiness attributes on trust, perceived value, and purchase intention: A conceptual framework for cross-cultural study on consumer perception of online auction*. Paper presented at 5th International Center for Electronic Commerce. Pittsburgh, PA.

Doney, P. M., & Cannon, J. P. (1997). An examination of the nature of trust in buyer–seller relationships. *Journal of Marketing*, *61*(2), 35–51. doi:10.2307/1251829

Fogg, B. J., Soohoo, C., Danielson, D. R., Marable, L., Stanford, J., & Tauber, E. R. (2003). *How do users evaluate the credibility of web sites? A study with over 2,500 participants.* Paper presented at 2003 ACM Conference on Designing for User Experiences. San Francisco, CA.

Grewal, D., Hardesty, D. M., & Iyer, G. R. (2004). The effects of buyer identification and purchase timing on consumers' perceptions of trust, price fairness, and repurchase intentions. *Journal of Interactive Marketing*, *18*(4), 87–100. doi:10.1002/dir.20024

Jarvenpaa, S. L., Tractinsky, N., & Vitale, M. (2000). Consumer trust in an internet store. *Information Technology Management*, *1*(1–2), 45–71. doi:10.1023/A:1019104520776

Karvonen, K. (2000). *The beauty of simplicity*. Paper presented at 2000 ACM Conference on Universal Usability. Arlington, VA.

King-Casas, B., Tomlin, D., Anen, A., Camerer, C. F., Quartz, S., & Montague, P. R. (2005). Getting to know you: Reputation and trust in a two person economic exchange. *Science*, *308*, 78–83. doi:10.1126/science.1108062

Lee, M. K. O., & Turban, E. (2001). A trust model for consumer internet shopping. *International Journal of Electronic Commerce*, *6*(1), 75–91.

Liu, C., Marchewka, J. T., Lu, J., & Yu, C.-S. (2005). Beyond concern — A privacy–trust–behavioral intention model of electronic commerce. *Journal of International Management*, *42*(2), 289–304.

McKnight, D. H., Choudhury, V., & Kacmar, C. (2000). *Trust in e-commerce vendors: A two-stage model*. Paper presented at Twenty First International Conference on Information Systems. Brisbane, Australia.

Morgan, R. M., & Hunt, S. D. (1994). The commitment—trust theory of relationship marketing. *Journal of Marketing*, *58*(3), 20–38. doi:10.2307/1252308

Riegelsberger, J. M., Sasse, A., & McCarthy, J. D. (2003). *Shiny happy people building trust? Photos on e-commerce websites and consumer trust*. Paper presented at ACM Conference on Human Factors in Computing Systems. Ft. Lauderdale, FL.

Schlosser, A. E., Barnett White, T., & Lloyd, S. M. (2005). Converting web site visitors into buyers: How web site investment increases consumer trusting beliefs and online purchase intentions. *Journal of Marketing*, *70*(2), 133–148. doi:10.1509/jmkg.70.2.133

Schoenbachler, D. D., & Gordon, G. L. (2002). Trust and customer willingness to provide information in database-driven relationship marketing. *Journal of Interactive Marketing*, *16*(3), 2–16. doi:10.1002/dir.10033

Smith, D., Menon, S., & Sivakumar, K. (2005). Online peer and editorial recommendations, trust, and choice in virtual markets. *Journal of Interactive Marketing*, *19*(3), 15–37. doi:10.1002/dir.20041

KEY TERMS AND DEFINITIONS

Benevolence: An inclination to do good.

Competence: The ability to do something efficiently and successfully.

E-Commerce: The buying or selling of products or services over the Internet.

Experience: Practical contact with and observation of facts or events.

Integrity: The quality of being honest and treating others with uprightness.

Internet Privacy: The desire for an individual's personal or transaction data to be protected and not sold to another party or used without the individual's permission.

National Culture: The general attitudes, belief systems, values, and traditions, particular to a nation.

Peer: A person of the same age, status, or ability as another specified person.

Trust: A confident expectation or a vulnerable party in the integrity, competence, and benevolence of another party.

ENDNOTES

1. Statistical Package for the Social Sciences.
2. Cheung and Lee's measurement instrument was developed in Hong Kong.

Chapter 3
Subject Recommended Samples:
Snowball Sampling

Pedro Isaías
Universidade Aberta, Portugal

Sara Pífano
Universidade Aberta, Portugal

Paula Miranda
Polytechnic Institute of Setúbal, Portugal

ABSTRACT

In a research project, the selection of the sample method is crucial, since it has repercussions throughout the entirety of the study. It determines how the population under scrutiny will be represented and with what accuracy. Hence, it has an important impact in terms of the reliability and validity of the research in general, and consequently, its conclusions. This chapter aims to explore snowball sampling as a chain-referral sampling method. An introductory review of the relevant literature highlights its main characteristics, benefits, and shortcomings, and provides a broader insight to circumstances where it can be successfully applied. This theoretical prologue is followed by the analysis of its employment in an online questionnaire and the presentation of the lessons learned from this sampling decision.

INTRODUCTION

In the methodology research arena, sampling precepts and processes have remained with a status of mere phases of studies and have not been carefully analysed per se (Noy, 2008). Nonetheless, the determination of the sampling method has a vital impact in the validity and reliability of a research. The potential to draw valuable conclusions from the data collected in a research lies also in the meticulousness of the sample method. Why and how researchers choose and apply a sample to the studies they conduct is a main element when assessing the value of their results.

There are multiple sampling methods and they should be selected and employed according to each research's particular objectives and characteristics. Studies targeting hidden populations, for example,

DOI: 10.4018/978-1-4666-2491-7.ch003

Copyright © 2013, IGI Global. Copying or distributing in print or electronic forms without written permission of IGI Global is prohibited.

tend to prefer non-probability and chain referral sampling methods rather than the conventional probability sampling approaches. This happens because these populations are often composed by subjects with no sampling frame, who express a concern for the safeguarding of their privacy. This derives from the fact that they belong to a peculiar group and solely represent a niche of the generality of the population (Heckathorn, 2002). Hidden populations usually include people affected by issues that are socially regarded as being sensitive, such as homosexuality, homelessness and drug use. The snowball sampling approach, which will be the focus of this chapter, was specifically designed to address the sampling challenges deriving from populations that are difficult to access (Brackertz, 2007).

As a method of chain referral, snowball sampling has an interactional nature compatible not only with the study of hidden populations, but also with the study of social networks (Noy, 2008). The sampling procedure can be initiated by researchers' contacts or by placing adverts in relevant locations (Noy, 2008). With difficult populations, involving sensitive matters, the researcher might first opt for a thorough scan of the locations where these populations meet and then approach some of them suggesting a preliminary interview. Later they are asked if they would like to recommend other individuals to participate in the study as well. When people are recommended and decide to participate, they are equally invited to suggest potential respondents. With this cycle of recommendations, the number of participants increases. So, snowball sampling resorts to an initial random selection of individuals who then recommend other subjects that in turn make their own suggestions of research subjects. This process continues until the size of the desired sample is reached (Kendall, et al., 2008). Only those who are connected to the individuals in the network of recommendations are used in the sample (Ahn, Han, Kwak, Moon, & Jeong, 2007). The starting sample is usually selected via a convenience system, making the entirety of the sample a non-probability sample. This sampling approach is used mostly in practice and less seen in statistics related literature (Handcock & Gile, 2011).

The rise of studies focusing on online populations has caused a continuous growth in the number of researcher using online surveys. They face the challenge of transposing conventional research methods to Web-based environments (Wright, 2005, p. 1). The authors have developed an online questionnaire to explore Web 2.0 users' preferences using the snowball sampling method. Since the population is Web-based, it was decided to use methods that can be applied over the Internet. The only condition that was required to participate in this questionnaire was that the respondents used Web 2.0 technologies or tolls. The only characteristic that the participants had in common was that they used the Social Web, independently of purpose, frequency or expertise and also regardless of participants' demographics. Such a study demands research and sampling methods that can encompass a large, diverse, and scattered population. The use of snowball sampling derives from the fact that Web 2.0 users represent a population that is widely spread throughout the world and the best solution to reach as many respondents as possible is to use Web 2.0 itself and its users to communicate with each other. Thus, allowing the population to define itself. The way this online questionnaire was applied and distributed by the researchers and then by the respondents, allows an empirical analysis of the snowball sampling method.

The first section of the chapter explores the method of snowball sampling, by presenting and discussing its core traits and the challenges and benefits it presents. The second section intends to provide insight on the use of the snowball sampling method in online questionnaires and the use of Web 2.0 to apply this method efficiently. This leads the way to the presentation of the online questionnaire developed by the authors, its results and the lessons learned from the way it was conducted. The chapter's contribution ends with a concluding section on future directions and prospective research on snowball sampling and sampling methods in general.

THE SNOWBALL EFFECT

Where a sampling frame is nonexistent, non-probability sampling methods are usually employed. Snowball sampling is one of these methods, being mainly defined as a of chain-referral sampling approach.

Subject Recommended Sampling

Chain-referral sampling tends to be used where a sampling frame is absent. More recently, respondent-driven sampling, an improved form of chain referral, which aims to minimise some of the shortcomings deriving from this approach, is equally being used. Respondent-driven samples aim to address issues such as the selection of the initial subject, volunteerism, and masking. Volunteerism refers to the over recruitment of particularly accommodating subjects and masking has to do with the opposite situation, i.e., the under recruitment of a reduced amount of obliging respondents. Heckathorn, Semaan, Broadhead, and Hughes (2002) argue in one of their studies that this approach can be improved with "steering incentives," in the sense that they can create a more representative sample and minimise volunteerism and masking. In the research they conducted, incentives were given to the subjects to recruit individuals within a specific age group (18-25) which was being underrepresented in previous research.

Time location sampling is a method that aims to estimate probability sampling by charting the entirety of the locations where the intended subjects exist in great quantity and then select at random a location, a day and a time to steadily engage with potential subjects. This approach is very limited to the same setting, which can introduce a geographical bias. While respondent-driven sampling, despite its still experimental status, tries to establish a more rigorous snowball sampling by investing in more extensive recruitment chains; creating recruitment boundaries; and having present, at the phase of data collection and analysis, the common bias that the participants usually have of choosing to recruit people with akin features (homophily). In spite of being still statistically questioned by some, there is an increasing number of researchers employing this approach (Kendall, et al., 2008).

Snowball Sampling

Snowball sampling addresses the needs of researchers who need to study hidden subjects. It is an advantageous method in cases where the individuals know more subjects of the target population and acknowledge them as its members. The process of sampling is initiated with a selected number of subjects who will be the starting point from where other contacts will be identified and engaged. Each new contact is asked to supply contacts and thus a chain of referrals is created. Depending on the method of referral, the numbers can increase and the sample could be more diverse. If the subjects can refer people that are not only limited to their vicinity and have communication technology at their disposal to do it, then the population can grow in scope and number (Heckathorn, 2002).

By employing this sample method, the researcher is giving power to its respondents and he/she is learning about their choices and behaviours. Ideally, snowball sampling will provide more than data. It has two main aspects: the data supplied by the participants and the patterns of the snowball effect (Noy, 2008). Handcock and Gile (2011) highlight a duality present in the literature in terms of the snowball sampling purpose: on the one hand it can be used to identify members of a hidden population through the recommendation of other individuals from that population until a certain sample is generated; and on the other hand it can be used to request individuals to signal other subjects for a pre-determined amount of phases to determine the number of common relations within that population.

The snowball sampling technique is a repetitive process. The term snowball employed in this method of sampling tries to depict the snowball effect to metaphorically describe its core trait: its accumulative nature (Noy, 2008). Snowball sampling can be the primary technique through which participants are recruited. However, it can also serve as a secondary method employed to help the access to other respondents or to boost the number of contacts that at certain points of the research may prove to be insufficient (Noy, 2008).

The real promise of snowball sampling lies in its ability to uncover aspects of social experience often hidden from both the researcher's and lay person's view of social life (Atkinson & Flint, 2001).

Benefits and Challenges

When dealing with hidden populations, those difficult to approach and reach, snowball sampling presents the great benefit of enhancing the identification and augmenting the number of participants by resorting to their target population to engage other people. Nonetheless, its validity in terms of representation is feeble. This happens because the sample is reliant on the choice of the initial participants and the people who were subsequently recruited by them (Kendall, et al., 2008).

Snowball sampling is advantageous when the population that the researcher wishes to study is not entirely evident and where the outline of a list of the population is intricate. This sampling approach is popular for the study of populations dealing with issues of sensitive or illegal nature such as drug abuse, but also with elite populations that are not as straightforwardly public as they may seem and tend to live surreptitiously (Tansey, 2007).

This sampling method, besides being the most effective in terms of unveiling hidden populations is also an excellent tool to study networks (Noy, 2008). In order to uncover the main characteristics of online social networks it is imperative to use a sampling method, due to their size. In most cases it is practically unattainable to conduct an analysis of the entirety of the network (Ahn, Han, Kwak, Moon, & Jeong, 2007).

The snowball approach requires some kind of system; it is not as arbitrary as one might think. The researcher is required to engage with the process since the beginning and make sure that the chain of referrals is kept within the limits of what is pertinent to the research aims (Tansey, 2007). Snowball sampling represents a peril of having participants that recommend subjects with too many similar traits, so the researcher needs to verify if the initial group is diverse enough to guarantee that the sample is not too distorted in any specific course (Tansey, 2007).

The snowball sampling method is often maculated by bias. Bias can be found right at the beginning of the research when it becomes necessary to select the primary subjects and then again in the subsequent individuals who are recruited: "friendship and other forms of affiliation tend to occur among persons who are similar in levels of education, income, ethnicity, and interests" (Heckathorn, Semaan, Broadhead, & Hughes, 2002, p. 57). This is a potential cause of bias since the patterns in the sample may often reflect this homophily. In practice, this would mean that the sample was composed with people that were very similar to the primary subjects, therefore making the remaining respondents seem redundant. The sample would be a near transcript of the initial participants.

A recurrent critique to snowball sampling is that, since it is dependent on its primary participants to increment the number of additional subjects by the means of referral, it makes it susceptible to bias and more defenseless against claims of lack of representativeness of the population (Fricker, 2008). Moreover, this dependence on the initial cluster of subjects to ignite and increase the snowball may not go according to the researchers' plan. There are reports of studies where the population had sensitive and illegal features and the initial respondents simply refuse to provide any other names (Duncan, White, & Nicholson, 2003). In situations like this, it is not possible to continue, since there are no referrals.

INVESTIGATING ONLINE POPULATIONS: WEB 2.0 USERS

On a previous study, the authors conducted an online questionnaire aimed to explore the online behaviour and precepts of Web 2.0 users. From design to analysis, the choice of the snowball sampling method was always determinant for this Web-based questionnaire, making it a valuable example of the feasibility of this sampling approach. Hence, to illustrate how snowball sampling can be used in practice, this section will focus on the online questionnaire that was designed to map Web 2.0 users' preferences online.

The choice of research methods requires careful consideration. The elements to have in account concern mainly the research objectives, time, and budget (Wright, 2005, p. 1). The research time and budget were very limited and its objectives involved accessing a population that geographically was scattered and numerically was a challenge. The target population was Web 2.0 users in general, that meant users from any platform, age, sex, or country. The objective was to distribute this questionnaire to as many users as possible in as many countries and as many Web 2.0 platforms as possible in order to have a global depiction of Web 2.0 users' preferences. The research had expectations of gathering a significant number of responses from different countries, which implied having to develop a method that would select a global sample from an uncountable population: Web 2.0 users. With a population of this proportion, the selection of a sample was conditional for its viability. The sample method that would be selected would have to address two seemingly contradictory requirements: on the one hand, limit the total population and on the other hand create a sample that would increase the number of respondents and diversify their characteristics. In sum, reduce the population, but at the same time constantly increase the sample.

The decision of using online questionnaires was based on many of its advantages, but mainly because they allow the research of populations that could not be accessed otherwise, either because they would be difficult or impossible to reach. They are able to reach respondents geographically distant or with whom it is improbable to establish contact. Furthermore, certain populations exist solely on the Internet (Wright, 2005), which is the case of the target population on this research. Nonetheless, like any other research method, online questionnaires present some disadvantages, namely they present sampling challenges (Wright, 2005). To address this challenge and given the characteristics of this study, snowball sampling was the chosen sampling method.

This sampling method, besides being the most effective in terms of unveiling hidden populations is also a valuable asset in the study of networks. By employing this sampling method, the researcher is giving power to its respondents and is learning about their choices and behaviours. Ideally, snowball sampling will provide more than data. There are two main aspects in snowball sampling: the data supplied by the participants and the patterns of the snowball effect (Noy, 2008). Snowball sampling was chosen also due to its interactional and viral nature, two features it shares with Web 2.0 itself. Web 2.0 also improves by peer recommendation and the more people are involved in it the better it will perform (O'Reilly, 2005). Moreover, the questionnaires were distributed online and the Internet is a venue where people need to trust the content they are sent. By using this method, each respondent would only receive a message with the questionnaire link from someone they knew. Some studies show that, to a tremendous extent, participants will recommend people with whom they already have trusting bonds (Heckathorn, et al., 2002). To increase adhesion to the questionnaire and ensure its reliability all design specificities for online versions of this method of research were as strictly followed as possible. All efforts were made to follow the main guidelines: a brief and informative welcome page; the guarantee of anonymity and confidentiality; the provision of instructions and clarifications where necessary; a closing page where the participants are thanked for the time spent participating in the study; and usability and accessibility as the core precepts of design (Ritter & Sue, 2007).

The initial participants were chosen through a sample of convenience. One of the authors used its own contacts in Web 2.0 websites to initiate the snowball sample. A questionnaire link and request to forward it was posted on Linked In profile and send as a private message to all contacts there; sent by email via Windows Live and posted on Windows Live blog and personal profile; on Facebook it was posted on the author's Wall and sent by private messages to some of the friends; and on Hi5 the link was also posted on the author's profile and sent by message to some of the friends. Additionally, one blog was set up just for updates on the questionnaire and results, which could be accessed by anyone on the Internet. This blog also contained the link to the questionnaire. The majority of the author's contacts on the above-mentioned Web 2.0 websites have Portuguese as their mother tongue, but because this study wanted to reach as many countries as possible, two links of the same questionnaire were created, one in Portuguese and one in English. Not everyone knows English, but most Web 2.0 users tend to engage in interactions with people from other countries and use English to do so, hence it was considered the best choice in terms of international languages.

RESULTS AND LESSONS LEARNED

The responses to the questionnaire enabled the attainment of the research objectives at the time, in the sense that they allowed the outline of a pattern of use of Web 2.0 that enabled some important conclusions in terms of Web 2.0's users' behaviours, preferences, and beliefs. This questionnaire will now permit an analysis focused on the determinant role of the chosen sampling method.

The online questionnaire had a total of 628 responses, from which 621 were considered valid and analysed. The process of data cleaning helped to eliminate some contributions that were not relevant for the study and its objectives. The data was processed into a workable form by using the statistical system SPSS 17. The existing techniques of data transformation assist in the configuration of the data collected for its highest usability (Ritter & Valarie, 2007). This allocates more time to the analysis of the data and avoids the great amounts of time spent on organising it.

The questionnaire was available online for completion from the 16th of June 2010 until the end of September 2010. Throughout this period, 487 initial invitations were sent individually using a convenience sample, at different times. These primary invitations were sent at different times via Facebook, Google Gmail, Hi5, Windows Live Hotmail, and LinkedIn, as was explained in the previous section. The link for the questionnaire was sent with a short friendly message that appealed to the link of familiarity between the author and the contact and explained the scope of the study and what was requested of them. A message in English for the Portuguese speaking contacts was also included, so that they could just use that template when forwarding the link to friends abroad. Familiar contacts were used first to potentiate the number of replies, but also to guarantee that the

Figure 1. Evolution of initial invitations vs. actual replies

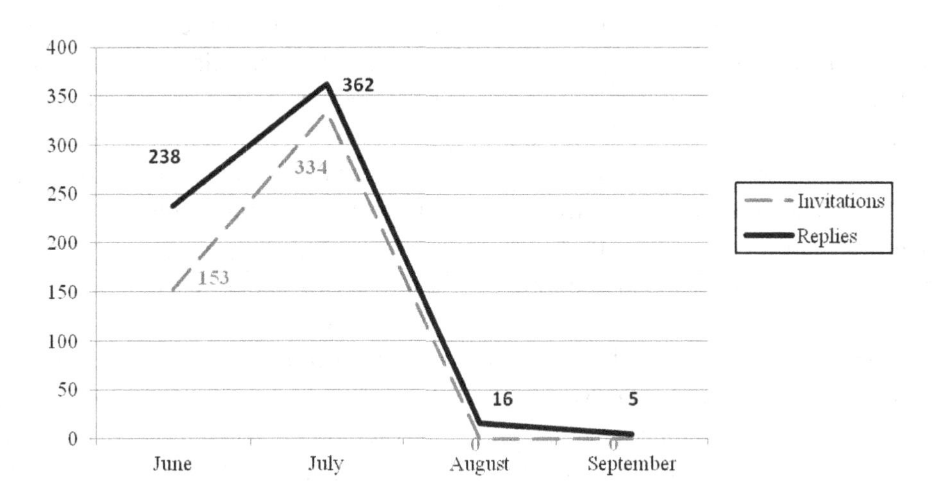

messages were actually read. Since the amount of messages from unknown contacts and spam is growing, it was important that the contacts knew the sender. The same was asked of the respondents who were asked to send the link to their contacts. This way there was a better probability of people trusting the questionnaire and participating in the study. As Heckathorn et al. (2002) once stated "... peer recruiters need not cultivate the trusting relationship required to access...recruitment, because overwhelmingly, respondents recruited those with whom they already had a trusting relationship" (p. 65). Thus, it was anticipated that trust in the sender would endow some extra reliability to the study and consequently an increased opportunity of participation in it.

A total of 53 first invitations were sent to the author's closer contacts, on the 16th, 17th, and 18th of June and by the end of that day there were already 75 questionnaires answered online. In June, on the 19th, 20th, and the 23rd, a total of 100 of new preliminary invitations were sent. This makes a total of 153 initial invitations from the convenience sample, sent in June which by the end of the month had already snowballed into 238 respondents. On the 4th and 11th of July the last invitations, which amounted to 334, were sent. July saw 362 new respondents complete the questionnaire on the Internet. After that, since all initial contacts had already been used, the snowball effect diminished in speed and number. August received 16 new answers and September only 5. In September, the questionnaire was closed, since a desired number of respondents had been reached. Figure 1 illustrates the evolution of the initial invitations send by the researcher and the replies obtained both by initial participants and the people they successfully recruited.

The graphic on Figure 1 portrays how the initial invitations were at every stage of the recommendation process, outnumbered by the participation of respondents online. In addition, it shows how the responses start to become scarce, once the waves of initial invitations by the researcher stopped. The evolution of this pattern of response shows that it is more beneficial to gradually send invitations, especially since the questionnaire was anonymous and sent using a snowball sampling with no way of tracking the respondents and send reminders to people directly, without taking the chance of harassing someone who had already completed the questionnaire. The number of responses was one of the positive aspects of using the snowball sampling method. With an opening convenience sample of 487 people, only snowball sampling would allow a response rate higher than 100%.

The snowball sampling method also allowed the accomplishment of the objective of reaching as many countries as possible, but only to a certain extent. The convenience sample comprised 24 different countries of residence, where Portugal was predominant with 376 residents in a sample of 487 and the United Kingdom followed with 55 residents. The remaining countries had between 1 and 7 residents and were: Australia, Austria, Brazil, Canada, The Czech Republic, Denmark, France, Germany, Greece, India, Indonesia, Italy, Montenegro, Poland, Romania, Russia, Spain, Switzerland, The Netherlands, Turkey, USA, and Venezuela. The online questionnaire was answered by participants from 33 different countries of residence. The results showed that these 33 countries did not include 6 countries of residence of the initial sample, meaning that the respondents residing in those countries did not answer the questionnaire themselves: Australia, Czech Republic, Montenegro, Romania, Russia, and Venezuela. This means that, in reality, from the initial 24 countries of residence only 18 participated. The initial sample of respondents recruited users residing in 15 other countries: Belgium, Bosnia and Herzegovina, China, Finland, Ireland, Japan, Latvia, New Zealand, Norway, Philippines, São Tome and Principe, South Africa, South Korea, Sweden, and Trinidad and Tobago. These recruited countries were represented by a reduced number of participants.

The general representation of Web 2.0 users' profile intended by the online questionnaire required a more geographically distributed user contribution, but what happened was the participation of an overwhelming majority of Portuguese users. The fact that this happened was predictable due to the high number of initial invitations sent to residents from this country, but nonetheless, more significant numbers from the other countries were expected. To potentiate that, a more targeted sampling could have been done and the subjects could have been instructed to forward the questionnaire in a way that would have maximised a more extensive reach to different countries. Alternatively, to have a more representative sample, the focus could have been placed on one country only or on one part of the world such as Europe. Even if the questionnaire did reach all 5 continents, in this case the sample was not representative of the target population, due to users' geographic distribution of participation. It was later learned that the expected international proliferation would hardly be obtained in a study of this dimension and design. Tansey (2007) alerts to the fact that snowball sampling is not, as the name might indicate, the type of sampling method that a researcher initiates and then abandons to its natural course. It is imperative that throughout the entirety of the process the researcher ensures that the string of recommendations is being contained inside of what is relevant and valuable to the objectives of the study. In this case, all responses were checked to see if the respondents met the criteria defined by the objectives. A great challenge of snowball sampling, already explained in previous sections, is homophily. As Tansey (2007) argues, it is crucial to confirm that the initial cluster of participants is sufficiently assorted to avoid creating a tainted sample. The initial group used to recommend other users was not diverse in terms of the country of residence and thus the sample illustrated just that and not the intended international diversity. The Web 2.0 questionnaire was affected by some of the bias that Heckathorn et al. (2002) argue. By using the author's contacts and then snowballing the contacts from there, a bias has been introduced not only in terms of interests, but mainly geographically and in terms of the Web 2.0 websites used.

The study aimed to have the users posting the questionnaire's link on several Web 2.0 platforms and websites, but when the users were asked where they accessed the questionnaire, most of them answered the same websites from where it had

Figure 2. Number of invitations sent and number replies received according to web 2.0 websites used

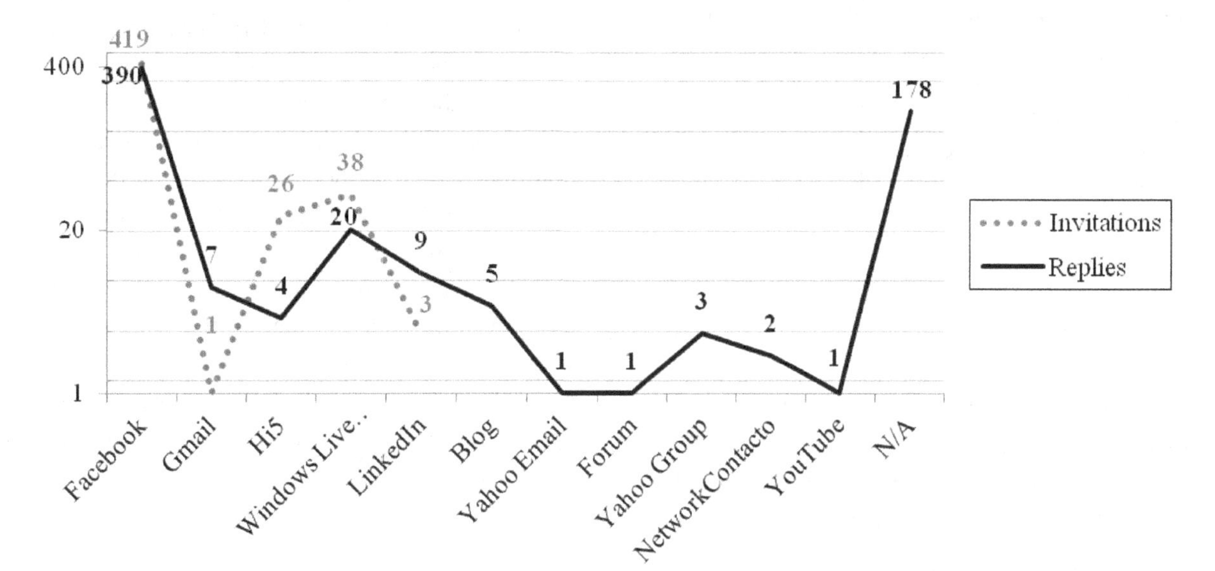

been initially sent. Figure 2 depicts the number of invitations sent per website and the number of replies received.

The participants were informed that they could use an unlimited number of Web 2.0 resources to forward the questionnaire, but the results demonstrated that they did not expand their choices. Although when counting the number of Web 2.0 applications and websites it is clear that they doubled with the snowball sampling, the number of replies received from them is not significant when compared with the main source of invitations, which was Facebook. Initially, only Facebook, Google Gmail, Hi5, Windows Live Hotmail, and LinkedIn were used in dissemination of the questionnaire. From these websites, the recruiting participants were able to add 6 new applications/websites to the list of venues for the questionnaire's diffusion: Blogs, Yahoo Email, Forums, Yahoo Group, NetworkContacto, and YouTube. Figure 2 also shows a significant number of users (178) which did not reply to this question or replied something different from what was requested, for example, instead of the name of the platform, the name of the friend that forwarded the questionnaire, invalidating their answers. Again, as argued by Tansey (2007), in this case, a greater diversity in the original Web 2.0 platforms from where the invitations were originally sent, would have made a difference and possibly contribute to the attainment of the objective of collecting the opinion of users from multiple platforms/websites.

Snowball sampling requires trust. The respondents need to be assured that any information provided to the researcher is protected. This is crucial, both practically and ethically speaking. The privacy and confidentiality of any data collected from the participants must be protected and they must know that it is being protected in order to collaborate (Atkinson & Flint, 2001). Although the questionnaire was designed with all the conditions to guarantee privacy and confidentiality to its respondents, concerns of privacy that involved suggesting a third party for a study, were ultimately left at the subjects' own common sense and will. All the used platforms and websites had the possibility of forwarding the link in a private way, without anyone's knowledge but the sender and the receiver's. The questionnaire was anonymous and there was no way to identify the

Table 1. Answers to the question: "How long have you been using social web internet sites?"

Frequency	Number of Users
Less than 6 months	19
More than 6 months but less than 1 year	37
1-3 years	164
Over 3 years	401
Total	621

subjects once they answered, unless they chose to, by leaving any comment or finishing it with their name or contact for further information from the researcher. In any other situation, their identity was fully protected.

Some of the questions were aimed at knowing more about the respondents experience in terms of Web 2.0. All answers are important, but respondents with more experience of Web 2.0 websites, who are active and frequent users, might have more to add than people who are starting to use these sites. This became vital because using snowball sampling implies directing some of the control to the initial participants in an important stage of the investigation (Noy, 2008), so it is important to understand who they are and what knowledge they have of the subject. Tables 1 and 2 illustrate the responses to questions 1 and 2 of the online questionnaire.

Table 2. Answers to the question: "How often do you use these websites?"

Frequency	Number of users
3 times a day or more	274
once a day	217
3 times a week	77
once a week or less	51
N/A	2
Total	621

These two questions provided the researcher with more information about the participant. Their answers endowed the researcher with important data regarding their suitability to actually be a participant. Table 1 shows an unequivocal predominance of respondents which have been engaging with Web 2.0 for over three years (401) and use it quite frequently, as shown in Table 2, with 274 users stating to use these sites three times a day or more and 217 stating that they participated once a day. These results show that the majority of the respondents had experience and knowledge in the field in terms of the questions that they were asked.

FUTURE RESEARCH DIRECTIONS

The literature on sampling methods is still scarce and deficient in practical cases that can assist researchers to make more informed decisions when it comes to selecting their samples. It seems to be fundamental to highlight the importance of the sample in a research. The sample represents the whole, it is through the chosen sample that a research will try to uncover patterns and create knowledge. An incorrect sampling method may invalidate the entirety of the study, since a misrepresented population will contaminate any conclusions with the suspicion of partiality, inescapably affecting the reliability of the study and its outcomes.

Future research on methodology should place an emphasis on sampling techniques in general. A representative sample establishes the difference between conducting a scientific study on a population that can be used by other social scientists and actually generate knowledge or the waste of time and budget in studies that provide insufficient data that can't really be re-used. Placing sampling methods on the sidelines of methodology is condoning years of misapplied samples and losing the criteria of scientific research.

Researchers should be better prepared to make decisions in their investigations and learn to attribute a sample technique to a research method and objective. It is important to highlight that each study is unique and requires much preparation in order to establish a plan to be properly executed. Objectives, design, research methods, samples, data collection, and analysis techniques are some of the factors to consider. Mainly, for research on methodology it is imperative to value each phase of the research process, because neglecting one of them might mean jeopardising the entirety of the study. It works in a domino effect, if one stage fails, then one by one the others will unavoidably be affected.

With regard to snowball sampling, prospective research should focus on its advantages and disadvantages, but having in mind a more practical application. It would be very useful for the development of this approach to understand under which research methods it works best and which ones should be avoided and why. The empirical use the snowball sampling is valuable to derive conclusions that assist researchers in the sampling decision making process and this should be potentiated.

CONCLUSION

This study sought to explore the method of snowball sampling and assess how it can be a valuable method in dispersed populations. For this purpose, an online questionnaire, which had been designed to explore the preferences of Web 2.0 users, was used here as an example of the application of the snowball sampling method.

One of the biggest advantages of using this method in this study was not only the abundant number of responses obtained, but also the subjects' behaviour, which provided valuable data itself. It not only provided the research with a chain of contacts, but it granted the possibility to understand more about Web 2.0 users' preferences and not only from their responses, but from the way they engaged with that chain and made efforts to continue it. One important limitation was also not being able to track, which one of the original contacts did reply to the questionnaire or not. The data proves that while some people forwarded the questionnaire and were able to persuade contacts of a third level of proximity to complete it and forward it, there were some initial contacts, the author's direct and closer contacts, who did not forward, nor completed the questionnaire. This also provides insight on how people create their networks of contacts and how easily it is to form a big cloud of online friends that sometimes people do not really know. Another explanation is the fact that the Social Web is so overwhelming sometimes, in terms of information and requests to engage, that people often get lost amid these networks.

The snowball sampling approach seems to, sometimes, be applied randomly, without a system of support to ensure that the aim with which it is being employed is attained. Unlike some other snowball effects, this sampling method is not entirely self-sufficient and it is not enough to initiate it, to roll the snowball. It needs to be supervised and guided by the researcher to provide the required insight.

REFERENCES

Ahn, Y.-Y., Han, S., Kwak, H., Moon, S., & Jeong, H. (2007). Analysis of topological characteristics of huge online social networking services. In *Proceedings of the 16th International Conference on World Wide Web - WWW 2007*, (pp. 835-844). New York, NY: ACM Press. Retrieved August 25, 2011, from http://www2007.org/papers/paper676.pdf

Atkinson, R., & Flint, J. (2001). Accessing hidden and hard-to-reach populations: Snowball research strategies. *Social Research Update, 33*. Retrieved August 28, 2011 from http://sru.soc.surrey.ac.uk/SRU33.pdf

Brackertz, N. (2007). Who is hard to reach and why? *Institute for Social Research.* Retrieved August 30, 2011 from http://www.sisr.net/publications/0701brackertz.pdf

Castillo, J. J. (2009). *Snowball sampling.* Retrieved July 30, 2011, from http://www.experiment-resources.com/snowball-sampling.html

Duncan, D. F., White, J. B., & Nicholson, T. (2003). Using internet-based surveys to reach hidden populations: Case of nonabusive illicit drug users. *American Journal of Health Behavior, 27*(3), 208-218. Retrieved September 10, 2011 from http://www.duncan-associates.com/hiddenpop.pdf

Fricker, R. (2008). Sampling methods for web and e-mail surveys. *The Sage Handbook of Online Research Methods, 1*, 195–217. Retrieved August 26, 2011 from http://faculty.nps.edu/rdfricke/docs/5123-Fielding-Ch11.pdf

Gile, K. J., & Handcock, M. S. (2010). Respondent-driven sampling: An assessment of current methodology. *Sociological Methodology, 40*, 285–327. Retrieved August 25, 2011, from http://arxiv.org/PS_cache/arxiv/pdf/0904/0904.1855v1.pdf

Handcock, M. S., & Gile, K. J. (2011). On the concept of snowball sampling. *Sociological Methodology, 1554*, 1–5.

Heckathorn, D. (2002). Respondent-driven sampling II: Deriving valid population estimates from chain-referral samples of hidden populations. *Social Problems*. Retrieved September 08, 2011 from http://www.jstor.org/stable/10.1525/sp.2002.49.1.11

Heckathorn, D., Semaan, S., Broadhead, R., & Hughes, J. (2002). Extensions of respondent-driven sampling: A new approach to the study of injection drug users aged 18 – 25. *AIDS and Behavior, 6*(1), 18-25. Retrieved September 02, 2011 from http://www.springerlink.com/index/00G4W675B7EVVH4T.pdf

Kendall, C., Kerr, L. R. F. S., Gondim, R. C., Werneck, G. L., Macena, R. H. M., & Pontes, M. K. (2008). An empirical comparison of respondent-driven sampling, time location sampling, and snowball sampling for behavioral surveillance in men who have sex with men, Fortaleza, Brazil. *AIDS and Behavior, 12*(4), 97–104. doi:10.1007/s10461-008-9390-4

Lee, S., Kim, P.-J., & Jeong, H. (2006). Statistical properties of sampled networks. *Physical Review E, 73*(1), 1-7. Retrieved August 24, 2011, from http://stat.kaist.ac.kr/~pj/sampling06.pdf

Noy, C. (2008). Sampling knowledge: The hermeneutics of snowball sampling in qualitative research. *International Journal of Social Research Methodology, 11*(4), 327–344. doi:10.1080/13645570701401305

O'Reilly, T. (2005). *What is web 2.0? Design patterns and business models for the next generation of software*. Retrieved July 25, 2011, from http://www.oreillynet.com/pub/a/oreilly/tim/news/2005/09/30/what-is-web-20.html

Ritter, L. A., & Sue, V. M. (2007). *The survey questionnaire. Using Online Surveys in Evaluation: New Directions for Evaluation* (pp. 37–45). New York, NY: Wiley.

Ritter, L. A., & Valarie, M. S. (2007). Managing online survey data. In Ritter, L. A., & Valarie, M. S. (Eds.), *Using Online Surveys in Evaluation: New Directions for Evaluation* (pp. 51–55). New York, NY: Wiley.

Tansey, O. (2007). Process tracing and elite interviewing: A case for non-probability sampling. *PS: Political Science & Politics, 40*(4), 765-772. Retrieved August 30, 2011 from http://www.springerlink.com/index/Q6Q17272M32757X6.pdf

Wright, K. B. (2005). Researching internet-based populations: Advantages and disadvantages of online survey research, online questionnaire authoring software packages, and web survey services. *Journal of Computer-Mediated Communication, 10*(3. Retrieved July 20, 2011, from http://jcmc.indiana.edu/vol10/issue3/wright.html

ADDITIONAL READING

Ahmed, N. K., Berchmans, F., Neville, J., & Kompella, R. (2010). Time-based sampling of social network activity graphs. In *Proceedings of the Eighth Workshop on Mining and Learning with Graphs (MLG 2010)*, (pp. 1-9). New York, NY: ACM Press.

Andrews, D., Nonnecke, B., & Preece, J. (2003). Electronic survey methodology: A case study in reaching hard-to-involve Internet users. *International Journal of Human-Computer Interaction, 16*(2), 185-210. Retrieved September 05, 2011 from http://www.google.co.uk/url?sa=t&source=web&cd=1&ved=0CCEQFjAA&url=http%3A%2F%2Fciteseerx.ist.psu.edu%2Fviewdoc%2Fdownload%3Fdoi%3D10.1.1.151.94%26rep%3Drep1%26type%3Dpdf&rct=j&q= Electronic%20survey%20methodology%3A%20A%20case%20study%20in%20reaching%20hard-to-involve%20Internet%20users&ei=v5R2TonLLciXhQfTmO2mDA&usg=AFQjCNGRqeiZ8j36XdIiJBQrSlRu-37Sxw&cad=rjt

Bernstein, M. S., Ackerman, M. S., Chi, E. H., & Miller, R. C. (2011). The trouble with social computing systems research. In *Proceedings of the ACM CHI Conference on Human Factors in Computing Systems, CHI 2011,* (pp. 389–398). Vancouver, Canada: ACM Press. Retrieved September 10, 2011 from http://portal.acm.org/citation.cfm?id=1979618

Cho, D. (2003). Email study corroborates six degrees of separation. *Scientific American.com*. Retrieved August 20, 2011 from http://www.scientificamerican.com/article.cfm?id=e-mail-study-corroborates

Etter, J. F., & Perneger, T. V. (2000). Snowball sampling by mail: Application to a survey of smokers in the general population. *International Journal of Epidemiology, 29*(1), 43–45. doi:10.1093/ije/29.1.43

Farquharson, K. (2005). A different kind of snowball: Identifying key policymakers. *International Journal of Social Research Methodology, 8*(4), 345. Retrieved September 10, 2011 from http://researchbank.swinburne.edu.au/vital/access/services/Download/swin:1568/SOURCE1

Fink, A. (2003). *How to sample in surveys* (2nd ed.). Thousand Oaks, CA: Sage.

Grupetta, M. (2006). Snowball recruiting: Capitalising on the theoretical 'six degrees of separation'. In *Proceedings of AARE 2005 International Education Research Conference: UWS Parramatta: Papers Collection*. AARE. Retrieved August 5, 2011 from http://www.aare.edu.au/05pap/gru05247.pdf

Korf, D. J., van Ginkel, P., & Benschop, A. (2010). How to find non-dependent opiate users: A comparison of sampling methods in a field study of opium and heroin users. *The International Journal on Drug Policy, 21*(3), 215–221. doi:10.1016/j.drugpo.2009.08.005

Krista, J. G. (2011). Improved inference for respondent-driven sampling data with application to HIV prevalence estimation. *Journal of the American Statistical Association, 106*(493), 135–146. Retrieved September 12 from http://arxiv.org/PS_cache/arxiv/pdf/1006/1006.4837v1.pdf

Magnani, R., Sabin, K., Saidel, T., & Heckathorn, D. (2005). Review of sampling hard-to-reach and hidden populations for HIV surveillance. *AIDS (London, England)*, *19*(2), S67–S72. doi:10.1097/01.aids.0000172879.20628.e1

Maiya, A. S., & Berger-Wolf, T. Y. (2010). Sampling community structure. In *Proceedings of the 19th international conference on World Wide Web (WWW 2010)*, (pp. 701-710). New York, NY: ACM Press.

Malekinejad, M., McFarland, W., Vaudrey, J., & Raymond, H. F. (2011). Accessing a diverse sample of injection drug users in San Francisco through respondent-driven sampling. *Drug and Alcohol Dependence*, *118*(2-3), 83–91. doi:10.1016/j.drugalcdep.2011.03.002

McLean, C. A., & Campbell, C. M. (2003). Locating research informants in a multi-ethnic community: Ethnic identities, social networks and recruitment methods. *Ethnicity & Health*, *8*(1), 41–61. doi:10.1080/13557850303558

Salganik, M. J. (2006). Variance estimation, design effects, and sample size calculations for respondent-driven sampling. *Journal of Urban Health*, *83*, 98–112. doi:10.1007/s11524-006-9106-x

Siah, C. Y. (2005). All that glitters is not gold: Examining the perils and obstacles in collecting data on the Internet. *International Negotiation*, *10*, 115–130. doi:10.1163/1571806054741155

Streeton, R., Cooke, M., & Campbell, J. (2004). Researching the researchers: Using a snowballing technique. *Nurse Researcher*, *12*(1), 35–47.

Stutzbach, D., Rejaie, R., Duffield, N., Sen, S., & Willinger, W. (2009). On unbiased sampling for unstructured peer-to-peer networks. *IEEE/ACM Transactions on Networking*, *17*(2). doi:10.1109/TNET.2008.2001730

Thompson, S. K. (2002). *Sampling* (2nd ed.). New York, NY: John Wiley and Sons.

Vervaeke, H. K. E., Korf, D. J., Benschop, A., & van den Brink, W. (2007). How to find future ecstasy-users: Targeted and snowball sampling in an ethically sensitive context. *Addictive Behaviors*, *32*(8), 1705–1713. doi:10.1016/j.addbeh.2006.11.008

Witte, J. C., Amoroso, L. M., & Howard, P. E. N. (2000). Research methodology: Method and representation in internet-based survey tools – Mobility, community, and cultural identity in survey 2000. *Social Science Computer Review*, *18*(2), 179–195. doi:10.1177/089443930001800207

Wright, R., & Stein, M. (2005). Snowball sampling. In Kempf-Leonard, K. (Ed.), *Encyclopedia of Social Measurement* (pp. 495–500). New York, NY: Elsevier. doi:10.1016/B0-12-369398-5/00087-6

KEY TERMS AND DEFINITIONS

Hidden Populations: Hidden populations, in the context of research, refer to populations either segregated by society or living at the margin of mainstream society due to issues of a sensitive nature, for example, prostitutes, drug users, homosexuals, or the homeless.

Population: A pre-defined group, often of large dimensions that is the focal point of a research investigation. The members of that population can have multiple differences among them, but are usually united by at least one common feature.

Respondent-Driven Sampling: Respondent-driven sampling is a sample method that is composed by individuals from a specific group and the people they have recommended to be part of the study.

Sample Frame: A sample frame consists in an itemisation of every element of the population, which is going to be researched, from which a sample will be drawn.

Sample: A sample is a section of the population and derives from the need to study certain populations that due to their dimension are extremely difficult or even impossible to explore. A representative sample allows the researcher to arrive at conclusions almost if he/she had studied each element of the population.

Sampling Methods: Methods used to extract parts of the population when the full population cannot be studied or isn't the aim of the research. There is a multiplicity of methods depending on the research aims, nature, and design.

Snowball Sampling: A sampling technique based on individual recommendation. A first set of individuals is selected and then invited to suggest other individuals who share the same characteristics, to also be part of the research. These suggested individuals are then asked the same and this process continues until a desired number of participants is reached.

Chapter 4
Practically Applying the Technology Acceptance Model in Information Systems Research

Reza Mojtahed
The University of Sheffield, UK

Guo Chao Peng
The University of Sheffield, UK

ABSTRACT

Explaining the factors that lead to use and acceptance of Information Technology (IT), both at individual and organizational levels, has been the focus of Information Systems (IS) researchers since the 1970s. The Technology Acceptance Model (TAM) is known as such an explanatory model and has increasingly gained recognition due to its focus on theories of human behaviour. Although this model has faced some criticism in terms of not being able to fully explain the social-technical acceptance of technology, TAM is still known as one of the best IS methodologies that contribute greatly to explain IT/IS acceptance. It has been widely used in different areas of IS studies, such as e-commerce, e-business, multimedia, and mobile commerce. This chapter discusses, describes, and explains TAM as one of the well-known information system research models and attempts to demonstrate how this model can be customised and extended when applyied in practice in IS research projects. In order to illustrate this, the chapter presents and discussed two case studies, respectively, applying TAM in the areas of mobile banking and mobile campus in the UK. It is also proposed that comparing with the traditional questionnaire approach, mixed-methods designs (that contain both a quantitative and a qualitative component) can generate more meaningful and significant findings in IS studies that apply the TAM model. The practical guidance provided in this chapter is particularly useful and valuable to researchers, especially junior researchers and PhD students, who intend to apply TAM in their research.

DOI: 10.4018/978-1-4666-2491-7.ch004

Copyright © 2013, IGI Global. Copying or distributing in print or electronic forms without written permission of IGI Global is prohibited.

INTRODUCTION

For the last four decades, the implementation of Information Systems (IS) in organizations has been known to be costly, frustrating and with a relative low success rate. Nonetheless, the literature in the field indicates that organizations continuously invest in IS in order to improve their performance, maintain customer satisfaction, increase the quality of their services and decline cost (Legris, et al., 2003).

Historically, low success rate and the failure to meet requirements, budgets and deadlines has been identified as expected outcome of investing on IS and Information Technology (IT). The frustration was apparent as early as 1979, when the US Government's Accounting Agency (1979) reported that less than 3% of the software that the US government had paid for, was actually used as delivered. More recently, the Standish Group (2001) reported that 31% of US software (SW) projects were failed in 1994 and 53% were only completed over their budgets and deadlines. Curiously, more than 30 years after the first report of failure, figures indicated by the Standish Group in 2009 show that the level of SW project success is still only at 32%. This apparent failure is usually linked to the "productivity paradox" (Brynjolfsson, 1993), that first put forward by researchers in the late 20 century, which challenges the expected benefit of using IT and IS.

Productivity is a simple concept. It is the amount of output produced per unit of input. While it is easy to define, it is notoriously difficult to measure, especially in the modern economy. In particular, there are two aspects of productivity that have increasingly defied precise measurement: output, and input (Brynjolfsson & Hitt, 1998).

The productivity paradox emerged during empirical studies of IT by the U.S. researchers during 80s and 90s (Brynjolfsson, 1993). However, this focus on input vs. output is rather reductionist in terms of understanding the effects of the adoption and implementation of an IS in organizations. Hitt and Brynjolfsson (1996) were among the first to propose that a shift in understanding the role and effects of IS in organisations was necessary. These authors suggested that IS provide have the potential of originating changes in quality, processes and work practices as well as in the nature and variety of products and services offered by the organization. IS may even have more drastic impacts in reforming the organizational structures and boundaries. However, these effects do not necessarily lead to increases in productivity, if measured strictly in terms of output vs. input. Nonetheless, all these impacts may increase competitiveness, organizational effectiveness, employee satisfaction and even provide extra value for the organisation's customers and business partners (Hitt & Brynjolfsson, 1996).

Therefore, the key question in IS successful adoption lies in exploitation of the system. That is how well the system is accepted and used internally in the organization. As recognized very early on, by authors such as Davis (1989), one of the main indications of IS success or failure is the level or degree of the acceptance of the system by the users. Identifying the reasons of acceptance or rejection of IS has been one of the main challenges of IS research ever since (Swanson, 1988). Sichel (1997) added to this argument by proposing that it is the low usage of installed system that is one of the main reasons for the failure of IS. The importance of IS usage has actually become one of the core concerns in modern organizational behaviour to such an extent that authors, such as Devaraj and Kohli (2003), consider it as one of the main determinants of organizational performance. This is confirmed by continued investment in IT by modern organizations that, even during financial crises, keep allocating a large portion of their assets to IT investment (Kanaracus, 2008). Hence, it is on intention to use, acceptance, and actual use of the system that will enable organizations to attain the expected benefits of IT/IS. Therefore, it is crucial to research the elements that lead to acceptance or rejection of technology by users.

As a result, identifying suitable models to explain and predict IS acceptance has been prioritised as one of the main goals in IS research. A number of social psychology theoretical frameworks have been introduced in IS research to attempt to explain the socio-technical phenomenon of IS acceptance and usage (Bhattacherjee & Premkamar, 2004). The Technology Acceptance Model (TAM) is such a framework that is based on social psychology theory and has been specifically developed to identify and explain the reasons for intention to use, acceptance, or rejection of IS. TAM model is widely used in various IS contexts such as banking (Wang, et al., 2003), online shopping (Grefen, et al., 2003), mobile devices (Pagani, 2004) and rich media (Kim & Forsythe, 2007). This chapter aims to describe and discuss the use of TAM, as well as to illustrate its use in practice through the discussion of two practical cases. It attempts to provide valuable practical guidance to help researchers, especially junior researchers and PhD students, to apply TAM more successfully and effectively in their studies.

TECHNOLOGY ACCEPTANCE MODEL (TAM)

Introduction of TAM

In the 1990s, despite the realization that usage and acceptance of technology was one of the main elements behind the gaining competitive advantages from IS and improve organizational efficiency and effectiveness. The existence of theories to explain and predict, the user tendency for acceptance and use of the technology was in short supply (Davis, 1989). Furthermore, there was an observed and increasing tendency for user resistance and lack of willingness to engage with technology (Young, 1984). Actually, a number of researchers had been previously exploring the influences of individual, organisational and technological variables on acceptance of IT (Benbasat & Dextter, 1986), however Davis (1989) criticised the situation by stating:

Despite the widespread use of subjective measures in practice, little attention is paid to the quality of the measures used or how well they correlate with usage behavior (Davis, 1989).

This realization led Davis (1989) to propose that measures of IS usage should be closely related to two factors: Perceived Ease of Use (PEOU) and Perceived Usefulness (PU). These two factors became the corner stone of TAM as new proposition and method to understand the user acceptance phenomenon.

Description of the TAM model

Davis (1989) and Davis et al. (1989) suggested TAM as a suitable model that can explain and characterise the reasons why users accept or reject IS. TAM is useful both as a predictive method, in order to assess the likelihood of people and organisations to adopt a particular new technology (Turner, et al., 2010), or as an evaluation technique to assess acceptance of technology already in use (Trevino & Webster, 1992). Despite the many additions and changes to TAM as a methodology (e.g. Venkatesh & Davis, 2000), the initial emphasis remains today, that is identifying the effective factors that influence user acceptance of IS. After more than a decade of use in the field of IS, IT, and Computer Science, TAM model is now recognized as one of the more efficient (Taylor & Todd, 1995), pervasive and continuously used model in measuring the adoption of IS (Venkatesh & Davis, 2000).

TAM is based on social psychology theory, in particular Theory of Reasoned Action (TRA) (Ajzen & Fishbein, 1980). TRA is based on the assumption that people, in performing of their organizational tasks, consider the impacts of their possible actions and this reasoning affects their decisions to undertake these actions (Ajzen & Fishbein, 1980). Therefore, based on TRA, TAM assumes that technology acceptance and usage is determined by users' reflections and reasoning, that in turn determine their attitudes, intentions, and internal beliefs.

As discussed above, the TAM model originally proposed by Davis (1989) contains two core elements, namely PEOU and PU. Specifically, Davis (1989) claims that the difficulties of using an IS can offset the usefulness and benefits of the system, as well as affecting user acceptance and satisfaction. PEOU is thus defined as "the degree to which a person believes that using a particular system would be free of effort" (Davis, 1989). On the other hand, if users perceive an IS to be useful to support their current and long-term job performance, they are more likely to adopt the system in their daily practice. PU is hence referred to as "the degree to which a person believes that using a particular system would enhance his or her [current and continuous] job performance" (Davis, 1989). A large number of IS studies confirmed and validated that both PEOU and PU can have direct impact on user's intention to use IS (e.g. Park, et al., 2009). Moreover, previous research also identified and supported that PEOU can in turn affect PU, since users' feeling on how easy or difficult the system can be used will shape their perception on the usefulness of the system (e.g. Wu & Wang, 2005; Park, et al., 2009).

The original version of TAM consists of combinations of these two core factors (i.e. PU and PEOU) and other essential elements, including External Variables, Attitude, Behavioural Intention, and Actual Use, as discussed by Davis et al. (1989) and shown in Figure 1. The user acceptance and actual use of technology is assumed to depend on the intention of users. In turn, the attitude of users, which influences the users' intention, is formed by the users' beliefs. According to Davis et al. (1989) the two base elements of perceived ease of use and perceived usefulness are the components of these users' beliefs. However, these beliefs are also constructed by development processes, personal experiences, professional experiences, organisational factors, social and political influences as well as the perceptions of the tasks to be performed using the technology. These latter influencing factors are known in TAM as external variables.

Although TAM has nowadays been recognised as a well-defined and widely used model in studying user intention for accepting IT/IS, it has some inherent limitations that IS researchers should consider when applying the model in their study. In particular, Wu (2009) highlighted that the popularity of TAM may be resulted from its simplicity and its efficiency in providing an initial road map for planning empirical IS research. However, it can be argued that when focusing on the limited set of six elements in the original TAM model, researchers may not be able to fully explore and explain the social-technical, cultural, and organizational dimensions embedded in the IS usage and operation environment (Wu, 2009). As a result, the original TAM model is rarely used by researchers as it is (Venkatesh, et al., 2003; Turner, et al., 2010). In fact, in order to satisfy the needs and contexts of specific studies, IS researchers very often need to establish their own research framework, by using the original TAM model as the core but extending it with new proposed elements and relations (e.g. Wu & Wang, 2005; Hill & Troshani, 2009). The following section presents two IS studies as exemplifications to illustrate further how TAM can be extended and applied in IS research.

EXPERIENCES IN APPLYING TAM IN IS RESEARCH

After giving an extensive introduction and description about the model, this section aims to provide further explanation and guideline regarding how to apply TAM in IS research projects. Two previ-

Figure 1. The original technology acceptance model (Davis, et al., 1989)

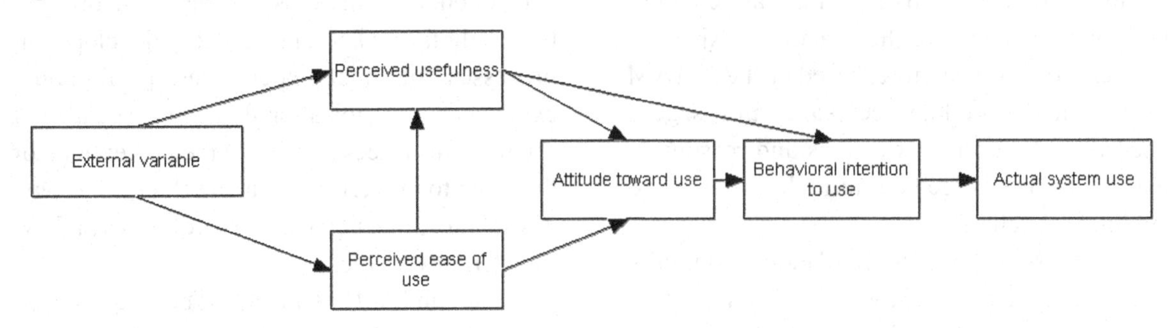

ous IS case study projects are used as examples to illustrate our discussion. The presentation of these IS studies is divided into four parts, namely introducing research aims and objectives, presenting the research design as well as the research model established based on TAM, discussing the data collection and analysis processes, and finally explaining the findings of research.

Case Study 1: Intention to Adopt Mobile Banking in the UK

Research Question and Research Aims

The International Telecommunication Union (ITU) estimated that by the end of 2012, the number of mobile subscriptions would reach around 4.5 billion in the world (ITU, 2008). In the UK, the number of mobile phone subscription reached 75 million in 2009, which accounted for around 1.6% of the world's total (Cellulr-news, 2009). Nonetheless, in a recent survey carried out by KPMG (2009), a global consulting and professional services firms, only 33 percent of the UK population felt comfortable using a mobile phone to access financial services, although the majority of the population also stated that having access to banking services through a mobile phone is important to them. This contradiction of choices and opinions led the researchers to carry out the research project reported in this chapter.

This research aimed at answering the question: *What are the factors that influence the intension of UK citizens to use mobile banking services?* In response to this predefined research question, the study aimed to identify the social and behavioural factors affecting mobile banking, as well as to explore the impact of these identified factors on the intention to use mobile banking.

Research Design

In order to gain a more in-depth understanding on the topic, an extensive literature review was conducted at the initial stage of the study. As a result of this literature review, the researchers identified and selected TAM as a highly suitable and valuable model to be used to understand and study user intention and acceptance towards using mobile banking services. Moreover, and as discussed earlier, the original TAM model, which consists of six main elements (External variable, PEOU, PU, BI, AT, and actual use) and a set of relations between them, is rarely used by researchers without changes (Legris, et al., 2003; Turner, et al., 2010). In order to satisfy the needs of specific subject and research questions, researchers very often need to customise the model by selecting and focusing on some of the originally proposed elements and relations, as well as to extend the model by added new elements and suggesting new relations between the elements (e.g. Wu &

Wang, 2005; Hill & Troshani, 2009). In light of this discussion, a new research model was established based on TAM, as further discussed below.

Furthermore, since TAM originated from positivistic quantitative study (Davis, 1989), quantitative methodology is generally adopted in research projects that apply the TAM model (Wu, 2009). In fact, by adopting TAM, researchers also often aim to capture a snapshot of the current situation and focus on a contemporary event, as well as to answer a "What" question. Yin (1994) highlights that when a research study has involved all of these three characteristics, it is highly appropriate to adopt a quantitative design, in particular a questionnaire survey.

Taking these factors into consideration, many previous studies using TAM have followed a common path of empirical investigation, as proposed by Wu (2009):

review previous literature → select relevant factors such as PU and PEOU for the study → propose hypotheses/model → collect empirical data from a quantitative survey → test the hypotheses and/ or validate the model (Wu, 2009).

The study reported here thus followed a similar research design. Specifically, a further literature review was carried out to propose a more specific and extended research model that was constructed based on TAM. A questionnaire survey was then conducted to test and validate the model. Based on the result of this survey, the researchers refined the model and proposed a set of factors that are crucial determinants to the use of mobile banking in the UK.

Research Model

The final research model of the study was established based on an extensive review of existing literature in the areas of mobile commerce and electronic banking. This research model consists of seven main factors that were deduced from the literature review. It includes the combination of TAM (Davis, 1989), perceived risk (e.g. Langendoerfer, 2002), demographic variables (e.g. Morris & Venkatesh, 2000; McKechnie, et al., 2006), perceived enjoyment (e.g. Davis, et al., 1989), and accessibility (e.g. Daniel, 1999).

Specifically, the TAM model indicated that PEOU and PU have direct influences on the intention of the use of technology, i.e. the use of mobile banking in this case. Moreover, PEOU was expected to have a moderating impact on perceived usefulness, as proposed by Davis (1989). On the other hand, PEOU was identified by Cyr et al. (2006) to have influential effect in users' enjoyment toward using the mobile commerce. Moreover, education (Laforet & Li, 2005), gender (Koivumaki, et al., 2008), and age (Wei, et al., 2009) were also identified as important demographic factors affecting the intention of users to use mobile technologies. In addition, the literature review also suggested perceived risk as an influential factor for using mobile banking (Wu & Wang, 2005). This factor can in turn impact on perceived usefulness of the system (Rose & Fogarty, 2006). Finally, the element of accessibility is identified by Kim et al. (2010) as an important factor to take into consideration. A full description and discussion of these elements and the hypotheses generated based on these factors are given below.

Perceived Ease of Use and Perceived Usefulness

Ease of use is an important element that can impact on the use of the Internet through mobile phones. However, inherent features of mobile phones, such as the small screen size and compacted keyboard, small bandwidth, and limited storage size and slow processor speed of mobiles, can affect their ease of use. In detail, Hoehle and Huff (2009) highlight that the small screen size and compacted keyboards on mobile phones make data entry more challenging, and therefore users may choose to use mobile banking only when no other banking methods are available. In addition, Laukkanen (2005) performed a comparison study between mobile banking, as a form of m-commerce, and Internet banking, as a form of e-commerce. It was concluded that the small screen size of mobile phones, which can only display limited amount of information, may make mobiles not the ideal device for transaction activities (Laukkanen, 2005). Moreover, Green (2000) reinforces that low Internet bandwidth, small screen size and simple functionality of mobile devices prevent the design of effective user interface for m-commerce. According to Carlsson and Walden's (2002) study in Finland, slow speed of Internet services offered through mobile phone is one of the main barriers of m-commerce diffusion. This finding was supported by Vrechopoulos et al. (2002), who carried out a wider study in the Europe and confirmed that high bandwidth is one of the critical success factors influencing the adoption and penetration of mobile commerce. Furthermore, Kim et al. (2007) argue that through mobile devices, Internet applications run on limited resources compared to computer systems (e.g. limited storage size, slower processor speed, and small screen size of mobiles). This can have negative impact especially for new adopters on using mobile Internet services (Kim, et al., 2007). The existence of these features may inevitably make the use of specific services, such as browsing banking websites through mobile phone, become more challenging, time consuming, and less enjoyable (Teo & Pok, 2003).

However, and despite these disadvantages, it should be highlighted that the use of m-commerce actually requires less technical skills and expertise than e-commerce (Vittet-Philippe & Navarro, 2000; Ropers, 2001). That is, computer applications are always more complicated (e.g. containing more functions and steps) than the mobile ones, and thus require users to have a higher level of computer literate skills to operate them. In contrast, mobile applications are designed with simpler layout and steps, clear commands, and less symbols (Condos, et al., 2002). As a result, users with a lower level of computer skills may perceive mobile phone as an easier and more enjoyable tool to use, than personal computers, for getting access to the Internet and engaging in banking transactions. These considerations led to the establishment of the following hypotheses:

- **H1:** Perceived ease of use has direct relationship with the Intention to use of mobile banking.
- **H2:** Perceived ease of use has direct relationship with the Perceived enjoyment of mobile banking.

On the other hand, it is widely acknowledged that mobile devices can offer more freedom over time and place (Wong & Hiew, 2005) than personal especially desktop computers. Kim et al. (2010) thus stated that the improved accessibility and mobility of mobile devices can increase the perceived usefulness of m-commerce, e.g. mobile banking. Moreover, the study of Wei et al. (2009) found that perceived usefulness in turn has an influence on the intention to use of mobile commerce. This finding has also been confirmed by other m-commerce researchers (e.g. Luarn & Lin, 2005; Lin & Wang, 2006; Lu, et al., 2003; Khalifa & Shen, 2008). Furthermore, a number of researchers (Wu & Wang, 2005; Liao, et al., 2007; Gu, et al., 2009) also identified that when users consider a specific mobile application to be easier to use, they are also more likely to perceive this application as useful, and are therefore more willing to use it. Consequently, the following hypotheses were formulated:

- **H3:** Perceived ease of use has direct relationship with the Perceived usefulness of mobile banking.

- **H4:** Perceived usefulness has direct relationship with the Intention to use of mobile banking.

Perceived Risk

The term 'perceived risk' was corned by Bauer, who proposed that, "Consumer behaviour involves risk in the sense that any action of a consumer will produce consequences which he cannot anticipate with anything approximating certainty, and some of which at least are likely to be unpleasant" (Bauer, 1960, p. 24). Originally, perceived risk is defined as "an uncertainty regarding the possible negative consequences of using a product or service" (Ko, et al., 2004). Bauer stressed that perceived risks are established from subjective views of consumers, and they may not necessarily be risks that will occur in the real world (Ross, 1975). The existence of these perceptions of risks can certainly affect consumer's acceptance of the related product or service.

With emergence of new technology, which offers new forms of trading and commerce, the definition of perceived risk has also consequently evolved. In particular, the perception of risk in technology is associated with the uncertainty about the capability of the specific technology in delivering the expected outcome (Im, et al., 2008). In terms of e-commerce, perceived risks of consumers/users may be related to four different categories (Forsythe & Shi, 2003), namely:

- **Financial Risks:** Which can lead to a net loss of money to a customer, e.g. in the case that credit card information is stolen online.
- **Product Performance Risks:** Which are related to the consumers' inability to accurately judge the quality of the product online.
- **Psychological Risks:** Which refer to disappointment, frustration, and shame experienced if one's personal information is disclosed.
- **Time/Convenience Risks:** Which are associated with the loss of time and inconvenience caused by difficulty of navigation and/or submitting order, finding appropriate websites, or delays receiving products.

In relation to mobile banking, there are some other specific risks perceived by customers. In particular, security and privacy issues are not a new concern in the field of m-commerce and have been widely discussed by previous researchers (Pikkarainen, et al., 2004; Fang, et al., 2005). Worries about the security and privacy aspects are the most crucial barriers preventing consumers from using mobile banking services. Luarn and Lin (2005) highlight that customers of mobile banking often worry that their personal information and financial data may be transferred to and illegally used by third parties without their permission. These perceived risks will inevitably have a negative impact on users' perceptions of using m-banking services. Moreover, these perceived risks are also expected to affect consumers' attitudes and perception towards the usefulness of m-banking technologies (Rose & Fogarty, 2006). Therefore, the following hypotheses were generated:

- **H5:** Perceived risk has direct relationship with the Intention to use of mobile banking.
- **H6:** Perceived risk has direct relationship with the Perceived usefulness of mobile banking.

Perceived Enjoyment

Perceived enjoyment is defined as "the extent to which the activity of using computers is perceived to be enjoyable in its own right, apart from any performance consequences that may be anticipated" (Davis, et al., 1992). When 'perceived ease of use' and 'perceived usefulness' are extrinsic factors, perceived enjoyment is an intrinsic element that motivates users to adopt m-banking services. PE has been frequently combined with the other elements of the TAM model to investigate user acceptance towards Internet and mobile technologies (Dabholkar, 1996; Moon & Kim, 2001; Bruner & Kumar, 2005; Liao, et al., 2007). In particular, findings of the survey study conducted by Liao et al. (2007) identified that perceived usefulness of 3G mobile services can be significantly affected by perceived enjoyment, which can also have very positive influence on users' attitudes towards the usage of these services. Similar findings were also found in other m-commerce studies (e.g. Pura, 2005; Kim, et al., 2007; Dai & Palvia, 2009). Therefore, the following hypotheses were proposed:

- **H7:** Perceived enjoyment has direct relationship with the Perceived usefulness of mobile banking.
- **H8:** Perceived enjoyment has direct relationship with Intention to use of mobile banking.

Socio-Demographic Factors

Traditionally, customers' socio-demographic backgrounds (e.g. gender, age, and education) can influence the type of banking information and services that they require (McKechnie, et al., 2006). These socio-demographic factors are also found to have direct impact on customers' intention to use of mobile banking services. For instance, in an m-banking study in China, Laforet and Li (2005) concluded that male and highly educated consumers are dominant users of mobile and online banking services. Koivumaki et al. (2008) reinforced that the majority of mobile services users are male and in relatively young age. Moreover, Howcroft et al. (2002) identified that the young generation is more likely to enjoy the features (e.g. convenient and time saving) of mobile technologies, and is thus more likely to accept and use mobile banking services. Based on the results of the literature review, the following hypotheses were established:

- **H9:** Gender has direct relationship with the Intention to use of mobile banking.
- **H10:** Age has direct relationship with the Intention to use of mobile banking.
- **H11:** Education has direct relationship with the Intention to use of mobile banking.

Accessibility

As discussed earlier, the TAM model was developed based on the Theory of Reasoned Action (TRA) (Ajzen & Fishbein, 1980). The Theory of Planned Behaviour (TPB) is an extended version of the TRA (Ajzen, 1991). When the TRA focuses mainly on the influence of individual attitudes and perceptions on their actions, TPB includes an additional aspect, namely Perceived Behavioural Control (PBC) (Ajzen, 1991). PBC is determined by the availability of facilities and resources that are required to perform an action, as well as the perceived importance of these facilities and resources to achieve outcomes (Ajzen, 1991). According to Ajzen (1991), the availability and accessibility of the required facilities and resources can substantially affect people's behaviour (e.g. their intention to use of a particular technology).

In light of this discussion, it can be expected that having access to the right tools can essentially influence users' intention to use mobile banking services. It is apparent that the use of m-banking requires the use of appropriate mobile phones, which is capable of accessing to the Internet. According to Daniel (1999), not having access to appropriate medium and tools is one of the main barriers preventing the use of e-banking in the UK. Moreover, Kim et al. (2010) identified that factors such as convenience, accessibility, and mobility, can impact users' perceived usefulness of mobile banking services. Therefore, following hypotheses were generated:

- **H12:** Accessibility to the Internet through mobile phone has direct relationship with the Intention to use of mobile banking.

Figure 2. The research model for mobile banking

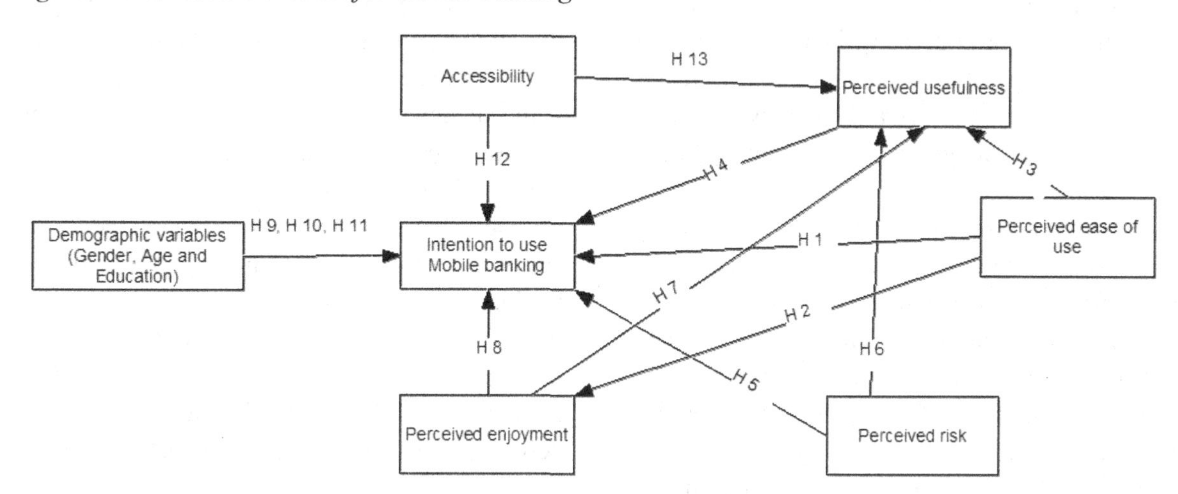

- **H13:** Accessibility to the Internet and mobile phone has direct relationship with the Perceived usefulness of mobile banking.

Overall, the seven factors contained in the research model, as well as the 13 hypotheses generated from the results of the literature review, are summarised in Figure 2.

Data Collection

As discussed above, questionnaire survey, which is the most commonly used method in TAM-based research (Wu, 2009), was selected and adopted as the data collection tool for this study. Moreover, it was considered that conducting a national survey covering the entire UK population is very difficult due to time and cost restrictions. Furthermore, not all citizens in the UK will use mobile phones or get access to the Internet by using mobile phones. These citizens may therefore be less interested in and/or less willing to getting involved in the study. In fact, the survey of Office for National Statistics (2007) shows that the use of mobile technologies has been particularly widespread and penetrated among young teenagers at universities, who consider mobile phones as a 'must-have' and even an integral part of their everyday life (Carter, et al., 2011). For these reasons, the researchers decided to carry out the questionnaire survey with students in their own institution, i.e. University of Sheffield, UK.

By following the guidance of Davis (1989) and fellow IS researchers (e.g. Cyr, et al., 2006; Wu & Wang, 2005; Rose & Fogarty, 2006), the questionnaire was designed by using multiple-item Likert scales. In fact, in his original paper, Davis (1989) introduced and attached the questionnaire that was used in his survey. Questions contained in this questionnaire are thus often re-used in other TAM-based studies. However, it was felt that since this set of questions was not originally developed to study mobile banking, they would not entirely suit this study. Also certainly, Davis' original questionnaire does not cover the additional elements that we established in the above research model. As a result, the questionnaire that we developed and used in this study was a revised version of Davis' work, with support from the literature. Specifically, this modified questionnaire consists of seven parts:

- In part one, the demographic information such as age, gender and education was gathered.
- Part two measured the scale of accessibility and the ease of access to the Internet through mobile devices.

- Parts three to seven contained questions that measured perceived ease of use, perceived usefulness, perceived risk in terms of security and privacy, perceived enjoyment and intention to use mobile banking.

Moreover, the designed questionnaire was sent to 11 students in the researchers' department for pilot testing. Subsequently, Cronbach's alpha test was used to examine the reliability of the data collected from the pilot study. The result of the test showed that Cronbach alpha for all variables are greater than 0.7. According to Kaplan and Saccuzzo (2008), when alpha is higher than 0.7, the data collected can be considered as internally reliable. Therefore, this pilot study concluded that the designed questionnaire was efficient in collecting reliable data, and was thus considered as the final one. The final questionnaire was then distributed to a randomly selected sample of 350 students at the university by email. In order to increase the response rate, some printed hard copies of the questionnaire were also distributed to students. Consequently, 140 valid and usable questionnaires were collected and analysed.

Data Analysis and Research Findings

After data collection, the questionnaire data was analysed by using the statistical software SPSS. A wide range of tests was carried out to analyse the data, including: Frequency table (to identify the distribution of the response to the questions), Mean comparisons (to examine the importance of each identified element), Correlation test (to identify level of dependency between the identified factors), and Regression test (to identify the direct impact of identified factors on intention to use of mobile banking). Nonetheless, since this chapter focuses mainly on how to apply TAM in IS research, the findings of this mobile banking study is not of particular importance to this chapter. These findings are thus only briefly discussed below.

Specifically, the most significant findings of this study came out from the regression test. The results of the regression test indicated that as the value of perceived risk increases the intention to use mobile banking declines. Moreover, if perceived usefulness and perceived enjoyment of mobile banking is effectively marketed by the financial sector, users' intention to adopt mobile banking will increase. These findings are supported by past studies, which identified direct influences of perceived risk (Dai & Palvia, 2009), perceived usefulness (Wu & Wang, 2005), perceived enjoyment (Hill & Troshani, 2009; Dai & Palvia, 2009) on intention to use mobile banking by customers.

On the other hand, perceived ease of use, perceived risk, and perceived enjoyment were found to have influences on perceived usefulness of mobile banking. These findings are consistent with the results of other studies, which suggests that perceived risk (Rose & Fogarty, 2006; Gu, et al., 2009), perceived enjoyment (Liao, et al., 2007), and perceived ease of use (Wu & Wang, 2005; Liu, et al., 2009) have direct impact on perceived usefulness of mobile commerce.

Furthermore, the study also confirmed the results derived by Cyr et al. (2006) and Liao et al. (2007), who identified that perceived enjoyment of using mobile commerce can be directly affected by perceived ease of use. Nonetheless, the survey also generated some unexpected findings. For example, the survey did not confirm the direct relationship between perceived ease of use and intention to use mobile banking. Moreover, direct relationships of demography and accessibility with other identified variables were not validated in our findings. Overall, the results of the survey led to the development of the following model, which was originally established based on TAM and then tested, validated and revised based on research findings (see Figure 3).

Case Study 2: Factors Leading to the Acceptance and Usage of Mobile Services in UK Universities

Research Question and Research Aims

UK has extremely high rate of mobile phone usage. According to the latest Oxford Internet Survey (OxIS, http://microsites.oii.ox.ac.uk/oxis/), mobile phones have been used by almost all UK households (90%) in 2011. Apart from traditional applications (e.g. telephone calls and sending messages), the OxIS report also shows a significantly increased rate in the use of mobile phones to access Internet-related applications in the UK in recent years (i.e. mobile Internet users increased from 11% in 2005 to 36% in 2011). In relation to this national context, the use of mobile technologies has been particularly widespread and penetrated (i.e. almost 100%) among young teenagers at universities, who consider mobile phones as a 'must-have' and even an integral part of their everyday life. Moreover, a recent survey report provided by Nielsen (2010), a global information and media marketing company, highlights that a significant amount (i.e. 46%) of British youth have been using their mobile phones to get Internet access. This figure is over 20% higher than those in other European countries (e.g. 24% in Italy, 21% in Spain, and 20% in Germany).

Owing to this widespread mobile phone penetration and the increasing mobile Internet usage, leading universities across the UK have been attempting to develop various Internet-based mobile applications to provide campus-wide information (e.g. location of printers/PCs) and administrative services (e.g. renewing books) to students. These advanced and value-added mobile campus applications are widely perceived by institutions as a means to enhance student satisfaction, as well as a strategic solution to maintain their leading positions in the Digital Age. However, and despite these potential values and importance, current development of Web-based mobile phone services in universities in the UK are still in infant stage. More importantly, anecdotal evidence shows that many initial mobile campus applications invested and introduced by UK universities have not been broadly accepted and/or used by students. Consequently, this project aims to answer the research question: *What factors can lead to the acceptance and usage of Internet-based mobile services in universities in the UK?* It particularly aims to explore what types of mobile services are actu-

Figure 3. Approved model on factors influencing the intention to use mobile banking

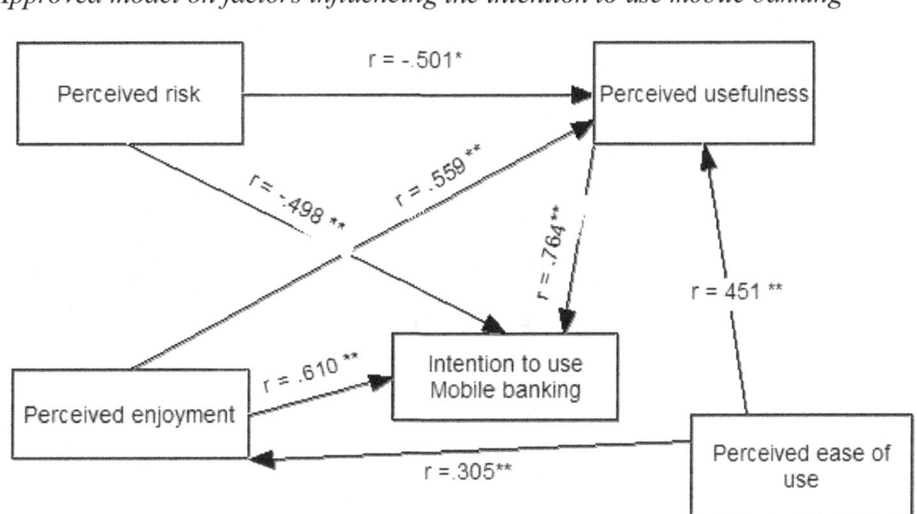

ally needed by students, as well as to investigate the factors leading to the acceptance and usage of these applications.

Research Design and Research Model

For similar considerations as the above mobile banking study, the TAM model is applied as the basic theoretical lens of this project. However, and as mentioned before, the TAM model only contains a limited set of elements, which need to be extended and customised based on the specific subject needs and research context. Consequently, a new research model was developed based on TAM and was extended by using results of an extensive literature review on mobile campus technologies. As shown in Figure 4, this new research model consists of seven elements.

Specifically, perceived usefulness and perceived ease of use, as two key factors of TAM, were expected to have direct influence on the intention to use of mobile campus services. As discussed above, perceived enjoyment and accessibility are another two elements that are frequently reported to have impact on the use of Internet and mobile technologies (Dabholkar, 1996; Daniel, 1999; Moon & Kim, 2001; Kim, et al., 2010). These two factors were thus considered as important to be included in the research model. Moreover, findings of previous researchers identified that perceived ease of use may have a strong impact on the perceived usefulness (Davis, 1989) and perceived enjoyment (Cyr, et al., 2006) of the use of a new technology, e.g. mobile campus tools. Furthermore, since potential participants of this project are all university students, they will fall in the same age group (i.e. age between 18 and 24) and have high educational level. Socio-demographic factors, such as age and education, were considered less important to this study, and were thus not included in the research model. Nevertheless, we were still interested in exploring whether gender would be a factor affecting the use of mobile campus applications among HE students. Finally, the cost of mobile Internet access was identified, in a recent study completed by Walsh (2010), as one of the key factors preventing students from using Internet-based mobile services provided by universities. Consequently, the following hypotheses were established based on the results of the literature review:

Figure 4. The research model for mobile campus

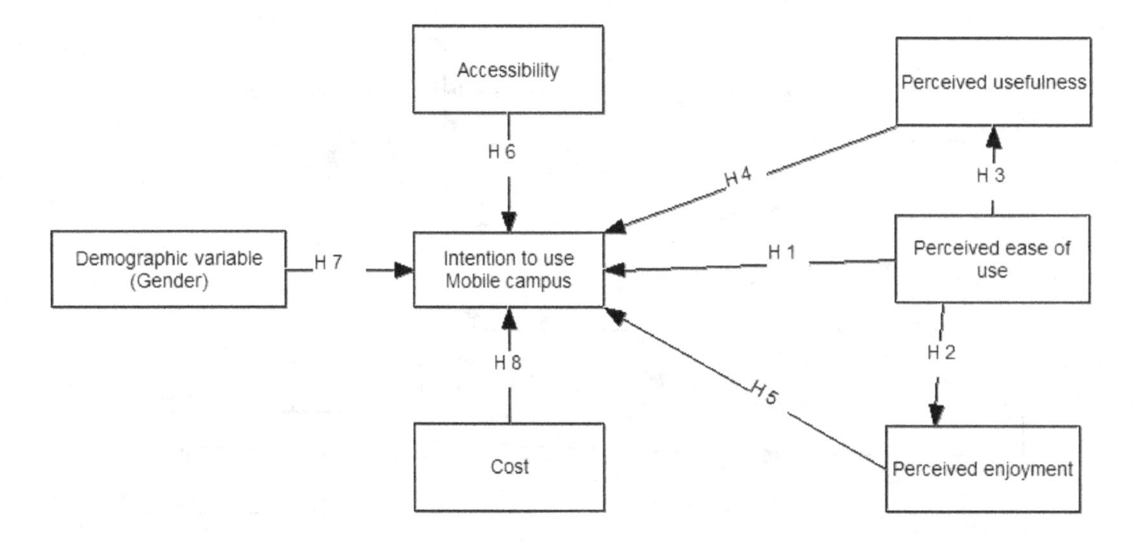

- **H1:** Perceived ease of use has direct relationship with the Intention to use of mobile campus applications.
- **H2:** Perceived ease of use has direct impact on the Perceived enjoyment of mobile campus applications.
- **H3:** Perceived ease of use has direct impact on the Perceived usefulness of mobile campus applications.
- **H4:** Perceived usefulness has direct relationship with the Intention to use of mobile campus applications.
- **H5:** Perceived enjoyment has direct relationship with Intention to use of mobile campus applications.
- **H6:** Accessibility to the Internet through mobile phone has direct relationship with the Intention to use of mobile campus applications.
- **H7:** Gender has direct relationship with the Intention to use of mobile campus applications.
- **H8:** Cost of mobile Internet access has direct impact on the Intention to use of mobile campus applications.

In order to test and validate the proposed research model and the established hypotheses, questionnaire survey was selected as the main data collection method of this study, as further discussed below.

Data Collection

It should be noted that this mobile campus project is currently still ongoing. At the preliminary stage of the project, the University of Sheffield was selected and used as a case study to generate initial findings and shed light on the next stage of the research. The University of Sheffield was chosen as the case study due to the following reasons:

- It is one of the pioneer universities in the UK to launch mobile campus applications, namely the CampusM. This is a free Internet-based mobile application, which provides students the opportunity of access to a wide range of personalised information (e.g. interactive campus map, library account, computer availability on campus, location of printers, and course timetables) through mobile devices. Students of the university were thus expected to have good opinions and perceptions towards the use of these CampusM services.
- The research team of the study is based at the University of Sheffield, which making data collection of this preliminary study easier and more efficient.

As discussed in the previous example, quantitative survey is the most commonly used research method for studies that apply the TAM model. Therefore, a questionnaire survey was selected as the data collection method for this preliminary study that involves the University of Sheffield as a case study. The questionnaire was designed and structured based on the research model presented above, and consists of eight parts:

- The first part of the questionnaire was designed to collect demographic data of respondents, such as gender, age, course undertaken, and year of study.
- The second part of the questionnaire asked students to rank a comprehensive list of mobile campus services provided by the University of Sheffield, from 1 = Not very useful to 5 = Very useful.
- Part three to seven of the questionnaire included a set of Likert-scale questions that measured PEOU, PU, PE, Accessibility, Cost, and Intention to use of mobile campus.

Moreover, the designed questionnaire was initially sent to 10 students at the University of Sheffield for pilot testing. Some minor changes were made to the questionnaire based on the results of the pilot test. Subsequently, the final questionnaire was widely distributed to student through using the University of Sheffield e-mail list. In order to enhance the response rate, 100 hard copies of the questionnaire were also printed out and distributed to students face-to-face. With these efforts, 176 valid and usable questionnaires were received and analysed.

Data Analysis and Research Findings

SPSS was used to analyse the quantitative data collected from the questionnaire. The demographic data showed that the respondents of the survey consisted of 51% of male students and 49% of female students. This indicated that the questionnaire received a good balance of opinions from students in both genders. A wide range of statistical techniques was used to provide a univariate analysis (e.g. frequency tables, bar/pie charts, and means) of the collected data. Based on results of the univariate analysis, students at Sheffield considered the following Internet-based mobile services provided by the university to be most useful to them (see Table 1).

Furthermore, the most significant part of the survey findings came out from a further bivariate analysis (i.e. Pearson's correlation), which aimed to examine empirical relationships between the established variables and thus test the proposed hypotheses in the context of Sheffield University. In particular, the results of the correlation analysis (as summarised in Figure 5) indicated that Perceived ease of use (r = 0.383**), Perceived usefulness (r=0.722**), and Perceived enjoyment (r=0.72**) have statistically significant relationship with the Intention to use of mobile campus services. Moreover, Perceived ease of use was also identified to have positive influence on Perceived usefulness (r = 0.425**) and Perceived enjoyment (r = 0.362**) of mobile campus applications. Furthermore, Cost (r = 0.27**) and Accessibility (r = 0.435**) were also found to be important factors that can affect the use of mobile campus services among university students. Nonetheless, the findings of the survey did not indicate gender as a crucial factor towards mobile campus usage. Therefore, universities should pay substantial attention to the influential factors identified from this preliminary study during the processes of designing, developing, and promot-

Table 1. The top 10 mobile campus services perceived by students at the University of Sheffield

Rank	Mobile campus services	Mean	Standard Deviation
1	Check university e-mail	4.24	1.21
2	Access to course timetable	3.78	1.47
3	Access to modules (lectures + tutorials) information	3.60	1.46
4	Access to exam results	3.52	1.62
5	Book a library PC or group room	3.44	1.53
6	Manage library account	3.43	1.43
7	Access to information about buildings in the campus (e.g. number of seats available in a lecture theatre or lab, location of building on a campus map)	3.40	1.39
8	Access to financial information	3.14	1.49
9	Access to latest news and events around campus	3.10	1.31
10	Access to a friend locator on the campus map	2.84	1.46

1 = Not useful at all; 5 = Very useful

Figure 5. Approved model on factors affecting the use of mobile campus applications

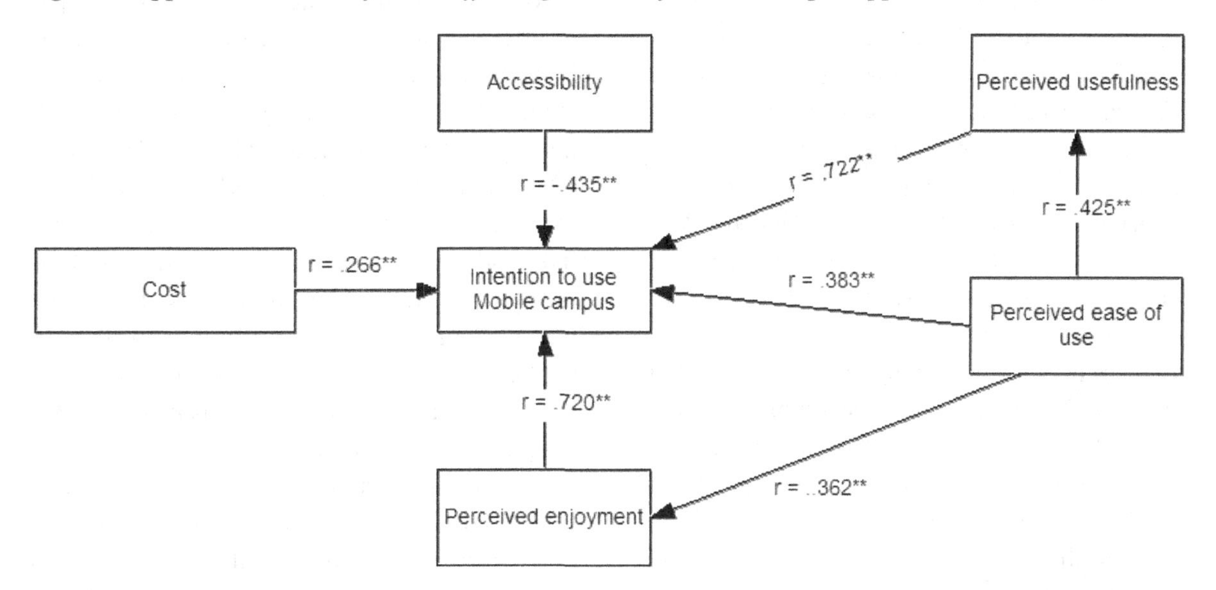

ing m-campus tools. It will also be beneficial if universities can provide free Wi-Fi which can be accessed by as many types of mobile phones as possible, in order to reduce cost and potential accessibility difficulties of using the implemented m-campus applications.

FURTHER PRACTICAL GUIDANCE FOR APPLYING THE TAM MODEL

As mentioned above, TAM has been widely used in IS research since 1989. However, when this model is very useful in providing an initial theoretical lens for studying user acceptance towards new technologies, it was criticised to be too simple to cover all important elements of the phenomenon being investigated (Wu, 2009; Bagozzi, 2007).

As a result, there is always a need to customise and extend the TAM model when applying it in IS studies, as proposed by previous researchers (Lee, et al., 2003; Legris, et al., 2003; Turner, et al., 2010). In the above mobile banking study, a number of factors (e.g. Accessibility, Perceived enjoyment, and Perceived risk) were added and combined together with the original TAM elements (e.g. Perceived usefulness and Perceived ease of use). On the other hand, in the mobile campus project, a number of new factors (e.g. Accessibility and Perceived enjoyment) were also added in the research model. It is clear from the above discussion that these new factors included in the extended research model were identified based on results of an extensive literature review. Moreover, Perceived risk is traditionally considered as an influential factor for using Internet and mobile banking which involves sensitive financial and data transactions (Pikkarainen, et al., 2004; Fang, et al., 2005). This factor was thus included in the final research model of the above m-banking study. However, since mobile campus applications will not normally involve sensitive data (e.g. only involve data such as library information, PC/printer location, and course timetables), perceived risk was considered less important in the context of mobile campus. This factor was therefore not included in the mobile campus study. This example clearly demonstrates that the customisation and extension of the TAM model should be done based on the nature of the research subject and the actual context of the study.

On the other hand, and as discussed above, quantitative survey is typically used in studies that apply TAM. However, there are actually some criticisms about the use of self-reported survey as the dominated method in TAM studies. In particular, and as argued by Straub et al. (1995), when measuring both dependent and independent variables subjectively by using self-reported survey, artifacts rather than truth are generated. As a result, the findings of a self-reported questionnaire may not accurately reflect users' actual intension for using the new technology (Straub, et al., 1995; Turner, et al., 2010). Moreover, Wu (2009) and Martins and Nunes (2009) argued that the use of quantitative surveys in TAM research may not allow researchers to generate rich and deep understanding of human intentions to use technology. Given these considerations, Wu (2009) proposed to use mixed-methods design in TAM studies. Specifically, it was suggested that a qualitative component may be carried out before the questionnaire, in order to explore and better understand the context and subject before the survey stage (Wu, 2009). In order words, Wu suggested a Qualitative + Quantitative approach to be used in TAM studies. When this is certainly one of the possible mixed-methods designs for TAM research, we would also like to propose that an alternative design can also be considered to use, namely the Quantitative + Qualitative approach. In particular, we would propose that a qualitative component may actually be conducted after the questionnaire survey. In fact, the above two IS studies, as well as many other TAM studies carried out by other researchers (e.g. Pikkarainen, et al., 2004; Park, et al., 2009), constantly prove that findings derived from a questionnaire survey may not always meet researchers' original expectations. That is, certain hypotheses proposed in the study may not be confirmed from the findings. If only a quantitative method is used in the study, it will not be possible for researchers to explore further these unexpected outcomes. Moreover, a questionnaire survey will also not be able to generate in-depth results to explore each of the identified variables in the model (e.g. if Perceived risk is important to the use of mobile banking, what particular risk items users are concerned about?). Therefore, after the questionnaire, it would be meaningful for researchers to carry out a follow-up qualitative study (e.g. an exploratory case study), in order to generate richer human insights on the phenomenon under investigation, as well as to explore any unexpected results derived from the survey. Given these reasons, we are actually planning to adopt a multi-case study at the next stage to further explore and validate the questionnaire findings derived from the above two IS studies. Figure 6 provides a more visual view on these two (i.e. Qualitative + Quantitative, and Quantitative + Qualitative) mixed-methods designs that can be potentially applied in TAM studies. Further

Figure 6. Two alternative mixed-methods designs to be applied in TAM research

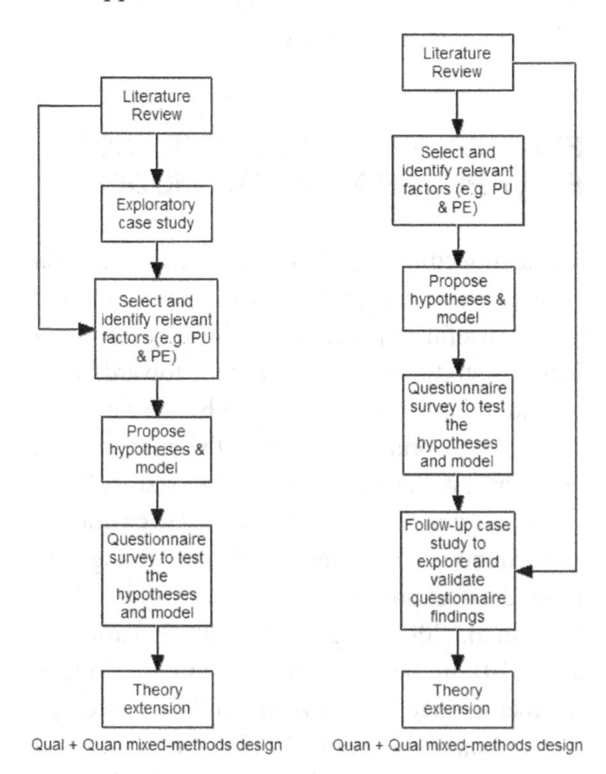

guidance in selecting and applying mixed-methods designs is given in another chapter of this book, namely *Experiences in Applying Mixed-Methods Approach in Information Systems Research.*

CONCLUSION

This chapter discussed the use of TAM as a methodology to investigate and explain factors that lead to use and acceptance of IT and IS. Two previous IS studies were used as exemplifications to illustrate the discussion. In conclusion, we would like to highlight that when the TAM model provides a useful and initial theoretical lens to understand user acceptance towards new information technologies, it needs to be earnestly modified and extended when applying in IS research, in order for researchers to fully explore and explain the social-technical environment in which the IT/IS usage takes place. As demonstrated in the above IS projects, decisions on customising and extending the TAM model should be made based on the actual context of the research and the nature of the research topic. An extensive literature review should be carried out to support the establishment of such an extended model. Moreover, apart from the traditional questionnaire approach, we suggest that more flexible research designs should be developed and adopted in TAM studies, in order to yield more comprehensive and significant findings. Two mixed-methods designs (i.e. Qualitative + Quantitative, or Quantitative + Qualitative design) are proposed in this chapter and can be applied immediately by researchers in their TAM studies. Overall, when these principles are kept rigorously and applied effectively, TAM can be a very useful theoretical tool in studying IS acceptance and usage.

REFERENCES

Ajzen, I. (1991). The theory of planned behaviour. *Organizational Behavior and Human Decision Processes, 50*, 179–211. doi:10.1016/0749-5978(91)90020-T

Ajzen, I., & Fishbein, M. (1980). *Understanding attitude and predicting social behavior*. Englewood Cliffs, NJ: Prentice-Hall.

Bagozzi, R. P. (2007). The legacy of the technology acceptance model and a proposal for a paradigm shift. *Journal of the Association for Information Systems*, *8*(4), 243–254.

Bauer, R. A. (1960). Consumer behavior as risk-taking. In Hancock, R. S. (Ed.), *Dynamic Marketing for a Changing World* (pp. 389–398). Chicago, IL: American Marketing Association.

Benbasat, I., & Dexter, A. S. (1986). An investigation of the effectiveness of color and graphical presentation under varying time constraints. *Management Information Systems Quarterly*, *10*(1), 59–84. doi:10.2307/248881

Bhattacherjee, A., & Premkumar, G. (2004). Understanding changes in belief and attitude toward information technology usage: a theoretical model and longitudinal test. *Management Information Systems Quarterly*, *28*(2), 229–254.

Bruner, G. C., & Kumar, A. (2005). Explaining consumer acceptance of handheld Internet devices. *Journal of Business Research*, *58*(5), 553–558. doi:10.1016/j.jbusres.2003.08.002

Brynjolfsson, E. (1993). The productivity paradox of information technology: Review and assessment. *ACM*, *36*(12), 66–77. doi:10.1145/163298.163309

Brynjolfsson, E., & Hitt, L. M. (1998). Beyond the productivity paradox. *Communications of the ACM*, *41*(8), 49–55. doi:10.1145/280324.280332

Carlsson, C., & Walden, P. (2002). Mobile commerce: Some extensions of core concepts and key issues. In *Proceedings of the SSGRR 2002s Conference*. L'Aquila, Italy: SSGRR.

Carter, M., Thatcher, J. B., Applefield, C., & Mcalpine, J. (2011). What cell phones mean in young people's daily lives and social interactions. *SAIS 2011*. Retrieved from http://aisel.aisnet.org/sais2011/29

Cellular-News. (2009, April 23). Vodafone sees loss of UK market share and lower ARPUs. *Cellular-News*. Retrieved from http://www.cellular-news.com/story/37159.php

Condos, C., James, A., Every, P., & Simpson, T. (2002). Ten usability principles for the development of effective WAP and m-commerce services. *Aslib Proceedings*, *54*(6), 345–355. doi:10.1108/00012530210452546

Cyr, D., Head, M., & Ivanov, A. (2006). Design aesthetics leading to m-loyalty in mobile commerce. *Information & Management*, *43*(8), 950–963. doi:10.1016/j.im.2006.08.009

Dabholkar, P. A. (1996). Consumer evaluations of new technology-based self-service options: An investigation of alternative models of service quality. *International Journal of Research in Marketing*, *13*(1), 29–51. doi:10.1016/0167-8116(95)00027-5

Dai, H., & Palvi, P. C. (2009). Mobile commerce adoption in China and the United States: A cross-cultural study. *ACM SIGMIS Database*, *40*(4), 43–61. doi:10.1145/1644953.1644958

Daniel, E. (1999). Provision of electronic banking in the UK and the Republic of Ireland. *International Journal of Bank Marketing*, *17*(2), 72–82. doi:10.1108/02652329910258934

Davis, F. (1989). Perceived usefulness, perceived ease of use, and user acceptance of information technology. *Management Information Systems Quarterly*, *13*(3), 319–340. doi:10.2307/249008

Davis, F., Bagozzi, R. P., & Warshaw, P. R. (1989). User acceptance of computer technology: A comparison of two theoretical models. *Management Science*, *35*(8), 982–1002. doi:10.1287/mnsc.35.8.982

Davis, F. D., Bagozzi, R. P., & Warshaw, P. R. (1992). Extrinsic and intrinsic motivation to use computers in the workplace. *Journal of Applied Social Psychology*, *22*(14), 1111–1132. doi:10.1111/j.1559-1816.1992.tb00945.x

Devaraj, S., & Kohli, R. (2003). Performance impacts of information technology: Is actual usage the missing link? *Management Science*, *49*(3), 273–289. doi:10.1287/mnsc.49.3.273.12736

Fang, X., Chan, S., Brzezinski, J., & Xu, S. (2005). Moderating effects of task type on wireless technology acceptance. *Journal of Management Information Systems*, *22*(3), 123–157. doi:10.2753/MIS0742-1222220305

Forsythe, S., & Shi, B. (2003). Consumer patronage and risk perceptions in internet shopping. *Journal of Business Research*, *56*(11), 867–875. doi:10.1016/S0148-2963(01)00273-9

GAO. (1979). *Contracting for computer software development--Serious problems require management attention to avoid wasting additional millions*. Washington, DC: General Accounting Office.

Gefen, D., Karahanna, E., & Straub, D. W. (2003). Trust and TAM in online shopping: An integrated model. *Management Information Systems Quarterly*, *27*(1), 51–90.

Green, R. (2000). The internet unplugged. *eAI Journal*, 82-86.

Gu, J. C., Lee, S. C., & Suh, Y. H. (2009). Determinants of behavioral intention to mobile banking. *Expert Systems with Applications*, *36*(9), 11605–11616. doi:10.1016/j.eswa.2009.03.024

Hill, S. R., & Troshani, I. (2009). Adoption of personalisation mobile services: Evidence from young australians. *BLED 2009 Proceedings*. Retrieved from http://aisel.aisnet.org/bled2009/35

Hitt, L. M., & Brynjolfsson, E. (1996). Productivity, business profitability, and consumer surplus: Three different measures of information technology value. *Management Information Systems Quarterly*, *20*(2), 121–142. doi:10.2307/249475

Hoehle, H., & Huff, S. (2009). Electronic banking channels and task-channel fit. *ICIS 2009 Proceedings*. Retrieved from http://aisel.aisnet.org/icis2009/98

Howcroft, B., Hamilton, R., & Hewer, P. (2002). Consumer attitude and the usage and adoption of home-based banking in the United Kingdom. *International Journal of Bank Marketing*, *20*(3), 111–121. doi:10.1108/02652320210424205

Im, I., Kim, Y., & Han, H. J. (2008). The effects of perceived risk and technology type on users' acceptance of technologies. *Information & Management*, *45*(1), 1–9. doi:10.1016/j.im.2007.03.005

ITU. (2008). ICT statistics news log - Global mobile phone subscribers to reach 4.5 billion by 2012. *ITU*. Retrieved from http://www.itu.int/ITUD/ict/newslog/Global+Mobile+Phone+Subscribers+To+Reach+45+Billion+By+2012.aspx

Kanaracus, C. (2008). Gartner: Global IT spending growth stable. *InfoWorld*. Retrieved from http://www.infoworld.com/t/business/gartner-global-it-spending-growth-stable-523

Kaplan, R. W., & Saccuzzo, D. P. (2008). *Psycho- logical testing: Principles, applications, and issues*. Monterey, CA: Brooks/Cole.

Khalifa, M., & Shen, N. K. (2008). Explaining the adoption of transactional B2C mobile commerce. *Journal of Enterprise Information Management*, *21*(2), 110–124. doi:10.1108/17410390810851372

Kim, C., Mirusmonov, M., & Lee, I. (2010). An empirical examination of factors influencing the intention to use mobile payment. *Computers in Human Behavior*, *26*(3), 310–322. doi:10.1016/j.chb.2009.10.013

Kim, H. W., Chan, H. C., & Gupta, S. (2007). Value-based adoption of mobile internet: An empirical investigation. *Decision Support Systems*, *43*(1), 111–126. doi:10.1016/j.dss.2005.05.009

Kim, J., & Forsythe, S. (2007). Hedonic usage of product virtualization technologies in online apparel shopping. *International Journal of Retail and Distribution Management*, *35*(6), 502–514. doi:10.1108/09590550710750368

Ko, H., Jung, J., Kim, J. Y., & Shim, S. W. (2004). Cross-cultural differences in perceived risk of online shopping. *Journal of Interactive Advertising*, *4*(2). Retrieved from http://jiad.org/article46

Koivumaki, T., Ristola, A., & Kesti, M. (2008). The perceptions towards mobile services: An empirical analysis of the role of use facilitators. *Personal and Ubiquitous Computing*, *12*(1), 67–75. doi:10.1007/s00779-006-0128-x

KPMG. (2009). UK consumers prefer to pay for digital content in time, not cash. *KPMG*. Retrieved from http://rd.kpmg.co.uk/mediareleases/15612.htm

Laforet, S., & Li, X. (2005). Consumers' attitudes towards online and mobile banking in China. *International Journal of Bank Marketing*, *23*(5), 362–380. doi:10.1108/02652320510629250

Langendoerfer, P. (2002). M-commerce: Why it does not fly (yet?). In *Proceedings of the SSGRR 2002s Conference*. L'Aquila, Italy: SSGRR.

Laukkanen, T. (2005). Comparing consumer value creation in Internet and mobile banking. In *Proceedings of the International Conference on Mobile Business (ICMB 2005)*. ICMB.

Lee, Y., Kozar, K. A., & Larsen, K. R. T. (2003). The technology acceptance model: Past, present, and future. *Communications of the Association for Information Systems, 12*, Retrieved from http://aisel.aisnet.org/cais/vol12/iss1/50

Legris, P., Ingham, J., & Collerette, P. (2003). Why do people use information technology? A critical review of the technology acceptance model. *Information & Management, 40*(3), 191–204. doi:10.1016/S0378-7206(01)00143-4

Liao, C. H., Tsou, C. W., & Huang, M. F. (2007). Factors influencing the usage of 3G mobile services in Taiwan. *Online Information Review, 31*(6), 759–774. doi:10.1108/14684520710841757

Lin, H.-H., & Wang, Y.-S. (2006). An examination of the determinants of customer loyalty in mobile commerce contexts. *Information & Management, 43*(3), 271–282. doi:10.1016/j.im.2005.08.001

Liu, Z., Min, Q., & Ji, S. (2009). An empirical study on mobile banking adoption: The role of trust. In *Proceedings of the 2nd International Symposium on Electronic Commerce and Security, ISECS,* (Vol. 2, pp. 7-13). Washington, DC: IEEE Computer Society.

Lu, J., Yu, C. S., Liu, C., & Yao, J. E. (2003). Technology acceptance model for wireless internet. *Internet Research, 13*(3), 206–222. doi:10.1108/10662240310478222

Luarn, P., & Lin, H. H. (2005). Toward an understanding of the behavioral intention to use mobile banking. *Computers in Human Behavior, 21*(6), 873–891. doi:10.1016/j.chb.2004.03.003

Martins, J. T., & Nunes, M. B. (2009). Methodological constituents of faculty technology perception and appropriation: Does from follow function? In *Proceedings of the IADIS International Conference on e-Learning*. Algarve, Portugal: IADIS.

McKechnie, S., Winklhofer, H., & Ennew, C. (2006). Applying the technology acceptance model to the online retailing of financial services. *International Journal of Retail & Distribution Management, 34*(4/5), 388–410. doi:10.1108/09590550610660297

Moon, J. W., & Kim, Y. G. (2001). Extending the TAM for a world-wide-web context. *Information & Management, 38*(4), 217–230. doi:10.1016/S0378-7206(00)00061-6

Morris, M. G., & Venkatesh, V. (2000). Age differences in technology adoption decisions: Implications for a changing workforce. *Personnel Psychology, 53*(2), 375–403. doi:10.1111/j.1744-6570.2000.tb00206.x

Nielsen. (2010). *Mobile youth around the world.* Retrieved from http://no.nielsen.com/site/documents/Nielsen-Mobile-Youth-Around-The-World-Dec-2010.pdf

Office for National Statistics. (2007). Use of ICT at home. *Office for National statistics*. Retrieved from http://www.statistics.gov.uk/cci/nugget.asp?id=1710

Pagani, M. (2004). Determinants of adoption of third generation mobile multimedia services. *Journal of Interactive Marketing, 18*(3), 46–59. doi:10.1002/dir.20011

Park, N., Roman, R., Lee, S., & Chung, J. E. (2009). User acceptance of a digital library system in developing countries: An application of the technology acceptance model. *International Journal of Information Management, 29*(3), 196–209. doi:10.1016/j.ijinfomgt.2008.07.001

Pikkarainen, T., Pikkarainen, K., Karjaluoto, H., & Pahnila, S. (2004). Consumer acceptance of online banking: An extension of the technology acceptance model. *Internet Research, 14*(3), 224–235. doi:10.1108/10662240410542652

Pura, M. (2005). Linking perceived value and loyalty in location-based mobile services. *Managing Service Quality, 15*(6), 509–553. doi:10.1108/09604520510634005

Ropers, S. (2001). New business models for the mobile revolution. *eAI Journal*, 53-57.

Rose, J., & Fogarty, G. (2006). Determinants of perceived usefulness and perceived ease of use in the technology acceptance model: Senior consumers' adoption of self-service banking technologies. in *Proceedings of the Academy of World Business, Marketing & Management Development Conference*, (vol 2, pp. 122-129). Paris, France: World Business, Marketing, & Management Development.

Ross, I. (1975). Perceived risk and consumer behavior: a critical review. In Schlinger, M. J. (Ed.), *Advances in Consumer Research* (*Vol. 2*, pp. 1–20). New York, NY: Association for Consumer Research.

Sichel, D. E. (1997). *The computer revolution: An economic perspective*. Washington, DC: Brookings Institution Press.

Standish Group. (2009). *Chaos report*. Boston, MA: Standish Group. Retrieved from http://www.standishgroup.com/newsroom/chaos_2009.php

Standish Group Inc. (2001). *Extreme chaos*. Retrieved from http://standishgroup.com/sample_research/extreme_chaos.pdf

Straub, D., Limayem, M., & Karahanna-Evaristo, E. (1995). Measuring system usage implications for IS theory testing. *Management Science, 41*(8), 1328–1342. doi:10.1287/mnsc.41.8.1328

Swanson, E. B. (1988). *Information system implementation: Bridging the gap between design and utilization*. Homewood, IL: Irwin.

Taylor, S., & Todd, P. A. (1995). Understanding information technology usage: A test of competing models. *Information Systems Research, 6*(2), 144–176. doi:10.1287/isre.6.2.144

Teo, T. S. H., & Pok, S. H. (2003). Adoption of WAP-enabled mobile phones among internet users. *Omega: The International Journal of Management Science, 31*(6), 483–498. doi:10.1016/j.omega.2003.08.005

Trevino, L. K., & Webster, J. (1992). Flow in computer-mediated communication: Electronic mail and voice evaluation. *Communication Research, 19*(2), 539–573. doi:10.1177/009365092019005001

Turner, M., Kitchenham, B., Brereton, P., Charters, S., & Budgen, D. (2010). Does the technology acceptance model predict actual use? A systematic literature review. *Information and Software Technology, 52*(5), 463–479. doi:10.1016/j.infsof.2009.11.005

Venkatesh, V., & Davis, F. D. (2000). A theoretical extension of the technology acceptance model: Four longitudinal field. *Management Science, 46*(2), 186–204. doi:10.1287/mnsc.46.2.186.11926

Venkatesh, V., Morris, M. G., Davis, G. B., & Davis, F. D. (2003). User acceptance of information technology: Toward a unified view. *Management Information Systems Quarterly, 27*(3), 425–478.

Vittet-Philippe, P., & Navarro, J. M. (2000). Mobile e-business (m-commerce): State of play and implications for European enterprise policy. *European Commission Enterprise Directorate-General E-Business Report*. Retrieved from http://www.ncits.org/tc_home/v3htm/v301008.pdf

Vrechopoulos, A. P., Constantiou, I. D., Mylonopoulos, N., & Sideris, I. (2002). Critical success factors for accelerating mobile commerce diffusion in Europe. In *Proceedings of the 15th Bled Electronic Commerce Conference*. Bled, Slovenia. IEEE.

Walsh, A. (2010). Mobile phone services and UK higher education students, what do they want from the library? *Library and Information Research, 34*(106), 22–36.

Wang, Y. S., Wang, Y. M., Lin, H. H., & Tang, T. I. (2003). Determinants of user acceptance of internet banking: An empirical study. *International Journal of Service Industry Management, 14*(5), 501–519. doi:10.1108/09564230310500192

Wei, T. T., Marthandan, G., Chong, A. Y. L., Ooi, K. B., & Arumugam, S. (2009). What drives Malaysian m-commerce adoption? An empirical analysis. *Industrial Management & Data Systems, 109*(3), 370–388. doi:10.1108/02635570910939399

Wong, C. C., & Hiew, P. L. (2005). Diffusion of mobile entertainment in Malaysia: Drivers and barriers. *Enformatika, 5*, 263–266.

Wu, J. H., & Wang, S. C. (2005). What drives mobile commerce? An empirical evaluation of the revised technology acceptance model. *Information & Management, 42*(5), 719–729. doi:10.1016/j.im.2004.07.001

Wu, P. F. (2009). Opening the black boxes of TAM: Towards a mixed methods approach. *ICIS* 2009. Retrieved from http:// aisle.aisnet.org/icis2009/101

Yin, R. K. (2003). *Case study research design and method* (3rd ed.). Thousand Oaks, CA: Sage Publisher.

Young, T. R. (1984). The lonely micro. *Datamation, 30*(4), 100–114.

Chapter 5
Designing Personalised Learning Resources for Disabled Students Using an Ontology-Driven Community of Agents

Julius T. Nganji
University of Hull, UK

Mike Brayshaw
University of Hull, UK

ABSTRACT

The exploration of social artifacts for the disabled is an important and timely issue. The affordances of new technologies like the Semantic Web allow more intelligent handling of educational learning resources that open up the potential of personalisation of services to individuals. Contemporary legislation calls for "reasonable adjustments" and "reasonable accommodation" to be made to services in order to accommodate the needs of disabled people. Here, the authors examine, from a design perspective, how this might be done in the context of higher education. Specifically, they advocate a design based upon an ontology-based personalisation of learning resources to deliver to students' real needs. To this end, so far little effort has been directed towards disabled students in higher education. The authors note some of the problems and issues with online assistive/adaptive technologies and propose a methodological fix. Here, they propose an ontology-based methodology for a Semantic Web community of agents that personalises learning resources to disabled students in higher education, specifically highlighting a disability-aware Semantic Web agency development methodology. The authors also present the results of usability evaluation of the implemented visual interface with some disabled and non-disabled students.

DOI: 10.4018/978-1-4666-2491-7.ch005

Copyright © 2013, IGI Global. Copying or distributing in print or electronic forms without written permission of IGI Global is prohibited.

INTRODUCTION

An increasing number of disabled students worldwide enter into higher education institutions every year. In the UK for instance, the Higher Education Statistics Agency (HESA) estimated the number of such students in the 2007-2008 academic year to be about 62,510 (HESA, 2009). This increasing number makes delivering education online more challenging as disabled students may have varying requirements based on their specific needs. To solve this problem, educational institutions utilise assistive technologies to assist disabled students in their learning. The problem however is that some of these technologies are not compatible with some digital learning systems, resulting in an exclusion of some disabled students if the institutions cannot handle the disability (Steyaert, 2005). The Disability Discrimination Act (DDA) was enacted in 1995 and extended in 2005 to prohibit discrimination against disabled people. Service providers are thus required to make "reasonable adjustments" to their services to meet the needs of disabled people. The Special Educational Needs and Disability Act 2001 (SENDA) introduces the right for disabled students to have equal opportunities to contribute and benefit from education and not to be discriminated against. The Americans with Disabilities (ADA) Act 1990 also prohibits discrimination against disabled people. Higher education institutions amongst other solutions have resorted to personalisation of e-learning and other services.

The problem of searching and retrieving online information continues because the current Web is not meaningful and hence using search engines to retrieve information is in some cases a difficult task as the results are often numerous and sometimes irrelevant. To solve this problem, there is need to provide students with personalised resources that meet their needs. To accomplish this, intelligent and semantically rich agents forming a community or society of agents (Minsky, 1986) which could empower the students based on their needs and requirements analyses are therefore needed. The semantic Web offers a solution as it is by its nature more meaningful. Hence, researchers are currently using semantic Web technologies to personalise learning and services (e.g. Brut & Braga, 2008; Henze, et al., 2004; Nganji, et al., 2011; Razmerita & Lytras, 2008) and to facilitate search and retrieval of information. However, as semantic Web technologies are used for personalisation, little seems to be done to consider the needs of disabled people and to personalise services for them thus inaccessibility problems continue, requiring more robust agents that understand their needs. The difficulties encountered in presenting learning resources using the current non-semantic Web which is based on HTML can now be resolved using the more meaningful semantic Web particularly with technologies such as Web ontologies. In order to successfully integrate assistive technologies with online learning environments in a way that will be suitable for disabled students, the integration must be based on an architecture that is disability-aware, using an approach that includes the needs of disabled students. In this light, this study proposes a personalisation approach based on a disability ontology containing information on various disabilities, which can be used to present disabled students with learning resources that are suitable for their specific needs, following a disability-aware Semantic Web agency development methodology. This chapter will therefore focus on presenting a disability-aware and agent-based methodology for designing personalised learning resources for disabled students. The fact that disabled students are emphasised does not mean that this methodology cannot be employed to personalise learning for non-disabled students. It will be shown in the results section that this methodology can be effectively employed to meet the needs of all students regardless of their disability status (disabled or non-disabled), thus being inclusive of all learners.

This chapter is organized as follows: first, the problems faced when using assistive technologies are discussed, and then related work in agent development methodology and system architecture is reviewed. The chapter then presents the research methodology and approach, which includes an agent-centric information systems approach and disability-aware semantic Web agency development. The context for this research is therefore a contribution to the overall notion of system design from a perspective of agents as important structuring entities upon which to drive the underpinning system architecture and philosophical motivation. As ontology is vital for providing personalisation of learning resources, the ADOOLES (Abilities and Disabilities Ontology for Online LEarning and Services) ontology employed in this personalisation is presented. The architecture of a semantic Web agency for personalising services for disabled students in higher education is then presented, discussing the various agents and the roles they play in personalisation. The results of testing the implemented visual interface of the ONTODAPS (Ontology-Driven Disability-Aware Personalised E-Learning System) model with 40 disabled and non-disabled students are presented and discussed. The chapter concludes with a summary of the contents of this work and future directions specifically highlighting the role of ONTODAPS in social learning for disabled students.

SHORTCOMINGS OF ASSISTIVE TECHNOLOGIES

Although assistive technologies greatly enable learning for students with disabilities, these technologies themselves can be a disabling factor if not properly integrated into student learning. As Tompsett (2008) notes, "poor integration of some e-learning and virtual learning environments with some assistive technologies such as screen readers can significantly degrade the productivity and achievement of some disabled students." With most course materials now being made available online, it is vital for course designers as well as Web designers to adhere to accessibility and usability standards in order to ensure that the online learning resources are truly accessible to disabled students. If a course designer uploads a visual learning resource say in the form of an image and fails to include the appropriate description for that image, a screen reader user may be unable to understand the information that is presented and hence will be left out; a form of discrimination.

Some screen readers at times are unable to read information contained in the ALT or LONGDESC attributes of the IMG tag due to incompatibility or lack of inclusion of such attribute. The problem arises from the fact that screen readers tend to read things rather superficially, at the level of HTML rather than at the deeper level of ontologies where knowledge is represented and machines can understand the information as well as humans. Thus, some assistive technologies like screen readers tend to depend on other applications such as windows and other applications in order to function properly, not being able to work independently. Consequently, these assistive technologies will depend on the structure of the applications and in the event where the structure is faulty, information will not be correctly conveyed; the case of screen readers and some HTML pages.

It is therefore necessary to look for new ways of integrating assistive technologies with e-learning environments and developing novel architectures, which are disability-aware and can help solve the problems associated with the shortcomings of existing architectures. This research will address these needs through a novel architecture for personalising learning resources for disabled students, which is ontology-driven and agent-based, following a disability-aware semantic Web agency development methodology.

RELATED WORK

One of the earliest adaptation models is AHAM (De Bra et al, 1999), Adaptive Hypermedia Application Model that is an extension of the Dexter model. In this model, the topmost layer, the run-time layer represents the user interface on top of a presentation specifications layer and a storage layer containing three models: an adaptation model that combines elements from a domain model and user model to describe an event, a domain model that contains a conceptual representation of the application domain, and a user model containing conceptual representations of all aspects of the user relevant for the adaptive hypermedia application. Whilst the user model represents aspects of the user, it does not specifically deal with disabled users, and does not show how such model can be employed to personalise or adapt services for disabled people.

The FOHM (Fundamental Open Hypermedia Model) model (Millard, et al., 2000) defines a common data model and set of related operations that are applicable for the three hypertext domains of navigational hypertext, spatial hypertext, and taxonomical hypertext.

Ontologies have been successfully employed to help provide effective solutions to the difficulties faced by disabled people. Coyle and Doherty (2008) for instance described the potential of using ontologies in developing interactive systems that can support mental health interventions. Lorenz and Horstmann (2004) on the other hand used the Resource Description Framework (RDF) to facilitate access to graphically represented information for blind people by semantically annotating the graphical information with RDF.

With the introduction of the Semantic Web, newer architectures and systems such as the Personal Reader (Henze & Herrlich, 2004) make use of Semantic Web technologies for personalisation of services. In this framework, different Web services cooperate with each other by exchanging RDF documents to form a Personal Reader instance. One of the key advantages with this framework is that it enables users to select services which provides extended functionality such as personalisation services, combining them into a Personal Reader instance. This framework as others is not specific for disabled people and does not describe how to deal with specific needs.

There is evidently a great gap in using ontologies to personalise services to disabled people, which necessitates an adaptation of existing architectures to provide such personalisation or newer architectures. Our architecture presented in this work although specific for disabled people, can be used for non-disabled people, thus being inclusive.

RESEARCH METHODOLOGY AND APPROACH

The work in this chapter is a practical study that aims to impact on the personalisation of Semantic Web services for Disabled Students. As such it could look to Software Engineering (e.g. Pressman, 2005; Somerville, 2007) to find its robust, practiced, reliable and well understood underpinning. This is a wide area of study that initially was conceived to take as its starting point the hard methodology of engineering as a way out of a perceived software crisis (Dijkstra, 1972). As a nascent discipline, it was an obvious step to look to more established cognate areas. As an aspiring science engineering (hence software engineering) was an obvious route to take out of this impasse. Engineers could solve major engineering problems like building bridges, so why could this methodology not be adapted to building software? Laterally this embrace was expanded to cover the whole software lifecycle to include not just design and implementation but also testing, reengineering, and maintenance. A critical notion here from the outset was software quality management (Sommerville, 2007) and linked to that, issues of modularization (e.g. Parnas, 1972). Should a system be broken down into parts based on models of its flow of control? Alternately maybe more abstract components might be a better level of dialog leading to higher level design descriptions involving Objects (e.g. Goldberg & Robson, 1980). However, an Object-Based Modularization may lead to different parts of the system that are closely related being distributed widely across different Objects. The help system is often held up as an example where it is implemented individually for a class of objects but really should be a system in its own right. This leads to other models of modularization based on these related concerns known as Aspect Orientated Programming (Kiczales, et al., 1996). To sum up how you modularize is a debatable issue strictly from an information systems viewpoint.

In the above, a critical assumption is made that the known a priori are sufficient to allow a requirement capture exercise to be undertaken and the roll out implementation expressed. These structuring principles are good from a software engineering perspective, but when we move to one of knowledge, then they are more problematic and other models warrant consideration.

Designing Information Systems from an Agent-Centric Viewpoint

It has been noted (e.g. Galliers, 1996, 1997) that approaches to information systems are often characterised by the parent discipline in which the reporters of this research are located e.g. library science, management science, or computer science. However, in the context of knowledge systems the use of the term agents and societies of agents has developed as a powerful structuring model (Minsky, 1986). There are clearly comparisons between Agents and Object orientated programming. The notions of abstract data, typing, hierarchy, and inheritance of both data (and knowledge) or procedures (or methods) is common. The emphasis in the world of Agents is, however, knowledge and the paradigm is one of knowledge engineering. In Knowledge Engineering, Knowledge is the King. With the growth of the Semantic Web and its wealth of knowledge as expressed via its associated markup languages, knowledge modelers and knowledge engineers turned to the world of Agents as an underpinning structure model. Using Agents as an Organizing Principle of the Semantic Web has already been well established (e.g. Luke, et al., 1997; Hendler, 2001; Hadzic, et al., 2009; Håkansson, et al., 2009) and as the basis of organising search (Hartung & Håkansson, 2010) including the fusion and aggregation of information (Nodine, et al., 2000). In the work reported here, we use agents as our organisational components and specifically address the social problem of educational resource access for those with disability using the Semantic Web.

When designing a system, structure and organisation is major a priori. Design rationale and methodology has been Holy Grail of this research enterprise (e.g. Somerville, 2007; Pressman, 2005). One approach has been to consider the problem in its entirety and context. The context is often considered ill structure and messy (Flood & Jackson, 1991) and a more holistic approach is called for. A classic view of information systems from this systems thinking perspective is provided by SSM (Checkland, 1982). In this perspective, a good model is considered to be one of considering the system as an entirety. The emphasis is one of the System as a whole. The concern of the observer is as an onlooker of the overall behaviour of the whole, not of the components. The observer is detached and the act of observation does not affect that which is being observed. The goal of the modeler is to develop a Rich Picture of the observed and that this zeitgeist is to be the guiding force in subsequent interaction/intervention. The critical notion is that this wholeist perspective gives a more rounded perspective. A critical notion here is "Does observation change what you are looking at?" This in essence is the critical rub between Psychology and Sociology. Psychology says that you can look at a system and observe. The act of looking does not change what you are looking at. A Sociologist would argue that the act of observation changes. If you know you are being watched then this changes your behaviour. A Systems Thinking approach would aspire to observe the whole. Indeed the most anthropologist methodology would encourage the observer to so totally embed themselves within the society to be observed that they become absorbed and therefore ignored so that their observations are unnoticed. For as Information Systems Designer this might mean going as a co-worker, sharing office/hot desking with others and spending time with fellow workers. The critical notion here is that after some time the observer (Information Systems Designer) will be considered to be part of the team, not an observer. Once one has "gone native" in anthropologist parlance then they can observe but their observation is not perceived by the observed. They can thus evaluate the Information System without the evaluation task getting in the way.

A different Information System Design perspective is that from a scientific—not a social science—perspective. This argues that we can scientifically observe the world. The act of observation does not radically change what is being observed such that we have to change what we are doing. We can empirically evaluate our Information Systems in a scientific way and report on the results. This is precisely what we do at the end of the chapter. Critically from an Information Scientist perspective, we can break what we are looking at into sub-parts and consider these notions as discrete components, an act of reductionalism.

A Software Engineering/ Information Systems Design reductionalist model considers how to break an overall software design problem down into subcomponents and by interoperability of the subcomponents solve the overall goal. This is a traditional software engineering trajectory of problem resolution but in precisely how we go about this lies the rub. We can break the problem down into real world entity/action steps, turn these into structure diagrams and network models before undertaking actualisation into lines of code (Jackson, 1983). A more modern approach would be to black box describe the behaviour of our target program and describe behaviours to check this model. This is test-driven development (Kent, 2003). An alternative paradigm is to think of the reductionalism into function components. These components can be in terms of objects that are represented in a hierarchical ordering with subclasses being specialisations of their parent (Kay, 1996). For some domains, this is a natural breakdown where the objects and their relations naturally speak to the designer. The work described here in terms of Agents is sympathetic to this perspective. Other possible breakdowns are into software components (e.g. Szyperski, 2002) which emphasize their inherent re-use and their independence irrespective of context; they are defined by a signature interface, and are an example of a service-orientation software philosophy (Bell, 2008). Objects are often language specific, subject to the architectural operational semantics of a particular language—components are free of that. A component that is written in JAVA should be able to talk to one written in Objective C with no issues. We will argue later in this chapter that both can be re-described in terms of general agency and play an important part in the Internet of Things.

Stepping back from direct Information System Design/Implementation issues we may wish to consider components as agencies in their own right that interact with others in an autonomous self-organising style. Such a society (Minsky, 1985) is a loose body of interacting components that can, but do not necessarily have, ordering relationships. As such, this does not necessarily limit them to being entirely software components or of fixed media. Within the context of the Semantic Web (Berners-Lee, et al., 2001) this may be a normal assumption, but within the context of Social Computing and the Semantic Web this assumption does not hold. In modern social computing the implementation of those who we interact with is a soft constraint—our interlocutors may be classical software agent sources (e.g. WWW), fellow users, or others of unknown origin like talking head agents (which may be software or people and will not readily give up their identity). This distinction only blurs as we move into the age of the Wisdom Web of Things (W2T, e.g. Zhong, et al., 2010) where the origin of agency blurs into a host of possible implementations. The world consists of one of "smart devices" be they embedded devices, autoIDs, computers, or people (e.g. Serbanati, et al., 2012). This chapter considers software agency as a basic architectural building block but in doing this we embrace the upward compatibility that this assumption affords for the future.

One of the reasons why information systems are inaccessible to some people with disabilities is because their needs are not taken into consideration during the design stage. The root of the problem could thus be traced to the methodology used. For such system to meet the needs of the end user, the methodology therefore needs to incorporate stages and processes that are particularly targeted to produce an outcome that will solve the problems of the end user and result in satisfaction with the product. Our methodology was thus designed to be disability-aware, thus anticipating that various disabled people will need to use the end product and incorporating their needs in the design process.

Disability-Aware Semantic Web Agency Development

It has been noted elsewhere that the education of disabled people needs to consider their needs, thus being disability-aware (Leicester, 2001).

Figure 1. Disability-aware semantic web agency development methodology

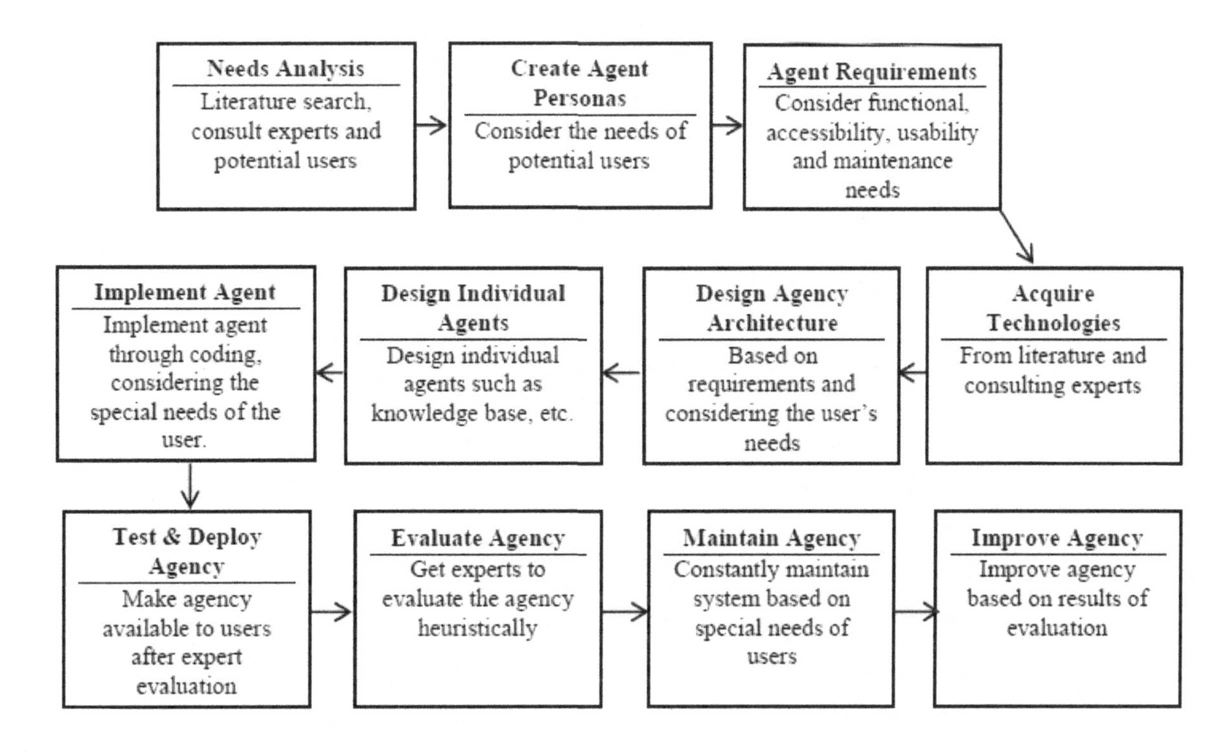

Similarly, in implementing agents or systems for disabled students in higher education, it is important to consider their specific needs, without which the system, which should include them, would exclude them. This follows some procedures common to software engineering such as the various models for software development (Boehm, 1988; Royce, 1987). Nevertheless, those models would be insufficient without considering the disabled person by incorporating their needs at crucial stages. By incorporating the needs of disabled people at each stage of development, the product will be both accessible and usable. We propose a methodology to follow in implementing a disability-aware agency (Figure 1), adapted from a similar approach to disability-aware software engineering (Nganji & Nggada, 2011). These are not separate stages that need to be completed before moving on, but constant review is needed for the agency to be improved during implementation.

Other researchers have considered design for people with special needs (e.g. Patrizia, et al., 2009; Senatore, et al., 2008). A disability-aware agency development begins with a consideration of the needs of the disabled user for which it is intended. Needs analysis is informed by literature, consultation of experts in the specific disability as well as potential users. Creating multi-agent personas further helps in having a clear idea of how the agency would be implemented, based on the needs of the user. These all inform the functional, accessibility, usability and maintenance needs of the agency. The developer will then need to acquire relevant technologies for implementation based on a review of related communities of agents and/or by consulting an expert in the field.

Figure 2. The semantic web community of agents for personalising services for disabled students in higher education is made up of the information translation and presentation agent, knowledge representation agent, information retrieval agent, and the management agent

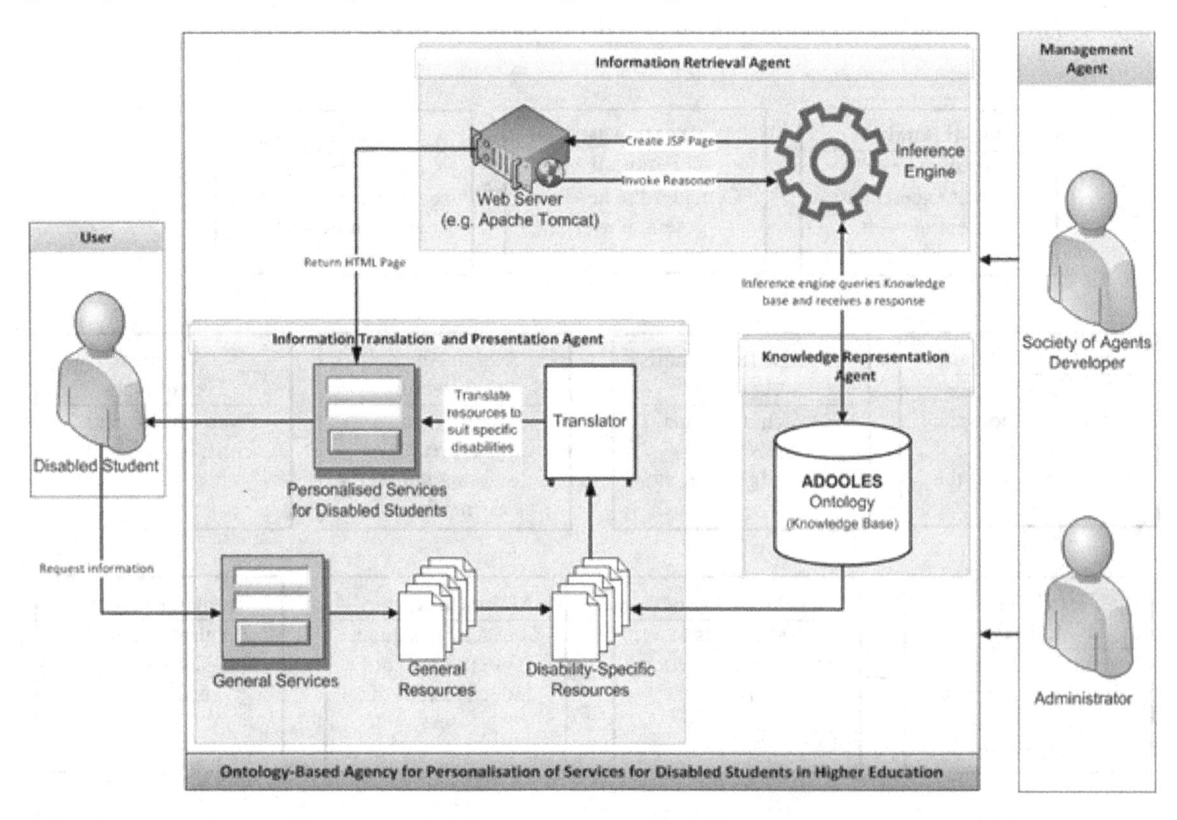

The agency architecture could then be designed, based on the requirements analysis and the needs of the user which will enable the developer to design various agents and thence to coding, testing, deploying and evaluating the agency, constantly maintaining it as need arises. Practical implementation of the agency into a visual interface requires an architecture, which should also be disability-aware. To accomplish this, agents still need to be employed. Other researchers have developed a cognitive architecture based on the "Society of Mind" (e.g. Davis & Vijayakumar, 2010). We have developed an ontology-driven architecture (Figure 2) for a Semantic Web agency based on a non-cognitively motivated "Society of Agents" metaphor; which personalises services for disabled students. This architecture is presented later in this chapter.

Outcomes of an Agent-Based Approach

Agents can be many things and vary in complexity. They can be as simple as switches or finite state machines (e.g. Rabin, 2002) to full knowledge system architectures. Here we use them as functional components to provide the basic structuring of our system. To judge the efficacy of our approach we consider the following to be advantages of an agent theoretic architecture:

- The decomposition of key features into knowledge components within a system.
- A consistent structure framework for all knowledge components.
- A hierarchical task decomposition with the potential to delegate tasks to other agents.
- Given a consistent structure agencies can consist of machines of a variety complexity. How it performs its agency may vary from simple automata at one end to machines given to sophisticated reasoning at the other.
- Consistent modular interfacing allowing for autonomous reconfiguration.
- Flexible Control Mechanisms for both horizontal and vertical (hierarchical) control mechanisms.
- Each agent can be regarded as a black box, therefore modern, agile, test driven software development paradigms would naturally fit onto this model.

There are however some assumptions here. Two potential critiques are as follows:

- Assumes that one size fits all and a domain can be represented by goal-orientated agents.
- Modularization can be hard and some goals/concepts do not break into subcomponents well (e.g. Fodor, 1983); however, for the Semantic Web to be tolerable some order must be possible.

We shall now try to take forward these ideas and demonstrate how they may be used to personalise services on the Semantic Web and later look at the results of their implementation and evaluation. Given that the implementation relies on an ontology, we will begin with a description of the ADOOLES (Abilities and Disabilities Ontology for Online LEarning and Services) ontology.

ADOOLES: AN ONTOLOGY FOR PERSONALISING SERVICES FOR DISABLED STUDENTS

Various disabilities exist amongst students in higher education. Some of these students with disabilities may require some form of assistance, which could be provided through assistive technologies, human support workers or through animals such as guide dogs for blind people. These assistive mechanisms could be vital for the disabled students accessing various services in the university such as the Study Advice Service, Disability Services, Library, etc. Paradoxically, the use of such assistive technologies depends on the extent of some abilities. For instance, the use of screen readers and guide dogs may be recommended for students with very low ability to see. Using the screen reader also depends on the user's ability to hear as screen readers read out content to the user who depends on such form of communication.

Modeling an ontology to personalise services for disabled students could use the following concepts: Abilities, Disabilities, Assistive Mechanisms, Person, Course and Services which contain subclasses. Disability could have sub classes such as Mental Disability, Physical Disability, Specific Learning Disability and Unseen Disabilities. Those subclasses are further divided into subclasses.

We have designed an ontology for personalising services for disabled students using OWL (Web Ontology Language), modeled with Protégé. Each concept of the ADOOLES ontology is represented as an OWL class. ADOOLES incorporates some concepts, classes and properties contained in the ADOLENA ontology (Keet, et al., 2008) but also includes the Services, Person, and Course classes which are not part of the ADOLENA ontology. Additionally, the classification of the ADOOLES ontology includes disabilities such as Specific Learning Disability (or Difficulty) which is not part of ADOLENA. The Services class of ADOOLES contains the following subclasses: *Accommodation*, *FinancialSupport*, *Leisure*, *Social*, and *Study*. These are the services offered in some higher education institutions that are accessed by disabled students. By accessing these services regularly, disabled students could receive support that would be helpful to their study and overall experience at the university. Accommodation for a student with mobility difficulties such as those using wheelchairs would need to be very accessible such as on the ground floor or accessible through lifts and ramps. A dyslexic student might need support from the Study Advice Service who will provide resources in alternative formats that could help them in their study while the Disability Services might provide alternative examination arrangements. The *AssistiveMechanism* class contains subclasses such as *Device*, *GuideDog*, and *SupportWorker*, with some subclasses. A human support worker may be required in some instances to provide physical support to some disabled students in lecture rooms while assistive technology devices could be employed to provide support in accessing online information.

The OWL classes also have properties such as Object properties linking two or more individuals. For instance, in our ontology, the Object property *isSuitableFor* links a *LowerLimbMobility* individual with a *GroundFloorAccommodation* individual. Various other Object and *DataType* properties are used in the ontology. The ontology constitutes a key aspect of the Knowledge Representation Agent of our architecture as seen in Figure 2.

ARCHITECTURE OF AGENCY FOR PERSONALISING SERVICES

The semantic Web agency architecture has four main agents as depicted in Figure 2. An agent is a basic unit or component within the community of agents, which performs specific functions such as information retrieval, presentation, representation, etc.; while an agency constitutes communities of distributed agents dynamically corresponding to deliver an effective user solution. The agents work together to present the user with their requested service when they interact with the system from its user interface on an electronic device (computer, mobile phone, PDA). The agents and their functions are as follows.

Knowledge Representation Agent

This agent represents domain knowledge in a format that could be easily understood and interpreted by the agency. Its inference engine can easily query the knowledge base and obtain specific results, which could be used to personalise the services that have been requested. The agent also provides a means for assistive technologies to interact with it, ensuring compatibility and interoperability.

This agent has a disability ontology specified in OWL (Web Ontology Language). The ontology contains various concepts such as Ability, Disability, Assistive Mechanisms, and Services, which all form the main classes. The ontology we have developed is known as Abilities and Disabilities Ontology for Online LEarning and Services (ADOOLES) and incorporates some concepts, classes, and properties contained in the ADOLENA ontology (Keet, et al., 2008) as already described in section 3.

Information Translation and Presentation Agent

This agent enables the disabled student to obtain content suitable for their specific needs. Initially, the student interacts with the semantic Web society of agents through the user interface from their device, which first presents them with content comprised of several general services. The agency identifies the student as one with special needs through an authentication mechanism and retrieves the services that are specific and suitable for their disability. This is facilitated by the translation module, which translates and presents the services to the student in an appropriate format. The translator for instance could accurately and fully describe an image to a blind user by semantically enriching it with comprehensive ontological descriptions, which are then translated into text for interpretation by a screen reader, thus facilitating understanding of any visual resource as represented in Figure 3.

Initially, an assistive technology such as a screen reader may be unable to describe the image to the blind user because it is not semantically enriched with accurate descriptions of the image and in some instances, the Web designer or developer may not have adhered to Web standards for accessibility. There is therefore a semantic description barrier, which needs to be overcome by semantically annotating the image with comprehensive descriptions of its content through

Figure 3. Image interpretation in the translator module

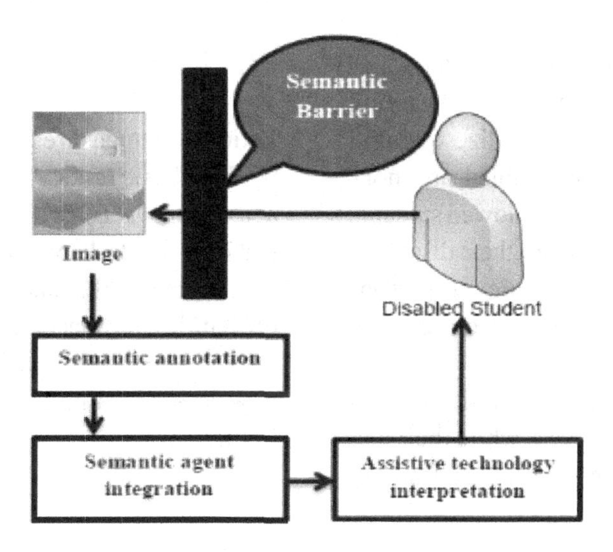

automatic processes. This image can then be integrated into the presentation agent for interpretation by the assistive technology, which conveys the meaning of the image to the user for better understanding.

Information Retrieval Agent

This agent is one of the key agents of the society of agents, which interacts with the knowledge representation agent to collect information on the services that are being requested, transmitting it to the information translation and presentation agent. It has an inbuilt inference engine which queries the ontology and a Web server (e.g. Apache Tomcat) interacting with the inference engine to present the information to the user.

Management Agent

This agent has a visual interface, which enables administrators such as lecturers, agency administrators, and developers to manage the student's accounts and to solve some problems they may encounter whilst interacting with the society of agents.

RESULTS OF VISUAL INTERFACE EVALUATION

The prototype known as ONTODAPS (Ontology-Driven Disability-Aware Personalised E-Learning System) was designed to personalize learning resources for students, both disabled and non-disabled. The e-learning system allowed students to select their learning goals and their disability type and then query the system to provide personalized learning resources based on their learning goals and disability type.

The learning goals of the students determined what modules they could access while their disability type determined what format the learning resources could be presented to them. A learner with severe hearing impairment such as complete deafness could thus be recommended learning resources in text and video formats, the video containing subtitles.

Some students (N=40) from the University of Hull were invited to participate in the research by interacting with the visual interface of ONTODAPS and then to fill in a questionnaire which contained a 10 point likert scale to evaluate the usability of the e-learning system.

According to Nielsen (n.d), usability has five quality attributes, which include: Learnability, Efficiency, Memorability, Errors, and Satisfaction.

Learnability

According to Nielsen (2012) mainly determines how easy it is for users to accomplish basic tasks the first time they encounter a design. The student evaluators of ONTODAPS rated the learnability of the system according to the results in Figure 4.

On a scale of 10, ONTODAPS scored 8.63 for learnability, which indicates that most users could easily use the system the first time they encountered it. This might have been facilitated by the fact that before interacting with the system, students were asked to read, watch, or listen to a tutorial on the system. The tutorial contained information on how to perform basic tasks.

Efficiency

Efficiency measures how quickly users can perform basic tasks. After the students had interacted with ONTODAPS, they were then asked to perform five tasks while the researcher watched them perform this and recorded any errors and difficulties they encountered as well as how long it took to perform the tasks. Generally, most tasks took less than a minute to perform as shown in Figure 5.

Figure 4. Results of ONTODAPS learnability evaluation

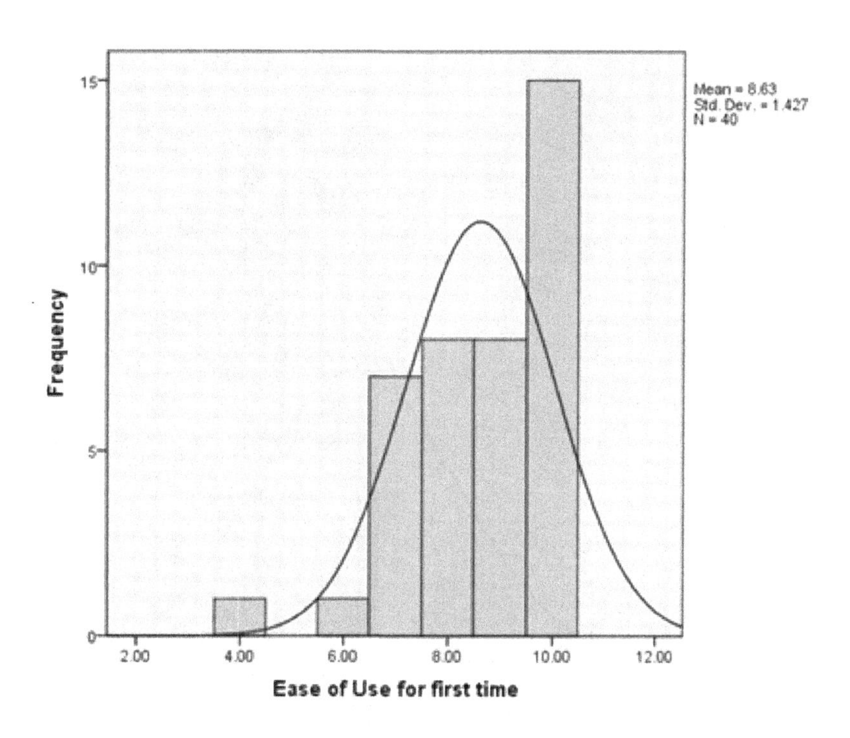

Figure 5. Results of users' efficiency in performing tasks

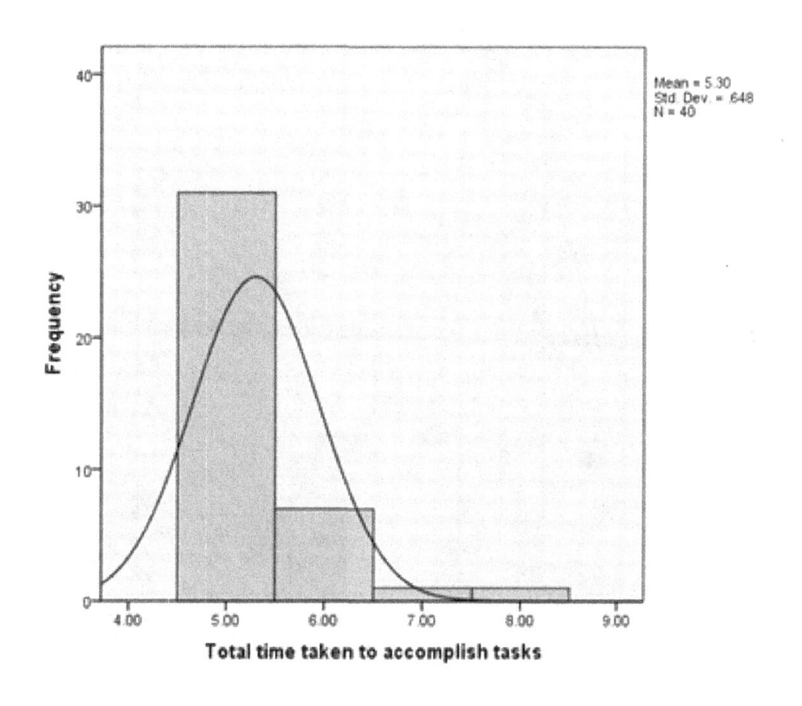

Figure 6. Memorability scores on ONTODAPS features

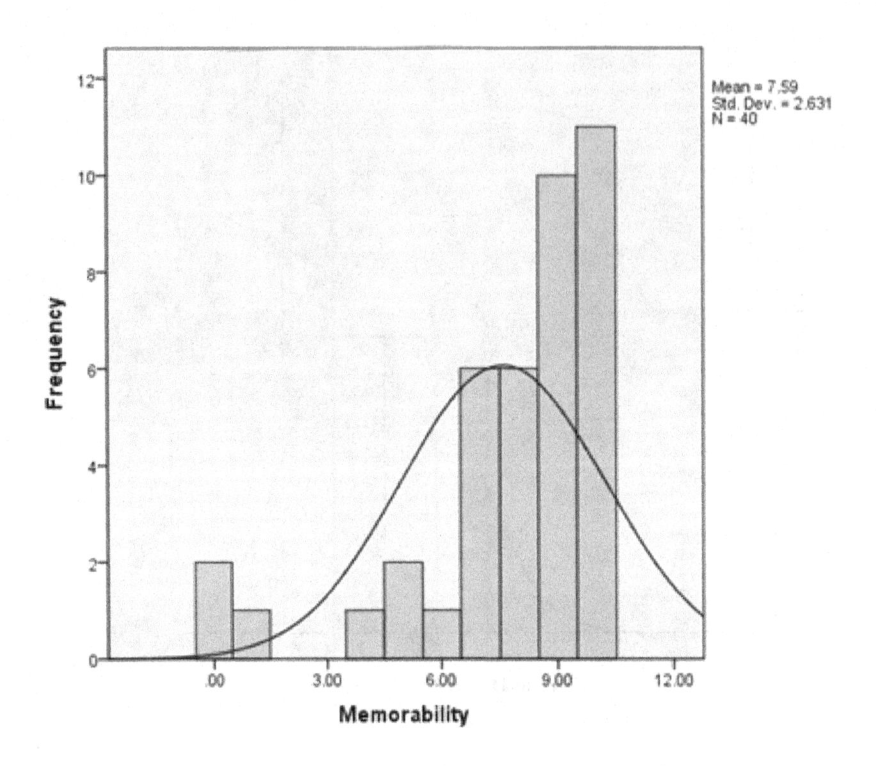

Figure 7. Number of errors made by students

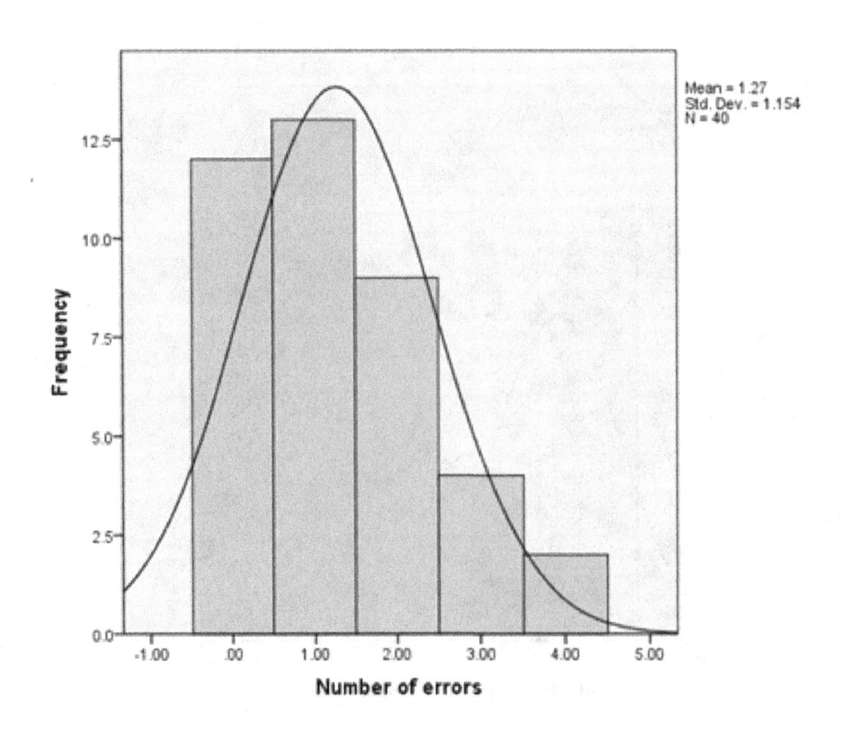

As could be seen in Figure 5, most users took 5 minutes to perform five tasks, an average of 1 minute per task. It could thus be inferred that the ONTODAPS design enables users to easily quickly perform tasks.

Memorability

An important usability quality of a design is the memorability: the ability for users to reestablish proficiency when they return to the system after a while. To measure memorability, after students had interacted with ONTODAPS and had logged out of it such that they had no access to the system, the questionnaire contained some four questions to test their memory on some of the system's functions. Each question was marked on a scale of 2.5 with a cumulative score of 10. The results of the memorability evaluation of ONTODAPS are presented in Figure 6.

As could be seen from Figure 6, the mean score for memorability was 7.59, which indicates that most users could remembers the main functions of the system. It will thus be easy for users to remember how to perform tasks on ONTODAPS after a while of not using it.

Errors

It is common for users to make errors whilst using a design. The ONTODAPS model was designed to minimize these errors. As the students interacted with the system, the number of errors they made were recorded as well as the type of errors. Generally, simple errors were made, which were basically a confusion of how to access resources. For example, most errors were due to double clicking to open a learning resource in the interface rather than selecting the resource and then clicking a button to open the resource. The number of errors performed by students is recorded in Figure 7.

As seen in Figure 8, the mean number of errors made by students whilst interacting with ONTODAPS was 1.27. This indicates that most users made very few errors. The normal distribution confirms this result. Thus, ONTODAPS helps to prevent errors. This is accomplished through its simplicity, ease of use and error messages which help users know how to recover from errors.

Satisfaction

For users to return to a design after encountering it for the first time, they must be satisfied with the design. The students rated their satisfaction with ONTODAPS as shown in Figure 8.

As could be seen in Figure 8, the mean satisfaction rating for ONTODAPS was 8.07. Overall, most of the students were satisfied with what ONTODAPS offers such as personalisation and the ability to manage vast amounts of learning resources through aggregation into a single location from which they could easily access the learning resources without having to go through each of them.

The above results of the user evaluation of ONTODAPS shows that the ontology-based and agent-centred methodology for designing learning environments and personalizing learning resources could be a very effective method for meeting learners' needs. This method is very effective with both disabled and non-disabled students.

FUTURE RESEARCH DIRECTIONS

The implemented visual interface is currently being evaluated heuristically by some experts (lecturers) in the e-learning and HCI domain at the University of Hull against Nielsen's (1994) ten interface design heuristics, some learning with software heuristics by Squires and Preece (1999) and Quinn's (1996) educational design heuristics, in a method similar to Granic and Cukusic (2011), thus addressing usability issues that may arise with the interface. By following accessibility and usability guidelines incorporated in the disability-aware semantic agency develop-

Figure 8. Students' satisfaction with ONTODAPS

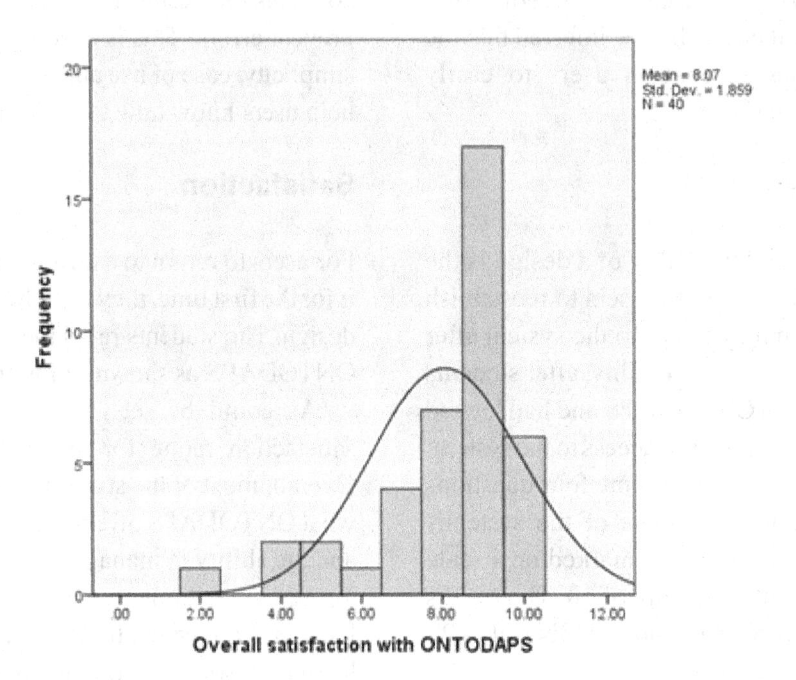

ment methodology, problems of accessibility and usability are expected to be greatly minimised. The heuristic evaluation will also give an indication of the strengths and weaknesses as well as the applicability of the ontology-based approach for personalising services for disabled people (Longpradit, et al., 2008).

The interface is also being evaluated by students. The aim is to have 30 disabled and 30 non-disabled students evaluate this from the user's perspective. The results presented herein are from the 40 students who have thus far evaluated it.

So far, this work has been applied to a specific domain, which is e-learning where we have control over the learning environment. Future research will apply the concepts described in this chapter to the wider Internet. Future work will also assess the success of the prototype in enhancing study for students with disabilities by assessing their performances pre and post interaction with the prototype over time, compared to other students with no such access.

An important progress in semantic Web and social artifacts is the social semantic Web (Breslin, et al., 2010) which opens endless opportunities to further this research for the benefit of disabled students as discussed below.

Social Learning ONTODAPS

With the increasing use and popularity of social networking sites amongst students and many people in the society today, social networking could be positively utilized to enhance study for disabled students. Future research could look at extending ONTODAPS to cater for the social interaction needs of disabled students for academic peer support in what we could describe as social learning ONTODAPS. This will be in line with the Social Semantic Web (Breslin, et al., 2010). A means of interaction for these students where they discuss academic matters and share experiences could possibly help improve their learning experience and results. In this light, specific autonomous agents could be developed which enable students to upload and collaboratively tag learning resources, allowing others to rate and review these resources in terms of their usefulness for specific disabilities. Consequently, highly rated resources will be listed at the top of recommended learning resources for different types of disabilities.

Social learning ONTODAPS will also need to provide the possibility of converting learning resources into different formats through the translation agent.

CONCLUSION

In this chapter, we have examined the difficulties faced by disabled students when accessing e-learning resources particularly when these are not accessible to them, including the shortcomings of some assistive technologies in e-learning or virtual learning environments. The need for novel methodologies and architectures that are disability-aware was raised; hence, we presented a novel architecture for a semantic Web community of agents, which personalises learning resources and services for disabled students in higher education based on a disability ontology. The agency is made up of four main agents which work together to present the user with the services they request, based on their specific needs as identified by the system.

The methodology for developing such agency was also presented which considers the needs of disabled students at crucial stages of development. The results of user evaluation of the visual interface have been presented, showing that the resultant ontology-based system can effectively personalise learning resources and services for disabled and non-disabled students.

From a methodological perspective, we can reflect on what role this has taken in the overall outcome of the reported work. This work was located within approaches to the Semantic Web; it also looked forward to the future Internet of Things. In such a future information space the world is populated by Things, with the Internet an implicit domestic technological transport device that is taken for granted. This raises the question "What are these Things?" They may be intelligent AI's, Social Communication Applications, Netbots, simple Finite State Machines, Object, or Component. Significantly their role is that of an Agent that does something, to which you can relate to (either or together) as a specific component, a social artifact, an intelligent device, or as one to which you can delegate action to. Thus, they are "Things." The inherent heart of this design methodology is to place Agency as the structural core of the Information System design. By identifying your Things you have a high level, top-down and bottom-up methodology for Information System Design. Things are neither objects nor components but both. It is how they fit into the overall society that is the key modeling notion. This allows you to mix Intelligent Artifacts, Social Artifacts, and AIs, in a considered design and that places new technological interaction at the center of the design rationale.

REFERENCES

Beck, K. (2003). *Test-driven development by example*. Reading, MA: Addison Wesley.

Bell, M. (2008). Introduction to service-oriented modeling. In *Service-Oriented Modeling: Service Analysis, Design, and Architecture*. New York, NY: Wiley & Sons.

Berners-Lee, T., Hendler, J., & Lassila, O. (2001, May 17). The semantic web. *Scientific American*, 35–43.

Boehm, B. W. (1988). A spiral model of software development and enhancement. *ACM SIGSOFT Software Engineering Notes*, *11*(4), 14–24.

Breslin, J. G., Passant, A., & Decker, S. (2010). *The social semantic web*. Berlin, Germany: Springer-Verlag.

Brut, M., & Buraga, S. C. (2008). An ontology-based approach for modelling grid services in the context of e-learning. *International Journal of Web and Grid Services*, *4*(4), 379–394. doi:10.1504/IJWGS.2008.022543

Checkland, P. B. (1981). *Systems thinking, systems practice*. Chichester, UK: John Wiley.

Coyle, D., & Doherty, G. (2008). Towards ontologies for technology in mental health interventions. In *Proceedings of the 2008 First International Workshop on Ontologies in Interactive Systems*. Washington, DC: IEEE Computer Society.

Davis, D. N., & Vijayakumar, M. V. (2010). A "society of mind" cognitive architecture based on the principles of artificial economics. *International Journal of Artificial Life Research*, *1*(1), 51–71. doi:10.4018/jalr.2010102104

De Bra, P., et al. (1999). AHAM: A dexter-based reference model for adaptive hypermedia. In *Proceedings of the ACM Conference on Hypertext and Hypermedia*. ACM Press.

Dijkstra, E. W. (1972). The humble programmer. *Communications of the ACM*, *15*(10), 859–866. doi:10.1145/355604.361591

Flood, R. L., & Jackson, M. C. (1991). *Creative problem solving*. New York, NY: John Wiley.

Fodor, J. A. (1983). *The modularity of the mind*. Cambridge, MA: MIT Press.

Galliers, R. D. (1995). A manifesto for information management research. *British Journal of Management*, *6*, 45–52. doi:10.1111/j.1467-8551.1995.tb00137.x

Galliers, R. D., Mylonopoudos, N. A., Morris, C., & Meadows, M. (1997). IS research agendas and practices in the UK. In D. E. Avison (Ed.), *Key Issues in Information Systems: Proceedings of the 2nd UKAIS Conference*, (pp. 143-172). New York, NY: McGraw-Hill.

Gershenfeld, N., Krikorian, R., & Cohen, D. (2004, October). The internet of things. *Scientific American*. doi:10.1038/scientificamerican1004-76

Goldberg, A., & Robson, D. (1983). *Smalltalk-80: The language and its implementation*. Reading, MA: Addison-Wesley.

Granic, A., & Cukusic, M. (2011). Usability testing and expert inspections complemented by educational evaluation: A case study of an e-learning platform. *Journal of Educational Technology & Society*, *14*(2), 107–123.

Håkansson, A., Nguyen, N. T., Hartung, R. L., Howlett, R. J., & Jain, L. C. (Eds.). (2009). Agent and multi-agent systems: Technologies and applications. *Lecture Notes in Artificial Intelligence, 5559*.

Hartung, R. L., & Håkansson, A. (2010). Meta agents, ontologies and search, a proposed synthesis. *Lecture Notes in Computer Science*, *6277*, 273–281. doi:10.1007/978-3-642-15390-7_28

Hazdic, M., Wongthongtham, P., Dillon, T., & Chang, E. (2009). Ontology-based multi-agent systems. In *Studies in Computational Intelligence*. London, UK: Springer.

Hendler, J. (2001). Agents and the semantic web. *IEEE Intelligent Systems*, *16*(2), 30–37. doi:10.1109/5254.920597

Henze, N. (2004). Reasoning and ontologies for personalized e-learning in the semantic web. *Journal of Educational Technology & Society*, *7*(4), 82–97.

Henze, N., & Herrlich, M. (2004). The personal reader: A framework for enabling personalization services on the semantic web. In *Proceedings of the Twelfth GI Workshop on Adaptation and User Modeling in Interactive Systems (ABIS 2004)*. Berlin, Germany: ABIS.

HESA. (2009). Students and qualifiers data tables. *Higher Education Statistics Agency*. Retrieved from http://www.hesa.ac.uk/index.php?option=com_datatables&Itemid=121&task=show_category&catdex=3

Jackson, M. A. (1983). *Systems development I*. Englewood Cliffs, NJ: Prentice Hall.

Kay, A. (1996). The early history of smalltalk. In Bergin, T. J., & Gibson, R. G. (Eds.), *History of Programming Languages II* (p. 511). Reading, MA: Addison-Wesley. doi:10.1145/234286.1057828

Keet, C. M., et al. (2008). Enhancing web portals with ontology-based data access: The case study of South Africa's accessibility portal for people with disabilities. In *Proceedings of the Fifth International Workshop OWL: Experiences and Directions (OWLED 2008)*. Karlsruhe, Germany: OWLED.

Leicester, M. (2001). A moral education in an ethical system. *Journal of Moral Education*, *30*(2), 251–260. doi:10.1080/03057240120077255

Longpradit, P. (2008). An inquiry-led personalised navigation system (IPNS) using multi-dimensional linkbases. *New Review of Hypermedia and Multimedia*, *14*(1), 33–55. doi:10.1080/13614560802316095

Lorenz, M., & Horstmann, M. (2004). Semantic access to graphical web resources for blind users. In *Proceedings of the 3rd International Semantic Web Conference (ISWC 2004)*. Hiroshima, Japan: ISWC.

Luke, S., Spector, L., Rager, D., & Hendler, J. (1997). Ontology-based web agents. In *Proceedings of the First International Conference on Autonomous Agents*. New York, NY: ACM.

Millard, D., et al. (2000). FOHM: A fundamental open hypertext model for investigating interoperability between hypertext domains. In *Proceedings of the Eleventh ACM Conference on Hypertext and Hypermedia HT 2000*, (pp. 93-102). ACM Press.

Minsky, M. (1986). *The society of mind*. New York, NY: Simon and Schuster.

Nganji, J. T., Brayshaw, M., & Tompsett, B. (2011). Ontology-based e-learning personalisation for disabled students in higher education. *ITALICS*, *10*(1), 1–11.

Nganji, J. T., & Nggada, S. (2011). Disability-aware software engineering for improved system accessibility and usability. *International Journal of Software Engineering and Its Applications*, *5*(3), 47–62.

Nielsen, J. (1994). Heuristic evaluation. In Nielsen, J., & Mack, R. L. (Eds.), *Usability Inspection Methods* (pp. 25–64). New York, NY: John Wiley.

Nielsen, J. (2012). *Usability 101: Introduction to usability*. Retrieved from http://www.useit.com/alertbox/20030825.html

Nodine, M., Fowler, J., Ksiezyk, T., Perry, B., Taylor, M., & Unruh, A. (2000). Active information gathering in InfoSleuth. *International Journal of Cooperative Information Systems*, *9*(1/2), 3–28. doi:10.1142/S021884300000003X

Parnas, D. L. (1972). On the criteria to be used in decomposing systems into modules. *Communications of the ACM*, *15*(12), 1053–1058. doi:10.1145/361598.361623

Patrizia, M., et al. (2009). A robotic toy for children with special needs: From requirements to design. In *Proceedings of the 11th International Conference on Rehabilitation Robotics*, (pp. 1070-1075). Kyoto, Japan: IEEE.

Pressman, R. S. (2005). *Software engineering: A practitioner's approach* (6th ed.). Boston, MA: McGraw-Hill.

Quinn, C. N. (1996). Pragmatic evaluation: Lessons from usability. In A. Christie & B. Vaughan (Eds.), *Proceedings of the 13th Annual Conference of the Australasian Society for Computers in Learning in Tertiary Education (ASCILITE 1996)*. Adelaide, Australia: Australia Society for Computers in Learning in Tertiary Education.

Rabin, S. (2002). *AI game programming wisdom*. Boston, MA: Charles River Media Inc.

Razmerita, L., & Lytras, M. D. (2008). Ontology-based user modeling personalization: Analyzing the requirements of a semantic learning portal. In *Proceedings of the 1st World Summit on the Knowledge Society (WSKS 2008)*, (vol 5288, pp. 354-363). Rhodes, Greece: WSKS.

Royce, W. W. (1987). Managing the development of large software systems. In *Proceedings of the 9th International Conference on Software Engineering*, (pp. 328-338). IEEE Computer Society.

Senatore, F., et al. (2008). Development of a generic assistive platform to aid patients with motor disabilities. In *Proceedings of the 14th Nordic-Baltic on Biomedical Engineering and Medical Physics*, (pp. 168-171). Riga, Latvia: IEEE.

Serbanati, A., Medaglia, C. M., & Ceipidor, U. B. (2012). *Building blocks of the internet of things: State of the art and beyond*. Retrieved from http://cdn.intechopen.com/pdfs/17872/InTech-Building_blocks_of_the_internet_of_things_state_of_the_art_and_beyond.pdf

Sommerville, I. (2007). *Software engineering* (8th ed.). Harlow, UK: Pearson Education. Retrieved from http://www.pearsoned.co.uk/HigherEducation/Booksby/Sommerville/

Squires, D., & Preece, J. (1999). Predicting quality in educational software: Evaluating for learning, usability and the synergy between them. *Interacting with Computers*, *11*, 467–483. doi:10.1016/S0953-5438(98)00063-0

Steyaert, J. (2005). Web-based higher education: The inclusion/exclusion paradox. *Journal of Technology in Human Services*, *23*(1), 67–78. doi:10.1300/J017v23n01_05

Szyperski, C. (2002). *Component software: Beyond object-oriented programming* (2nd ed.). Boston, MA: Addison-Wesley Professional.

Tompsett, B. C. (2008). Experiencias de enseñanza a estudiantes de informática con discapacidad en universidades del reino unido. *Technologías de la Información Y las Comunicaciones en la Autonomía Personal, Dependencia Y Accesibilidad*. Fundación Alfredo Brañas. *Colección Informática Número*, *16*, 371–398.

Videria Lopes, C., Maeda, C., & Mendhekar, A. (1996, December). Aspect-oriented programming. *ACM Computing Surveys*.

Warren, L., Hitchin, L., & Brayshaw, M. (1997). IS: The challenge of neo-disciplinary research. In D. E. Avison (Ed.), *Key Issues in Information Systems: Proceedings of the 2nd UKAIS Conference*, (pp. 187-194). New York, NY: McGraw-Hill.

Watt, S., Zdrahal, Z., & Brayshaw, M. (1995). A multi-agent approach to configuration and design tasks. In *Proceedings of Articial Intelligence and Simulation of Behaviour Conference*. IEEE.

Zhong, N., Ma, J. H., Huang, R. H., Liu, J. M., Yao, Y. Y., Zhang, Y. X., & Chen, J. H. (2010). Research challenges and perspectives on wisdom web of things (W2T). *Journal of Supercomputing*. Retrieved from http://kis-lab.com/zhong/open_publish/Research_challenges_and_perspectives_on_Wisdom_Web_of_Things.pdf

ADDITIONAL READING

Deline, G., Lin, F., Wen, D., & Gasevic, D., & Kinshuk. (2009). A case study of ontology-driven development of intelligent educational systems. *International Journal of Web-Based Learning and Teaching Technologies*, *4*(1), 66–81. doi:10.4018/jwltt.2009010105

Gladun, A., & Rogushina, J. (2007). An ontology-based approach to student skills in multiagent e-learning systems. *International Journal Information Technologies and Knowledge*, *1*, 219–225.

Hadzic, M., Wongthongtham, P., Dillon, T., & Chang, E. (2009). *Ontology-based multi-agent systems*. Berlin, Germany: Springer-Verlag. doi:10.1007/978-3-642-01904-3

Nganji, J. T., & Brayshaw, M. (2011). Towards an ontology-based community of agents for personalisation of services for disabled students. In K. Blashki (Ed.), *Proceedings of the IADIS International Conference Interfaces and Human Computer Interaction 2011*, (pp. 193-200). IADIS.

Nganji, J. T., Brayshaw, M., & Tompsett, B. (2011). Ontology-based e-learning personalisation for disabled students in higher education. *ITALICS*, *10*(1), 1–11.

Nganji, J. T., Brayshaw, M., & Tompsett, B. (2011). Describing and assessing image descriptions for visually impaired web users with IDAT. In *Proceedings of the Third International Conference on Intelligent Human Computer Interaction*. Prague, Czech Republic: IEEE.

Quynh-Nhu, N. T., & Low, G. (2008). MOBMAS: A methodology for ontology-based multi-agent systems development. *Information and Software Technology*, *50*, 697–722. doi:10.1016/j.infsof.2007.07.005

Santos, O. C., & Boticario, J. G. (2011). Requirements for semantic educational recommender systems in formal e-learning scenarios. *Algorithms*, *4*, 131–154. doi:10.3390/a4030131

Schmidt, A., & Schneider, M. (2007). Adaptive reading assistance for dyslexic students: Closing the loop. In *Proceedings of the 15th Workshop on Adaptivity and User Modeling in Interactive Systems*. Hildesheim, Germany: IEEE.

Vargas-Vera, M., & Lytras, M. D. (2008). Exploiting semantic web and ontologies for personalised learning services: Towards semantic web-enabled learning portals for real learning experiences. *International Journal of Knowledge and Learning*, *4*(1), 1–17. doi:10.1504/IJKL.2008.019734

Zhu, F., & Yao, N. (2009). Ontology-based learning activity sequencing in personalized education system. In *Proceedings of the International Conference on Information Technology and Computer Science*, (pp. 285-288). IEEE.

KEY TERMS AND DEFINITIONS

Agent: A basic unit or component within a community or society of agents which performs specific functions. The Image Description Assessment Tool (IDAT) is an example of an agent in the ONTODAPS system, which translates written information into an audio format for a visually impaired individual.

Assistive Technology: Any hardware or software product or service that helps compensate for a loss in function for individuals with disability, enabling them to function independently.

Community of Agents: An aggregation of coordinated agents working together to achieve a specific goal such as presenting a disabled student with a specific learning resource in a specific format suitable for their needs.

Disability: A physical or mental condition that affects an individual's ability to perform certain functions and hence affects equal participation in the society. Lower limbs mobility difficulty for instance affects the way an individual could move and hence may require a wheelchair for mobility.

Ontology: A set of concepts in a specific domain and the relations between them. For instance, blindness is a visual impairment, which affects the sense of sight, as expressed in the ADOOLES disability ontology.

Personalisation: Presenting specific services, resources, or products to individuals based on their needs or preferences that ensure it is relevant to them.

Screen Reader: A software application that enables people with visual impairments to read information on an electronic device.

Semantic Web: A new form of Web that can define things more meaningfully such that machines can understand.

Chapter 6
Taxonomy of IT Intangible Assets for Public Administration Based on the Electronic Government Maturity Model in Uruguay

Helena Garbarino
Universidad ORT Uruguay, Uruguay

Bruno Delgado
Universidad ORT Uruguay, Uruguay

José Carrillo
Universidad Politécnica de Madrid, Spain

ABSTRACT

This chapter presents a taxonomy of IT intangible asset indicators for Public Administration, relating the indicators to the Electronic Government Maturity Model proposed by the Uruguayan Agency for Electronic Government and Information Society. Indicators are categorized according to a consolidated intellectual capital model. The Taxonomy is mapped at the indicator level against the EGMM subareas covering all of the relevant aspects associated with the intangible IT assets of the Public Administration in Uruguay. The main challenges and future lines of work for building a consolidated maturity model of IT intangible assets in Public Administration are also presented.

DOI: 10.4018/978-1-4666-2491-7.ch006

Copyright © 2013, IGI Global. Copying or distributing in print or electronic forms without written permission of IGI Global is prohibited.

INTRODUCTION

When analyzing assets related to Information Technology (hereafter IT), the importance of intangibles that originate competitive advantage comes into play. The purpose of this article is to provide a taxonomy of IT intangible asset indicators for Public Administration as defined by Garbarino and Delgado (2011) that may allow to build, in the future, a valuation model of IT intangible assets. The objective is to support the effective management of IT resources in the Public Administration.

At this stage, the taxonomy is expanded and completed by traceability with the Electronic Government Maturity Model (hereafter EGMM) proposed by the Uruguayan Agency for Electronic Government and Information Society (hereafter AEGIS, AGESIC in Spanish), in order to consider the factors related to the evolution and maturity of electronic government in Uruguay.

The original taxonomy is based on a group of indicators previously defined by Zadrozny (2005). Indicators are categorized according to a consolidated model for intellectual capital where the indicators are grouped into two categories (appropriate and inappropriate). The indicators are then classified according to a consolidated intellectual capital model presented below. According to this taxonomy, it is established that each one of the six intellectual capital classes proposed in the model exists within the Uruguayan Public Administration.

The expanded taxonomy proposed is an extension and adaptation of the model proposed by Merino-Rodríguez *et al.* (2003), which emerged from models such as Intellectus (Trillo & Sánchez, 2006), after being adapted and revised by external experts.

With the taxonomy obtained, each indicator is mapped to EGMM subareas so that it considers all relevant aspects regarding the intangible IT assets of the Uruguayan Public Administration.

The model defined in García de Castro *et al.* (2004) and Medina (2003) for public Spanish companies is used as a basis. Then, this model is adapted taking into account the characterization of the evolution of the Uruguayan Public Administration carried out by Garbarino and Delgado (2011) and the need for identifying the intangibles of the Public Administration, measuring their potential, directing public policies toward a change in the focus and meaning of public service (Merino-Rodríguez, et al., 2003; AGESIC, 2011) and transforming it into a tool that supports Public Administration IT Governance. Finally, a taxonomy of intangible IT indicators is built according to the reality of the Public Administration of Uruguay (Garbarino & Delgado, 2011).

INTANGIBLE ASSETS

Definitions

According to IASC (2009), intangible assets "*are characterized as identifiable assets, without physical substance, and that are allocated for use in the production or supply of goods and services to be lent to third parties, or for administrative ends.*" Baruch Lev (2001) defines intangible assets in the following manner: "*an intangible asset is a claim to future benefits that does not have a physical or financial (a stock or a bond) embodiment. A patent, a brand, and a unique organizational structure (for example, an Internet-based supply chain) that generate cost savings are intangible assets.*"

Classification Models for Intangible Assets

There are several models whose purpose is to serve as tools for identifying, structuring, and to a lesser degree, assessing intangible assets. Some of these are: Balanced Business Scorecard (Kaplan & Norton, 1996), Intellectual Assets Monitor (Sveiby, 1997), Skandia Navigator (Euroforum,

1998), Intellect Model (Euroforum, 1998), Intellectus Model (CIC-IADE, 2003), the AIE model for assessing intangibles (Hubbard, 2007), and MERITUM (Kaplan & Norton, 1996), among others.

Our taxonomy will be based particularly on three models: the proposal by the MERITUM project, the Intellect model and its evolution, Intellectus. MERITUM proposes the need of a common, internationally accepted framework of reference. This framework would serve as a basis for companies to identify, to measure and to track their intangibles, since it is difficult for competitors to imitate these processes, which then become an important sustainable advantage for companies (Kaplan & Norton, 1996). With regard to Intellectus, it is especially relevant to consider the structure that it proposes for intellectual capital.

As in Lev (2001), the terms intangible assets, intangible resources, and knowledge assets or intellectual capital shall be used indiscriminately, since they are widely used in the literature. However, when the focus is accounting, the term "intangible resource" is found to be the most accepted; if the perspective is economic, the term used will be "knowledge" and if the perspective is from a company or human resources, the term is "intellectual capital."

Cañibano *et al.* (2002) propose a classification of intangibles in two groups: static and dynamic. Within the first group (static), intangible resources are divided into assets and capabilities. In the second group (dynamic), intangible activities are classified into:

- Activities that lead to the development or acquirement of the new intangibles.
- Activities that allow the value of current intangibles to increase.
- Activities that enable the evaluation and oversight of said intangible activities.

Both knowledge and intellectual capital are intangible resources. There are other intangibles that do not belong to intellectual capital (Sánchez-Medina, et al., 2004), such as company reputation or clients loyalty. The link between intellectual capital and intangible skills/assets is established by intangible activities, which transform the former into the latter (Figure 1).

When considering intellectual capital, there is a certain consensus to categorize it into three capitals: human, structural, and relational (Cañibano, Covarsí, & Sánchez, 1999; Euroforum, 1998; Trillo & Sánchez, 2006), although said model has evolved.

Human capital is the knowledge of the people who make up the organization. It includes capabilities, experience, attitudes, and aptitudes both generic to the entire organization or exclusive

Figure 1. Classification of intangibles

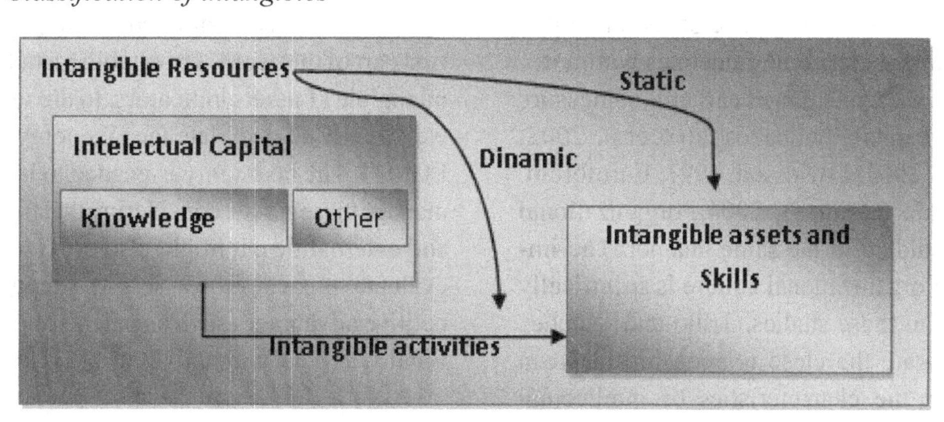

to people. From the strategic perspective of the company, according to Kaplan and Norton (2004), human capital is measured by focusing on the capabilities, education, and knowledge of the people that belong to the company.

The structural capital referred to by Euroforum (1998) is the accumulated knowledge that remains in the company, such as business routines, information technologies, business culture, procedures and systems. The Intellectus model further divides this capital into organizational capital and technological capital. Organizational capital is classified into business philosophy and its management. Technological capital is further divided into two categories: technological infrastructure and the knowledge that it creates and manages.

Finally, what is referred to as relational capital in the Intellect model is divided into business capital and social capital. Business capital refers to the value for the organization of relationships to the other basic participants of the business (clients, suppliers, shareholders, among others). Social capital refers to the value of relationships with social agents of the business environment (for instance, communications media, public institutions, and environmental conservation). We add to these five capital categories the one proposed by Trillo and Sánchez (2006): cultural capital (comprised by values, organizational culture, innovative and creative culture, and relational culture).

Trillo and Sánchez (2006) cite Deshpandé and Webster, his definition of culture as "the shared group of beliefs and values that help individuals to understand the performance of the organization, providing behavioral guidelines within it."

In the models mentioned earlier that measure intellectual capital (Cañibano-Calvo, et al., 2002; CIC-IADE, 2003; Edvinsson, 1997; Euroforum, 1998; Kaplan & Norton, 2004), organizational culture is studied in the same manner. The importance of organizational culture is specifically considered in these studies. Trillo and Sánchez (2006) indicate the close relationship between culture and the characteristics of intellectual capital indicators and propose an extension of the model proposed in CIC-IADE (2003), where cultural capital is removed from the rest of the proposed categories. Therefore, according to this line of analysis, the intellectual capital would be made up of the following six elements: human capital, organizational capital, technological capital, business capital, social capital, and cultural capital, where cultural capital is defined as the set of the company's cultural values, organizational culture, creative and innovative culture and relational culture.

THE IMPORTANCE OF IT INTANGIBLE ASSETS

Intangible assets are relevant when performing financial evaluations. Investors need to know the value of the firm's assets and evaluate if the expenses associated with those assets are really productive (Brealey, et al., 1996). Organizations define a set of basic competences and assets that are needed to achieve corporate goals. Intangible assets are part of those organization basic assets as mentioned in MERITUM (above-mentioned).

Within the overall set of intangible assets are those related to IT, which are the subject of this study. It is essential to identify and measure IT intangible assets, to establish the relationships between them, to identify the intangible activities that add value to these assets and to assess the coherence of investments.

As previously mentioned, the objective of the first part of this work is to establish a taxonomy of intangible IT assets indicators. In the second part, we extend and complete the taxonomy regarding EGMM. The taxonomy is needed to identify and manage these assets, considering their interactions and externalities. It might also provide a basis to evaluate impact of the assets in strategy or competitive advantage (such as patents and licenses). Finally, as per Cañibano (2001), it is important to have visible information about intangible assets (including IT) because:

- Intangibles may be value drivers that do not appear in financial statements.
- The information may prevent mistakes in price formation.
- Reduces uncertainty.
- Reduces the volatility.
- Prevents investors and managers contemplating only the short-term.
- Reduces the cost of capital.
- Lowers interest rates.
- Reduces the risk of hostile takeovers.
- Improves corporate governance mechanisms.
- Strengthens corporate image.

All of these reasons are valid, both in contexts of competition and also when there is no competition, as in the case of Public Administration which we study in this chapter.

According to this extended model, a list of 90 intangibles proposed by Zadrozny (2005) based on a varied and accepted bibliography will be used as a starting point. First, the relevance within the proposed area of study, the Central Administration of Uruguay, will be studied. Then, the assets are classified into the categories proposed by Trillo and Sánchez (2006).

BUILDING A TAXONOMY OF INTANGIBLE IT ASSETS

As mentioned in the previous section, the proposed model extends and adapts the one proposed by Merino Rodríguez *et al.* (2003), resulting from the study of models such as Intellectus (CIC-IADE, 2003), its subsequent adaptation and revision by external experts, and its restriction to intangible IT assets.

Given the boundaries of the study, it is postulated that all of the six capitals that are proposed in the model exist in the Uruguayan Public Administration. The main problem becomes, therefore, to find out which indicators are appropriate for each one of the proposed IT assets. We shall take as a basis the study of Zadrozny (2005).

Wlodek Zadrozny is an IBM researcher who has been working with IT intangible assets since 2005. In this work we refer to his article titled "Making the Intangibles Visible: How Emerging Technologies will Redefine Enterprise Dashboards" (Zadrozny, 2005). This author has published other related works.

The Zadrozny taxonomy is not IT oriented. In fact, is a general work that analyzes how to systematically access information about intangibles. In this work, we focus on IT intangibles as a way to improve management of organizations. In particular, we are interested in enhancing IT governance in Public Administration, i.e. to improve the effectiveness and efficiency of electronic government, starting from an improvement in the governance of Public Administration IT intangible assets.

Previously, in the first phase (Garbarino & Delgado, 2011), the work was related to adapting the taxonomy presented by Zadrozny (2005) to one representative to the study domain (the Public Administration of Uruguay). The work was based in the most representative frameworks of Intellectual Capital (cited above). In a second phase, this taxonomy was re-adapted to consider the maturity model of IT Governance framework for e-government presented by AGESIC, in order to define a taxonomy aligned with the practice of maturity management of IT intangible assets in Public Administration (Garbarino, Delgado, & Carrillo Verdún, 2011).

Preliminary Concepts

- Within the context of the Uruguayan Central Administration, the client may be a public or private individual, a legal entity or other type of entity (for instance, non-government organizations).

- All of the citizens and organizations that are able to receive services from the Administration may be considered as clients.
- Citizens and organizations that are subjects to oversight by the Central Administration are also considered clients.

Application Environment

At the end of the 20th century, public organizations participated in an organizational trend that changed their management models, centering their processes on the citizen. This trend is based on concepts associated with services and a focus on the client. Along with this process, which is constantly evolving and far from over, the managers of Uruguayan public organizations need models that enable them to enter the so-called Information Society. These transformations imply a change in roles, agents and managing processes that need to redirect their objectives in order to focus on knowledge. The changes also point to the need to clarify the intangible assets possessed by the Public Administration as a source for creating value. Clarifying the intangible assets of the Public Administration and measuring their potential should help to change the focus and purpose of public service (Merino Rodríguez, et al., 2003). Using the model defined by García de Castro *et al.* (2004) and Sánchez Medina *et al.* (2004) for public Spanish businesses as a basis, a model compatible with the reality of the Uruguayan Public Administration is built.

To clarify the scope of this work it is necessary to describe the different types of organizations that exist in the Uruguayan Public Administration.

Uruguayan Public Administration organizations include government offices and other State organizations that may be defined as industrial undertakings, whose mission is the delivery of services associated with economic earnings and/or public services of a business nature. The Central, Regional and Local Administrations are those organizations inherent to public functions that develop their own policies, under competencies assigned by law and the Constitution of the country. The Uruguayan Constitution (ROU, 2009) divides these organizations into:

- **National Autonomous Entities:** BCU (Central Bank), BROU (Development, Commercial, and Retail Bank), BSE (Insurance), BHU (Mortgages), BPS (Social Security), UTE (Electrical utilities), ANTEL (Telecommunications), ANCAP (Oil products), ANP (Port authorities), OSE (Water), AFE (Railroads), PLUNA (Airways), INC (Rural settling). Autonomous public education organisms and cultural organizations also fall into this category.
- Regional Governments (there are 19 local administrative units, named "Departamentos") and Local Councils.
- Central Administration, divided into Executive (Presidency, Ministries and Planning Office), Legislative (two Chambers) and Justice Administration, which includes a number of different branches.

National Autonomous Entities of an industrial or commercial nature, in the first group, have a ROI (Return on Investment) to fulfill, including a social ROI where appropriate, associated with Corporate Social Responsibility policies. Their management is more or less oriented by criteria similar to those of private business. Therefore, they may eventually assimilate private management, whose main objective is to attain benefits and being currently an area of development, research, and specific discussion. Some of these entities compete in the market although there are some situations where monopolies exist. Within these organizations, competitive strategies associated to

Table 1. Inappropriate indicators

Inappropriate indicator	Reason of elimination
Ratio of large clients	This indicator measures the ratio of large clients in the target portfolio or market (indicator for competitive markets, outside scope).
Cash spending rate	The Public Administration has general budgets voted by Parliament and an acceptable fiscal deficit within public policies. Furthermore, this is not an IT indicator, but rather a financial one.
Frequency of Repeated Requests	This indicator measures the ratio of clients in the portfolio to which a product or service could be sold again. It is an indicator for competitive markets, therefore outside of scope.
Right to offer, right to sell	Applies to business context (competitive market, outside scope)
Supplier/Client Integration	Improves the Supplier-Business-Client relationship in order to create loyalty and improve competitiveness (indicator of competitive market, outside scope)
Right to design, Productivity Gains	Applies to the business context and improves competitive conditions (repetitive market, outside scope)
Relationship with the Environment	Company improvement in the production of goods and services with regard to respect and improvement of the environment. The Central Administration is a regulatory entity that is not subject to oversight (outside scope).
Client Acquisition Cost	The clients are acquired with regard to the services requested or duties required or fulfilled.
Clinical Trials, FDA Approval	Indicator for health industries in competitive markets, outside scope.
Debt Capacity	The Public Administration has general budgets voted by Parliament for the fulfillment of duties and an acceptable fiscal deficit within public policies, all of which fix debt capacity.
Records (e.g. Owner Databases)	If this indicator is interpreted as measuring ownership of database capabilities for competition, and therefore looking for comparative advantages, the indicator would be intended for a competitive market, outside scope. Another possible reference is to the Law for the Protection of Personal and Business Information, which limits the use of personal information and affects both public organizations and private businesses.
Alliances	Useful in a context where a better position in a market is sought or where economies of scale are to be built for both products and services (including software and the knowledge applied) in such a way that improves basic efficiency. This is an indicator for a competitive market, outside scope.
Possibility of Selling Facilities	In a context where the maintenance and/or improvement of physical assets is sought by the possibility of resale in an effort to improve the comparative advantages or the level of competitiveness. This is an indicator of a competitive market and is outside scope.
Patent Royalties / Know-how	The Central Administration does not generate patents.

continued on following page

Table 1. Continued

Inappropriate indicator	Reason of elimination
Employee Loyalty	The availability and operational mobility of the Public Administration is not contingent upon loyalty, but rather the labor conditions and salaries of the Public Administration. The different units are not under a competitive policy, and therefore it is impossible to buy or sell specific capabilities that allow for the improvement of a determined competitive position. Indicator for competitive markets, outside scope.
Customer Loyalty/Satisfaction, Subscriptions, Ratio of Loyal Clients, Market Potential/Growth, Online Revenue, Client Strengthened by Competition, Professional/Client Benefit, Franchise Agreements, License Agreements, Marketing Alliances, Revenue Increases by Segment, Online Supply Channels, Revenue by means of Alliances, Alliances R + D / Joint Ventures, Eyeballs (traffic of use), Relative Pay – Salaries, Competition, Classification/Volume of Business, Strength of support from shareholders, including opinion leaders, Client Contracts, Quality of Gains, Market Share / Growth, Credit Assessment, Benefit by Employee, Formal Alliances, Organization for strengthening clients, Trademark (Research, Level, Support), Regulatory Deposits, Residual Value of Inverse Engineering,	All those indicators are intended for competitive markets, outside scope.

IS/IT should be consistent with the state of practice and eventually the state of the art in their market, in order to compete effectively and efficiently. In the first group, there are also oversight agencies such as the BCU (Central Bank), the retirement and pensions agency (BPS) and the rural settling institution, INC, which have service and oversight duties that are highly specialized and more difficult to standardize.

Because of these very heterogeneous characteristics, we chose not to concentrate in this first group when developing our group of indicators.

The second group is associated with the implementation of local and regional government. Our work could be adapted to this group in future stages.

In this work we concentrate on the organizations of the third group. In particular the General Directorate of Commerce (www.comercio.gub.uy), within the Ministry of Economy and Finance was used as the base case for construction of the model.

Revision of Available Indicators

Once the set of indicators proposed by Zadrozny (2005) and the area of application described in the previous section were analyzed, the first step was to eliminate from the expanded taxonomy those indicators that were inappropriate for our purpose. The eliminated indicators were either those that did not assess IT assets, or those that were intended for companies in competition in the market. These indicators value the competitiveness of businesses and their capability of return on investments, which is outside the scope of

Table 2. Types of intellectual capital

Type	Name
1	**Human Capital**
2	**Organizational Capital**
3	**Technological Capital**
4	**Business Capital**
5	**Social Capital**
6	**Cultural Capital**

Table 3. Indicators and grouping by type of capital

Type of Capital	Appropriate indicators	
	Indicator	**Reason**
1	**Employee Experience**	An universal indicator. Applies in the context related to IT Knowledge Management.
1	**Upper Management Experience**	Abilities of upper management for fulfilling the mission of the organization; ability of upper IT management in governing the IT of the organization.
1	**Quality of Upper Management**	The value contributed by the quality upper IT management significantly influences the quality of IT products and services. This indicator might be drilled down in order to assess the quality of middle management.
1	**Know-How**	The final quality of the product or service delivered is associated with IT capabilities, training, and knowledge regarding the operation being carried out and its objectives.
1	**Capabilities**	Indicators should enable the measurement of different capabilities in order to fulfill goals. These indicators should refer to: adequately satisfying the information demands, providing a quality service or following the status of good e-government practices.
1	**Employee Added Value**	Basic indicator for obtaining employee performance with regard to specific operations. Examples: amount of requests received, and amount of service instances carried out at the Service Desk.
1	**Rotation of support personnel**	This indicator is relevant because of the volatility of technical support personnel due to labor market, contracting methods of companies that provide services to the Public Administration, and horizontal operational movements between different public offices.
1	**Training**	Indicator that measures the efficiency of the continuous corporate governing of IT formation and human resources training.
2	**Ability to attract talented employees** **Loss of talent**	There are different perceptions and strategies for recruiting and relationship management for both Public Administration service providers and Public Administration workers.
2	**Investments in Internal Structure**	Considering this indicator as an investment in the IT organizational structure, it is an improvement indicator that applies within the context.
2	**IT Acquisitions**	Refers to the acquisition processes of IT solutions and services.
2	**Loss of Professionals**	Measures the impact of the loss of IT professionals within the organization.
2	**Lost Assets**	The capability of measuring lost assets regarding IT (hardware such as general infrastructure, and software such as code, executables, technical documentation of users, and databases). A study of this indicator might signal how many, which, and what percentage may or may not be accepted as a loss. This is a measurement of the quality of asset management that should be evaluated in strategic-monetary terms in order to support the decision-making process.
2	**Formalized Processes**	Once the processes have been formalized and implemented, the next step is to evaluate them with tools (for example under an ISO, CMMI, or COBIT framework)
2	**Structural Adaptation**	This indicator measures the relationship between the capabilities of the IT structure of the organization and its operations. When the indicator value is not unique, it could measure the additional capabilities to provide IT services to other units or third parties (such as Web hosting, server space for databases, applications and training rooms).
2	**Costs of Education and Training**	This is relevant for all organizations, including Public Administration organizations. A possible indicator could be the cost per employee in education and IT training, comparing it with international standards.
2	**Informal Processes**	There may be informal IT processes for various reasons: lack of resources; lack of process-oriented culture, including quality processes; tradition, organizational culture, and informal hierarchical relationships, among others.
2	**Quality of Processes, Products or Services**	These indicators are currently measured based on international standards derived from institutions like ISO, in a way that enables benchmarking.

continued on following page

Table 3. Continued

Type of Capital	Appropriate indicators	
	Indicator	**Reason**
2	**Investment in IT**	This indicator is a fundamental element in the strategy for IT government and should consider the different aspects referring to IT mission and principles, IT architecture and the defined infrastructure, strategic relevance of investments (alignment with the Administration strategy) and decision-making processes for investment.
2	**Staff renewal ratio**	Measures the Administration's capability to renew IT staff.
2	**Quality of Corporate Government**	Often, the practical restriction on Corporate IT Government with existing models for measuring, analyzing and improving (e.g. COBIT)
3	**Software (Applications)**	Portfolio of the company's IT applications focused on offering specific services.
3	**Purchase of Technology**	Applies within a context referring to the focus and participating processes of acquiring and updating technology.
3	**Data Access Capability**	Availability of data for metrics.
3	**IT Development**	Capability of developing and managing IT solutions and services.
3	**Research and Development**	Measurement of the innovation capability regarding improvements in effectiveness (more and better products and services) and efficiency (lower cost, time or effort)
3	**Systems**	This is a systemic indicator that should be measured at two levels: one regarding the existence of information systems and processes supported by information technology, and the other regarding the measurement of the efficiency and effectiveness of these systems and processes in carrying out their operations. Parker and Benson (Parker and Benson, 1988) may be used, even from an economic point of view.
4	**Lending of Services**	The evaluation of incoming or outgoing products/services to and from third parties may be an indicator of the organization's IT capabilities. Within the current framework of electronic government trends and the complementing of Public Administration services (with or without economic relationship), this could be an indicator of maturity, and therefore increase the IT assets.
5	**Hierarchical knowledge**	This indicator is needed for measuring the satisfaction of citizens with the Public Administration services within the G2C context (Government to Citizen), the performance of the hierarchies in the fulfillment of their duties, and the development of services based on electronic government. In this sense the development of electronic government agencies (AGESIC for example) implements standardized evaluation models for measuring the dimensions of this attribute.
5	**Quality of Supplier contracts**	In Uruguay, a unit under the Ministry of Economy and Finance that specializes in purchases for the Public Administration (UCA and MEF, 2011) was created, in search of economies of scale.
6	**Organic Growth**	The measurement of the evolution of products and services offered is an indicator of improvement and added value within the context of innovation.
6	**Organizational Reputation**	The Public Administration of Uruguay, by means of AGESIC, is currently immersed in a transformation process with the objective of improving its Public Administration IT capabilities, aligning them with better practices and international standards. Organizational reputation should be measured independently. Later on it may be consolidated with the different EG actors: Citizens, Businesses, Government and Employees.
6	**Innovative culture**	This indicator should be considered and measured from the point of view of the different actors. Even if there is no competition in the market, it may be necessary to innovate in order to improve efficiency and effectiveness in the deployment of services.
6	**Flexibility of the Physical Plant**	The volume, quality, and flexibility of the physical plant influences the development of IT capabilities for improved performance, which should be measured in terms of product and software service productivity. This is an indicator that should be supported by several measurements such as the amount of function points/developed monthly, customer service satisfaction, the capacity of reactive response to IT demands and the capability of proactive proposals for IT solutions.
6	**Plant Infrastructure**	
6	**Modernity of the Plant**	

this study. The list of the eliminated indicators is shown in Table 1.

The six capitals mentioned in Table 2 gather the intangible assets associated with IT that exist in the Uruguayan Central Administration in terms of the nature of knowledge and its relations. These capitals break down into a set of elements that conceptually identify and describe homogeneous groups of intangible assets.

In Table 3, the proposed indicators are classified according to the type of capital they correspond to.

Each component is broken down into a set of variables that allow for measurement, improvement, and tracking of the IT intangible assets according to the reality of the office, branch, or ministry in question.

Analysis of the Electronic Government Maturity Model in Uruguay AGESIC (AEGIS, Agency for Electronic Government and Information Society)

AGESIC is the "Agency for the Development of Electronic Government Management and Information Society." Created in 2005, it reports to the President of the Republic and operates with technical autonomy. Its mission is to lead the strategy and implementation of Electronic Government in the country as a basis for an efficient Public Administration that is centered on the citizen; fostering the Information Society and promoting the inclusion of Information and Communication Technology, and the quality of its use.

Its main actions are centered on defining and disseminating the computer guidelines and overseeing their fulfillment, analyzing technological trends, developing Information and Communication Technology projects, advising public institutions with regard to computer material, disseminating material regarding Electronic Government, and supporting the transformation and transparency of the Public Administration.

EGMM: Electronic Government Maturity Model

EGMM is an assessment guide that enables the analysis of the capabilities of services management for the citizenry, as well as techniques and strategies (AGESIC, 2008a). It also enables the fulfillment of objectives for efficiency and effectiveness in the use of Information Technology by

Table 4. Breakdown of areas and subcategories (AGESIC, 2008a, 2008c, 2011a)

Dimension	Area	Subcategory
Organization	**Strategy**	Strategy; Governing; Value Management; External Analysis and Benchmarking
	People	Competence; Culture; Structure; Change Management; Recognition and Rewards
	Performance	Framework for measuring performance; Indicators
IT	**Technology**	Infrastructure; Networks and Connectivity; Integration; Architecture; Security; Standards
	Information	Content; Architecture; Security; Standards; Privacy/Access to public information
	Operations	Management; Execution (processes and projects); Origin of Resources; Fulfillment; Financing Process
Relationships with the public	**Services**	Channels; Rendering of Services
	Citizens	Understanding of the citizens' perspective and their needs; Satisfying the citizens; Participation/Adoption/Involvement with citizens
	Communications	Internal and External strategies; Execution; Promotion of services

Table 5. Indicators and grouping by type of capital

Type	No.	Indicator	Description
1	1.1	**Employee Experience**	Universal indicator. Applies in the context related to IT Knowledge Management.
	1.2	**Upper Management Experience**	Abilities of upper management for fulfilling the mission of the organization; ability of upper IT management in governing the IT of the organization.
	1.3	**Quality of Upper Management**	The value contributed by the quality of upper management significantly influences the quality of the products and services. It may be broken down in order to assess the quality of middle management.
	1.4	**Aligning of middle management**	The level of alignment of decision-makers in the middle management chain conditions the effectiveness of implemented strategies.
	1.5	**Know-How**	The final quality of the product or service is associated to the capabilities, training, and knowledge regarding the operation carried out, and its objectives.
	1.6	**Capabilities**	Indicators should enable the measurement of different capabilities in order to fulfill goals. These indicators should refer to: adequately satisfying the information demands, providing a quality service or following the status of good e-government practices.
	1.7	**Employee Added Value**	Basic indicator for obtaining employee performance with regard to specific operations. Examples: amount of requests received, and amount of service instances carried out at the Service Desk
	1.8	**Rotation of support personnel**	This indicator is relevant because of the volatility of technical support personnel due to labor market, contracting methods of companies that provide services to the Public Administration, and horizontal operational movements between different public offices.
	1.9	**Training**	Indicator that measures the efficiency of continuous corporate governing of IT formation and human resources training.
	1.10	**Employee satisfaction**	Measures the level of satisfaction regarding strategy, organization, operation, and work conditions.
2	2.1	**Ability to attract talented employees**	There are different perceptions regarding the provision of services to the Public Administration, and working for the Public Administration.
	2.2	**Investments in Internal Structure**	Considering this indicator as an investment in organizational structure, it is an improvement indicator that applies within the context.
	2.3	**IT Acquisitions**	Refers to the processes for the acquisition of IT solutions and services.
	2.4	**Loss of Professionals/Talent**	The capability of measuring the impact of the loss of professionals within the organization.
	2.5	**Lost Assets**	The capability of measuring lost assets regarding IT (hardware such as general infrastructure, and software such as code, executables, technical documentation of users, and databases). A study of this indicator might signal how many, which, and what percentage may or may not be accepted as a loss. This is a measurement of the quality of asset management that should be evaluated in strategic-monetary terms in order to support the decision-making process.
	2.6	**Formalized Processes**	The evaluation of formalized processes is the step following the formalization and implementation of said processes (for example under a referential ISO, CMMI, COBIT framework)
	2.7	**Structural Adaptation**	Measurement of the relation between the IT capabilities of the organization and its operations. When the value is not unique, it can measure the additional capabilities of offering IT services to other units, and should be related to adequate demand management.
	2.8	**Costs of Education and Training**	This is relevant for all organizations, including Public Administration organizations. One possible indicator is the operation cost in comparison with international standards.

continued on following page

Table 5. Continued

Type	No.	Indicator	Description
	2.9	**Informal Processes**	There may be informal IT processes for various reasons: lack of resources; lack of process-oriented culture, including quality processes; tradition, organizational culture, and informal hierarchical relationships, among others.
	2.10	**Quality of Processes, Products, or Services**	These indicators are currently measured based on international standards derived from institutions like ISO, in such a way as to enable benchmarking.
	2.11	**Investment in IT**	This indicator is a fundamental element in the strategy for IT government, and should consider the different dimensions referring to the following of IT principles, IT architecture of the business and the defined infrastructure, strategical relevance of investments (alignment with the administration strategy), and decision-making processes.
	2.12	**Staff renewal ratio**	Indicator that represents the Administration's capability to renew the IT staff.
	2.13	**Quality of Corporate Government**	The new public management based on transparency in managing, external and internal auditing, focus on results, and public accountability.
	2.14	**Defined strategy**	Explicit definition of the mission and vision of the IT organization, as well as plans for carrying them out.
	2.15	**Quality of the system for recognition and rewards**	Existence, effectiveness, and efficiency of the organization's systems for recognition and rewards.
	2.16	**Performance - strategy level**	Existence of a model for performance indicators.
	2.17	**Performance - alignment**	Alignment of the performance indicator model with the defined strategy.
	2.18	**Performance - level of operation**	Effective use of the model.
	2.19	**Information Security**	Security management of the information, including policies for information security, information risk management, standards followed, respect for data privacy and access to public information, as well as auditing procedures
	2.20	**Capabilities**	Management of capabilities based on a capability plan with a defined management cycle.
	2.21	**External financing**	Effective mechanisms for obtaining external resources in order to finance Electronic Government initiatives.
	2.22	**Scalability of EG Services**	Capabilities associated with integration, adaptation, and maintenance of integrated services on an EG platform.
	2.23	**Multichannel GE Services**	Capabilities regarding the rendering of services by means of different channels (on-site, electronic, and by telephone).
	2.24	**Communication strategy**	Existence of an organic communication strategy for dissemination and awareness of services, as well as management.
	2.25	**Multichannel communication**	Alignment of communications management with the multichannel strategy.
3	3.1	**Software (baseline)**	Quality and adaptation of the support systems portfolio of the organization.
	3.2	**Software (Applications)**	Quality and adaptation of the organization's applications portfolio focused on rendering specific services.
	3.3	**Purchase of Technology**	This applies within a context referring to the focus on acquiring and updating technology.
	3.4	**Data Access Capability**	Availability of data for constructing measurements that support indicators.
	3.5	**Development of in-house IT**	This applies within the context referring to the management and development of IT solutions and services on the part of the organization with its own resources.

continued on following page

Table 5. Continued

Type	No.	Indicator	Description
	3.6	**Development of outsourced IT**	This applies within a context referring to the management and development of IT solutions and services by third parties. Outsourcing carried out at the physical limits of the organization may be distinguished from pure outsourcing for the purpose of better supervision.
	3.7	**Research and Development**	Measurement of the innovation capability regarding improvements in effectiveness (more and better products and services) and efficiency (less costs-time-effort).
	3.8	**Systems**	At the IT portfolio management level, the strategic relevance, effectiveness, and efficiency of the information systems and the processes supported by IT regarding the operations that they carry out shall be measured. Parker and Benson (1988) may be used for evaluation.
	3.9	**Flexibility of the Physical Plant**	The volume, quality, and flexibility of the physical plant influences the development of IT capabilities for improved performance, which should be measured in terms of product and software service productivity. This is an indicator that should be supported by several measurements such as the amount of function points/developed monthly, customer service satisfaction, the capacity of reactive response to IT demands, and the capability of proactive proposals for IT solutions.
	3.10	**Plant Infrastructure**	
	3.11	**Modernity of the Plant**	
4	4.1	**Lending of Services**	The evaluation of incoming or outgoing products/services to and from third parties may be an indicator of the organization's IT capabilities. Within the current framework of electronic government trends and the complementing of Public Administration services (with or without economic relationship), this could be an indicator of maturity, and therefore increase the IT assets.
	4.2	**Operations**	Operations management, based on an operations plan with a defined management cycle.
	4.3	**Projects**	Project management, methodology, integration, organizational structure, and benefit management.
5	5.1	**Hierarchical knowledge**	This indicator is needed for measuring the satisfaction of citizens with the Public Administration services (within the G2C5 context – Government to Citizen), the performance of the hierarchies in the fulfillment of their duties, and the development of services based on electronic government. In this sense the development of electronic government agencies (AGESIC for example) implements standardized evaluation models for measuring the dimensions of this attribute.
	5.2	**Quality of Supplier contracts**	The integration of products and services with different suppliers is a characteristic of IT-type architecture. Monitoring the relationship and fulfillment of supplied services by means of Service Level Agreements implies service quality assurance.
	5.3	**Deployment of communication**	Monitoring and control of communications while keeping in mind the public objective (G2C, G2B, G2G, and G2E).
6	6.1	**Organic Growth**	The measurement of the evolution of products and services offered is an indicator of improvement and added value within the context of innovation.
	6.2	**Organizational Reputation**	The Public Administration of Uruguay, by means of AGESIC, is currently immersed in a transformation process with the objective of improving its Public Administration IT capabilities, aligning them with better practices and international standards.
	6.3	**Innovative culture**	This indicator should be considered and measured from a citizen satisfaction point of view. Even if there is no competition in the market, it may be necessary to innovate in order to improve efficiency and effectiveness in the deployment of services.
	6.4	**Focus on service and the citizen**	Capability and focus on IT in satisfying the demands for services on the part of citizens, companies, and government.

continued on following page

Table 5. Continued

Type	No.	Indicator	Description
	6.5	**Focus on citizen participation**	Capability and focus on IT in order to encourage and facilitate citizen participation.
	6.6	**Encouragement of multichannel EG Services**	Focus on the promotion and encouragement of multichannel EG services, with the objective of improving access, deployment, and universality of services.

the Public Administration. In Uruguay, the Public Administration shows uneven development in its sectorial operational capabilities, in particular regarding capabilities for using Information and Communication Technology efficiently and effectively (AGESIC, 2009).

The objective of the maturity model is to support the transformation of the Public Administration: to give assurances of the quality of management, to oversee Public Administration projects, to improve and to increase the number of Public Administration transactions and services that are available online, and to strengthen Public Administration from the point of view of organization and capabilities, by means of the implementation of maturity models and continuous training.

National and international experts in Public Administration and different technical areas participated in defining the model with the objective of obtaining a solution that is in line with the best international practices, and adapted to the possibilities of the Public Administration of Uruguay. The model was used as an assessment tool for current capabilities and, since it contains a list of good practices, it also serves as a behavioral model. Based on the assessments that were carried out, with a definitive model and a tool for self evaluation supporting improvement in the level of maturity in the Executive Units and the rest of the organisms of the Public Administration, progress was made in the drafting of the roadmaps to improve capabilities (AGESIC, 2009).

As seen in Table 4, the model is based on 3 dimensions, 9 related sub areas, and 35 dependent subcategories.

RELATIONSHIP MODEL: TAXONOMY – EXPANDED TAXONOMY INDICATORS

Each one of the six types of capital mentioned in Table 1 includes intangible assets associated with IT that are related to the defined EGMM. However, the set of intangible assets previously described in Table 3 is not enough to cover the EGMM model presented by AGESIC. Therefore, we propose an extended set of indicators (Table 5).

The indicators are sorted according to the intellectual capital type (Table 2) of the assets.

Taxonomy Traceability to EGMM Subareas

Table 6 displays the complete relationship between EGMM at the subarea level and the different indicators. This mapping should enable the assesment of strengths and weaknesses in each subarea by evaluation of the indicators proposed in the taxonomy.

CURRENT AND FUTURE RESEARCH DIRECTIONS

A construction and validation process is currently underway for a quantitative model based on the taxonomy proposed. In this sense, the most relevant lines of work would be:

- To build a quantitative model for each indicator, taking into consideration the exist-

Table 6. Taxonomy traceability: EGMM

Area	Subcategory	Related indicators
Strategy	**Strategy**	2.14
	Government	2.2; 2.3; 2.5; 2.9; 2.11; 2.13; 3.5; 3.6; 5.2
	Value management	2.5; 2.9; 2.10; 3.5; 3.6
	External analysis and benchmarking	2.6; 2.10; 3.7
People	**Competition**	1.1; 1.2; 1.3; 1.4; 1.5; 1.8
	Culture	6.4
	Structure	6.5
	Change management	1.3; 1.7; 2.1; 2.4; 2.12
	Acknowledgment and rewards	1.10; 2.15
Performance	**Framework for measuring performance**	1.6; 3.7
	Indicators	2.16; 2.17; 2.18
Technology	**Infrastructure**	2.7; 3.1; 3.2; 3.8; 3.9; 3.10; 3.11; 4.1
	Networks and Connectivity	2.7; 3.1; 3.9; 3.10; 3.11; 4.1
	Integration	2.7; 3.1; 3.2; 3.8; 3.9; 3.10; 3.11; 4.1
	Architecture	2.7; 3.1; 3.2; 3.8; 4.1
	Security	3.1; 3.2; 4.1
	Standards	3.1; 3.2; 4.1
Information	**Content**	3.4
	Architecture	3.4
	Security	2.19
	Standards	2.19
	Privacy/Access to public information	2.19
Operations	**Management**	2.20; 4.2
	Execution (processes and projects)	4.2; 4.3
	Origin of resources	3.3
	Compliance	3.3; 4.2
	Financing process	2.21
Services	**Channels**	2.22; 2.23; 6.6
	Rendering of services	5.1; 6.1
Citizens	**Understanding of the citizen's perspective and needs**	6.2; 6.4
	Citizen satisfaction	5.1; 6.2; 6.3
	Participation/Adopting/Involving citizens	5.1; 6.1; 6.5
Communications	**Internal and external strategy**	2.24
	Execution	2.25
	Promotion of services	6.6

The taxonomy presented in previous tables (Tables 5 and 6) was tested in a survey applied in different offices of the Ministry of Economy and Finance.

ing models based on qualification models of organizational maturity and practices in different areas (CMMI, ISO, and COBIT, among others).
- To build a consolidated quantitative model for measuring the maturity of intangible IT asset management in the Public Administration in order to compare IT management performance of different Public Administration organizations and their relationships with society.

With regard to the future, taxonomy will evolve based on comparisons with other existing Electronic Government Maturity Models until the model is balanced, using parameters of volatility-stability for indicators in terms of inclusion/exclusion. In this way, the model may be transported to other countries and it would be possible to compare maturity management of intangible IT assets in different Public Administrations in different countries.

CONCLUSION

From the preceding analysis, a taxonomy proposal emerges that incorporates new indicators in order to cover all of the dimensions of the model proposed by AGESIC, without being limited to it. Starting from a consolidated taxonomy of 38 indicators, a total of 58 are proposed, representing an increase of 52%. The taxonomy obtained covers all the areas of EGMM. The more highly covered EGMM areas are Technology (33%) and Strategy (18%), while the least covered is Communications (3%). The final result is a completely traceable taxonomy, in which every subarea of EGMM is covered by different indicators.

Regarding the taxonomy reference (Garbarino & Delgado, 2011), a relevant percentage increase of coverage is observed in business capital (67%), organizational capital (19%) and technological capital (12%), with no change in social capital. On the other hand, the proportion of indicators related to cultural and human capital decreases significantly, 40% and 23% respectively. The percentage decrease occurs because of the increment of management-related indicators.

In summary, this taxonomy should help to build a set of representative IT intangible assets indicators in the Uruguayan Public Administration. This set of IT intangible assets indicators must allow, in the future, the definition of a quantitative model, allowing measurement, comparison and improvement of performance of IT intangible assets management.

REFERENCES

AGESIC. (2008a). *Memoria anual 2008*. Retrieved from http://www.agesic.gub.uy/innovaportal/file/267/1/Memoria_Anual2008.pdf

AGESIC. (2008b). *Modelo de madurez de gobierno electrónico*. AGESIC.

AGESIC. (2009). *Memoria anual 2009*. Retrieved from http://www.agesic.gub.uy/innovaportal/file/267/1/Memoria2009.pdf

AGESIC. (2011). *Plan estratégico 2011 - 2015*. Retrieved from http://www.agesic.gub.uy/innovaportal/file/265/1/planestrategico2011-2015.pdf

Brealey, R. A., Marcus, A. J., & Myers, S. C. (1996). *Principios de dirección financiera*. Madrid, Spain: McGraw-Hill.

Cañibano, L. (2001). *La relevancia de los intangibles en el análisis de la situación financiera de la empresa*. Paper presented at the IVIE Universidad Autónoma de Madrid. Madrid, Spain.

Cañibano, L., Covarsí, M. G.-A., & Sánchez, M. P. (1999). *The value relevance and managerial implications of intangibles: A literature review*. Retrieved from http://www.oecd.org/industry/industryandglobalisation/1947974.pdf

Cañibano Calvo, L., Sánchez Medina, A. J., García-Ayuso, M., & Chaminade, C. (Eds.). (2002). *Directrices para la gestión y difusión de información sobre intangibles*. Madrid, Spain: Fundación Airtel Móvil.

CIC-IADE. (2003). *Modelo intellectus. CIC-IADE. CMMi_Product_Teem. (2002). Capability maturity model® integration (CMMISM), version 1.1*. Pittsburgh, PA: Carnegie Mellon University.

Edvinsson, L. (1997). Developing intellectual capital at Skandia. *Long Range Planning, 30*(3), 366–373. doi:10.1016/S0024-6301(97)90248-X

Euroforum. (1998). *Medición del capital intelectual: Modelo intelect*. Madrid, Spain: IUEE.

Garbarino, H., & Delgado, B. (2011). *Taxonomía de indicadores de activos intangibles de TI para la administración pública: Caso república oriental del Uruguay*. Paper presented at the CISTI 2011. Madrid, Spain.

Garbarino, H., Delgado, B., & Carrillo Verdún, J. (2011). *Taxonomy of IT intangibles based on the electronic government maturity model in Uruguay*. Paper presented at the MCCSIS 2011. Madrid, Spain.

García de Castro, M. A., Merino Moreno, C., Plaz Landaeta, R., & Villar Mártil, L. (2004). *La gestión de activos intangibles en la administración pública*.

Hubbard, D. W. (2007). *How to measure anything: Finding the value of intangibles in business* (2nd ed.). Hoboken, NJ: John Wiley & Sons, Inc.

IASC. (2009). *NIC 38 activos intangibles*. Retrieved from http://www.iasb.org/NR/rdonlyres/8C28675D-FE12-468C-A7B7-66F8C1628673/0/IAS38.pdf

Kaplan, R., & Norton, D. (2004). La disponibilidad estratégica de los activos intangibles. *Harvard Business Review, 122*, 38–51.

Kaplan, R. S., & Norton, D. P. (Eds.). (1996). *The balanced scorecard*. Boston, MA: Harvard Business School Press.

Lev, B. (Ed.). (2001). *Intangibles: Management, measurement, and reporting*. New York, NY: Brookings Institution Press.

Medina, A. J. S., González, A. M., & Falcón, J. M. G. (2003). *El capital intelectual: Concepto y dimensiones*. Retrieved from http://www.aedem-virtual.com/articulos/iedee/v13/132097.pdf

Medina, A. S. (2003). *Modelo para la medición del capital intelectual de territorios insulares: Una aplicación al caso de Gran Canaria*. Gran Canaria, Spain: Universidad de las Palmas de Gran Canaria.

Merino Rodríguez, B., Merino Moreno, C., Plaz Landaeta, R., & Villar Mártil, L. (2003). *Capital intelectual en la administración pública: El caso del instituto de estudios fiscales*. Madrid, Spain.

Parker, M., & Benson, R. (1988). *Information economics: Linking business performance to information technology*. Englewood Cliffs, NJ: Prentice Hall.

RAE. (2010). *Diccionario de la lengua española*. Retrieved 28/09/2010, from http://www.rae.es/RAE/Noticias.nsf/Home?ReadForm

ROU. (2009). *Constitución de la república: Constitucion 1967 con las modificaciones plebiscitadas el 26 de noviembre de 1989, el 26 de noviembre de 1994, el 8 de diciembre de 1996 y el 31 de octubre de 2004*. ROU.

Sánchez Medina, A. J., Melián González, A., & García Falcón, J. M. (2004). *El capital intelectual: Concepto y dimensiones*. Gran Canaria, Spain: Campus Universitario de Tafira Las Palmas de Gran Canaria.

Sveiby, K. E. (1997). The intangible assets monitor. *Journal of Human Resource Costing and Accounting, 2*(1), 73–97. doi:10.1108/eb029036

Trillo, M. A., & Sánchez, S. M. (2006). Influencia de la cultura organizativa en el concepto de capital intelectual. *Intangible Capital*, *11*(2), 164–180.

UCA & MEF. (2011). *Unidad centralizada de adquisiciones*. Retrieved from http://uca.mef.gub.uy/portal/web/guest/portada

Zadrozny, W. (2005). *Making the intangibles visible: How emerging technologies will redefine enterprise dashboards. On-Demand Innovation Services*. New York, NY: IBM Research.

ADDITIONAL READING

Bismuth, A., & Kirkpatrick, G. (2006). *Intellectual assets and value creation: Implications for corporate reporting*. Paris, France: OCDE. Retrieved from http://www.oecd.org/dataoecd/2/40/37811196.pdf

Carol, P., & Aileen, C.-S. (2009). *Justifications, strategies, and critical success factors in successful ITIL implementations in U.S. and Australian companies: An exploratory study*. Retrieved from http://eprints.usq.edu.au/5287/2/Pollard_Cater-Steel_AV.pdf

Dressler, M. (2007). Tecnologías de la información para la gestión del conocimiento. *Intangible Capital*, *15*(3), 31–59.

Gray, C. (2006). *e-Governance issues in SME networks*. IKD Working Paper 10. London, UK: The Open University Business School.

ISACA. (2010). *E-commerce and consumer retailing: Risks and benefits*. ISACA.

ITGI. (2009). *Enterprise value: Governance of IT investments: Getting started with value management*. ITGI.

ITGI. (2011). *Global status report on the governance of enterprise it (GEIT)—2011*. ISACA.

Kaplan, R. S., & Norton, D. P. (2004). *Mapas estratégicos: Cómo convertir los activos intangibles en resultados tangibles*. Barcelona, Spain: Mapas.

Monteverde, F. (2010). *The process of e-government public policy inclusion in the governmental agenda: A framework for assessment and case study*. Hershey, PA: IGI Global.

Oh, W. (2006). The moderating effect of context on the market reaction to IT investments. *Journal of Information Systems*, *20*(1), 27. doi:10.2308/jis.2006.20.1.19

Penfold, M. (2010). *Fortalecimiento institucional y gobierno corporativo: Retos para el desarrollo empresarial*.

Ponce, H. G., & Calderón, E. P. (2009). Un cuadro de mando integral para la gestión táctica y estratégica del patrimonio tangible e intangible. *Revista del Instituto Internacional de Costos*, *4*, 37–52.

Rama Rao, T. P. (2004). *E-governance assessment frameworks*. New Delhi, India: Government of India.

Srivastava, S. C., & Teo, T. S. H. (2010). E-government, e-business, and national economic performance. *Communications of the Association for Information Systems*, *26*(14), 23.

Van Grembergen, W. (2000). The balanced scorecard and IT governance. *Information Systems Control Journal ISACA, 2*.

Walton, J. (2006). *Maturing your PMO - A partner in the planning process*. New York, NY: Pacific Edge Software, Inc.

Weill, P., & Ross, J. (2004). *IT governance: How top performers manage IT decision rights for superior results*. Boston, MA: Harvard Business School Press.

Weill, P., & Ross, J. (2011). *Four questions every CEO should ask about IT*.

KEY TERMS AND DEFINITIONS

Classification: Action or effect of classifying, relationship between what is classified in a determined situation (RAE, 2010).

Electronic Government: The possibility of accessing information from the relevant Public Administration, 24 hours a day, 365 days a year and at the moment that it is needed. The citizen should have access to procedures and services online in order to avoid travel, waiting and set schedules. Adapted from AGESIC (2008b).

G2B: Government-to-Business.

G2C: Government-to-Citizen.

G2E: Government-to-Employee.

G2G: Government-to-Government.

Intangible Asset: A measurable and identifiable asset with no physical substance that is used in the production or supply of goods or services. Adapted from IASC (2009) and Lev (2001).

Intellectual Capital: A knowledge group that remains in the company, such as routines, information technologies, business culture, procedures and systems. Many models generalize the above in reference to all intangible assets. Adapted from Medina *et al.* (2003).

Maturity Model: A maturity evaluation tool for processes regarding the fulfillment of a set of good practices (defined and accepted as a general framework, e.g. COMMI and COBIT, among others). Traditionally, according to a model defined by the Software Engineering Institute, the models respond to a scale of 0 to 5 where the level 0 implies the absence of practices and level 5 implies the optimization of processes in a cycle of continuous quality improvement. Adapted from CMMi_Product_Teem (2002).

Quantitative Model: The model whose main symbols represent numbers. These are the most common and useful in business.

Taxonomy: The science that covers the principles, methods and purpose of classification (RAE, 2010).

Section 2

Interpretivist and Social Constructivist Approaches

Chapter 7
Online Ethnographies

Hannakaisa Isomäki
University of Jyväskylä, Finland

Johanna Silvennoinen
University of Jyväskylä, Finland

ABSTRACT

Various new approaches of ethnographic research have been developed for inquiries in online settings. However, it is not clear whether these approaches are similar, different from each other, and if they can be used to study the same phenomena. In this chapter, the authors compare different suggested methodological approaches for conducting ethnographic research in online environments. Based on a literature review, 16 approaches, such as netnography, webnography, network ethnography, cyber-ethnography, and digital ethnography, are analysed and compared to each other. The analysis of the online ethnography proposals is conducted through a case comparison method by a data-driven framework. The framework provides structures for the analysis in terms of relations between Information and Communication Technologies (ICT), research object, researcher position, and research procedures. The results suggest that these approaches disclose fundamental differences in relation to each other and in relation to the basic idea of traditional ethnography. Finally, the authors discuss the advantages and disadvantages of the analysed approaches.

INTRODUCTION

A pertinent recent change in various daily activities in workplaces, homes, and schools concerns networked social practices, such as the snowballing use of social media. While new information technologies are increasingly used as promoters of various types of social practices, the study of online communities has become of particular interest. The key idea is to share information and create knowledge together with other users, in line with the view of knowledge sharing as a

DOI: 10.4018/978-1-4666-2491-7.ch007

Copyright © 2013, IGI Global. Copying or distributing in print or electronic forms without written permission of IGI Global is prohibited.

socially constructed and mediated phenomenon (Vygotsky, 1978). A contemporary shift in the practice of online communities is the transformation of users' roles from mere consumers to producers of information: many applications enable content sharing by offering environments that allow the integration of pictures, audio and videos, with written expressions (Iivari, Isomäki, & Pekkola, 2010).

Essential in this shift is that the users generate contents by which interactive relations between users in online communities function, forming unique online cultures or communities of practice (Wenger, 1999). Online communities are characterized by feelings of belonging, empathy and support (Rheingold, 1993), and a particular purpose and policies for community life (Preece, Abras, & Maloney-Krichmar, 2004). Online communities' designated cultures include social rules, norms and joint understandings, which are developed and learned as becoming and being a member of a community (Lave & Wenger, 1991). These practices are the key to online communities' meaning and functions, for example, social rules enact special types of communication (Yates & Orlikowski, 1992).

Online communities and cultures interact in online environments, nowadays often referred to as social media, which refers to Web-based applications that are constructed on the ideological and technological foundation of Web 2.0 that allows the creation and changing of user-generated contents (Kaplan & Haenlein, 2010). The inquiry regarding the use of such applications has been increasing during the past decades, and is currently intensively studied in numerous contexts, such as knowledge sharing and diffusion across gaming spaces (Fields & Kafai, 2008, 2009), information sharing with listserv application (Pomson, 2008), participation in a community of learning in a workplace (e.g., Stacey, Smith, & Barty, 2004), viability and veracity of online learning (Song, Singleton, Hill, & Koh, 2004), online communities as part of blended learning (e.g., Garrison & Kanuka, 2004), informal learning in an online community of practice (Gray, 2004), and social technologies in higher education (Hemmi, Bayne, & Landt, 2009), to name a few. It is worth noticing that methodological approaches in these studies represent variations of how traditional research methods have been applied to online settings.

To study these specific online communities forming a unique online culture a suitable research method is needed. In particular, to understand online communities, a suitable research method should offer conceptual means to shed light to the specific social and cultural nature of those communities and facilitate the bridging of relevant concepts and researchers' 'raw' experiences in a way that fulfils the demands of valid inquiry (cf. Weick, 1996).

In the current situation, it is difficult to select an appropriate research method to study online communities: research literature includes several approaches that aim to apply traditional ethnography to the study of online communities, to their social and cultural dynamics while creating and sharing knowledge. However, it is not clear whether these approaches are similar to or different from each other, or whether they share the same idea of the research object. It is especially unclear how these approaches facilitate online communities' research in terms of knowledge sharing, and how can they be employed within novel Information and Communication Technologies (ICT)?

In this chapter, we present a review of research methods that are presented suitable for conducting research in online communities seen as sociocultural phenomena. Based on a literature review, we analysed various new approaches found in the current research literature. The different approaches to online ethnography are compared by an explanatory framework depicting the essential factors that illustrate the features of the methods. The results suggest that the online ethnography approaches differ from each other, and thus, may be deployed in a different manner. The variation in the current online ethnographic methods is also perceptible in the terminology, which involves different methodological designations. There are many advantages in the current online ethnographies but also some undeveloped features, which call for future work.

The rest of the chapter is organised as follows: first, we describe the literature review method; second, the explanatory framework is illustrated; third, we present the findings; and finally, the discussion and conclusions.

BACKGROUND: FROM ETHNOGRAPHY TO ONLINE ETHNOGRAPHY

The study of communities and cultures has traditionally involved ethnography as a research method. Besides Bronislaw Malinowski (1884–1942), one of the main creators of the method is Clifford Geertz, who focuses on the symbolic dimensions of social action—e.g., art, religion, ideology, science, law, morality, common sense—as the essential mission of ethnography. In so doing, he also identifies communities with cultures by defining the concept of culture as Webs of significance that people have built to themselves (Geertz, 1973). During times, the widely adopted ethnographic method has undergone various intellectual developments (cf. Marcus, 1995; Schlecker & Hirsch, 2001). The development of ethnographic approach, however, entails given discipline specifity due to the method's inherent contextuality. In the field of information systems, the potential of ethnography is addressed as holistic, semiotic, and critical schools of thought; the semiotic school being further divided into thick description and ethnoscience (Myers, 1999).

The holistic school of thought stresses empathic, participant perspective on informants members (Schlecker & Hirsch, 2001), and necessitates identification with the social grouping under study (Myers, 1999). The researcher is assumed to bracket away any prior definitions of the observed social action in order to fully understand the social and cultural practices of a community, particularly to reveal the issues behind the taken-for-granted assumptions. The holistic approach insists that the researcher should 'go native' and live just like the local people (Myers, 1999).

The semiotic school of thought, as argued by Geertz (1973), does not prioritize empathy but emphasizes the analysis of symbolic forms such as words, images, institutions, and behaviors. The main idea is to analyse and describe a culture in order to find meaning in the 'Webs of significance' (Myers, 1999). Geertz (1973, pp. 3-30) highlights elaborated description of cultural categories in studying the significance and meaning of social action by referring to Gilbert Ryle's (1968) account of thick description. The mission of this school of thought is to observe, analyse, and understand human action in a very detailed manner taking into account the nuances of human behavior and its intentionality in context.

Ethnoscience or ethnomethodology, in turn, gears ethnography towards sociologically informed study of everyday practices and means by which people (re)produce social order. Sharrock and Randall (2004) maintain that ethnomethodology is particularly useful in systems design and may even provide grounds for standardisation of practical purposes for systems development and use.

Moreover, the critical school of thought sees ethnography as an emergent research process, involving dialogue between researchers and informants members. The idea of this school of thought is in line with emancipatory notions of research: critical ethnography describes, analyzes, and opens for inspection otherwise hidden agendas, power relations, and assumptions that repress and constrain organisational action (Lee & Myers, 2004).

The above mentioned ethnographic approaches are most often applied and developed for the design and evaluation of information systems (e.g., Orlikowski, 1991; Grudin & Grinter, 1995; Star,

1999; Karasti, 2001), and recently also for the design of online communities (e.g., Preece, Abras, & Maloney-Krichmar, 2004). Often ethnography has also been taken into use to study organisational life in offline settings (e.g., Ruhleder & Jordan, 1997; Schultze, 2000) to promote the understanding of information systems' "object systems," which need to be seen in terms of their underlying concept structure, representation forms, ontology and epistemology (Hirschheim, Klein, & Lyytinen, 1995, p. 16). However, the ethnographic method has not received as much attention regarding the ongoing change of the traditional method for the study of online communities.

The current wave of technological development, particularly the exponentially growing use of social media, diverts us to look for new applications of ethnography for the study of online communities. In the following, we present a comparison of current proposals of online ethnographies to find prominent suggestions for the study of online communities.

METHOD

To study online communities and cultures many different versions of online ethnography have recently been suggested. To clarify the nature of these recent suggestions, a stand-alone literature review (Okoli & Schabram, 2010) was conducted in spring 2011. The method of the literature review combines elements of snowball sampling (Arber, 1993) and forward and backward searches (Webster & Watson, 2002; Levy & Ellis, 2006). As a starting point, we chose Robert Kozinets' (2010) Netnography, which is frequently cited in the Internet as a suitable method to study online communities and cultures. Employing backward search, we found in this book more references of different approaches of online ethnography.

To complete this search result we conducted a keyword search based on the scope of the study and the literature in the main electronic databases (Science Direct, EbscoHost, Inderscience, ACM Digital Library, and Springer Link) and accomplished also further searches with Google Scholar (Levy & Ellis, 2006). The search words used were netnography, online ethnography, virtual ethnography, ethnography of Internet, cyberethnography, Webnography, and digital ethnography. These keywords had to occur in the title, as keywords, and/or in the abstract section of the articles in the databases. Some papers were screened out from the final set of resulting papers because they used non-novel methodologies, for example, presented a method in a similar manner proposed earlier by other authors. Eventually, we identified 14 approaches suggested as new applications of the ethnographic method deployed in various disciplines. In autumn 2011, we updated the review and added two approaches.

Analysis

To carry out comparisons, Cunningham (1997) suggests a case comparison method. In this approach, the researcher constructs an explanation for one case or set of cases and then replicates this process with a similar case or a set of cases. In doing the comparison, the researcher develops an understanding of why certain factors did or did not occur and offers interpretations. In this chapter through the comparison of online ethnographic approaches third level interpretation is conducted. First, we studied how the approaches are defined, what are the similarities and differences between them, and are they employed to study same phenomena. Second, we explored how these approaches are defined in relation to the methodological shift regarding ICT as a field in ethnography, and third, we interpreted through these co-ordinations a bigger picture of the multi-voiced arena of online ethnographic approaches.

In our analysis, the construction of the framework for comparing the different approaches was data driven, i.e., the key factors of the framework were developed from central factors of the different approaches found in the literature searches. These key factors also reflect the change process from traditional ethnography to online ethnography in that the core of this change concerns the ethnographic field's new condition to 'go ICT.' Therefore, the approaches' relation to ICT is considered along with the research object. Moreover, paying attention to the validity of data collection and analysis, researcher position is included to the framework. Finally, research procedures are considered to clarify how ethnographic approaches are employed within utilisation of novel ICT applications.

RESULTS

In line with the purpose of this chapter, in the following methodological approaches suggested for conducting ethnographic research in online environments are analysed. Based on the literature review 16 approaches, such as netnography (Kozinets, 2010), three versions of connective ethnography (Dirksen, Huizing, & Smit, 2010; Fields & Kafai, 2008, 2009; Hine, 2007), mediated ethnography (Bealieu & Estalella, 2009; Bealieu, 2004), three versions of digital ethnography (Wesch, 2009; Murthy, 2008; Masten & Plowman, 2003), cyberethnography (Rybas & Gajjala, 2007), cyber-ethnography (Ward, 1999), Webnography (Puri, 2007), virtual ethnography (Hine, 2000), network ethnography (Howard, 2002), and multi-sited ethnography (Wittel, 2000; Green 1999) are compared. Comparisons are made with the aid of four factors that structure the framework: relation to ICT, research object, researcher position and research procedures (Table 1). In the following, the results are briefly depicted.

First, traditional ethnography is mentioned to both clarify the original form of ethnography and to contrast the new constructions of the method. In traditional ethnography, there is no specific relation to ICT; research is usually conducted in face-to-face (F2F) settings in studying cultures and communities. The researcher position is determined by the demands of F2F interactions and usually the researcher pursues to immerse in the culture under study and aims to the position of unobtrusively observing, interviewing, and otherwise collecting data. In this way traditional ethnography aims to combine different data collection and analysis methods.

In Netnography (Kozinets, 2010) Computer Mediated Communication (CMC) is used as a source of data by which ethnographic understanding and representation of cultural or communal phenomena under study are pursued. A netnographic research process can be conducted in F2F settings in offline worlds, online in network field sites, or through combination of both of these environments. In research settings, data collection should be done as apparent as possible by gaining informed consent from the informants. Netnography follows phased research process through six steps adapted from traditional ethnography, including research planning and entrée, data collection, data analysis, ensuring ethical standards, representation, and evaluation. Netnography offers explicit and practical guidelines for conducting ethnography when studying online communities and cultures.

Dirksen, Huizing, and Smit (2010) describe connective ethnography as a modern form of ethnography which can be used to study social dynamics in both offline and online contexts. It emphasizes connectivity between offline and online spaces. The research object is studied through particular combination of offline and online methods through which understanding of specific dynamics of online social practices is pursued. Traditional ethnographic methods are blended to online ethnographic methods in order to add "layers of understanding," which requires researcher's active engagement in online and offline contexts through participant observation and interviews. Connective ethnography (Dirksen, Huizing, & Smit, 2010) aims at describing how technology appropriation can be comprehended in more detail by moving back and forth between online and offline research sites.

Table 1. Comparison of the approaches of online ethnographies

Approach	Notable examples	Relation to ICT	Research Object	Researcher Position	Research Procedures
Traditional Ethnography	For instance, Geertz (1973)	F2F, offline	Cultures and communities	Participant -observer	Combining different methods
Netnography	Kozinets (2010)	Online & offline	Online (offline) cultures and communities	"Apparent" participant-observer	Phased method through 6 steps
Connective Ethnography	Dirksen, Huizing and Smit (2010)	Online embedded to offline and vice versa	Social dynamics in local physical context with online context(virtual communities)	Interviewer and participant observation online and offline; active engagement	Traditional methods blended with online research methods, combination of online &offline methods, SNA to online (log file) data
Connective Ethnography	Fields and Kafai (2008, 2009)	Online and offline as a hybrid space	Virtual communities, knowledge diffusion, support groups	Active, towards holistic understanding; long term	Multiple sources of data through connection of different spaces
Connective Ethnography	Hine (2007)	Moves between online and offline through connections	e-science in different disciplines, diverse forms of connected activities	Holistic understanding, participant observation	Various sites through connections and diverse methods both online and offline
Mediated Ethnography	Beaulieu and Estalella (2009), Beaulieu (2004)	Online and offline through technologies	Internet, traces of links, hits and hyperlinks	Participant-observer	Contiguity and traceability
Digital Ethnography	Wesch (2009)	Online (Offline)	(experiences of self-awareness through) Vlogs	Long term active participation and observation, "friending"	Creation of videos, discussions with vloggers and among research group, interviews
Digital Ethnography	Murthy (2008)	Added to offline research or only online	Digital Video, Social networking websites, Blogs	Covert participant-observer	Questionnaires and email interviews
Digital Ethno	Masten and Plowman (2003)	Online (offline) Mobile devices	Mobile communities connected to sites	Lurker; swift analyser of multiple digital data with permission	Many digital techniques; participant observation by participants
Cyberethno-graphy	Rybas and Gajjala (2007)	Online and offline world (intersection)	Cyberculture, Social network environments and virtual communities, identity production	Long term involved participant-observer	Epistemology of doing; doing technology living online, observing the physical environment
Cyber -Ethnography	Ward (1999)	Online and offline as hybrid space	Online communities' interactions, for instance in chatrooms	Opportunist, participant-observer, possibility of lurking	Reflexivity, semi-structured interviews
Webnography	Puri (2007)	Online (offline)	Blogs, chatrooms, discussion boards	Lurker, "a part of furniture"	Digital collection of text-based consumer data

continued on following page

Table 1. Continued

Approach	Notable examples	Relation to ICT	Research Object	Researcher Position	Research Procedures
Virtual Ethnography	Hine (2000)	Online and offline	Social shaping of virtual communities	Mediated participant-observer, shaped by mediated interactions	Field connections, intermittent engagement
Network Ethnography	Howard (2002)	Online and offline	Network field sites, hypermedia organisations	Active or passive participant -observer	Synergetic research design, amalgam of traditional ethnography and Social Network Analysis
Multi-sited Ethnography	Wittel (2000)	F2F, traditional and virtual ethnography combined	"Real people" and virtual space, overlappings, connections	Participant-observer online and offline	Multiple objects and fields of study
Multisited Ethnography	Green (1999)	Offline and virtual reality technologies	Virtual reality technologies, mobile connections between multiple sites, stories, objects, texts, relationships	Involved fluent and multisited participant-observer	Multisited research fields following stories, people and objects. Multiple objects of study, data gathering and approaches to analysis.

To study how knowledge sharing and diffusion spreads in informal groups, and to trace youth participation in online and in offline social contexts, particularly collaborative peer-to-peer learning of norms and practices in online gaming world, Fields and Kafai (2008; 2009) used an integrative approach of connective ethnography that combines multiple data sources including online and offline interactions. Connective ethnography approach is necessary in terms of the complexity caused by moving between online and offline environments and the large number of participants. In this approach, the researcher actively strives to a holistic understanding of the phenomena under study through connection of spaces. This approach integrates traditional ethnography and online ethnography across physical and virtual environments, avoiding the dichotomy between online and offline interactions. Hine (2007) emphasizes extensive online fieldwork and long-term engagement through connections of various linked activities in online and offline environments by which a holistic understanding is pursued. Online and offline settings are crossed through the connections of sites that inform one another in diverse ways. From this point of view, transferring ethnographic research approaches to online settings does not require overall redefining of methodology or practice.

According to Beaulieu (2004) and Beaulieu and Estalella (2009), mediated ethnography deals with ethnographic fieldwork in which researchers mainly articulate their presence and interactions through technology, for instance recording videos, writing blogs or chatting. Traceability and contiguity are determined as important aspects of mediated ethnography concerning fieldwork ethics in addition to, for instance, privacy, harm, and alienation. These concepts deal with the anonymity of informants and with traceability of information of researcher and informants in online settings through Internet traces, such as links and log files. What kind of traces the researcher leaves in the Internet, how visible they are to others and how extensive they are, are also essential viewpoints in mediated ethnography. These aspects lead to the question of responsibilities of the researcher for instance in relation to future implications of the material on the online field. This research approach to ethnography in online settings seems to have similar definitions of the transferability of the approach between offline and online spaces as do Christine Hine´s (2007) approach to connective ethnography.

Wesch (2009) deploys digital ethnography onto the study of vlogs. Vlogs are video diaries shared popularly through YouTube. By creating, viewing and answering to vlogs experiences of self-awareness can be attained through reflecting one´s self in relation to others. Every new ICT application provides novel ways of self-expression, self-awareness, and contemplation. Digital ethnographic research mainly functions in online environments, but also in physical environments when members of research group, for example, discuss their procedures of interacting with vloggers. Typical to this approach is a deep longitudinal involvement in the vlog community: the researchers create videos, leave comments and practice "friending" in the vlog community under study. They also utilize various types of technologies to support their online data collection and analysis, which also includes self-analysis.

Murthy's (2008) version of digital ethnography involves digital data collection techniques in the study of social networking sites. This approach is beneficial for a socially informed study in that it enables researchers to demarginalise the voice of the informants. Moreover, a balanced combination of physical and digital ethnography provides researchers a novel array of data collection tools, such as blogs and online questionnaires. Murthy also points out a limitation: access to new technologies is not equal to all people. Digital ethnography may also be conducted only online and with covert participant observation.

Masten and Plowman (2003) call their approach Digital Ethno, which refers to a convergence of traditional methods with digital technology by updating traditional ethnographic methods with digital technology. Remote sensing devices, e.g., cell phones, PDAs, email, Webcams, SMS, GPS, and digital cameras, form the platform for applying traditional ethnography in a novel way. Digital Ethno is assumed to produce rich insights to the informants' lives due to multiple data collection through different technologies that enable the use of written and audio-visual data. Particularly, by putting the power of participant observation in the participants' own hands through the use of technology (e.g., sending pictures with cell phones to researchers), this approach is considered to convey the real-time richness of informants' lives and environments.

Cyberethnography (Rybas & Gajjala, 2007) is mainly aimed at studying the production of identities in social networking sites and online communities through multifaceted social settings in the intersection of online and offline worlds, in which ICT, for instance the Internet, is used. The cyberethnographic approach is based on an epistemology of doing, such as emphasizing the production of subjectivity in online spaces, consumption of technological artifacts, sustainable interactions, and understanding of every day practices of the community under study. This approach demands long-term commitment and immersion by becoming a part of the online community in combining the observation and participation of both online and offline, at the intersection of these worlds. Cyberethnographic research practice is many-sided aiming towards nuanced and diverse understanding of identity creation (Rybas & Gajjala, 2007).

Ward (1999) argues that Cyberethnography is the most suitable approach to study interactions in online communities as through reflexive qualities of cyber-ethnography the characteristics of online communities can emerge, allowing the participants to define their own borders. Semi-structured interviews are used to allow the subjects to talk back rather than just being observed during the process. In this approach, online communities are seen as hybrid spaces that cannot be separated into physical and virtual worlds. The cyberethnographer has the possibility of lurking in online communities in order not to affect too much on the interaction of the members in a specific online community. Despite the opportunity of lurking, Ward (1999) emphasizes consideration of ethical research practices and researchers responsibilities. The nature of participation differs considerably between online spaces and F2F settings in terms of continuity. According to Ward (1999) there is no time for 'friending' in online environments and the manners of meeting people are often one-off occasions. Ward does not consider cyberethnography merely as a tool, but it can be utilized in conducting ethnographic research in online settings rather than making a throughout separation between cyberethnography and traditional ethnography.

Webnography (Puri, 2007) is developed for market and consumer studies in order to examine network sites to find new ideas and insight from conversations in various discussion areas, emphasizing interactions and discussions in their natural contexts. It is mainly developed to study consumers' viewpoints of products through blogs. Blogs are seen as connective factors between private and public spaces providing relative anonymity to bloggers. The researcher participates in online environments as a lurker trying to emerge as a part of a specific community collecting online data. Speed and cost-efficiency are seen as benefits of the method.

Instead of directly participating in a certain online community's network site, virtual ethnography studies the connections between specific fields that are shaped in and by the Internet, which is seen to produce multiple orderings of time and space that cross the online/offline boundary (Hine, 2000). Intensive commitment is required for accomplishing reflexive dimensions to ethnographic research. Interaction is seen mediated and mobile rather than direct and multi-sited, and immersion in the research setting is only intermittently achieved, thus, a holistic description is not achieved. According to Hine (2000, p. 64), the object of virtual ethnography can be reshaped by focusing on flow and connectivity rather than location and boundary as the organizing principle. Here online communities are seen through social theory, i.e., social shaping of technology, without emphasis on concrete technology but on the shaping of interactions with informants by the technology. Hine (2000, p. 64) emphasizes that technologies are used and interpreted differently in different contexts, and have to be acquired, learnt, interpreted and incorporated into context. The shaping of interactions with informants by technology is essential part of research, as is the ethnographer's engagement with technology, which is seen as an important source of insight. However, this engagement with ICT is solely viewed as a social practice, and Internet is considered to acquire its sensibility in use.

In network ethnography (Howard, 2002) in-depth interviews, extended involvement, and active or passive observation are conducted in network field sites. These sites are selected through comparison by Social Network Analysis (SNA). Network ethnography is a synergistic multidisciplinary approach that is primarily suited to study communications of hypermedia organisations in the area of new media combining research in real and online worlds. According to Howard (2002), hypermedia organisations are seen as social groups that are consisted through CMC from the viewpoint of networking and companies. However, the transmission of human, social, cultural, or symbolic capital may be difficult in some organisational forms causing obstacles for the ethnographer's observation in technological environments.

Wittel (2000) approaches ethnographic research in online environments by concentrating to study what kind of challenges arise when F2F research field is moved to online environments. Through four challenges, uncertainty about information of users' real identities, one-sided observation in CMC, one-sided information of social relations between users through for instance hyperlinks, and the lack of field sites' physical location multi-sited ethnography is suggested as an approach to solve these ambivalences' of an online field site. Wittel (2000) argues that research conducted in online settings can only be based on ethnographic method if the study is multi-sited in that it combines aspects of traditional and online ethnographies. This approach also requires thorough participant-observation of research objects in various sites. Besides focusing on the differences of these spaces new perspective could be achieved by concentrating on overlappings and similarities of offline and online worlds.

Green (1999) discloses a stance emphasising fluency and long-term involvement studying a diverse field of study, which requires a flexible methodological approach. Multisited ethnographic research was conducted by following stories across diverse sites, which led on following people in communities that engage with virtual reality technologies representing connections between diverse social worlds. The following of people led to following of objects, which provided access to different spaces and connections of multiple social worlds. Dirksen, Huizing and Smit (2010) and Green (1999) have defined their connective ethnographic approaches partly on the basis of Marcus' (1995) Multi-Sited ethnography, which is a thorough and well delineated approach but does, however, not explicitly express connections to online environments.

The above-mentioned approaches are prominent for applying ethnography into the study of online communities, and have been developed in the context of different disciplines or schools of thought. This broad range of contexts is not eclectic in a negative sense in our view, but rather interesting because ethnography has always been adaptive to the conditions it has been introduced to (Hine, 2000, p. 65). In the following, we discuss the advantages and disadvantages of the approaches as we have interpreted from the documents used as data in this study. Our discussion is limited to both the view of online communities as knowledge sharing phenomena and our own disciplinary backgrounds.

ADVANTAGES AND DISADVANTAGES OF THE APPROACHES

In this study, we compared different novel applications of the ethnographic method into the study of online communities. The comparison was made in relation to a framework originating from the central interconnected ethnographic tenets of research object, approaches' relation to ICT, researcher position, and research procedures. All the approaches include valuable insights to online ethnography, but imply also advantages and disadvantages. Some issues may also be seen as both advantage and disadvantage depending on the viewpoint in question. In the following, we present our ideas of these issues.

Research Object

The emergence of online communities and the need for their research is highlighting a totally new research object for traditional ethnography. The nature, meaning and functioning of online communities can be clarified only by focusing on ICT-based social and cultural activities, which may include both online and offline components. The objects of research varied in terms of different ICT applications in diverse contexts and separate disciplines. In addition, the unit of analysis varied from individuals' interactions in a single social media application to a network of different sites.

An obvious disadvantage of the methodological situation of current research is that there is a lack of well-delineated stances of the online communities' ontology and epistemology, which is shown in the approaches as thin conceptual contextualisation of research method. The approaches should take a much more profound stance in describing effects of methodological changes. In accordance with Sharrock's and Randall's (2004) claim regarding traditional ethnography, also the new online methods provide sparingly conceptual tools for researchers to "go out and look," which is a suitable guideline for researchers trained to apply anthropological or sociological imagination (Mills, 1959) but insufficient for researchers with other academic backgrounds. For example, concentrating solely on mediated immersion, Hine (2000, p. 64) points out with good reason that the object of ethnographic research can be reshaped by focusing on flow and connectivity rather than location and boundary as the organizing principle. However, when studying communities from the viewpoint of both online and offline locations, the researcher may experience a need to know what specific conceptualisations he or she could deploy when observing the flow in online-offline environments. Moreover, in order to make a difference between online and offline situations, explicit conceptualisations of location and boundary are also needed. It seems that the applications of online ethnography tend to embrace the idea of online communities' unique social practices and cultures as "Webs of significance," but the level of conceptualisation varies, which, in turn, appears to be connected to the different purposes that different disciplinary orientations pursue to see or achieve by the ethnographic study of online communities.

To overcome the scarcity of well-grounded theoretisations, researchers need to develop these themselves to find what they see as essential in the adaptation of ethnography to their object of research. Disciplinary foundational analyses of the research objects in question facilitate in this endeavor if taken into account that this analysis is serving also the fundamental issue of overturning prior assumptions. In this sense, methodological conceptualisations need to maintain a certain theoretical level of both flexibility and accuracy to meet the demands set for a useful methodological 'lens' to analyse complex cultural actions in online or combined online-offline settings. As such conceptualisation, we think much of the approach referred to as the ethnography of infrastructure (Star & Ruhleder, 1996; Neumann & Star, 1996; Star, 1999, 2002). This approach is flexible as it may be deployed with an "infrastructural inversion" (Bowker, 1994), which means foregrounding the backstage elements of cultural practices, and vice versa, turning the focus back on the non-infrastructural cultural actions. This way the conceptualisations of the approach may be used to study a continuum of objects including a flow from rigid technical to adaptive human issues. On the other hand, the approach is accurate and useful by describing diversely infrastructure by both specific conceptual definitions and emergent properties, which appear to people in practice and are connected to activities and structures (Star & Ruhleder, 1996). Star (1999) defines infrastructure by the following properties:

- **Embeddedness:** Infrastructure is embedded in other structures, social arrangements, and technologies, which people may not distinguish.
- **Transparence:** Infrastructure is transparent to use in that it does not have to be reinvented each time for each task but invisibly supports those activities.
- **Reach or Scope:** Infrastructure has both spatial and temporal reach beyond a single event or on-site practice.
- Infrastructure is learned as part of membership forming an indispensable action taken-for-granted by members but encountered by strangers.
- Infrastructure both shapes and is shaped by conventions of practice.
- **Embodiment of Standards:** Modified by scope and perhaps also conflicting conventions, infrastructure becomes transparent by plugging into other infrastructures and tools in a standardised manner.

- Infrastructure is built on an installed base; becomes visible upon breakdown.
- Because infrastructure is big, layered and complex, and because it means different things locally, it is never changed from above.

In addition, Star (1999) emphasizes the scalability of infrastructure. Her conceptualisation of infrastructure is accurate but theoretical enough to stimulate researchers own imagination without setting any prior assumptions of the online community under study. The level of theorisation is also suitable to include a study's relation to ICT uniquely.

The Approaches' Relation to ICT

The relation that the online ethnographies hold towards ICT is an important issue, because the traditional ethnographic methodology is pursued to apply to the study of online communities due to the global implementation of ICT. The compared approaches took different stances towards ICT, most often technology was discussed on a concrete practical level from a tool perspective. Some of the approaches offered abstracted theorisations of technology (e.g. connective ethnography). Nearly all approaches regarded the study of online communities important both in online and offline situations. Clear advantages in the approaches' relation to ICT are:

- A broad range of different and interesting novel research objects found in virtual worlds.
- A myriad of new ICT applications introduced to collect and analyse data.
- The use of ICT tools may be used to empower the informants to participate in finding the meaningful situations in their daily life and capture it in a digital form.
- Some of the approaches clarify in detail why and how to study online communities both in online and offline settings (e.g. virtual ethnography and cyber-ethnography).

Some disadvantages follow from the predominantly tool-centered stance towards ICT. For example, the analysis guidelines are embedded in the use of ICT applications, and are affected by the nature of different data. Further, ICT was seen only as a practical organisation of social relations in knowledge sharing communities. Moreover, only a few proposed approaches developed the role of ICT to reach a theoretical level. Thus, a more diverse meaning of the technology's significance may not become explicit. The disadvantages may be surmounted by defining technology on appropriate levels as part of the research object.

Researcher Position

Inherently significant in the ethnographic process of "going there – being there – being here" is researcher position. Essential in this current shift is that users generate communication contents by which interactive relations between users in online communities function, forming a unique online culture or community of practice. This way the study of online communities concentrates on mediated interactions, and researcher position should include the delineation of the researcher's relation to both to the community and to ICT by which the contents are created and interaction is mediated (cf. Wesch, 2009). A disadvantage in the approaches is that often the approaches described researcher positions by traditional ethnographic concepts concentrating on researcher's relation to the community without explication of the researcher's own experience on ICT in the study. Thus, the focus of mediated interactions remained often only in the investigation of other users' relation to ICT and omitted the significance of researcher's own relation to ICT. To maintain validity in collecting data for understanding ICT mediated interactions, researcher position should be explained in the studies from the viewpoint of acquisition, learning, interpretation, and context of ICT as the researcher's own experience (cf. Hine, 2000, p. 64). A useful conceptualisation here may be the concept of User Experience (UX), which refers to user's experiential, affective and meaningful aspects of ICT use in addition to utility, ease of use, learnability and efficiency of the technologies employed in the study (e.g., Isomäki, 2009). These features of UX affect the understanding of communication and other functionalities of the infrastructure within online communities. By disclosing researcher's UX of the ICT the making of online ethnography could offer accounts of validity to the readers.

Research Procedures

Finally, research procedures presented by the analysed approaches consisted of various advantages regarding the application of ethnographic procedures. Multiple sources of data through connections to different online and offline sites were usual. Moreover, resourceful implementation of novel ICT applications was used to promote online communities' interactions and involvement in research. For example, by putting the power of participant observation in the participants' hands through the use of mobile technology, e.g., sending pictures of one's own current environment with cell phones to researchers (Masten & Plowman, 2003). In many approaches visual data (e.g., video) were brought easily as parts of online ethnographical data. A disadvantage is that the approaches are rich only in data collection accounts but thin in explicating ways of analysis.

Further, the use of personal information requires high ethical standards. Yet in some approaches the ethical aspects of including ICT into online ethnography were not explicated. This may mirror the minor backwardness in the novel proposals' research procedures. Most voluminous issue of disadvantages in research procedures concerned "being there" in terms of intensity of immersion, i.e., "going native," the way of living in the online-offline community: the approaches varied a lot in terms of presence in the community. For example, some approaches favored efficiency and utilised archives of the community discussions, some deployed lurking, and some stressed active immersion and comprehensive understanding of sociality and identity formation in cyber world. Lurking was seen as appropriate method of being there by three proposals. The ethicality of lurking - how different approaches justify lurking and how ethical viewpoints are taken into consideration varied, too.

The validity of observations in online settings was often expressed as a problem. Particularly, how observations change in online environments when all the gestures, social cues and facial expressions cannot be perceived similarly as in face-to-face settings when people are expressing themselves and communicating via videos, text, animations, detached voice, and still pictures? The approaches did not offer means to observe and analyse textual gestures. Howard (2002) also pointed out that the transmission of human, social, cultural, or symbolic capital may be difficult in some organisational forms causing obstacles for the ethnographer's observation in technological environments. To overcome these problems new means to observe and analyse textual gestures and visual presentations need to be developed.

An exception in this sense is Wesch´s (2009) digital ethnography, in which the interpretation of textual gestures is not in a crucial role because the study concentrates on moving images, vlogs, in which people are describing their lives. Gestures and facial expressions can be observed through vlogs as in F2F settings. However, can their experiences and gestures be affected of the medium through which they are represented? Definition of research field in digital ethnography is complex, for instance various fields are opened through Webcam. Functionalities of Webcams represent huge distinction between private and public spaces.

FUTURE RESEARCH DIRECTIONS

The development of contemporary online ethnography requires further explication of its connections to traditional ethnography in terms of theory and its operationalisation into the approaches' relation to ICT, research object, researcher position, and research procedures. Particularly, in further research attention should be paid to how different technologies shape the nature of online communities including researcher's position to both ICT and mediated content creation within online communities. Further, novel disciplinary theoretisations should be developed in order to facilitate more profound stances in describing effects of methodological changes. Finally, the maintenance of validity of observations needs support from further studies clarifying ways to observe and analyse textual gestures.

CONCLUSION

In this chapter, we have presented a literature review based comparison of cases regarding contemporary online ethnography approaches. It seems that, on one hand, the need for online ethnographic methodology is increasingly urgent, and on the other hand, it is difficult to select an appropriate online ethnographic method due to proliferation of different proposals for novel approaches. The relation that the online ethnographies hold towards ICT is an important issue, because the traditional ethnographic methodology is pursued to apply to the study of online communities due to the global implementation of ICT.

The results suggest that the current various online ethnography approaches differ fundamentally from each other, and thus, may be deployed in a different manner. The variation in the current online ethnographic methods is also perceptible in the terminology, which involves different methodological designations. A predominant disadvantage concerns theorisation of the research object: the new online methods provide sparingly conceptual tools for researchers to "go out and look," which is a suitable guideline for researchers trained to apply anthropological or sociological imagination but deficient for researchers with other intellectual backgrounds. It would be reasonable to assume that the online ethnographic approaches would take a much more profound stance in describing effects of methodological changes on diverse levels. The methodological conceptualisations need to maintain a certain theoretical level of both flexibility and accuracy in order to meet the demands for a useful methodological 'lens' to analyse complex cultural actions in online or combined online-offline settings. As such conceptualisation we value the ethnography of infrastructure (e.g., Star & Ruhleder, 1996; Star, 1999). This approach is flexible in that it may be deployed with an "infrastructural inversion" (Bowker, 1994), which means foregrounding the backstage elements of cultural practices, and vice versa, turning the focus back on the non-infrastructural cultural actions.

Thin level of theorisation concerns also the approaches' relation to ICT. The compared approaches took different stances towards ICT, most often technology was discussed on a concrete practical level from a tool perspective. Nearly all approaches regarded the study of online communities important both in online and offline situations. Clear advantages in the approaches' relation to ICT are a broad range of interesting novel research objects found in virtual worlds, a myriad of new ICT applications introduced to collect and analyse data and the empowering use of ICT tools to engage informants in finding the meaningful situations in their daily life and capture it in a digital form. The prevailing view of ICT as a practical tool affects also the research procedures employed by the approaches: the approaches are rich only in data collection accounts but thin in explicating the ways of analysis, which have been often adapted to the functionalities of ICT applications.

A regularly expressed problem concerned the validity of observations in online settings. On the one hand, how observations change in online environments when all the gestures, social cues, and facial expressions cannot be perceived similarly as in face-to-face settings when people are expressing themselves and communicating via videos, text, animations, detached voice, and still pictures; the approaches did not offer means to observe and analyse textual gestures. On the other hand, the focus of mediated interactions was often solely in the investigation of other users' relation to ICT and omitted the significance of researcher's own relation to ICT. To maintain validity in collecting data for understanding ICT mediated interactions, researcher position should be explained in the studies from the viewpoint of acquisition, learning, interpretation, and context of ICT as the researcher's own experience. The concept of user experience could serve as a useful method of conceptualisation in explicating researcher position in online settings.

ACKNOWLEDGMENT

The authors would like to thank the editors and the three anonymous reviewers for their valuable comments and suggestions, which helped to improve the chapter.

REFERENCES

Arber, S. (1993). Designing samples. In Gilbert, N. (Ed.), *Researching Social Life* (pp. 68–93). London, UK: Sage.

Beaulieu, A. (2004). Mediating ethnography: Objectivity and the making of ethnographies in the internet. *Social Epistemology, 18*(2-3), 139–163. doi:10.1080/0269172042000249264

Beaulieu, A., & Estalella, A. (2009). Rethinking research ethics for mediated settings. In *Proceedings of 5th International Conference on e-Social Science,* (pp. 1-15). Cologne, Germany: IEEE.

Bowker, G. (1994). Information mythology and infrastructure. In Bud-Frierman, L. (Ed.), *Information Acumen: The Understanding and Use of Knowledge in Modern Business* (pp. 231–247). London, UK: Routledge.

Cunningham, B. (1997). Case study principles for different types of cases. *Quality & Quantity, 31*, 401–423. doi:10.1023/A:1004254420302

Dirksen, V., Huizing, A., & Smit, B. (2010). Piling on layers of understanding: The use of connective ethnography for the study of (online) work practices. *New Media & Society, 12*(7), 1045–1063. doi:10.1177/1461444809341437

Fields, D., & Kafai, Y. (2008). Knowing and throwing mudballs, hearts, pies, and flowers: A connective ethnography of gaming practices. In V. Jonker et al. (Eds.), *8th International Conference on Learning Sciences.* Utrecht, The Netherlands: University of Utrecht.

Fields, D., & Kafai, Y. (2009). A connective ethnography of peer knowledge sharing and diffusion in a tween virtual world. *Computer-Supported Learning, 4*(1), 47–68. doi:10.1007/s11412-008-9057-1

Garrison, D. R., & Kanuka, H. (2004). Blended learning: Uncovering its transformative potential in higher education. *The Internet and Higher Education, 7*, 95–105. doi:10.1016/j.iheduc.2004.02.001

Geertz, C. (1973). *The interpretation of cultures: Selected essays*. New York, NY: Basic Books.

Gray, B. (2004). Informal learning in an online community of practice. *Journal of Distance Education, 19*(1), 20–35.

Green, N. (1999). Disrupting the field: Virtual reality technologies and "multisited" ethnographic methods. *The American Behavioral Scientist*, *43*(3), 409–421. doi:10.1177/00027649921955344

Grudin, J., & Grinter, E. R. (1995). Ethnography and design. *Computer Supported Cooperative Work Journal*, *3*(1), 55–59. doi:10.1007/BF01305846

Hemmi, A., Bayne, S., & Landt, R. (2009). The appropriation and repurposing of social technologies in higher education. *Journal of Computer Assisted Learning*, *25*, 19–30. doi:10.1111/j.1365-2729.2008.00306.x

Hine, C. (2000). *Virtual ethnography*. London, UK: Sage.

Hine, C. (2007). Connective ethnography for the exploration of e-science. *Journal of Computer-Mediated Communication*, *12*(2). doi:10.1111/j.1083-6101.2007.00341.x

Hirschheim, R., Klein, H., & Lyytinen, K. (1995). *Information systems development and data modeling: Conceptual and philosophical foundations*. Cambridge, UK: Cambridge University Press. doi:10.1017/CBO9780511895425

Howard, P. N. (2002). Network ethnography and the hypermedia organization: New media, new organizations, new methods. *New Media & Society*, *4*(4), 550–574. doi:10.1177/146144402321466813

Iivari, J., Isomäki, H., & Pekkola, S. (2010). The user – The great unknown of systems development: Reasons, forms, challenges, experiences and intellectual contributions of user involvement. *Information Systems Journal*, *20*, 109–117. doi:10.1111/j.1365-2575.2009.00336.x

Isomäki, H. (2009). The human modes of being in investigating user experience. In Saariluoma, P., & Isomäki, H. (Eds.), *Future Interaction Design II* (pp. 191–207). London, UK: Springer-Verlag. doi:10.1007/978-1-84800-385-9_10

Kaplan, A., & Haenlein, M. (2010). Users of the world unite! The challenges and opportunities of social media. *Business Horizons*, *53*(1), 59–68. doi:10.1016/j.bushor.2009.09.003

Karasti, H. (2001). *Increasing sensitivity towards everyday work practice in system design*. (Unpublished Doctoral Dissertation). University of Oulu. Oulu, Finland.

Kozinets, R. V. (2010). *Netnography – Doing ethnographic research online*. London, UK: Sage.

Lave, J., & Wenger, E. (1991). *Situated learning: Legitimate peripheral participation*. Cambridge, UK: Cambridge University Press. doi:10.1017/CBO9780511815355

Lee, J. C., & Myers, M. (2004). Dominant actors, political agendas, and strategic shifts over time: A critical ethnography of an enterprise systems implementation. *The Journal of Strategic Information Systems*, *13*, 355–374. doi:10.1016/j.jsis.2004.11.005

Levy, Y., & Ellis, T. J. (2006). A systems approach to conduct an effective literature review in support of information systems research. *Informing Science Journal*, *9*, 181–212.

Marcus, G. E. (1995). Ethnography in/of the world system: The emergence of multi-sited ethnography. *Annual Review of Anthropology*, *24*, 95–117. doi:10.1146/annurev.an.24.100195.000523

Masten, D., & Plowman, T. (2003). Digital ethnography: The next wave in understanding the consumer experience. *Design Management Journal*, *14*(2), 74–81.

Mills, C. W. (1959). *The sociological imagination*. London, UK: Oxford University Press.

Murthy, D. (2008). Digital ethnography: An examination of the use of new technologies for social research. *Sociology*, *42*(5), 837–855. doi:10.1177/0038038508094565

Myers, M. (1999). Investigating information systems with ethnographic research. *Communications of AIS, 2*.

Neumann, L., & Star, S. L. (1996). Making infrastructure: The dream of a common language. In Blomberg, J., Kensing, F., & Dykstra-Erickson, E. (Eds.), *Computer Professionals for Social Responsibility* (pp. 231–240). Palo Alto, CA: ACM.

Okoli, C., & Schabram, K. (2010). A guide to conducting a systematic literature review of information systems research. *Sprouts: Working Papers on Information Systems, 10*(26). Retrieved February 13, 2011, from http://sprouts.aisnet.org/10-26

Orlikowski, W. J. (1991). Integrated information environment of matrix of control? *Accounting. Management and Information Technologies*, *1*(1), 9–42. doi:10.1016/0959-8022(91)90011-3

Pomson, A. (2008). Look who's talking: Emergent evidence for discriminating between differences in listserv participation. *Education and Information Technologies*, *13*(2), 147–163. doi:10.1007/s10639-008-9056-x

Preecen, J., Abras, C., & Maloney-Krichmar, D. (2004). Designing and evaluating online communities: Research speaks to emerging practice. *International Journal of Web Based Communities*, *1*(1), 2–18. doi:10.1504/IJWBC.2004.004795

Puri, A. (2007). The web of insights – The art and practice of webnography. *International Journal of Market Research*, *49*(3), 387–408.

Rheingold, H. (1993). *The virtual community: Homesteading on the electronic frontier*. Reading, MA: Addison-Wesley.

Ruhleder, F., & Jordan, B. (1997). Capturing complex, distributed activities – Video-based interaction analysis as a component of workplace ethnography. In A. S. Lee, J. Liebenau, & J. DeGross (Eds.), *The IS and Qualitative Research Conference,* (pp. 246-275). London, UK: Chapman & Hall.

Rybas, N., & Gajjala, R. (2007). Developing cyberethnographic research methods for understanding digitally mediated identities. *Forum Qualitative Sozial Forschung*, *8*(3), 1–33.

Ryle, G. (1968). Thinking and reflecting. *Royal Institute of Philosophy Lectures*, *1*, 210–226. doi:10.1017/S0080443600011511

Schleckler, M., & Hirsch, E. (2001). Incomplete knowledge: Ethnography and the crisis of context in studies of media, science and technology. *History of the Human Sciences*, *14*(1), 69–87. doi:10.1177/09526951010101400104

Schultze, U. (2000). A confessional account of an ethnography about knowledge work. *Management Information Systems Quarterly*, *24*(1), 3–41. doi:10.2307/3250978

Sharrock, W., & Randall, D. (2004). Ethnography, ethnomethodology and the problem of generalisation in design. *European Journal of Information Systems*, *13*, 186–194. doi:10.1057/palgrave.ejis.3000502

Song, L., Singleton, E. S., Hill, J. R., & Koh, M. H. (2004). Improving online learning: Student perceptions of useful and challenging characteristics. *The Internet and Higher Education*, *7*, 59–70. doi:10.1016/j.iheduc.2003.11.003

Stacey, E., Smith, P. J., & Barty, K. (2004). Adult learners in the workplace: Online learning and communities of practice. *Distance Education*, *25*(1), 107–123. doi:10.1080/0158791042000212486

Star, S. L. (1999). The ethnography of infrastructure. *The American Behavioral Scientist*, *43*(3), 377–391. doi:10.1177/00027649921955326

Star, S. L. (2002). Infrastructure and the ethnographic practice: Working on the fringes. *Scandinavian Journal of Information Systems*, *14*(2).

Star, S. L., & Ruhleder, K. (1996). Steps toward an ecology of infrastructure: Design and access for large information spaces. *Information Systems Research*, *7*(1), 111–134. doi:10.1287/isre.7.1.111

Vygotsky, L. S. (1978). *Mind in society: The development of higher psychological process* (Cole, M., Lopez-Morillas, M., Luria, A. R., & Wertsch, J., Trans.). Cambridge, MA: Harvard University Press.

Ward, K. J. (1999). The cyber-ethnographic (re) construction of two feminist online communities. *Sociological Research Online*, *4*(1). doi:10.5153/sro.222

Webster, J., & Watson, R. T. (2002). Analyzing the past to prepare for the future: Writing a literature review. *Management Information Systems Quarterly*, *26*(2), 13–23.

Weick, K. (1996). Drop your tools: An allegory for organizational studies. *Administrative Science Quarterly*, *41*(2), 301–313. doi:10.2307/2393722

Wenger, E. (1999). *Communities of practice: Learning, meaning and identity*. Cambridge, UK: Cambridge University Press.

Wesch, M. (2009). YouTube and you –Experiences of self-awareness in the context collapse of the recording webcam. *Explorations in Media Technology*, *8*(2), 19–34.

Wittel, A. (2000). Ethnography on the move: From field to internet. *Forum Qualitative Sozial Forschung*, *1*(1).

Yates, J., & Orlikowski, W. J. (1992). Genres of organizational communication: A structurational approach to studying communication and media. *Academy of Management Review*, *17*(1), 299–326.

KEY TERMS AND DEFINITIONS

CMC: Computer Mediated Communication.

Ethnography: Qualitative research methodology to understand cultures and communities.

F2F: Face-to-face interaction.

ICT: Information and communication technology.

Online Communities: Groups of people interacting, sharing, and creating information in virtual communities.

Online Ethnography: Ethnographic research conducted in online environments, or in combination of online environments and face-to-face settings.

Researcher Position: The relationship between researcher´s subject and research problem, data and methods.

Vlog: A video diary shared in social media.

Chapter 8
Grounded Theory in Practice:
A Discussion of Cases in Information Systems Research

Jorge Tiago Martins
The University of Sheffield, UK

Miguel Baptista Nunes
The University of Sheffield, UK

Maram Alajamy
The University of Sheffield, UK

Lihong Zhou
Wuhan University, China

ABSTRACT

A growing number of Information Systems (IS) research is drawing upon Grounded Theory (GT), as an inductive research methodology particularly suited to the development of theory that is grounded in rich socio-technical contexts that are understood through the analysis of data collected directly from those specific environments. Studies in IS are eminently applied research projects; therefore, GT seems the ideal approach to enable rich understandings and inductive explanatory theories on the socio-technical human activity systems studied by the discipline. Nevertheless, GT is still an underutilised methodology in IS and several scholars critique the scarcity of theories developed specifically to account for IS phenomena. This chapter makes a contribution to IS research through providing a discussion of GT's application in the field by presenting an outline of the method's modus operandi and a case-based overview of its use in three different IS research projects in the fields of patient knowledge sharing in healthcare environments, IS strategic planning in academic libraries, and E-Learning adoption. As the potential of GT for IS research remains to be maximized, the objective of this chapter is to counter the lack of explicit and well-informed views of GT use in this discipline, whilst sharing lessons learned from the use of the method in context-dependent interpretive studies.

DOI: 10.4018/978-1-4666-2491-7.ch008

Copyright © 2013, IGI Global. Copying or distributing in print or electronic forms without written permission of IGI Global is prohibited.

INTRODUCTION

If complex organizational behaviors are modeled as if they are simple, well understood, deterministic systems, or even as stochastic systems, then the resulting models will tend to be insignificant (Daft & Wiginton, 1979, p. 187).

The opening quotation of this chapter establishes the rationale for theory development as a response to emergent IS phenomena. Existing theories may be partly applicable or hold an explanatory capacity of some degree. However, because emergent IS phenomena are essentially ever-evolving novel socio-technical topics (Fernandez, et al., 2002), prior theoretical studies may not fully account in scope and richness of meaning for what is going on in the field. Consequently, deductively extending existing frameworks may seem like a parsimonious application of cumulative scientific knowledge, but it can be equally frustrating if, as Daft and Wiginton (1979) suggest, established models prove to be insignificant when applied to a context alien to the one they originally respond to. It can also happen, as Lehmann (2010) argues, that the available "body of literature is narrow, mostly conjectural [or] often weak in terms of detailed predictive power" (p. 23), therefore limiting its use for the formulation and verification of hypotheses. Nonetheless, this would be true for any social science and, in particular, to any discipline concerned with socio-technical phenomena. This would also be true even if the research were of a more quantitative and positivist nature due to the continuously changing and unpredictable nature of human-activity systems.

In IS, "what is going on" is usually a composite social entity, as information systems are hybrids of human, social and technical research objects (Kroenke, 1992). In other words, the development of information systems is a human-oriented process that involves multiple stakeholders, hence the two-layered position advocated in this chapter. Firstly, we concur with Hansen and Kautz's (2005) argument that "it is impossible to separate the action itself from the organizational or social context in which it takes place," which invites the IS researcher to ground the inquiry at the heart of the social context where the interaction of actors builds up a given social reality. Secondly, the previous assertion leads us to propose that IS research should seek to "generate empirically valid theory by systematically exploring the new phenomena and its players in non-simulated environments" (Fernandez, et al., 2002, p. 111).

Therefore, in IS theory building usually means the systematic extraction of the factors that explain and characterise an empirical phenomenon. The data that accounts for that phenomenon should be grouped in clusters of meaning, aggregated into categories that form the basis of a descriptive and explanatory framework that is fit to context. According to Dey (1999), this implies focusing on "how individuals interact in relation to the phenomenon under study" (pp. 1-2), because a theory should assert "a plausible relation between concepts and sets of concepts," whilst being derived from data acquired through data collection strategies such as interviews, observations or document analysis.

The GT method, initially proposed by Glaser and Strauss (1967), incorporates iterative interaction with the social-technical environment under study, through direct contact with either human informants or other resources. This interaction results in a closely linked process of data collection and analysis, and is operated through coding, memoing, and constant comparison at each stage of the analysis. The theory construction in the methodology is based on the construction of analytical codes and concepts from data (not from logically deducted hypotheses). These procedures are well explained and defined and offer the IS researcher a sense of assurance by means of a concrete set of methods that promise validity (theoretical saturation) and lead to the emergence of theory. Nonetheless, as outlined by Hughes and Jones (2003), there "can be a difficulty in effectively introducing such methods into a technically dominant field such as information systems" (p. 57).

Despite this difficulty, there are growing calls for a shift in IS research and for a refocusing on "contextual and processual elements as well as the action of key players associated with organizational change elements that are often omitted in IS studies" (Orlikowski, 1993). Similarly, Watson (2001), Weber (2003) and Seidek and Recker (2009) claim for more theory development-focused IS research, whilst Myers (2000) and Urquhart (2001) highlight GT's usefulness in developing context-based, process-oriented descriptions and explanations of IS phenomena.

More recently, Fendt and Sachs (2007) posited that the GT method is "engaged with the world and helps, especially with the constant comparison and theoretical sampling techniques, to come skin close to the lived experience and incidents of the management world and make sense of them" (p. 448). This is particularly relevant for IS research, in which "it is often that organizational cases are the dominant unit of analysis" (Lehman & Fernandez, 2007, p. 7).

However, GT is still an underutilised method for IS research, being scarcely employed in comparison to more positivist and quantitative studies, as discussed by a variety of studies devoted to understanding the use of GT in the discipline. According to Lehmann (2010), "only 3 out the 7372 papers in 1st tier journals [published between 1985 and 2005] contain 'Grounded Theory' as a keyword" (p. 53). Urquhart (2007) presents a more optimistic figure, reporting a total of 32 GT-based IS studies between 1996 and 2005, despite the incongruent application of the method, its reported adaptation, or its coupling with other methods. To this regard, a later study conducted by Matavire and Brown (2008) yielded interesting results concerning the diversity of understandings and approaches to GT in IS research between 1985 and 2007: besides the traditional divergence between Glaserian (8% of studies) and Straussian (17% of studies) approaches, there is GT used in mixed-methods research designs (13% of studies), and an overwhelming tendency to simply employ GT data analysis techniques (67% of studies).

Furthermore, there is a growing body of literature that is descriptive about the method's implementation in a number of particular IS research projects (Toraskar, 1991; Orlikowski, 1993; Urquhart, 1997; Esteves, et al., 2002; Llundel & Lings, 2003; Lehmann & Gallupe, 2005; Allan, 2007; Coleman & O'Connor, 2007; Rodon & Pastor, 2007; Montoni & Rocha, 2010). Additionally, there are also a few reflective contributions providing a constructive analysis of GT, connecting it to epistemological issues and addressing the wider IS disciplinary identity (Hughes & Howcroft, 2000; Urquhart, 2001, 2007; Bryant, 2002; Fernandez, et al., 2002; Hughes & Jones, 2003; Fernandez, 2005; Urquhart & Fernandez, 2006; Lehmann & Fernandez, 2007; Martins & Nunes, 2009a). A recent example of the later is provided by Urquhart et al.'s (2010) in the form of a comparative study, where the authors advance guidelines for the conduct and evaluation of GT studies in IS, and where Orlikowski's (1993) study into the adoption of CASE tools is still considered "the high-water mark of Grounded Theory in IS research" (p. 378).

This chapter aims at exactly such a reflective contribution by providing insights of the method usage and to illustrate its application through the presentation and discussion of three IS research cases. Thus, the remainder of the chapter is structured as follows: the next section offers an overview of the most distinctive features of GT. This is followed by lessons learned about the use of GT in three distinct research projects, which are discussed to illustrate the operationalisation of the method and to help raise the profile of GT studies in information systems.

GROUNDED THEORY METHOD

GT is a systematic and flexible qualitative research method, specifically suited for theory generation and development. The GT approach is defined as "a qualitative research method that uses a systematised set of procedures to develop and inductively derive grounded theory about a phenomenon" (Strauss & Corbin, 1990, p. 24).

GT was first introduced by Glaser and Strauss in their book *The Discovery of Grounded Theory* published in 1967. Despite having different backgrounds (Strauss was a qualitative researcher from the University of Chicago and Glaser was a quantitative researcher from Columbia University), the co-founders of the method challenged the long-established verification paradigm by proposing a theory generation-centred paradigm for the advancement of social science. Together, they arrived at the position that inductive theory generation needs to be grounded in data systematically collected and analysed (Corbin & Strauss, 1990; Strauss & Corbin, 1998; Pickard, 2007).

Notwithstanding a convergence towards an understanding of theory as labour of the ground, Glaser and Strauss developed conflicting understandings regarding the application of GT, which produced two streams of systematic approaches, namely Straussian and Glaserian (Fernández, 2004), advocated respectively in the key publications *Basics of Qualitative Research* (Strauss & Corbin, 1990), and *Basics of Grounded Theory Analysis* (Glaser, 1992).

The Straussian stream advocates using a more structured approach to collecting and analysing data (Pickard, 2007, p. 156). However, the Straussian approach has been criticised for having "moved too far from the original concepts" (Pickard, 2007, p. 156) of GT, and for imposing an interpretive structure that may be "forcing data" (Glaser, 1992, p. 122). By contrast, the Glaserian approach advocates that the researcher should stand at a passive position, free from preconceptions, not forcing structure onto data, and trusting that theory will emerge (Rodon & Pastor, 2007).

Despite the decades of dispute between the two approaches, both Straussian and Glaserian researchers adopt an identical philosophical view, namely that theory should emerge from or be "grounded" in the data (Van Niekerk & Roode, 2009). Moghaddam (2006) adds that even though the two approaches have very distinctive paths to develop a theory, both Straussian and Glaserian have a conjugate definition for the main processes that form the operational core of GT, namely: use of literature, theoretical sampling, coding processes, comparative analysis, and theoretical saturation.

These issues are addressed in the subsequent subsections, following the Straussian approach to GT, which provides more pragmatic rigour and clearer techniques, especially for the novice researcher who is less familiarized with the processes of data collection and analysis (Rodon & Pastor, 2007).

Use of Literature

In GT projects, the literature review needs to be practiced cautiously. Its function is not to produce any a priori theoretical framework, which is commonly adopted as the theoretical foundation and starting point for data collection and analysis in deductive research designs.

Strauss and Corbin (1998) argue that in a GT-based inquiry there is no need to review all the literature at the beginning of the research project, since early steeping in the literature may "constrain," "stifle," or even "paralyze" (p. 49) the researcher's analytical senses, and may possibly introduce strong biases in the emerging theory. Therefore, an objective stance is advisable for the GT researcher, which means setting aside personal knowledge and experience, and listening to the "voice" from the data.

Nonetheless, as also discussed by Strauss and Corbin (1998, p. 43), although familiarity with the relevant literature may interfere with the researchers' analytical capacity, it can also enhance their theoretical sensitivity. Theoretical sensitivity is crucial for theory development, as it refers to the ability to capture subtle nuances in data, generate concepts from data, and relate them according to models of theory in general (Glaser, 1978). Consequently, a general review of literature may be recommended at the beginning of the research project to provide background knowledge for the development of theory and enhance the researcher's theoretical sensitivity. An additional literature review should be carried out at the end of the research project, after the emergence of theory, to confirm the research findings and to identify where the literature is incorrect or overly simplistic (Strauss & Corbin, 1998, p. 52).

Such a procedure keeps with the precepts of GT: instead of heavily relying on any a priori framework to guide the process of data collection and analysis, it entrusts the researcher with the mission of collecting data according to the analysis and the needs of theory development (Strauss & Corbin, 1998, pp. 204-205). This strategy is called theoretical sampling, and will be introduced in the following subsection.

Theoretical Sampling

Theoretical sampling is valued both as a basic principle and as a distinctive feature of GT. Theoretical sampling is defined as "data gathering driven by concepts derived from the evolving theory and based on the concept of "making comparisons," whose purpose is to go to places, people, or events that will maximize opportunities to discover variation among concepts and to densify categories in terms of their properties and dimensions" (Strauss & Corbin, 1998, p. 201).

The theoretical sampling process closely connects the processes of data collection and analysis, between which there is an ongoing interplay: the collection of data is driven by the analysis, which starts as soon as the first element of data is gathered. On the basis of the data analysis, the researcher articulates and derives indications for further data collection (Corbin & Strauss, 1990). Ideally, the researcher should enter the investigation site with an unbiased mind (i.e. free of pre-established conceptual and interpretive devices), and make use of the theoretical sampling strategy to "maximize opportunities to compare events, incidents, or happenings to determine how a category varies in terms of its properties and dimensions" (Strauss & Corbin, 1998, p. 202).

Nevertheless, it is very important to mention that Strauss and Corbin (1998) do not suggest that the researcher should be completely "empty-minded" (p. 205). In fact, they argue that the researcher needs to decide whether to "develop a list of interview questions or areas for observation. [...] Initial interview questions or areas of observation might be based on concepts derived from literature or experience or, better still, preliminary field work" (p. 205). However, and because any predefined concepts are not rooted in the "real" data, they must be considered as provisional, and abandoned as soon as the data is elicited.

Coding Process

GT demands microanalysis of data, which means detailed line-by-line examination of data in order to identify incidents and concepts, to generate categories with properties and dimensions, and eventually to formulate a theory (Strauss & Corbin, 1998, p. 57). Strauss and Corbin (1998) advise that the microanalysis is practiced by applying three types of coding to the data: open coding, axial coding, and selective coding (p. 58).

Open coding is "the analytic process through which concepts are identified and their properties and dimensions are discovered in data" (Strauss & Corbin, 1998, p. 101). It entails breaking data down into discrete fragments, thereafter thoroughly examined for similarities and differences. Through these processes, events, objects, actions and interactions that are identified as conceptually similar or related in meaning are grouped to form categories (Corbin & Strauss, 1990).

Axial coding is "the process of relating categories to their subcategories" (Strauss & Corbin, 1998, p. 123). This means interconnecting properties, concepts, and subcategories around the axis of a category, and checking this back again against data.

Selective coding represents "the process of integrating and refining the theory" (Strauss & Corbin, 1998, p. 143). In selective coding, "the major categories are finally integrated to form a larger theoretical scheme in which the research findings take the form of theory" (Strauss & Corbin, 1998, p. 143). Moreover, this step offers the opportunity for spotting and refining poorly developed categories until a full integration is achieved.

When approaching coding, the researcher must note that although the practice of open, axial, and selective coding seems sequential, structured and static, researchers do not need to follow a rigidly circumscribed sequence (Pandit, 1996). Strauss and Corbin (1998) argue in favour of analytical flexibility, claiming that data analysis is a free-flowing process, in the sense that analysts move quickly back and forth between all three types of coding (p. 58).

Constant Comparison

Constant comparison is the result of comparative analysis, and is a symbol of social science research and an essential feature of GT analysis. GT requires the researcher to adopt the comparative analysis technique throughout all data analysis and theory development processes.

More specifically, Strauss and Corbin (1998) propose two fundamental types of comparison. The first type of comparison "pertains to the comparing of incident to incident or of object to object, looking for similarities and differences among their properties to classify them" (p. 94). Goulding (2002) further explains that this type of comparison is mostly used when exercising open and axial coding.

The second type of comparison is practiced at an abstract level, as it compares "categories (abstract concepts) to similar or different concepts to bring out possible properties and dimensions when these are not evident to the analyst" (Strauss & Corbin, 1998, p. 94). This type of comparative analysis is mostly used in axial and selective coding processes (Goulding, 2002).

In order to support the practice of comparative analysis, it is advisable to adopt a set of practical tools, namely a code definition list (to support the comparison between individual open codes), a quotation list (to support the comparison between individual quotations), and concept maps (to support comparisons between properties, concepts, sub-categories, and categories).

Theoretical Saturation

The theoretical saturation is an extremely important stage as it is a signpost for the completion of data gathering, theoretical sampling, and data analysis. Additionally, and more importantly, the theoretical saturation indicates the completion of theory generation. Strauss and Corbin (1998) formulate three essential rules which need to be used when determining whether the theoretical saturation has been achieved: (1) "no new or relevant data seem to emerge regarding a category"; (2) "the category is well developed in terms of its properties and dimensions demonstrating variation"; and (3) "relationships among categories are well established and validated" (p. 212).

In practical terms, and according to the rules presented above, theoretical saturation occurs when (1) no new open codes emerge from data; (2) properties and dimensions of individual categories have been examined as explainable to the social phenomenon reflected from data; (3) the relationships between individual categories have been examined and confirmed by checking with data.

A DISCUSSION OF CRITICISMS OF GT

GT is characterised by theory generation through interaction with data collected directly from human-activity systems instead of the traditional and well-established positivist method of testing hypotheses from existing theory. This is often subject to criticism, particularly in relation to the prevalent hypothesis driven deductive practices that have traditionally characterised science, leading to misinterpretation of the aims and analytical methods proposed by GT and a misunderstanding of its findings. In fact, the qualitative nature of GT focuses on the search for meaning to build theory that is not speculative or extracted from preliminary hypotheses. However, this does not mean that the GT inductive theory building endeavour invites the researcher to incur in a lax or less rigorous behaviour: "the analyst should take as much time as necessary to reflect and carry his own thinking to its most logical (grounded in the data, not speculative) conclusions" (Glaser & Strauss, 1967, p. 107).

In practical terms, this means that the methodology is not interested in predefining the composition of samples for the collection of data, in the verification of variables or in the development of probabilistic relations, as expressed by Katz (1983):

If we view social life as a continuous symbolic process, we expect our concepts to have vague boundaries. If analytical induction follows the contours of experience, it will have ambiguous conceptual fringes (...). For the statistical researcher, practical uncertainty is represented by statements of probabilistic relations; for the analyst of social processes, by ambiguities when trying to code border line cases (...) (Katz, 1983, p. 133)

Consequently, we argue that GT's close proximity with qualitative data affords the understanding of complex socio-technical phenomena, but it cannot eliminate the conceptual crossing and inter-relation of meanings that are characteristic of any context-bound enquiry. Furthermore, the inductive generation of concepts entails a cumulative process - not a sequentially fragmented process - whereby the researcher:

1. Breaks down data into descriptive categories.
2. Thinks theoretically rather than descriptively and evaluates interrelationships.
3. Subsumes concepts into higher order categories that indicate an emergent theory.

This poses the GT researcher an additional challenge, because unlike quantitative methods, the reliability and validity of findings does not lie with inserting statistical analysis or appending copies of questionnaires, but on achieving theoretical saturation and conceptual integration.

Therefore, we concur with Bryant and Charmaz's (2010) argument that GT has sought to "produce outcomes of equal significance to those produced by the predominant statistical-quantitative, primarily mass survey methods (...)" (p. 33). In fact, three main stages should be observed in the process of developing a grounded theory. These include, according to Goulding (1999): (1) "demonstrating how, why and from where early concepts and categories were derived" by making them traceable to the data; (2) abstracting the concepts and looking for theoretical meaning—"concepts should be sufficiently developed as to warrant an extensive re-evaluation of literature to demonstrate the fit, relationship and where applicable the extension of that literature through the research findings"; and finally (3) by presenting the theory, "uniting the concepts and integrating them into categories which have explanatory power within the specific context of the research" (p. 17).

Finally, a recurrent source of criticism is the idea that GT methodology recommends eliminating the researcher's exposure to literature to prevent bias. Urquhart (2006) classifies this assumption as the "researcher as a blank slate" myth, since as previously discussed the constant comparison principle determines the obligation to engage emergent theory with the literature. Furthermore, and as argued by Bowen (2006), an initial literature review can "provide starting points for building analysis to produce a grounded theory," and "give the researcher a sense of how observed instances of a phenomenon might fit within conceptual categories."

In any case, and since it has become the most preeminent inductive methodology in use today, many of the criticisms thrown at GT, are criticisms of inductive theory in general, rather than of the methodology itself. There will never be an agreed consensus on the validity and benefits of inductive theory in general. However, accepting that inductive research enables the production of new theoretical propositions, closely grounded in complex social environments, where there were none before, then GT emerges as an ontologically solid, easily applicable and rigorous research methodology.

CASE STUDY EXAMPLES OF GT IN IS RESEARCH

In this section, we present three IS research projects that used GT as their overarching methodology. We draw implications of using GT in each case, offering details about the main issues addressed: patient knowledge sharing in Chinese healthcare environments, IS strategic planning for Syrian academic libraries, and e-learning adoption in Portuguese Higher Education (HE) institutions. The discussion aims at discussing how each investigator aligned their study objectives with the research design of GT, and with how the methodological features of GT were handled throughout each study. Essentially, what transpires across the cases presented is the voice of the ordinary novice researcher. The presentation of cases follows a common structure, namely initial research objectives; use of GT methodology; and conclusions.

Identifying Barriers to Sharing Patient Knowledge between Healthcare Professionals from Traditional and Western Medicines in Chinese Hospitals

Initial Research Objectives

The Chinese healthcare system incorporates two entirely different medical philosophies, namely Traditional Chinese Medicine (TCM) and Western Medicine (WM). In the 1950s and by the request of the central government, the two medical professional communities have been both required and encouraged to collaborate with each other. In order to ensure the successful implementation of a patient-centred healthcare policy, also imposed by the central government, these two very different healthcare professional groups are required to communicate and share knowledge about individual patients. This tacit Knowledge Sharing (KS) aims at protecting the needs, interests and benefits of patients, as well as guaranteeing that the patient is at the centre of the collaborative processes. However, the two medical communities do not coexist harmoniously and do not readily communicate and share knowledge with each other. There are barriers hindering the processes of KS between TCM and WM healthcare professionals. Therefore, this research project aims to identify barriers to the sharing of patient knowledge between the two types of healthcare professionals in the context of Chinese hospitals.

Due to the very specific characteristics of the Chinese healthcare environment, it was very difficult to establish a base theory for the problem. In fact, there are other examples of coexistence between WM and traditional medicine (e.g. homeopathy, herbal medicine, and chiropractic). However, in no other circumstances are these different health professionals working together by political design and in the same hospital, sometimes even in the same building. Therefore, the research approach selected for the project is inductive in nature and essentially aiming at building a theory.

To achieve the theory building aim, a GT approach was adopted. The main reason is that GT is very useful for theory building and development, and is particularly advantageous when, like in this research project, a theoretical foundation cannot be established from the existing body of literature. Moreover, without a preliminary framework to guide the processes of data collection and analysis, GT offers a systematised set of strategies, procedures and techniques, which can be applied not only to direct the collection and analysis of data, but also to explore insights for generating theory closely related to the context of the phenomena being studied.

Use of GT Methodology

Considering that China is one of the largest countries in the world, with a population exceeding 1.3 billion and with 56 ethnic groups and 34 provinces, it would be virtually impossible to generate a theory that would encompass the whole nation. Consequently, since this project aimed at generating a first set of insights into this problem, and in order to enable an in-depth investigation, a single case-study design was adopted to provide a social context for the application of GT. A public hospital in central China was selected for the case study. This hospital was chosen for two main reasons. Firstly, it provides both WM and TCM services to patients and has done so for several decades. Secondly, the researcher obtained guaranteed and management supported access to the informants and the project.

More specifically, the application of GT in the context of this case study was planned to have three stages: a pilot study, a main study, and a follow-up study.

The pilot study stage aimed to confirm whether the KS problems anticipated did present themselves in the actual practice of a Chinese hospital, and to identify early results and insights to guide the remaining research stages. Overall, 7 healthcare professionals and workers were purposively approached and interviewed. These participants were 2 WM doctors, 1 WM nurse, 2 TCM doctors, 1 ICT manager and 1 hospital administrator. Findings from this stage suggested that different departments in the hospital exhibit very different patterns of KS behaviour between the two medical communities. The study also showed that very different levels of integration of complementary treatments may take place in different departments. This resulted in the decision to choose one specific department, namely the Department of Neurosurgery. This department has a proven history of using WM and TCM compound treatments for rehabilitating patients after craniotomies.

The purpose of the main study stage was to identify KS barriers between the two communities when dealing with problems from neurosurgical patients. At this stage, 11 neurosurgeons, 8 neurosurgical nurses, and 6 TCM doctors were selected as participants, approached, and interviewed. Data collected was analysed using open and axial coding through the constant comparison method of analysis. The findings indicated a need for further study on the external influences on KS.

Finally, a follow-up study stage was aimed at exploring these external influences on KS. Furthermore, it aimed at exploring and studying those categories for which theoretical saturation could not be achieved in the previous stage. Theoretical sampling, further coding and constant comparison analysis were used at this stage. This exercise of gaining a deeper understanding involved the interviewing of specific informants, namely 2 neurosurgeons, 2 TCM doctors, 3 neurosurgical nurses, 1 TCM educator, 1 public administrator in the local healthcare department, and 8 patient carers.

Overall, 46 informants were interviewed in a total number of 49 interviews. As required by the theoretical sampling strategy, these informants were sampled by the emerging theory and were interviewed by using semi-structured and evolving question scripts.

Following the GT methodology, the processes of data collection and analysis were operationalised interactively. That is, immediately after each individual interview, the collected data were transcribed and analysed. Results from this analysis were used not only for the theory building, but also to direct following interviews. Furthermore, the practice of data analysis adopted two GT analytical techniques, namely, coding (open, axial and selective) and constant comparative analysis. Consequently, data collection and analysis coexisted until the theoretical saturation was achieved, that is, until no new open codes emerged from the data analysis.

Conclusion

As the result of the analytical process described in the earlier subsection, the emerging theory saturated on five main categories: contextual influences, philosophical issues, Chinese healthcare education, interprofessional training, and hospital management.

The research findings show that the sharing of patient knowledge is mainly obstructed by the divergent philosophical and conceptual perspectives adopted by the two types of healthcare professionals. These divergences in conceptual systems, theoretical foundation, and diagnosis and treatment methods lead to severe difficulties in understanding each other's diagnosis, clashes in indications for treatment, difficulties in interpreting requirements for complementarity of treatments and difficulties in understanding interpretations of patients' problems, and thus made the communication of patient knowledge extremely difficult. Moreover, the data collected also reflect that WM professionals have a higher professional standing and have almost dominant power over patients. Therefore, they often explicitly instruct and regulate TCM doctors in what to do about the patient. These inequalities of status and power not only create untrusting, uncooperative and even resenting relationships between TCM and WM professionals, but also effectively demotivate them from actively and voluntarily sharing knowledge about individual patients.

After further conceptualising the research findings, it was identified that KS is mostly prevented by philosophical and professional tensions between the two medical communities. Therefore, to improve KS and reduce the effects of the identified barriers, efforts should be made targeted at resolving both types of tensions by establishing national policies and hospital management strategies aimed at maintaining equality of the two medical communities and putting in place an interprofessional common ground to encourage and facilitate communication and KS.

The Role of Academic Librarians in Information Systems Strategic Planning: The Case of Syrian Governmental Universities

Initial Research Objectives

This research investigated academic librarian involvement in Information Systems Strategic Planning (ISSP) in Syrian Universities and was aimed at generating theoretical propositions about the role of Academic Librarians (AL) in University-wide ISSP, as perceived both intra-professionally and inter-professionally, namely the relationship with the Information Technology/ Information Systems (IT/IS) departments. This study takes on particular relevance in the Syrian context, as AL's traditional roles are increasingly being questioned and expanded, with impact on librarians' self-perceptions, institutional mission, and professional relations with students, staff, and planning stakeholders. Additionally, a theoretically-sensitising literature review indicated this to be an unexplored area in research, with no previous studies investigating IT/IS staff professional views towards AL and their role in the ISSP process in the context of the Middle East.

The fact that there is no satisfactory theory addressing the subject of investigation has contributed greatly to shape an inductive, qualitative research design, the goal being investigating professional worldviews towards an actor that operates in a complex process—i.e. AL role in ISSP—which develops across different departments.

Considering the objectives of this inquiry and that a major strength of GT lies in the flexibility and openness to facilitate grasping "shifts in the understanding of dynamic situation in complex organizational arenas" (Vasconcelos, 2007, p. 127), this methodology was elected to conduct the research. Since the aim was to explore the untackled relationship between different professionals' views about roles and capacities, it was considered that applying GT would lead to an in-depth understanding of the practices, discourses and interactions taking place between the different social actors: AL and IT/IS staff. Furthermore, and as Shannak and Aldhmour (2009) appreciate, GT allows "the inclusion and investigation of key organizational elements, such as the impact of ICT on the employees' and managers' behaviours, skills, experiences, performance, organization structures, activities, and so on" (p. 35).

From an operational point of view, GT also offered a structured guideline for data collection, analysis and inductive theory building, which reinforces the reliability of the resulting theory. In GT, data collection and analysis is performed in "successive" steps, which helps to focus on the interpretation of data (Esteves, et al., 2002). The systematic way of analysing data proposed by Strauss and Corbin (1998) was favoured in this research project, as this approach is helpful to guide the beginner researcher, who may encounter difficulties in making sense of the data and in understanding how to move towards developing concepts and, ultimately, theories.

Use of GT Methodology

GT requires an open-minded researcher, available to detect different parties' ideas and knowledge base. The investigation of this knowledge, patent in feelings, beliefs, and organisational behaviour patterns, demands a data collection technique that is able aiming to elicit and capture participants' accounts.

Therefore, semi-structured interviews were administered to IT/IS senior managers and AL. In addition to having interviews recorded and transcribed, memos have been used as a tool to enhance reflexivity and to keep track of the researcher's developing interpretation of emerging findings. Interviews were conducted in two main stages of data collection, the first stage being a pilot study and the second stage being the main study.

A pilot study was conducted to explore the emerging organisational identity of AL across the Syrian Higher Education ecology. Since the number of HE public organisations in Syria is relatively small, the pilot aimed at identifying groups of informants, if the groups and subgroups to be studied existed homogenously across the national scenario and if there were sufficient numbers of informants to enable meaningful data collection and analysis. Informants from all governmental Syrian Universities (Damascus University, Aleppo University, Tishreen University, and Al-Baath University) were identified using a snowballing technique, initiated by an initial holistic meta-inquiry exercise conducted with the Higher Education Ministerial Department.

Data collection for the main study followed the principle of theoretical sampling. The pilot study stage contributed to fine-tuning the analytical focus, directing it into the processes sustaining the unstructured and non-formalised contribution of AL in ISSP. Therefore, theoretical analysis directed the search for relevant social professional arenas, where the interplay between IT/IS senior managers and AL could be scrutinised. A pool of senior managers in information system departments and strategic planning departments, university senior managers, and academic librarians from Syria's four governmental universities (Damascus University, Aleppo University, Tishreen University, and Al-Baath University) has been approached, resulting in a total of 23 interviewees.

Data analysis proceeded through constant comparison, which means that every time the researcher selected a passage of text and coded it, she would compare it with all previous coded passages. Constant comparative analysis was enhanced through the use of three analytical tools, namely a code definition list, concept maps, and a quotation list.

The code definition list contains codes that emerged from the data. For each code, there is a simple definition explaining the meaning of this code. As the categories emerged, codes were reorganised and placed into appropriate categories.

The main emergent themes and structures of categories were represented diagrammatically by means of concept maps. Additionally, for each quotation added to the quotation list, a label demonstrating its specific location in the data was assigned. The process of data collection and analysis stopped when no new concepts emerged from data, and therefore no new codes and relationships could be established.

Conclusion

Conclusions extracted after iterative conceptualization of data come in the form of propositions that reflect organisational ambiguity concerning the recent role of AL in Syrian universities' ISSP process. The data reveals a situation where AL faces a situation of exclusion from ISSP. The proposed theory suggests that this situation may be caused by several factors, namely different perceptions regarding AL inclusion in ISSP; AL feeling undervalued, not heard and ignored; AL feeling burdened by multiple tasks; and inconsistency in stakeholders' views of AL job attributions and effective contribution to ISSP. Data analysis further revealed a strong emphasis of stakeholders on the need to remain "open" in the ISSP process. However the intention of remaining open and flexible is hampered by non-existent or inefficient communication channels, and by the lack of a framework that explicitly caters for the inclusion of AL in Syrian universities ISSP process.

On Trust and Assurance: Towards a Risk Mitigation Normative Framework for e-Learning Adoption in Portuguese Higher Education Institutions (HEI)

Initial Research Objectives

The methodology chosen for this study was GT because the main objective of the research was to investigate how Portuguese academics enact trust in e-learning and, consequently, how they face vulnerability and potential risks involved in e-learning adoption. The objective was not to investigate if academics' cognitions or practice pertaining to e-learning developed in a predetermined way, following an established model of trust. This decision was taken to ensure contextual fit, relevance of results, and adherence to the Portuguese HEI landscape.

Additionally, little research addressing Portuguese academics' perceptions of e-learning adoption has been conducted to date, hence the interest in understanding how practice behaves instead of investigating if practice behaves in a specific anticipated way, determined by a theoretical explanatory model of technological diffusion, infusion and adoption.

As Walz argues (2005), "individuals negotiate, choreograph, describe, perform, and thus cultivate close and distant relationships not only with one another, but also with everyday artefacts and their everyday environments" (p. 123). Therefore, after considering the objectives of this research, it became clearer that only a methodology that is qualitative and inductive in nature could help the researcher understand how academics' individual negotiations of meaning occur, and generate contextually-valid knowledge about the trust/ risk micro-environment within which e-learning appropriation takes place. Essentially, these are human relations issues, not easily apprehensible with the use of quantifiable parameters.

E-learning adoption develops within a human activity system, it is a context-bound process situated at the crossways of organisational, cultural and pedagogical continuums. Studying this process using a research methodology that gives the researcher some type of control over pre-set verifiable hypotheses and the possibility to extract testable moderating factors exclusively from the literature would misrepresent what is going on in the field (Martins & Nunes, 2009a).

As an alternative, an open approach that invites data to speak and academics' subjective manifestations to come forth holds the potential to generate locally-significant robust theories, with results being sufficiently "grounded directly in the observed data, instead of being produced by deduction or other intellectual experiments" (Hansen & Kautz, 2005, p. 4). Unequivocally, the results will not point towards scientific generalisability, reflecting otherwise a contribution to the accumulation of insights that feed into a vaster body of knowledge.

Use of GT Methodology

Considering that a strength of GT is its structured approach to "the gathering and analysis of human experiences and the associated interrelations with other human actors, coupled with situational and contextual factors" (Coleman & O'Connor, 2008, p. 775), opportunities for theory-building were construed from academics' perceptions concerning the links between decision making, action and self-reported e-learning adoption processes. The discovery of a multi-level ontology encompassing the structured social context of Portuguese HEI was concomitant to handling meanings, capturing respondents' situational definitions, and iterative attempts at developing an interview-based causal theory.

In terms of data collection, the study operated through semi-structured in-depth interviews with academics. Interview script analysis developed through constant comparison, following the steps described by Glaser and Strauss (1967): "(1) comparing incidents applicable to each category; (2) integrating categories and their properties; (3) delimiting the theory; and (4) writing the theory" (p. 105). The process led to the unearthing of conceptual categories and their properties from data, substantiating grounded theory's precept that systematic abstraction of empirical data, followed by comparison-centred conceptualisation, constitute the core methodological steps of theory generation.

Sampling efforts focused on the identification and recruitment of a relevant social arena of action, composed of academics in Portuguese public HEI, teaching at BA/ BSc Level, and affiliated with Faculties where e-learning appropriation manifested itself in considerable depth. Moreover, studying this community of academics allowed the researcher to examine a specific intra-organisational dynamics (Strauss, et al., 1985, p. 158), reinforced by a common professional and occupational world.

The sampling technique employed in this research therefore required selecting informants who were knowledgeable practitioners and who were willing to share their experiences with the researchers. Data collection efforts developed in two stages. A first interview round comprised 13 interviews, being part of a pilot study stage designed to develop "contextual sensitivity" (Nunes, et al., 2010). Additionally, a second interview round comprised 47 interviews.

A purposeful approach to preliminary informant selection was deemed necessary during the first round of interviews to, as Glaser (1978) admits, gain rapport with "knowledgeable people to get a line on relevancies and leads to track down more data and where and how to locate oneself for a rich supply of data," whilst maximizing "the possibilities of obtaining data and leads for more data in their question" (p. 45). During this stage, 13 academics of Portuguese Higher Education Institutions were interviewed.

In the course of data analysis conducted during the first interview round, the researcher found that emergent theoretical propositions related to academics' e-learning appropriation pathways could be refined and modified through comparison with other cases. This acknowledgement consequently dictated the decision to refine and extend the sampling strategy, basing the procedure on analytic grounds.

As the study developed into a second round interviews, theoretical sampling—employed as an inductive, systematic approach to extract theoretical formulations out of informants' disclosed cognitions followed by deductive validation—was used to guide further data collection, which involved recruiting additional informants for interview and collecting additional data to examine provisional categories, understand their interrelation and to ensure their fit and representativeness. As Glaser (1978) points out, theoretical sampling occurs when "the analyst jointly collects codes and analyses his data and decides what data to collect next and where to find it, in order to develop his theory as it emerges" (p. 36). On the other hand, in the words of Strauss and Corbin (1990), theoretical sampling occurs "on the basis of concepts that have proven theoretical relevance to the evolving theory" (p. 177). During this second stage of data collection, a total of 47 academics were interviewed.

The interviews were conducted until no new conceptualisations emerged from the interview data and when, in the concomitant process of data analysis (i.e. coding), no new properties of data were uncovered and no new relationships between categories were manifested.

Conclusion

The confines of the data set directed the researcher into understanding (1) the way academics face vulnerability and risk involved in e-learning appropriation; (2) institutional responses to augment trust and achieve higher levels of confidence; (3) the trustworthiness diffused by the social system in which academics are embedded, composed of rules, roles and routines; (4) and academics' response and enactment of trust in e-learning. Iterations in interview script analysis suggested that trust as a behavioural response is relevant but not a self-fulfilling condition for sustainable e-learning appropriation by academics. Therefore, it needs to be supplemented by an additional standard that enables the theoretical conceptualisation of risks and vulnerabilities and the calculative reason necessary for sound appropriation of e-learning.

In terms of theory-building, preliminary findings suggest that the emergent core category of 'risk mitigation normativity' can operate as a defining standard, able to generate efficiencies by lowering transaction costs associated with adopting e-learning, whilst conserving academics' cognitive resources. The 'risk mitigation normativity' refers to the institutional inculcation of norms and values by means of regulatory mechanisms that explore the capacity of academics to act as self-actualising agents. It is composed of pay-off structures, through which academics (1) have a voice in a system of joint planning and problem solving alongside with management (Martins & Nunes, 2009b); (2) become aware of e-learning adoption outcomes; (3) take cost-effective actions to mitigate risk and enhance benefit (Martins & Nunes, 2010b); (4) proceed with appropriation when expected gains are anticipated and sustained through recognition and reward (Martins & Nunes, 2010a).

CONCLUDING REMARKS

The examples offered in this chapter illuminate the use of GT as a valid and defensible methodology to conduct social and organisational context-rich IS research. The method's capacity to generate trustworthy, data-centred, and locally applicable theory is perhaps its major strength, intimately connected to the scaffolding provided by its very specific set of procedures, such as theoretical sampling, constant comparison, and the accurate process of breaking down interview data into units of meaning and generative concepts through the different stages of coding. This is not to be taken as a mechanical routine, rather as a highly creative process, manifest in "the ability of researchers to aptly name categories, ask stimulating questions, make comparisons, and extract an innovative, integrated, realistic scheme from masses of unorganized raw data" (Strauss & Corbin, 1998, p. 13).

The different research designs presented show how the general principles of GT can be customised according to the needs of different research questions and objectives. GT is therefore seen more as a flexible research framework rather than as rigid and structured methodology. In truth, due to its elliptic nature, GT research projects could conceivably be designed in very different ways. Nonetheless, this does not mean that GT is a mere overarching framework. GT actually offers prescriptive procedures that are described and illustrated with cases throughout this chapter. Furthermore, GT also offers guidelines or pointers for a process that is eminently interpretive in its nature, ultimately leading to a conceptual understanding of a research problem and to the unveiling of "underlying assumptions, contexts and experiences of those involved in the IS phenomenon under study" (Hughes & Jones, 2004).

However, as discussed above, the GT method is not without critique. The very tension between the founding fathers of the method as discussed in Section 2 may be considered inimical to consistency and mainstreaming. Therefore, we advise for an informed choice between a flexible, creative application of the method (following Glaser's unstructured emphasis on emergence) and a more scaffolded approach to analytical induction (embracing Strauss and Corbin's formalism).

Critics of the method such as Thomas and James (2006) are concerned with GT's oversimplification of the complexity contained in data. In this chapter, we counter-argue that GT researchers are not satisfied with superficial analysis and always aim at accounting for the structural core of social phenomena, through offering an understanding of the relationships between main concepts. This understanding comes in the form of a narrative, which is developed from iterations in data, from the interplay of analytical memos and, *aposteriori*, from the literature. To those who condemn the GT method either by focusing excessively on procedure at the cost of a richer interpretation (Layder, 1993) or by pushing researchers into "look[ing] for data, rather look[ing] at data" (Robrecht, 1995, p. 175), we offer the argument advanced by Fendt and Sachs (2007): GT presents "a useful systematic approach to handling and analyzing data that, if applied with courage and creativity, may lead to innovative perspectives. (...) [It] offers processual theory in the Strauss and Corbin sense. It produces plausible propositions of relationships among concepts and clusters of concepts that can be traced back to the data" (p. 448).

We further add that Strauss and Corbin's contributions in terms of procedural guidelines are beneficial for the novice researcher. This was apparent in the three studies presented above. These guidelines should not be taken as self-limiting rules that interfere with the capacity to advance interpretations and ideas, but helped the researchers define stages and milestones in their research design. Considering the turn in the IS research field from a technical view of systems to seeking rationalisation for non-materialistic issues such as cognition, culture and motivation (Bryant, 2002), we recommend the use of GT as a method that can help researchers:

- Confidently achieve rigour and relevance.
- Enhance the transferability of IS research to improve organisational processes.
- Affirm the disciplinary identity and specificity of IS by overcoming the ingrained habit of borrowing base theories from neighbouring fields, through reinstating the possibility of developing theories from inside the field.

More than a coding technique, GT is a method of theory generation particularly sensitive to the vitality of experiences emanating from the field, which gives the resulting theories the "form of [an] experience map" (Coleman & O'Connor, 2008, p. 781) and a more contextualised and dynamic depiction of IS phenomena. Again, this was evident in the findings in the three studies. In all of these, the theory proposed enabled the proposition of practical applications of that theory, recommendations to handle the phenomena studied and actually link the theory with the reality of practice.

ACKNOWLEDGMENT

This research was supported by FCT – Portuguese National Foundation for Science and Technology (grant SFRH/BD/39056/2007, awarded to co-author Jorge Tiago Martins).

REFERENCES

Allan, G. (2007). The use of grounded theory methodology in investigating practitioners' integration of COTS components in information systems. In *Proceedings of the 28th International Conference on Information Systems,* (pp. 1-10). Montreal, Canada: IEEE.

Bowen, G. A. (2006). Grounded theory and sensitizing concepts. *International Journal of Qualitative Methods*, *5*(3), 1–9.

Bryant, A. (2002). Grounding systems research: Re-establishing grounded theory. *The Journal of Information Technology. Theory and Application*, *4*(1), 25–42.

Bryant, A., & Charmaz, K. (2010). Grounded theory in historical perspective: An epistemological account. In Byant, A., & Charmaz, K. (Eds.), *The Sage Handbook of Grounded Theory* (pp. 31–57). London, UK: Sage.

Coleman, G., & O'Connor, R. (2008). Investigating software process in practice: A grounded theory perspective. *Journal of Systems and Software*, *81*(5), 772–784. doi:10.1016/j.jss.2007.07.027

Corbin, J., & Strauss, A. (1990). Grounded theory research: Procedures, canons, and evaluative criteria. *Qualitative Sociology*, *13*(1), 3–21. doi:10.1007/BF00988593

Daft, R., & Wiginton, J. (1979). Language and organization. *Academy of Management Review*, *4*(2), 179–191.

Esteves, J., Ramos, I., & Carvalho, J. (2002). Use of grounded theory in information systems area: An exploratory analysis. In *Proceedings of the 1st European Conference on Research Methodology for Business and Management Studies,* (pp. 129-136). Reading, UK: Business and Management Studies.

Fendt, J., & Sachs, W. (2007). Grounded theory method in management research: Users' perspectives. *Organizational Research Methods*, *11*(3), 430–455. doi:10.1177/1094428106297812

Fernandez, W. (2004). Using the Glaserian approach in grounded studies of emerging business practices. *Electronic Journal of Business Research Methods*, *2*(2), 83–94.

Fernandez, W. (2005). The grounded theory method and case study data in IS research: Issues and design. In Hart, D., & Gregor, S. (Eds.), *Information Systems Foundations: Constructing and Criticising* (pp. 43–60). Canberra, Australia: The Australian National University.

Fernandez, W., Lehmann, H., & Underwood, A. (2002). Rigour and relevance in studies of information systems innovation: A grounded theory methodology approach. In *Proceedings of the 10th European Conference on Information Systems,* (pp. 110-119). Gdansk, Poland: IEEE.

Glaser, B. (1978). *Theoretical sensitivity.* Mill Valley, CA: Sociology Press.

Glaser, B. (1992). *Emergent vs. forcing: Basics of grounded theory analysis.* Mill Valley, CA: Sociology Press.

Glaser, B., & Strauss, A. (1967). *The discovery of grounded theory.* London, UK: Aldine Transaction.

Goulding, C. (1999). *Grounded theory: Some reflections on paradigm, procedures and misconceptions.* Technical Working Paper. Wolverhampton, UK: University of Wolverhampton.

Goulding, C. (2002). *Grounded theory – A practical guide for management, business, and market researchers.* London, UK: Sage.

Hansen, B., & Kautz, K. (2005). Grounded theory applied – Studying information systems development methodologies in practice. In *Proceedings of the 38th Hawaii International Conference on System Science,* (pp. 1-10). Waikoloa, HI: IEEE.

Hughes, J., & Howcroft, D. (2000). Grounded theory: Never knowingly understood. *Information Systems Research, 4*(1), 181–197.

Hughes, J., & Jones, S. (2003). Reflections on the use of grounded theory in interpretive information systems research. *Electronic Journal of Information Systems Evaluation, 6*(1).

Katz, J. (1983). A theory of qualitative methodology: The social system of analytical fieldwork. In Emerson, R. (Ed.), *Contemporary Field Research: A Collection of Readings* (pp. 127–148). Boston, MA: Little Brown Company.

Kroenke, D. (1992). *Management information systems.* New York, NY: McGraw-Hill.

Layder, D. (1993). *New strategies in social research: An introduction and guide.* Cambridge, UK: Polity Press.

Lehmann, H. (2010). *The dynamics of international information systems: Anatomy of a grounded theory.* New York, NY: Springer. doi:10.1007/978-1-4419-5750-4

Lehmann, H., & Fernandez, W. (2007). Adapting the grounded theory method for information systems research. In *Proceedings of the 4th QUALIT Conference Qualitative Research in IT & IT in Qualitative Research,* (pp. 1-8). Wellington, New Zealand: QUALIT.

Martins, J., & Nunes, M. (2009a). Methodological constituents of faculty technology perception and appropriation: Does form follow function? In *Proceedings of the IADIS International Conference eLearning 2009,* (pp. 124-131). Algarve, Portugal: IADIS Press.

Martins, J., & Nunes, M. (2009b). Translucent proclivity: Cognitive catalysts of faculty's preference for adaptable e-learning institutional planning. In *Proceedings of the World Conference on E-Learning in Corporate, Government, Healthcare & Higher Education ELEARN 2009,* (pp. 1359-1366). Vancouver, Canada: ELEARN.

Martins, J., & Nunes, M. (2010a). Grounded theory-based trajectories of Portuguese faculty effort-reward imbalance in e-learning development. In *Proceedings of the 9th European Conference on Research Methodology for Business and Management Studies,* (pp. 310-119). Madrid, Spain: IEEE.

Martins, J., & Nunes, M. (2010b). Polychronicity and multipresence: A grounded theory of e-learning time-awareness as expressed by Portuguese academics. In *Proceedings of the 9th WSEAS International Conference on Advances in e-Activities, Information Security and Privacy,* (pp. 32-40). Merida, Venezuela: WSEAS.

Matavire, R., & Brown, I. (2008). Investigating the use of grounded theory in information systems research. In *Proceedings of the 2008 Annual Research Conference of the South African Institute of Computer Scientists and Information Technologists in IT research in developing countries: riding the Wave of Technology,* (pp.139-147). ACM Press.

Moghaddam, A. (2006). Coding issues in grounded theory. *Issues in Educational Research, 16*(1), 52–66.

Montoni, A., & Rocha, A. (2010). Applying grounded theory to understand software improvement implementation. In *Proceedings of the 7th International Conference on the Quality of Information and Communication Technology,* (pp. 25-34). Porto, Portugal: IEEE.

Nunes, M., Alajamy, M., Al-Mamari, S., Martins, J., & Zhou, L. (2010). The role of pilot studies in grounded theory: Understanding the context in which research is done! In *Proceedings of the 9th European Conference on Research Methodology for Business and Management Studies,* (pp. 34-43). Madrid, Spain: IEEE.

Orlikowski, W. (1993). Case tools as organizational change: Investigating incremental and radical changes in systems development. *Management Information Systems Quarterly, 17*(3), 309–340. doi:10.2307/249774

Pandit, N. (1996). The creation of theory: A recent application of the grounded theory method. *Qualitative Report, 2*(4), 1–14.

Pickard, A. (2007). *Research methods in information.* London, UK: Facet Publishing.

Robrecht, L. (1995). Grounded theory: Evolving methods. *Qualitative Health Research, 5*(2), 169–177. doi:10.1177/104973239500500203

Rodon, J., & Pastor, A. (2007). Applying grounded theory to study the implementation of an inter-organizational information system. *The Electronic Journal of Business Research Methods, 5*(2), 71–82.

Seidek, S., & Recker, J. (2009). Using grounded theory for studying business process management phenomena. In *Proceedings of the 17th European Conference on Information Systems,* (pp. 1-13). Verona, Italy: IEEE.

Shannak, R., & Aldhmour, F. (2009). Grounded theory as methodology for theory generation in information system research. *European Journal of Economics. Finance and Administrative Sciences, 15*, 33–50.

Strauss, A., & Corbin, J. (1990). *Basics of qualitative research: Grounded theory procedures and techniques.* Newburry Park, CA: Sage.

Strauss, A., & Corbin, J. (1998). *Basics of qualitative research: Grounded theory procedures and techniques.* London, UK: Sage.

Strauss, A., Fagerhaugh, S., Suczek, B., & Wiener, C. (1985). *Social organization of medical work.* Chicago, IL: University of Chicago Press.

Thomas, G., & James, D. (2006). Re-inventing grounded theory: Some questions about theory, ground and discovery. *British Educational Research Journal, 32*(6), 767–795. doi:10.1080/01411920600989412

Toraskar, K. (1991). How managerial users evaluate their decision-support: A grounded theory approach. *Journal of Computer Information Systems, 7*, 195–225.

Urquhart, C. (1997). Exploring analyst-client communication: Using grounded theory techniques to investigate interaction in informal requirements gathering. In Lee, A., Liebenau, J., & DeGross, J. (Eds.), *Information Systems and Qualitative Research* (pp. 149–181). London, UK: Chapman and Hall.

Urquhart, C. (2001). An encounter with grounded theory: Tackling the practical and philosophical issues. In Trauth, E. (Ed.), *Qualitative Research in IS: Issues and Trends* (pp. 104–140). Hershey, PA: IGI Global. doi:10.4018/978-1-930708-06-8.ch005

Urquhart, C. (2007). The evolving nature of grounded theory method: The case of the information systems discipline. In Charmaz, K., & Bryant, T. (Eds.), *The Handbook of Grounded Theory* (pp. 311–331). Thousand Oakes, CA: Sage Publications.

Urquhart, C., & Fernandez, W. (2006). Grounded theory method: The researcher as a blank slate and other myths. In *Proceedings of the ICIS 2006 International Conference on Information Systems*. Milwaukee, WI: ICIS.

Urquhart, C., Lehmann, H., & Myers, M. (2010). Putting the 'theory' back into grounded theory: Guidelines for grounded theory studies in information systems. *Information Systems Journal*, *20*(4), 357–381. doi:10.1111/j.1365-2575.2009.00328.x

Van Niekerk, J., & Roode, J. (2009). Glaserian and Straussian grounded theory: Similar or completely different? In *Proceedings of the 2009 Annual Research Conference of the South African Institute of Computer Scientists and Information Technologists*, (pp. 96-103). Johannesburg, South Africa: SAICSIT.

Vasconcelos, A. (2007). The use of grounded theory and of arenas/social worlds theory in discourse studies: A case study on the discursive adaptation of information systems. *The Electronic Journal of Business Research. Methods (San Diego, Calif.)*, *5*(2), 125–136.

Walz, S. (2005). Constituents of hybrid reality: Cultural anthropological reflections and a serious game design experiment merging mobility, media and computing. In Buurman, G. (Ed.), *Total Interaction: Theory and Practice of a New Paradign for the Design Disciplines* (pp. 122–141). Boston, MA: Birkhauser. doi:10.1007/3-7643-7677-5_9

Chapter 9
Creating Interpretive Space for Engaged Scholarship

Gabriel J. Costello
Galway-Mayo Institute of Technology, Ireland

Brian Donnellan
National University of Ireland – Maynooth, Ireland

ABSTRACT

The purpose of this chapter is to argue that the approach of engaged scholarship provides interpretive space for practitioners who are introducing change in their organization. In this case, the change involved implementation of process innovations, which continue to be an important challenge for business and public sector bodies. The research domain was a subsidiary of APC by Schneider Electric located in Ireland and involved a two-year study where the principal researcher had the status of a temporary employee. A new form of Action Research (AR) called dialogical AR was tested in this study. Key finding from an analysis of the interviews showed that the approach was both helpful and stimulating for the practitioner.

INTRODUCTION

The purpose of this chapter is to argue that the approach of engaged scholarship (Mathiassen & Nielsen, 2008; Van de Ven, 2010, 2007) provides interpretive space (Lester & Piore, 2004) for practitioners who are introducing change in their organization. In this case, the change involved implementation of process innovation which continues to be a salient challenge for business and public sector bodies (Baldwin & Curley, 2007; Brynjolfsson & Saunders, 2009; Chesbrough, 2006; Dodgson, Gann, & Salter, 2008; Pavitt, 2005; Smith, 2006; Tidd, Bessant, & Pavitt, 2005; Vanhaverbeke & Cloodt, 2006; von Hippel, 2005). The research domain is APC

DOI: 10.4018/978-1-4666-2491-7.ch009

Copyright © 2013, IGI Global. Copying or distributing in print or electronic forms without written permission of IGI Global is prohibited.

Ireland, formerly a subsidiary of the American Power Conversion (APC) Corporation. APC entered a major period of transition in the first quarter of 2007 with completion of its acquisition by Schneider Electric and the formation of a new subsidiary called APC (by Schneider Electric). The work involved a two-year study of innovation where the principal researcher had the status of a temporary employee.

The research question can be outlined as follows: how can engaged scholarship assist the provision of interpretive space for a firm undergoing a process of change? The reason for undertaking this work is to provide an empirical example of an action research study that employed the theoretical concepts of interpretive space and engaged scholarship. It aims to be pertinent to other researchers that are grappling with the perennial challenge of meeting the dual objectives of rigor and relevance in their work. Furthermore a new form of Action Research (AR) called dialogical AR proposed by Mårtensson and Lee (2004) is tested in this study. Such work we believe is an important addition to these debates which continue to be high on the research agenda (Benbasat & Zmud, 1999; Davison, Martinsons, & Kock, 2004; Lee, 1989; Zmud, 1996). The main objective of the chapter is to make the following contributions: providing evidence that Engaged Scholarship (ES) provides a milieu where researchers and practitioners can work together to develop an innovative organization, testing a novel form of action research and exploring the concept of interpretive space in an empirical setting. The chapter is organized as follows. Firstly a literature review is presented on the topics of interpretive space, engaged scholarship and dialogical action research. Then an overview is given of the case in which the study is based. After this the research approach, data collection and analysis is outlined. Finally, there is a discussion of the implications of the work for research and practice together with the main conclusions of the study.

BACKGROUND

This section of the chapter will explain the main concepts used in this study namely: interpretive space, engaged scholarship, and dialogical action research.

Interpretive Space

Lester and Piore (2004) undertook a series of field studies of new product development in cellular telephones, medical devices and clothing as part of a research program at the MIT Industrial Performance Centre from 1994 to 2002. A central conclusion of their studies was that the two dominant theses on the reason for the American boom of the 1990s did not adequately explain the economic renaissance. One of these theses argued that the expanding reach of market competition and the role of entrepreneurship fuelled the economy. A contrary view was that the radical changes in organisational structures and management practices were the chief drivers of the boom. However, they propose that it is "necessary to understand what actually happens when firms innovate" (p. 5). They concluded from their studies that the ability to innovate in the threefold manner of generating new products, improving existing ones and implementing more efficient supply chains depends on two fundamental processes which they term analysis and interpretation. The analytical process is essentially rational decision making and works best in situations where alternative outcomes are well understood and can be clearly defined. The approach is essentially problem solving where the problem is divided into a number of discrete and separable components each of which are assigned to a designated specialist. However, from their research they concluded that innovation was not all about problem solving but about a process, which they call interpretation. This process less tangible and identifies the role of the manager as "initiating and guiding conversations between individuals and groups" (p. 8) rather than problem solving or negotiation. Furthermore they content that the analytical viewpoint dominates both in the academic debates of the business and engineering schools and in managerial practice. Their cited aim is not to replace the analytical perspective but to develop the interpretive approach so that it stands alongside analysis in the cultivation of innovation. They propose that innovative firms must "continually seek out and participate in exploratory, interpretive conversations with a variety of interlocutors" (p. 9). Furthermore, they are concerned that such interpretive spaces are more difficult to find in a globalised, unregulated, and technologically changing environment. These interpretive spaces do not appear naturally in market economies but must be continually "created, cultivated, renewed, and enriched." In their field studies, they found that interviewees had difficulty in characterising and describing the non-analytical aspects of the innovation process. This activity sounded "an awful lot like a conversation" (p. 51) which were inherently ambiguous but from this ambiguity new ideas emerged. It is worth quoting their description of this process:

Interpretation is an open-ended process, on-going in time, perhaps with a beginning but with no natural end. Unlike people engaged in problem solving, the participants in a conversation often have no idea where their discussion is going when it starts; and even if they do, the actual direction may turn out to be quite different (Lester & Piore, 2004, p. 53).

The main differences between the analytical and interpretive views of product development are summarised in Table 1.

In new product development these analytical aspects and interpretive aspects "exist in perpetual tension" (Lester & Piore, 2004, p. 121) and resolving this tension is a central challenge for innovation managers. The answer to this dilemma, they contend, is the creation of public spaces, which in their research involved four areas:

- The interior of the firm itself.
- Industrial districts.
- The regulatory process.
- The university.

Lester and Piore (2004) reiterate their belief that the university is "unquestionably the most important public space for research and development" (p. 153). They quote Rosalind Williams former dean of students at MIT:

In a quite hard headed way, we argued that the sources of creativity, necessary to engender change, technological or otherwise, flourish only in a setting with time and space for the intense social interactions that are at the heart of both research and learning (p. 154).

Table 1. Analysis and interpretation from Lester and Piore (2004, p. 97)

Analysis	Interpretation
The focus is a project with a well defined beginning and end The thrust is to solve problems Managers set goals Managers convene meetings and negotiate to resolve different viewpoints and eliminate ambiguity Communication is the precise exchange of chunks of information (bits and bytes) Designers listen to the voice of customers Means and ends are clearly distinguished, and linked by a causal model	The focus is a process which is ongoing and open-ended The thrust is to discover new meanings Managers set directions Managers invite conversations and translate to encourage different viewpoints and explore ambiguity Communication is fluid, context-dependant, undetermined Designers develop an instinct for what customers want Means and ends cannot be clearly distinguished

Engaged Scholarship

The discipline of information systems has been considered to have certain failings in its effort to impact on practice (Kawalek, 2008). There have been numerous research studies identifying failures in IS in its attempts to achieve desired outcomes and disappointments in assessments of return on investment (Lam & Chua, 2005; Pan, 2005). The analyses in these studies often yield recommendations that operate at a high level of abstraction and lack the detail and specificity to lead to action-oriented solutions. Examples of such recommendations include (Public Accounts Committee Report, 2000):

- Commitment of senior management is critical.
- End-user must be identified and involved in the development process.
- Lack of clarity in the project specification can lead to lead to expensive misunderstandings subsequently.
- Organizations must learn lessons from previous projects undertaken.
- Training must address the needs of users, as well as those operating and maintaining the system

Such findings, while offered in a constructive spirit of helpfulness and concern for continuous improvement, do little to advance either (1) the capability of practitioners to achieve their goals or (2) the theoretical knowledge underpinning information systems academic research. One of the requirements for a more helpful approach is a greater sensitivity to the contextual complexity of the organizational problem-solving environment where IS practitioners work.

Van de Ven describes engaged scholarship as a participative form of research for obtaining the views of key stakeholders to understand a complex problem (Van de Ven, 2007, 2010). By exploiting differences between these viewpoints, he argues that engaged scholarship produces knowledge that is more penetrating and insightful than when researchers work alone. Engaged scholarship has a number of facets; a form of inquiry where researchers involve others and leverage their different perspectives to learn about a problem domain; a relationship involving negotiation, mutual respect, and collaboration to produce a learning community and an identity of how scholars view their relationships with their communities and their subject matter. In Van de Ven's view, you can increase the likelihood of advancing knowledge for science and practice by engaging with practitioners and other stakeholders in four steps;

1. Grounding the problem or research question in a real-world scenario.
2. Address the situation by developing a range of theories.
3. Collection of evidence.
4. Application and dissemination of the findings.

According to Van de Ven's (2007, pp. 10-11) schema, there are four stages in an Engaged Scholarship project. The stages can happen in any sequence. The stages are:

- **Problem Formulation:** Situate, ground, diagnose, and infer the research problem by determining who, what, where, when, why, and how the problem exists up close and from afar.
- **Theory Building:** Create, elaborate, and justify a theory by abductive, deductive, and inductive reasoning where abductive reasoning is for conceiving or creating a theory, deductive reasoning is for constructing or elaborating a theory and inductive reasoning is for justifying or evaluating a theory.
- **Research Design:** Develop a variance or process model for empirically examining the alternative theories.

- **Problem Solving:** Communicate, interpret, and apply the empirical findings on which alternative models better answer the research question about the problem.

Mathiessan and Nielsen (2008) see engaged scholarship as a grand opportunity to address key challenges within the IS discipline in a novel and constructive way. They applied the principles of engaged scholarship to analyze Scandinavian IS research through the lens of Scandinavian Journal of Information Systems (SJIS). After reviewing all the research papers published in SJIS over the past 20 years; they advocated a role for engaged scholarship in shaping the future of Scandinavian IS research and IS research and practice in general.

Van de Ven locates Action Research within the scope of Engaged Scholarship (Van de Ven, 2007, p. 27) and identifies four forms of the approach:

1. Informed basic research is undertaken to describe, explain, or predict social phenomenon.
2. Collaborative basic research entails a greater sharing of power and activities among researchers and stakeholders than informed research.
3. Design and evaluation research is undertaken to examine normative questions dealing with the design and evaluation of policies, programs, or models for solving practical problems of a profession in question.
4. Action/intervention research takes a clinical intervention approach to diagnose and treat a problem for a specific client.

Now we will provide an overview of the action research carried out in this study.

Dialogical Action Research

Action Research (AR) originated from the work of Kurt Lewin during the 1940s and has been summarised as an approach that "combines theory and practice (and researchers and practitioners) through change and reflection in an immediate problematic situation within a mutually acceptable ethical framework" (Avison, Lau, Myers, & Nielsen, 1999, p. 94). The application of AR has not been without controversy particularly in debates with positivist science on the justification and generation of knowledge. These arguments were addressed by Susman and Evered (1978) in their influential description of AR as consisting of a cyclical process involving five phases: diagnosing, action planning, action taking, evaluating, and specifying learning. The focus of AR is to address real-life problems through intervention together with the research objective of making a contribution to knowledge. Coghlan and Brannick (2005, p. 125) emphasise the importance of the social and academic context in which action research is carried out.

Dick (1993), an academic working in the field of psychology, proposes that the AR methodology has the twofold aim of action and research:

- Action designed to bring about change in some community, organization or program.
- Research to increase understanding on the part of the researcher or the client, or both—and in many cases some wider community.

Reason and Bradbury aim to "draw together some of the main threads that form the diverse practices of action research" and propose an almost lofty vision of AR contributing to the world's wellbeing and sustainability; in areas ranging from the economic and political to the psychological and spiritual. The following quotation with its emphasis on understanding and reflection is of particular relevance to this study.

So action research is about working towards practical outcomes, and also about creating new forms of understanding, since action without reflection and understanding is blind, just as theory without action is meaningless. Now a recent addition to the action research portfolio which was tested in this study will be presented (Reason & Bradbury, 2001, p. 2).

Mårtensson and Lee (2004) have suggested and described a new form of action research called dialogical AR. In dialogical action research, the scientific researcher does not "speak science" or otherwise attempt to teach scientific theory to the real-world practitioner, but attempt to speak the language of the practitioner and accepts him as the expert on his/her organization and its problems. Mårtensson and Lee (2004) propose that "reflective one-to-one dialogues" between the practitioner and the researcher; that take place at regular intervals in a location removed from the organisation; can help the manager to "reflect on, learn from, and remedy managerial problems in the organization." In their schema, the role of the researcher consists in suggesting actions based on one or more theories taken from their discipline.

The implementation of these suggestions is left to the judgment of the practitioner based on his experience, expertise, and tacit knowledge together with his reading of the organisational situation that confronts him. Furthermore, the ongoing dialogue is presented as an interface between the scientific world of the researcher, marked by theoria and everyday world of the practitioner, which is marked by praxis.

The overall aim of dialogical AR is to bring about some improvement to the real-world problem of the practitioner while at the same time contributing to the development, confirmation, or disconfirmation of theory by the researcher. Mårtensson and Lee draw heavily on Schön's model of professional inquiry (p. 510) consisting of a pattern of five features: situation requiring attention; a surprising response; reflection-in-action; critical examination and restructuring; and an "on-the-spot experiment." They make a fundamental distinction between traditional forms of consulting and dialogical action research in that the latter always involved reflection and learning. Furthermore action research-unlike consulting- involves someone who has academic expertise rooted in some scientific discipline; where teamwork takes place between researcher and practitioner and where "negative feedback" is seriously taken on board.

It is incumbent on the researcher, according to Mårtensson and Lee (2004, p. 514) to "explicitly and intentionally acquire an understanding of the social and historical context of the organization and its problems." This was carried out in the first year of the study undertaken in this work. Mårtensson and Lee take two concepts: the scientific attitude and the natural attitude of everyday life to form four features which differentiate dialogical AR from existing forms of action research. They are: adopting the scientific attitude; adopting the natural attitude of everyday life; accepting the role played by social and historical context; understanding the role played by social and historical context. As regards the philosophical underpinnings, they classify dialogical AR as viewing reality through social constructionist lens and the phenomenology of Schutz (1962) in Mårtensson and Lee (2004, p. 514).

In their vision of dialogical AR, the scientist makes suggestions to the practitioner but the practitioner remain the "agent of action" using his or her explicit and tacit knowledge (p. 515). Furthermore, they see the role of the researcher having the following attributes in the one-on-one dialogues: firstly, to listen in order to identify the problem that requires some action; secondly, to gather the facts to form the basis of deciding what suitable theory can be applied to the problem area; and thirdly, to suggest and monitor appropriate actions to the practitioner.

Interestingly, for this study they use the analogy of an anthropologist spending a yearlong ethnography to understand the world of the natives i.e. the practitioner. Mårtensson and Lee insist on the distinction between the practitioner and the scientific researcher and posit that ultimately it is

Figure 1. Dialogical action research (adapted from Mårtensson & Lee, 2004)

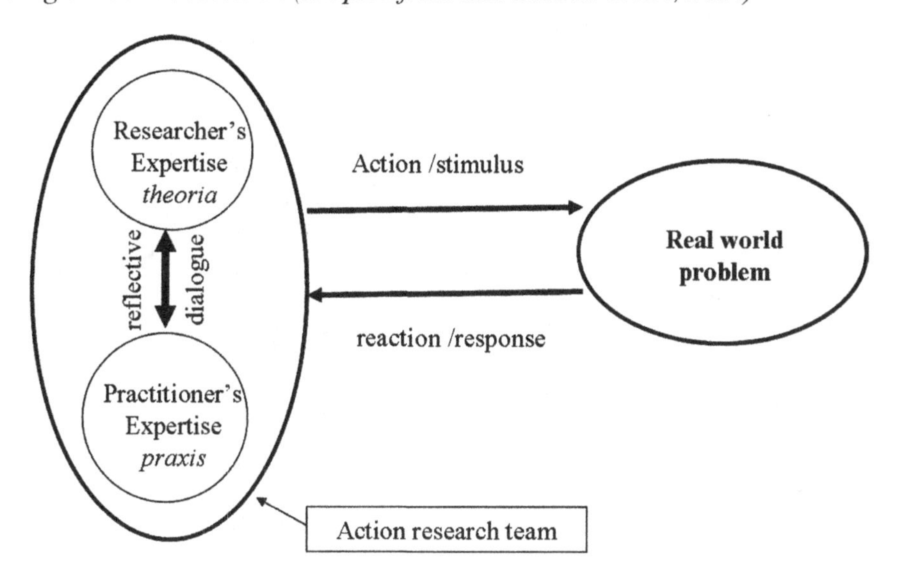

up to the practitioner to decide on the effectiveness of the action in solving or remedying the problem while it is up to the researcher to decide if the theory been tested is conformed or not. Importantly the authors contend that the theoria of the researcher and the praxis of the practitioner are "simply different forms of knowledge" and cannot be labeled as better or worse. (p. 517). The dialogical action research process is presented in Figure 1.

In order to evaluate dialogical AR they suggest three criteria that are outlined in Figure 2 (p. 519):

- The practitioner considers the real world problem to be solved or remedied satisfactorily.
- There had been an improvement in the practitioner's expertise.
- There has been an improvement in the researcher's expertise.

STUDY OVERVIEW

Kurt Lewin is famous for his assertion that "there is nothing as practical as a good theory." However we will take the aphorism of his student Bronfenbrenner (2005, p. 48) who reversed the classical Lewinian maxim to read: "There is nothing like the practical to build a good theory" (p. 48). This is part of a tradition that goes back to Aristotle "who made frequent reference to concrete example to illustrate his theoretical points" (Kenny, 2010). Consequently, we will outline the empirical study as we grappled with the topic of innovation in a multinational company and reflected on the role of the practitioner and researcher in the process.

This study is based in APC Ireland, formerly a subsidiary of the American Power Conversion (APC) Corporation. APC entered a major period of transition in the first quarter of 2007 with completion of its acquisition by Schneider Electric and the formation of a new subsidiary called APC (by Schneider Electric). The strength of the MIS function in APC was viewed as an important advantage by Schneider in their acquisition analysis and APC's "intimacy with information technology" was identified as central to the creation of synergies with Schneider's power solutions subsidiary MGE.

As the main part of this study was developed before the acquisition, this section will focus on providing a background to the APC context in which the work was carried out (APC, 2011). APC designs, manufactures, and markets back-

Figure 2. Improvements over time (adapted from Mårtensson & Lee, 2004)

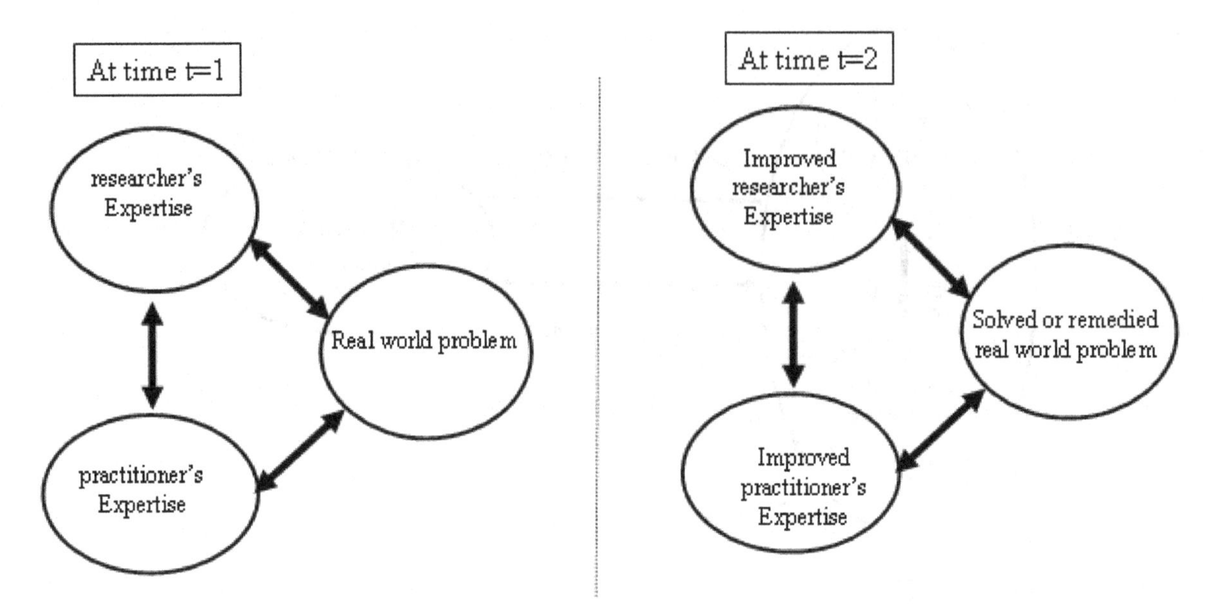

up products and services that protect hardware and data from power disturbances. The explosive growth of the Internet has resulted in the company broadening its product offerings from Uninterruptible Power Supplies (UPS) to the high-end InfraStruXure™ architecture in order to meet the critical availability requirements of Internet Service Providers (ISP) and data-centers. This modular design integrates power, cooling, rack, management, and services, which allows customers to select standardized modular components using a Web-based configuration tool. The Corporation reported sales of $2 billion in 2005, globally employs approximately seven thousand people and is a Fortune 1000 company. APC aims to set itself apart from the competition in three areas: financial strength, innovative product offerings and efficient manufacturing (Results APCC, 2006).

However, financial reports stressed that the company needs to implement significant improvements in manufacturing and the supply chain (Results APCC, 2005, 2006). According to these published reports, the company must work to develop a "lean, customer-centric, ambidextrous organization" in order to reach "optimal efficiencies in our processes." APC had two locations in the West of Ireland that serve EMEA region. The Manufacturing Operations site, based in Castlebar, employed approximately 100 people; and a number of functions including sales, information technology, business support and R&D are situated in Galway with workforce of approximately 300.

The widening of a focus from the manufacturing of discrete products, such as UPS, to the delivery of customized InfraStruXure™ solutions provides both challenges and opportunities for the Operations function. Responding to the supply chain challenge, a Lean Transformation Project was set up in the Castlebar campus in February 2006 with a cross-functional team of twelve members drawn from Management, Engineering, Manufacturing, Materials Planning, and Quality and Logistics functions. The Lean Project team set an objective to quickly deliver the message that Ireland is responding to, and leading, the corporate initiative and to provide a platform for the Irish subsidiary to obtain a reputation as an innovative location. An initial corporate feedback is that this project is "ahead of the curve" in terms of the other regions. A major requirement from corporate executives was that any innovations resulting from the initiative could be replicated in other regions.

APC Ireland is keen to take the leadership role in enhancing its global competitiveness by becoming a knowledge leader in the area of supply chain expertise. The manufacturing practices and processes used within the corporation have come under increased pressure from global competition. In addition, building up a lean enterprise is treated as a corporate-wide task. Now we will outline the research approach undertaken in the study.

RESEARCH APPROACH

The research approach utilized in the longitudinal study of innovation management in the Irish subsidiary is now presented.

Benbasat and Zmud (1999) conclusions concerning the lack of relevance in management research were, to put it mildly, a criticism of the discipline. Consequently, the initial approach to the case study was closely related to the following recommendation in their paper:

Researchers should look to practice to identify research topics and look to the literature only after a commitment has been made to a specific topic (p. 8).

However, the linear nature of their recommendation does not sufficiently accommodate the dynamics of a real-world corporate environment so this study adopted a more iterative approach, going from practice to literature in a continuous cycle. The study is presented from the perspectives of a researcher undertaking a longitudinal study of innovation management in the Irish subsidiary with the back up of colleagues in the research area. Slappendel (1996) recommends using a research team approach to overcome limitations when examining innovation in organizations from the interactive process perspective.

The innovation project consisted of two main phases: an ethnographic study during 2006 followed by dialogical Action Research in 2007. It is notable that Mårtensson and Lee (2004, p. 515) advocate that the researcher, akin to an anthropologist, spends a yearlong ethnography to understand the world of the practitioner. Data collection methods during the ethnographic phase involved: maintaining a log book, reviewing documents and information systems, records, interviews, observations (direct and participant), artifacts and surveys in order to develop a database and body of evidence (Gillham, 2000; Yin, 1994). A total of 29 unstructured or open interviews were undertaken that involved approximately 60 hours of interview time and 24 days spent in the company sites. The interviews were conducted across a wide area of the organization that included: Senior Managers with global, EMEA (Europe, Middle East, and Africa) and site responsibilities, Middle-Managers, Team Leaders, Engineers and a number of people in general planning roles. Furthermore, the researcher had the status of a temporary employee with his own email address and intranet access.

Dialogical Action Research Phase

The approach of Mårtensson and Lee (2004) described earlier was adopted during the second phase of the research. One persistent bone of contention has been the "paucity of methodological guidance" for conducting and evaluating action research studies. This resulted in the authors adopting a number of principles proposed by Davison, Martinsons, and Kock (2004) to enhance the rigor of the research process. These were as follows: the Principle of the Researcher-Client Agreement (RCA), the Principle of the Cyclical Process Model (CPM), the Principle of Theory, the Principle of Change through Action, and the Principle of Learning through Reflection.

Data Collection

There was an agreement in January 2007 to move forward using dialogical Action Research with meetings every two weeks. In their paper, Mårtensson and Lee (2004) propose that "reflective dialogues outside the organization can help the manager to reflect on, learn from, and remedy managerial problems in the organization." In particular, the discipline of having to take regular timeout in a time-pressured manufacturing environment was a major incentive for the Plant Manager to agree to this approach. The Plant Manager also considered the framework advantageous since it allowed him to retain control and responsibility for all decisions, implementations and communications. However, there are a number of practical risks with this type of longitudinal research in a dynamically changing corporate environment, such as the realities of reorganizations and relocations that are not pointed out by Mårtensson and Lee.

In addition to the above there were 11 meetings with the main point of contact for the project, which totaled seventeen hours in duration. These meetings became the basis for the dialogical AR approach during the second phase of the project. Data collection during the dialogical AR period involved recording of the meetings, which were subsequently transcribed verbatim by the researcher. Given the rich nature of the data, this was considered the optimum way of capturing the reflective meaning and ensuring consistent interpretation. Analysis was done manually through the examination of each meeting transcript and providing a summary of the topics discussed in the transcripts. This then was sent to the plant manager for evaluation and agreement that it was an accurate portrayal of the meeting. In total these

Table 2. Data collection summary

Number of Formal Interviews Estimated hours	22 34.5
Meetings with main point of contact (additional to above) Estimated Hours	11 17
Dialogical Action Research Meetings Estimated Hours	16 22.5
TOTAL INTERVIEW HOURS	74
TOTAL DAYS ON SITE	42
Additional detailed discussions Estimated Hours	8 18.5

transcripts ran to over 60,000 words. A profile of the interviews is set out in Table 2.

The data gathered from the interviews were by their nature subjective and hence open to interview bias. However, the broad range of interviewees was an attempt to get various perspectives across the organisation. As pointed out by Howcroft (1998) in a similar situation, this was not a positivistic study that wished to claim scientific objectivity but that "any values that are invoked are those that inform the theoretical perspective" (p. 123).

Data Analysis

Qualitative data focuses on words rather than the numbers of quantitative data and there has been a major expansion of qualitative enquiry over the last twenty years (Miles & Huberman, 1994). Furthermore, it is having an increased influence on the management discipline (Kaplan & Duchon, 1988; Lee, 2001; Myers, 1997; Trauth, 2001). A number of different methods can be adopted to analyse interviews such as "content, narrative and semiotic strategies" (Denzin & Lincoln, 2008, p. 34) while Myers and Newman (2007) caution that the interview has remained a relatively "unexamined craft" (p. 2).

Gummesson (2000) points out that deductive research "primarily tests existing theory" while inductive research "primarily generates new theory" (p. 64). Glaser and Strauss development of grounded theory was part of a significant growth in qualitative analysis during the 1960s and 1970s. Indeed Locke (2000) concurs that grounded theory was one of a number of methods that "attempted to bring more formalization and systematization to qualitative methods" (p. 12). In grounded theory the theory emerges during the research study and is "the product of continuous interaction between analysis and data collection" (Goulding, 2002, p. 42). Charmaz (2004) describes the fundamental premise of grounded theory as letting the "key issues emerge rather than to force them into preconceived categories" (p. 516). Also the approach in grounded theory is to let the "codes emerge as you study your data" (p. 506). Robson (2002, p. 59) describes qualitative analysis as being much closer to "codified common sense" than to the "complexities of statistical analysis" associated with quantitative data and provides the following typology from the work of Crabtree and Miller (1992, p. 457):

1. Quasi-statistical methods.
2. Template approaches.
3. Editing approaches.
4. Immersion approaches.

This follows a progression from a more structured approach to a less formal approach. In fact, there is a debate whether the immersion approach can be classified as a scientific method *per se*. The editing approach which is closest to the method employed in this study is characterised by being interpretive and flexible with no or few *a priori* codes. In this method "codes are based on the researcher's interpretation of the meanings or patterns in the texts" (p. 458) and it is typified in grounded theory approaches.

The analysis techniques adopted in this research consisted in a number of mechanisms that dovetail with the following methods described by Miles and Huberman (1994, p. 51):

- Contact summary sheet.
- Memoing.
- Interim case summary.

However, the basic approach of phenomenology, as these authors point out (p. 8), involves working with the entire interview transcripts, where coding is not normally used, in order to reach the "*Lebenswelt*" or life-world of the interviewee.

A contact summary sheet is "a single sheet focusing or summarising questions about a particular field contact" in order to develop a synopsis "of the main points in the contact" (p. 51). In this study it involved the transcription of the interview with the practitioner and then summarising the main themes by placing them in the following "bins": purpose of the meeting, summary of the main points of the meeting, actions arising out of the meeting and finally the agreed agenda for the next meeting. The direct tape recordings of the field events were transcribed into text and then the process involved making "notes, selecting excerpts, and making judgements" (p. 51). The contact summary sheet was placed as the cover sheet of each transcribed interview so the information was available for review. In the action research report, the excerpts from the dialogical research were presented in terms of the topics that emerged from the analysis.

Memoing is a method that took a classic form in the work of Glaser (1978) and involves attempting to stand back and to "make deeper and more conceptually coherent sense of what is happening." They are "primarily conceptual in intent" and strive to "tie together different pieces of data into a recognisable cluster" (Miles & Huberman, 1994, p. 72). In this study the researcher used "memoing" to gather together some of the concepts that were emerging from the interviews. These conceptual memos were of the more "elaborate variety" (p. 74) and formed the basis for writing the data into academic papers that involved crystallising of ideas emanating from the research.

An Interim Case Summary "provides a synthesis of what the researcher knows about the case" and "also indicates what needs to be found out." It involves pulling together what is known about the case (p. 79). This was done at various stages of the study—during and after the pilot study, during the various phases of the action research cycles- and together with the memoing discussed above became the basis for academic papers reporting the research.

A Supplementary Case Study of Engaged Scholarship

The discipline of Information Systems (IS) has been considered to have certain failings in its effort to impact on practice (Kawalek, 2008). Additionally, Sambamurthy and Zmud (2000) noted that there is a growing gap between scholarly research and the need of practitioners. There have been numerous research studies identifying failures in IS in its attempts to achieve desired outcomes and disappointments in assessments of return on investment (Lam & Chua, 2005; Pan, 2005). The analyses in these studies often yield recommendations that operate at a high level of abstraction and lack the detail and specificity to lead to action-oriented solutions. Such findings, while offered in a constructive spirit of helpfulness and concern for continuous improvement, do little to advance either (1) the capability of practitioners to achieve their goals or (2) the theoretical knowledge underpinning Information System academic research. One of the requirements for a more helpful methodology is a more systematic approach with greater sensitivity to the contextual complexity of the organizational problem-solving environment where IS practitioners work.

The development of the IT-CMF (The Information Technology Capability Maturity Framework) (Curley, 2004, 2007) is a response to the need for a more systematic, comprehensive approach to managing IT in a manner that meets the requirements of practicing IT professionals. This research is being undertaken by the Innovation Value Institute (www.ivi.ie). Applying the principles Design Science Research (DSR) (Hevner, March, Park, & Ram, 2004), IT Management is being investigated using a design process with defined review stages and development activities based on the DSR guidelines advocated by Hevner *et al.* During the design process, researchers participate together with practitioners and subject matter experts within research teams to capture the working knowledge, practices and views of key domain experts.

Developing innovative artifacts is a central activity in DSR (Vaishnavi & Kuechler, 2004). Such artifacts can be in the form of constructs, models, methods or instantiations. For the construction of such artifacts, two basic activities can be differentiated: build and evaluate where building "is the process of constructing an artifact for a specific purpose" and evaluation "is the process of determining how well the artifact performs" (p. 254). The construction of an artifact is a heuristic search process. Within this process an extensive use of theoretical contributions and research methodologies stored in the knowledge base should be made. On the one hand theoretical contributions can come from governance, value based management, risk management, compliance management, etc. to build an artifact, i.e. the situational method. The IT-CMF uses the following DSR patterns proposed in Vaishnavi and Kuechler (2007).

Different Perspectives: The research problem is examined from different perspectives, e.g. conceptual, strategic, organizational, technical, and cultural.

Interdisciplinary Solution Extrapolation: A solution or solution approach (i.e. methods, instructions, guidelines, etc.) to a problem in one discipline can be applied in or adapted to the integrated IT CMF.

Building Blocks: The complex research problem of IT Management is broken into thirty three critical competencies that are examined in turn.

Combining Partial Solutions: The partial solutions from the building blocks are integrated into the overall IT CMF and the inter-dependencies between the building blocks are identified and highlighted. In order to rigorously demonstrate the utility of the developed artifact, different evaluation methods can be used. Amongst others, the "informed argument" is suggested as an appropriate evaluation method.

DISCUSSION AND FINDINGS

The rigor versus relevance debate continues to be a matter of lively discussion in the information systems discipline (Avison, Fitzgerald, & Powell, 2004; Benbasat & Zmud, 1999; Cranefield & Yoong, 2007; Davison, et al., 2004; Dubé & Paré, 2003; Lee, 1989; Zmud, 1996). Dialogical action research provides a new approach to the challenge of engaged scholarship. It is especially suitable when the practitioner seeks to retain control of the implementation of the project. However, dialogical AR is relatively untested and this study is intended to contribute to debate on the approach. Referring to Van de Ven's alternative forms of engaged scholarship this research fits the quadrant "action research for a client." It can also be argued due to the publications that arose from the study that it also fits the quadrant "co-produce knowledge with collaborators."

As the purpose of this chapter is to demonstrate the value of engaged scholarship to a practitioner, we will now provide excerpts from the dialogues, which illustrate the value of providing interpretive space. The dialogues allowed the practitioner space to reflect on projects that were not suitable for the short-term bounded tasks typically associated with consultancy. Mårtensson had used his office as the interpretive space completely away from the operational environment. This would be the ideal situation but was not feasible in the project described in this study. However, it was possible to find a virtual interpretive space within the four walls of a location. Here are the words of the practitioner in this regard:

'I see a great value in this research by forcing me to take time out for reflection.' He then told me about someone who left a leading multi-national company to become a Six Sigma trainer because he had 'no time to think' in his last job.

He saw value in the framework for taking regular time out as this is very difficult in a manufacturing environment. He also considered the framework suitable as he retained control and responsibility for all decisions, implementations, and communications

The practitioner had this to say on the question of learning from the project.

So do we want to talk about what the learning has been since the start of the project?

For me if I reflect back on the last twelve months I would say that before we started this an innovative culture was something that you could almost not define: it was just an airy fairy type of concept. After having gone through the process over the last while being able to define a structure that helps support an innovative culture has been a key point. That gives you some concrete steps that can be taken to develop this innovative culture rather than floundering around hoping that something will develop because you don't seem to know what you are trying to develop.

Furthermore, the engaged scholarship provided the practitioner with the filtering of information.

At the basic level, the real benefit of bringing the literature to me is that: people from the academic world read lots—that's what they do. However, being able to filter it down to the likes of me who might want to read but does not have all that time is important. [This] is of huge value because it means that I do not have to read through the hundred pages of stuff to get the five pages of real value.

The most significant one was this book 'Managing Innovation' and you pointed me to chapter 11 in it. Therefore, I did not have to read through the first 466 pages which meant I got straight into this. It was one hundred per cent the type of stuff I was looking for.

A particular advantage was the discipline of regular meetings.

Another thing I find of benefit is that this stuff is the long-term benefit that we are working on. We will not see the benefit next week or next month and therefore oftentimes you get so consumed with the day to day issues on the floor that you just neglect this: so that having the regular meetings I find it puts a wee bit of pressure on me at least to have some updates to report. That is a positive thing, and if that wasn't there it would get put off and put off so it never gets done. I am going to show you the excel spreadsheet. The only reason that got done was that I did it yesterday because [the researcher] would be here tomorrow.

One of the outcomes of the project was the development of a conceptual model of innovation in the form of a Dolmen. A "Dolmen" (from the Breton for "stone table") is an ancient monument, found in many areas of Europe, consisting of two or more upright large stones that support a horizontal capstone. A well-known Irish example is found in Poulnabrone, Co.Clare that dates from the Neolithic period, about 3,000 B.C. The model shown in Figure 3 was developed from the engaged scholarship that took place in the interpretive spaces established during the project.

- **Cultural Context:** the "Dolmen" analogy is familiar in the West of Ireland and also it gave the message of putting in place a system that will endure.
- The concept seeks to impart the message that the innovation "climate" is supported by a number of critical dimensions.
- Some dimensions are more important than others e.g. strategy and processes being critical.
- Some attributes closely depended on each other e.g. strategy and organisation.

Figure 3. The innovation "dolmen" developed from the preliminary concept

Abbreviations

C = Innovation Culture/Climate
O = Organisation
S = Strategy
L_k = Links
L_r = Learning
I_{st} = Information Systems/ Technology
E_p = Environmental Performance
P_{sb}= Supply Chain /Business processes

- The IS/IT dimension was replaced by the more general term; technology management.

It was also considered significant that developing the "Dolmen" was aligned with Mintzberg's idea of "strategy as craft" and consistent with his identification of the multiple facets of strategy (Mintzberg, 1987a, 1987b).

The practitioner noted:

[This is what]I've said about the Dolmen concept: how it gives me a tangible idea to work on when dealing with the very intangible concept of innovation.

Let me see if I can remember them (the main pillars of the Dolmen). They are strategy, learning, linkages, organisation, and there was process, and we also had IS [This showed that the Dolmen concept allowed the practitioner to carry around the main concepts in his head]

End of project reflections

On reflection for me the strategy part was the most important part of the Dolmen. [It] gave direction and clear focus to people - the other pillars can follow afterwards (and in a defined order as well).

[The] 5 pillars do not need to be built simultaneously (in fact it may be impossible). Step 1 has to be strategy. After that comes defining the organization followed closely by the processes for how the organization should work. After that becomes making sure you are a learning organization (processes in place to support) and part of that is developing linkages which defines a clear sequence for how the Dolmen can get built. We did the strategy but did not do a good job on the others, as I believe we were trying to do too much all together.

Overall, I guess I would have liked to have been further along in building the Dolmen but I guess with the day to day pressures on numbers etc. it is difficult to take time out to develop this properly. That is where I found our meetings most beneficial in that it forced the time out for us to at least consider the direction we are heading in. You could say why don't I just schedule time out on my own but you know yourself unless you have made a commitment to somebody else it is simply too easy to let the day to day item take priority over the time out. Maybe that can be my new year's resolution—commit to taking time to reflect and evolve the development of an innovative culture using our Dolmen model (one pillar at a time!).

The work outlined in this paper provides empirical evidence of the utilization of engaged scholarship recently compiled in a detailed form by Andrew Van de Ven (2007). We believe the Van de Ven's thesis is an important contribution to the rigor vs. relevance debate. Consequently this work makes a contribution by: presenting evidence that Engaged Scholarship (ES) provides an environment where researchers and practitioners can explore and develop an innovative organization, by testing a novel form of action research (dialogical AR), and exploring the concept of interpretive space in an empirical setting. We believe it is an original approach to a real-life phenomenon of interest to researchers and practitioners. The work is limited in that it is a single case study. However the argument for the relevance of a single case study has been defended in the literature (Lee, 1989; Mintzberg, 1979). The work could be improved by further case studies that consider both analysis and interpretation in a company's innovation journey.

Advantages and Disadvantages of the Research Approach

The following advantages of the research approach were identified:

- The methodology meets the ongoing call for more relevant and pragmatic research programs that engage with practitioners (Ågerfalk, 2010; Baskerville & Myers, 2004; Benbasat & Zmud, 1999; Maurer & Githens, 2010; Zmud, 1996).
- The dyadic interaction provides a conduit for academic theory and literature to a real world scenario. Such a model can help the perception that much academic activity is not relevant to practice.
- The methodology allows the practitioner to control many of the practical aspects of the operations surrounding the research such as communications and implementations. This is a very significant factor for some practitioners who otherwise would be reticent to engage in a research program.
- Dialogical action research, with its continuous long-term access, provides the opportunity of carrying out a longitudinal investigation of a contemporary phenomenon and building a case-based narrative.
- Developing a relationship with a practitioner provides a milieu for testing and developing theory.
- The dialogical interaction can be very enjoyable for the researcher who has the opportunity to apply the fruits of his studies in a real-world context.
- The interaction with academia can provide the practitioner with a very motivating experience and an interaction with new ideas and fresh thinking.
- The discipline of regular meetings guards against the perpetual research problem of keeping momentum and focus during the sometimes arduous journey.
- The joint development of the Dolmen concept was beneficial to the practitioner. In particular, it allowed him to put some structure on the rather fuzzy notion of innovation. Also, because of its local and contextual nature it could be easily called into memory in the course of a discussion on the topic.

In the course of this research, a number of disadvantages of the approach became evident:

- The research approach is new with only one published paper concerning the topic. More field studies are required in order for the methodology to mature.
- The research was confined to a single case and is open to criticisms concerning generalization. However as Kelly points out, the emphasis in this tradition is on "theoretical (or analytical) generalisation as opposed to the statistical variety" (Kelly, 2004, p. 51).
- The question of research and data bias is very salient in connection with this approach. The data gathered from the interviews were by their nature subjective and hence open to interview bias. However, the broad range of interviewees was an attempt to get various perspectives across the organization.
- The reflective one-to-one dialogue with a single individual makes triangulation more difficult but it is possible to search for supporting external evidence. For example, the enthusiasm of the practitioner for a certain process innovation was collaborated by the organization winning an external health and safety award.
- The milieu of a multinational in a time of transition is that of rapid change and organizational restructuring. The commitment to dialogue with a single individual runs the risk of the practitioner moving to another role and was the reason that the timeframe was capped at one year.
- The approach requires the building of an empathetic relationship between researcher and practitioner. It is considered that

the researcher undertaking dialogical AR requires a certain ecological validity. This consists of some familiarity with the culture and sub-cultures of the practitioner's world.

- There is a danger that the focus of the research becomes the dialogue rather than the action. The approach by its nature is demanding requiring researcher to open up and demonstrate a certain sense of humility.

CONCLUSION

This chapter has presented the case for engaged scholarship between researchers and practitioners using the example of an action research study carried out "in the field." The approach used a new form of action research called dialogical AR, which has not been used extensively. Its advantages include providing the setting for researcher practitioner interaction that allows input from *theoria* (the world of research) to *praxis* (the world of practice). Results from an analysis of the extensive interviews over the period of one year indicated that it was beneficial to a plant manager who had taken on the project of making his subsidiary more innovative. Future work was suggested to further examine field studies using the engaged scholarship model.

REFERENCES

Ågerfalk, P. J. (2010). Editorial: Getting pragmatic. *European Journal of Information Systems, 19*, 251–256. doi:10.1057/ejis.2010.22

APC. (2011). *American power conversion corporation*. Retrieved from http://www.apcc.com/

Avison, D., Fitzgerald, G., & Powell, P. (2004). Editorial. *Information Systems Journal, 14*(1), 1–2. doi:10.1111/j.1365-2575.2004.00163.x

Avison, D. E., Lau, F., Myers, M. D., & Nielsen, P. A. (1999). Action research. *Communications of the ACM, 42*(1). doi:10.1145/291469.291479

Baldwin, E., & Curley, M. (2007). *Managing IT innovation for business value: Practical strategies for IT and business managers*. New York, NY: Intel Press.

Baskerville, R., & Myers, M. D. (2004). Special issue on action research in information systems: Making IS research relevant to practice—Foreword. *Management Information Systems Quarterly, 28*(3), 329–335.

Benbasat, I., & Zmud, R. W. (1999). Empirical research in information systems: The practice of relevance. *Management Information Systems Quarterly, 23*(1), 3–16. doi:10.2307/249403

Bronfenbrenner, U. (2005). Lewinian space and ecological substance. In Bronfenbrenner, U. (Ed.), *Making Human Beings Human: Bioecological Perspectives on Human Development* (pp. 41–49). Thousand Oaks, CA: Sage Publications.

Brynjolfsson, E., & Saunders, A. (2009). *Wired for innovation: How information technology is reshaping the economy*. Cambridge, MA: MIT Press.

Charmaz, K. (2004). Grounded theory. In Hesse-Biber, S. N., & Leavy, P. (Eds.), *Approaches to Qualitative Research: A Reader on Theory and Practice. Oxford, UK*. Oxford: Oxford University Press.

Chesbrough, H. W. (2006). Open innovation: A new paradigm for understanding industrial innovation In H. Chesbrough, W. Vanhaverbeke, & J. West (Eds.), *Open Innovation: Researching a New Paradigm*, (pp. 1-14). Oxford, UK: Oxford University Press.

Coghlan, D., & Brannick, T. (2005). *Doing action research in your own organization* (2nd ed.). London, UK: Sage Publications.

Crabtree, B. F., & Miller, W. L. (1992). Primary care research: A multi-method typology and qualitative road map. In Crabtree, B. F., & Miller, W. L. (Eds.), *Doing Qualitative Research*. Thousand Oaks, CA: Sage.

Cranefield, J., & Yoong, P. (2007). To whom should information systems research be relevant: The case for an ecological perspective. In *Proceedings of the 15th European Conference on Information Systems (ECIS 2007)*. ECIS.

Curley, M. (2004). *Managing information technology for business value*. New York, NY: Intel Press.

Curley, M. (2007). *Introducing an IT capability maturity framework*. Paper presented at the International Conference for Enterprise Information Systems, ICEIS. Madeira, Portugal.

Davison, R. M., Martinsons, M. G., & Kock, N. (2004). Principles of canonical action research. *Information Systems Journal*, *14*(1), 43–63. doi:10.1111/j.1365-2575.2004.00162.x

Denzin, N. K., & Lincoln, Y. S. (2008). *The landscape of qualitative research*. Thousand Oaks, CA: Sage Publications.

Dick, B. (1993). *You want to do an action research thesis?* Retrieved from http://www.scu.edu.au/schools/gcm/ar/art/arthesis.html

Dodgson, M., Gann, D., & Salter, A. (2008). *Management of technological innovation: Strategy and practice*. Oxford, UK: Oxford University Press.

Dubé, L., & Paré, G. (2003). Rigor in information systems positivist case research: Current practices, trends, and recommendations. *Management Information Systems Quarterly*, *27*(4), 597–635.

Gillham, B. (2000). *Case study research methods*. London, UK: Continuum.

Glaser, B. G. (1978). *Theoretical sensitivity: Advances in the methodology of grounded theory*. Mill Valley, CA: Sociology Press.

Goulding, C. (2002). *Grounded theory a practical guide for management, business and market researchers*. London, UK: SAGE.

Gummesson, E. (2000). *Qualitative methods in management research* (2nd ed.). Thousand Oaks, CA: Sage.

Hevner, A. R., March, S. T., Park, J., & Ram, S. (2004). Design science in information systems research. *Management Information Systems Quarterly*, *28*(1), 75–105.

Kaplan, B., & Duchon, D. (1988). Combining qualitative and quantitative methods in information systems research: A case study. *Management Information Systems Quarterly*, *12*, 571–586. doi:10.2307/249133

Kawalek, J. P. (2008). *Rethinking information systems in organizations: Integrating organizational problem solving*. New York, NY: Routledge.

Kelly, S. (2004). *ICT and social/organisational change: A praxiological perspective on groupware innovation.* (Ph.D. Thesis). University of Cambridge. Cambridge, UK.

Kenny, A. (2010). *A new history of western philosophy*. Oxford, UK: Oxford University Press.

Lam, W., & Chua, A. (2005). Knowledge management project abandonment: An exploratory examination of root causes. *Communications of the AIS*, *16*(23), 723–743.

Lee, A. (1989, March). A scientific methodology for MIS case studies. *Management Information Systems Quarterly*, 33–50. doi:10.2307/248698

Lee, A. (2001). Challenges to qualitative researchers in information systems. In Trauth, E. M. (Ed.), *Qualitative Research in IS: Issues and Trends* (p. 240). Hershey, PA: IGI Global. doi:10.4018/978-1-930708-06-8.ch010

Lester, R. K., & Piore, M. J. (2004). *Innovation-The missing dimension*. Boston, MA: Harvard University Press.

Locke, K. D. (2000). *Grounded theory in management research*. London, UK: SAGE.

Mårtensson, P., & Lee, A. S. (2004). Dialogical action research at omega corporation. *Management Information Systems Quarterly*, *28*(3), 507–536.

Mathiassen, L., & Nielsen, P. A. (2008). Engaged scholarship in IS research. *Scandinavian Journal of Information Systems*, *20*(2).

Maurer, M., & Githens, R. P. (2010). Toward a reframing of action research for human resource and organization development: Moving beyond problem solving and toward dialogue. *Action Research*, *8*(3), 267–292. doi:10.1177/1476750309351361

Miles, M. B., & Huberman, M. A. (1994). *Qualitative data analysis: An expanded sourcebook*. Thousand Oaks, CA: Sage.

Mintzberg, H. (1979). An emerging strategy of "direct" research. *Administrative Science Quarterly*, *24*(4), 582–589. doi:10.2307/2392364

Mintzberg, H. (1987a, July/August). Crafting strategy. *Harvard Business Review*.

Mintzberg, H. (1987b). The strategy concept I: Five Ps for strategy. *California Management Review*, 11–24.

Myers, M. D. (1997). Qualitative research in information systems. *Management Information Systems Quarterly*, *21*(2), 241–242. doi:10.2307/249422

Myers, M. D., & Newman, M. (2007). The qualitative interview in IS research: Examining the craft. *Information and Organization*, *17*, 2–26. doi:10.1016/j.infoandorg.2006.11.001

Nijhoff, M., & Slappendel, C. (1996). Perspectives on innovation in organizations. *Organization Studies*, *17*(1), 107–129. doi:10.1177/017084069601700105

Pan, G. (2005). Information system project abandonment: A stakeholder analysis. *International Journal of Information Management*, *25*(2), 173–184. doi:10.1016/j.ijinfomgt.2004.12.003

Pavitt, K. (2005). Innovation process. In Fagerberg, J., Mowery, D., & Nelson, R. R. (Eds.), *The Oxford Handbook of Innovation*. Oxford, UK: Oxford University Press.

Public Accounts Committee Report. (2000). *Public accounts committee report, 1st report*. London, UK: House of Commons.

Reason, P., & Bradbury, H. (2001). Introduction: Inquiry and participation in search of a world worthy of human aspiration. In Reason, P., & Bradbury, H. (Eds.), *Handbook of Action Research: Participative Inquiry and Practice*. Thousand Oaks, CA: Sage.

Results, A. P. C. C. (2005). *American power conversion reports record revenue for the fourth quarter and full year 2005*. Retrieved from http://www.apcc.com/

Results, A. P. C. C. (2006). *American power conversion reports first quarter 2006 financial results*. Retrieved from http://www.apcc.com/

Robson, C. (2002). *Real world research: A resource for social scientists and practitioner-researchers*. Oxford, UK: Blackwell.

Sambamurthy, V., & Zmud, R. W. (2000). Research commentary: The organizing logic for an enterprise's IT activities in the digital era – A prognosis of practice and a call for research. *Information Systems Research*, *11*(2). doi:10.1287/isre.11.2.105.11780

Schutz, A. (1962). Concept and theory formation in the social sciences. In Nijhoff, M. (Ed.), *Collected Papers* (*Vol. I*, pp. 48–66). The Hague, The Netherlands.

Smith, D. (2006). *Exploring innovation*. Maidenhead, UK: McGraw-Hill Education.

Susman, G. I., & Evered, R. D. (1978). An assessment of the scientific merits of action research. *Administrative Science Quarterly*, *23*(4), 582. doi:10.2307/2392581

Tidd, J., Bessant, J., & Pavitt, K. (2005). *Managing innovation: Integrating technological, market and organizational change*. Chichester, UK: John Wiley & Sons.

Trauth, E. M. (2001). *Qualitative research in IS: Issues and trends*. Hershey, PA: IGI Global.

Vaishnavi, V., & Kuechler, W. (2004). *Design science research in information systems*. Retrieved from http://www.desrist.org/desrist

Vaishnavi, V., & Kuechler, W. (2007). *Design science research methods and patterns: Innovating information and communication technology*. New York, NY: Auerbach Publications. doi:10.1201/9781420059335

Van de Ven, A. H. (2007). *Engaged scholarship: A guide for organizational and social research*. Oxford, UK: Oxford University Press.

Van de Ven, A. H. (2010). *Reflections on engaged scholarship*. Paper delivered to Carlson School of Management, University of Minnesota. Minneapolis, MN.

Vanhaverbeke, W., & Cloodt, M. (2006). Open innovation in value networks. In H. Chesbrough, W. Vanhaverbeke, & J. West (Eds.), *Open Innovation: Researching a New Paradigm*, (pp. 258-284). Oxford, UK: Oxford University Press.

von Hippel, E. (2005). *Democratizing innovation*. Boston, MA: The MIT Press.

Yin, R. K. (1994). *Case study research: Design and methods*. London, UK: Sage Publications.

Zmud, R. (1996). Editor's comments: On rigor and relevancy. *Management Information Systems Quarterly*, *20*(3).

Chapter 10
Exploring Higher Education Students' Technological Identities using Critical Discourse Analysis

Cheryl Brown
University of Cape Town, South Africa

Mike Hart
University of Cape Town, South Africa

ABSTRACT

This chapter applies a critical theory lens to understanding how South African university students construct meaning about the role of ICTs in their lives. Critical Discourse Analysis (CDA) has been used as a theoretical and analytical device drawing on theorists Fairclough and Gee to examine the key concepts of meaning, identity, context, and power. The specific concepts that inform this study are Fairclough's three-level framework that enables the situating of texts within the socio-historical conditions and context that govern their process, and Gee's notion of D(d)iscourses and conceptualization of grand societal "Big C" Conversations. This approach provides insights into students' educational and social identities and the position of globalisation and the information society in both facilitating and constraining students' participation and future opportunities. The research confirms that the majority of students regard ICTs as necessary, important, and valuable to life. However, it reveals that some students perceive themselves as not being able to participate in the opportunities technology could offer them. In contrast to government rhetoric, ICTs are not the answer but should be viewed as part of the problem. Drawing on Foucault's understanding of power as a choice under constraint, this methodological approach also enables examination of how students are empowered or disempowered through their Discourses about ICTs.

DOI: 10.4018/978-1-4666-2491-7.ch010

Copyright © 2013, IGI Global. Copying or distributing in print or electronic forms without written permission of IGI Global is prohibited.

INTRODUCTION

Globally, there is a paucity of research from the perspective of students as participants embedded in the setting of the higher education institution (Selwyn, 2006). It has been argued that such a situated approach is essential to understanding how Information Systems (IS) are used within an organisation (Johnson & Aragon 2003) as it avoids universalistic assumption of knowledge and value and instead examines meaning, legitimacy and value at an individual level (Avgerou & Madon, 2004).

The motivation for this research arose from a study of South African university students' access to and use of Information and Communication Technologies (ICTs) for learning at university. In this project, students' attitudes towards ICTs were overwhelmingly positive with quantitative data showing students thought ICTs were essential for education and a positive benefit to their learning proving to be an enabler in the take up of ICTs for e-learning (Czerniewicz & Brown, 2009). However, in a society such as South Africa with a history based on inequality and a growing and increasingly diverse student body (Cooper & Subotzky, 2001) in a sector only recently restructured towards transformation and operating for the most part under significant resources constraints (Steyn & de Villiers, 2007), it seemed curious that students attitudes could be so similar and so positive.

Consequently, research was undertaken to better understand what meaning (conceptualized broadly as values, assumptions, perceptions, attitudes, opinions, experience) ICTs have for students within their local context and the broader national higher education community.

BACKGROUND TO CRITICAL DISCOURSE ANALYSIS IN IS RESEARCH

The research described in this chapter is situated within the sphere of Critical Theory. Although critical IS research is characterized by a "diversity of topics, objectives, methods and philosophical roots," it does have certain basic assumptions (Cecez-Kecmanovic, 2005, p. 20). In describing these, many authors have drawn on Alvesson and Deetz's three concerns, namely insight, critique, and transformative redefinitions (Cecez-Kecmanovic, 2005; Howcroft & Trauth, 2005; McGrath, 2005).

Whilst Critical Theory approaches are starting to take a modest but firm hold within Information Systems, Critical Discourse Analysis (CDA) is still in its infancy (Alvarez, 2005). Discourse analysis plays a role in understanding people's interaction with ICTs, and can aid in interpreting the hidden meaning about ICTs and in understanding what ICTs are, how they can be used and how different interpretations affect use (Stahl, 2004).

Discourse analysis becomes critical when it seeks to analyze power relationships in society and it is often used in IS to criticize the status quo (particularly exclusion), for example the digital divide (Kvasny & Trauth, 2002). The dominant approach in IS, is perhaps more aptly described as a critical analysis of discourse, as it draws directly on critical theorists (usually Habermas, but to some extent Foucault) and does not form part of the more linguistically-oriented field of CDA..

Two researchers who are key in operationalizing a Habermasian approach to critical discourse analysis within IS are Cukier (Cukier, Bauer, & Middleton, 2004; Cukier, Ngwenyama, Bauer, & Middleton, 2009) and Stahl (2004, 2008a). Cukier and her colleagues have developed an approach to CDA which draws explicitly on Habermas' validity claims. They present an analysis of media discourses around a Canadian technology project called the Acadia Advantage (AA) (Cukier, et al., 2004, 2009). They used critical hermeneutics and content analysis to analyze a large corpus comprising 173 media texts. They focused largely on the content of the articles, i.e. statements about the advantages and disadvantages of the technology. They also looked at some of the language being used, i.e. adjectives and metaphors, to describe the project. They also examined the empirical analysis in terms of Habermas' four validity claims, i.e. truth, legitimacy, comprehensibility and sincerity. Findings showed the discourse around this project to be distorted and influenced by powerful players, such as suppliers and university administration.

Stahl and his colleagues (Stahl, et al., 2005) used Habermasian Discourse Analysis to identify contradictions between rhetoric and reality in Egyptian ICT policy. They drew on Cukier's pioneering method of operationalizing Habermas' validity claims. Drawing on questions posed by Cukier, the frequency of each validity claim was noted. They then used Foucault's theory at a micro level to explore the empowering effect of ICTs at the organizational level by looking at a particularly widely-used application, the Decision Support System. They demonstrated that, whilst the rhetoric is one of empowerment, the reality showed that for whatever reason these were not followed through.

This approach has also been applied in the South African context (Chigona & Chigona, 2008; Chigona, Mjali, & Denzl, 2007). In the first study, the authors analyzed 24 articles over a two-year period to determine the media discourse about a popular South African mobile instant messaging system, MXit. They utilized the guiding questions of Cukier and Stahl as a means of analyzing the corpus of media texts. The analysis showed that the media discourse was highly distorted and that amongst the voices of the educators, parents, MXit management, and even politicians, the voices of the youth using the technology were conspicuously absent. The authors likened parents' concerns to a sign of moral panic and their sense of loss of control over their children's activities. The second paper utilized the same theoretical and methodological approach. Here they sought to investigate the South African government's views on the role of ICTs in national development. They analyzed 18 speeches from ministries related to ICT and peripheral ministries. The conclusion they come to was that whilst the government does show a lot of technological optimism, it is not devoid of human development concerns. There are strong links across all levels of government in terms of what they say, and there are signs of influence by global trends and a commitment to globalization.

The level of Habermasian critical analyses of discourse is somewhat macro in its focus, foregrounding the societal level of government (Chigona, et al., 2007; Stahl, et al., 2005) or media (Chigona & Chigona, 2008; Cukier, et al., 2009). All of these authors focus on the "event" under scrutiny and how it is represented in a public arena of media reports or speeches. Whilst background is provided to the corpus under analysis, the inquiry is really related to the here and now, and is not contextualized historically.

The other main approach to CDA within IS draws more directly on the CDA "school," particularly, the concept of discursive practices and genres. Thompson sought to "draw attention to ICT's dual roles as both medium and subject of discursive power relations" (Thompson, 2004, p. 1), and to demonstrate to IS researchers the usefulness of an adapted form of Fairclough's CDA. In order to do this, he analyzed a single speech given by the President of the World Bank Group in 2000 (Thompson, 2004). Thompson selected an order of discourse (namely "development") and looked for an identifiable configuration of "discursive practices," idioms, references, inferences or phrases. He then distinguished between the speech genres, which can apply across various orders of discourse, and the discursive types of themes, which apply only within a particular order of discourse. Using a semi-grounded approach, Thompson organized the data according to concepts and then identified recurrent devices or themes. These were then reduced into six higher-level concepts or discursive types. Conscious of the subjectivity of this process, Thompson has explicitly linked his interpretations to the source's text so as to enable the reader to make independent judgments concerning the analysis. Thompson's approach to CDA was drawn upon by South African researchers in their analysis of South African government officials' speeches about ICT development in Africa (Roode, Speight, Pollock, and Webber, 2004). They analyzed three speeches made in 2002 and utilized the same speech genre and discursive types as Thompson. The clear technocentric approach enabled a further discussion and reflection on the socio-techno divide manifest at the societal level in South Africa.

Ng'ambi (2008) utilized the same approach to CDA, but in a completely different context, namely that of artefacts (text messages) from an anonymous knowledge-sharing environment. Ng'ambi used the texts to provide insight into the social practices of the community in which the students were located. It helped him to understand the assumptions of the traditional practices of institutions and the practicalities experienced by the students. In addition to redefining Thompson's discursive types in terms of the context of his study, Ng'ambi identified a further two new text genres namely panic and apologetic.

Alvarez has examined the realm of CDA in the field of Information Systems. In describing the "linguistic turn" she has observed in IS research, she provides a useful overview of the range of approaches IS researchers have taken to the study of language (Alvarez, 2005). Alvarez describes how researchers are starting to look at speech acts to grasp the meanings of everyday conversations about IS genres to examine socially constructed meaning attached to different forms of communication and, more directly, discourse analysis, which adopts a hermeneutic view to technology implementation and use.

Alvarez's personal approach is linked more directly with that of the CDA "school" and she draws on linguistic features proposed by van Dijk as her analytical tools. Specifically, Alvarez has used a variety of linguistic analytical strategies of including narrative analysis and rhetorical figures of speech in various ways (Alvarez, 2008). In the longitudinal investigation of the relationship between Enterprise Systems, organizational structures and identity, Alvarez undertakes a linguistic analysis of three dominant themes that emerge from the data, namely loss of control, arbiter of fairness and acts of resistance. She shows how before adoption there was overwhelming support for a new system with favourable characteristics ascribed to it. However, once implemented these characteristics caused shifts in power balance, resulting in loss of autonomy. The rules and routines inscribed within the system limited the way individuals undertook their work and caused a shift in professional identity. This in turn generated resistance and resulted in staff finding alternative "workarounds," enabling them to reshape and, in some cases, re-established their identity. The study was successful in uncovering the hidden aspects of power relations during the ES implementation, exposing both the intolerable consequences of the power shift as well as the new possibilities enabled by the resistance.

Other IS researchers are also drawing on Foucault in their critical investigations. For example both Doolin also draws on an understanding of power influenced by Foucault in studies of the implementation of IS in the public health sector in new Zealand (Doolin 2009). Stahl et al. draw on Foucault to investigate the micro level of organizational practice in the use of ICTs in Egypt (Stahl, McBride, & Elbeltagi, 2005). As Stahl notes, Habermasian methodology focuses more at the macro level of greater social structures whereas Foucault theories lend themselves to more micro-level analysis because of Foucault's interest in the individual (Stahl, et al., 2005). Stahl et al. tried to balance this perceived weakness by drawing on a Foucauldian lens to explore the empowering effect of ICTs on the organizational level, and were able to show that the rhetoric and reality were mismatched.

Theoretically, the concept of power is also hotly contested (Hindess, 1996). In this chapter, we take a Foucauldian view of power. Foucault believes there is no such entity as power. "Power exists only as exercised by some on others, only when it is put into action" (Foucault, 1994a, p. 340). He views power as something exercised over those who are in a position to choose, although power influences what those choices will be. On the one hand, power is a "strategic game between liberties" where people can engage in the exercise of power on their own account—i.e. the element of choice. On the other hand, domination is where the person has little room to manoeuvre because the margin of their liberty is extremely limited. The distinction between power and domination allows Foucault to condemn domination but not power. This view of power has enabled researchers to examine power at both a macro and micro level and to still view the individual as having agency.

In this chapter we adopt a critical view of students' use of technology and have sought to go beyond just understanding meaning to uncovering hidden power dynamics, critique the status quo, and challenge technological determinism. Our emancipatory intentions are focused on oppositions, conflicts and contradictions, and by uncovering these, we seek to provide information and a perspective on students' meanings of ICTs that can be used to eliminate the causes of alienation.

Critical Discourse Analysis as a Methodological Framework

One of the strengths of CDA is that it is multidisciplinary and essentially diverse (Van Dijk, 2001). In fact, Van Dijk, who is one of its original proponents, says that good CDA scholarship seldom follows just one person or one approach but is enriched through the integration of the "best work of many people, famous or not, from different disciplines, countries, cultures and directions of research" (Van Dijk, 2001, p. 95). In terms of this research, two CDA researchers have proved particularly useful, namely Norman Fairclough and James Paul Gee.

We use Critical Discourse Analysis (CDA) as the mode of analysis to understand more about the social relationships and identity students maintain in terms of ICTs drawing on a combination of Fairclough and Gee's approaches to Critical Discourse Analysis (particularly the theoretical concept of "recontextualisation") to examine how students draw on existing Discourses of globalisation to inform their personal technological identity.

Fairclough is regarded as one of the grandfathers of CDA, with his landmark publication in 1989, seen as starting CDA as a field in its own right (Blommaert, 2005; Myers, 2002; Van Dijk, 1997; Weiss & Wodak, 2003). Fairclough's (1995, 2001) models of CDA consisted of three inter-related processes of analysis tied to three dimensions of discourse. The three dimensions of discourse are the object of analysis (text), the process by which the object is reproduced (practice) and the socio-historical conditions, which govern this process. Each of these dimensions requires a different type of analysis, namely "description of the text, interpretation of the relationship between text and interaction and explanation of the relationship between interaction and social context" (Fairclough, 2001, p. 91).

Gee's work is largely situated in the movement of New Literacy Studies (NLS) which is based around the "idea that reading, writing and meaning are always situated within specific social practices within specific discourses (Discourses)" (Gee, 1999, p. 8). Gee describes two types of ways meanings are interpreted in texts. The general meaning for a word/phrase and the way that word or phrase is used within a particular situation that gives it a more specific meaning in terms of its context of use (Gee, 2005). This led to Gee's notion of D(d)iscourse (Gee, 1996). Gee's concept of "little d" discourse he describes as language in use whereas Big D Discourses Discourse encompasses more than just the use of language it includes ways of being (thinking, acting and interacting) (Gee, 2005) that take on socially meaningful identities in various situations or contexts. Another of Gee's concepts that is a useful tool of analysis is what he terms "Big C" Conversations. Gee notes that most of us are aware of "societal Conversations going on around us like abortion, creationism, global warming, terrorism" (Gee, 2005). Gee terms these Big C Conversations, as he views these as grand societal conversations. He uses Conversations explicitly as a tool of inquiry to examine what external Conversations a piece of text refers to and what it does not and to examine how these Big C Conversations shaping the Discourse?

Whilst these two proponents of CDA do clearly have different approaches, they are not incompatible (Gee, 2004). Recently, the two theorists have been represented together as chapter authors in a book "Critical Discourse in Education" (Rogers, 2004a). Rogers uses Gee's theory of Discourse and Fairclough's concept of orders of discourse in her research; whilst others in the book, Young (2004) and Rowe (2004), also combine various aspects of the two approaches.

Whilst Gee and Fairclough's terminology and process differ, Rogers has noted that there are common intersections between their two theoretical and methodological models (Gee, 2004; Rogers, 2004b). Both agree that language involves textual, interpersonal and ideational functions, and go beyond just describing practice to explore the relationship between language and social structure. Rogers also asserts that the tension between these two frameworks is productive as "it allows for the theory and methods of CDA to be reformulated and applied to important educational issues" (Rogers, 2004c).

Within CDA there have been calls for researchers to be more explicit in defining what they mean by various key components of CDA, (Rogers, 2004c) and more explicit in providing overviews of their conceptual and analytical frameworks (Weiss & Wodak, 2003). Within IS, critical theory, research has been criticized for its vagueness in terms of methodology (Pozzebon, 2004).

I have utilized Fairclough's (2001) dimensions of discourse as an analytical guide to move between the text, discursive level and social practice. My analysis begins at the textual level where I undertake linguistic analysis of components that are evident in the genre of text, drawing on Fairclough and the way he understands linguistic features that elucidate agency and modality, as well as questions from Gee's analytical process that relate directly to context, situated meaning, power and identities. At the discursive level, I draw on Gee's conceptualization of grand societal "Big C" Conversations (2005) to examine the process by which this object is reproduced before situating this within the socio-historical conditions and context which governs these process. I employ Gee's notion of Big D Discourses as a way of explaining a social group's way of being in the worlds, their "form of life" their very identity.

Positioning Discourse within a context is essential in Critical Discourse Analysis as the theory is based on the premise that everything we say is socially and culturally produced. All the interpretation and understanding within CDA is the result of a contextualisation process in which texts are made to fit a particular set of contexts by the participants in an interaction. In undertaking a CDA it is important not to restrict the notion of context to what happens in a specific communicative event. Not just because the researcher is performing the interpretation of meaning as a matter of post hoc recontextualisation but also because of intertexuality. This chapter draws heavily on these two concepts. The first—recontextualisation—is explained by Fairclough (2006) as the process where outside entities (ideas) come inside (the text) and by Blommaert (2005) as a process that extracts text, signs or meaning from its original context (decontextualisation) in order to introduce it into another context. It grounds discourse analysis firmly within histories of use, histories that are socially, culturally, and politically constituted. This concept helped the researchers look beyond the boundaries of a particular communicative event to see where the expressions used come from, what there sources are and whom they speak for.

Application of Critical Discourse Analysis in a Research Project

The primary data for this research is drawn from open-ended answers to a survey of students' access to and use of ICTs carried out in 2007. The survey was conducted amongst 3533 students from 6 universities across South Africa (located in 4 provinces). 88% of students responded to the open-ended questions providing an extensive corpus of text (Brown & Czerniewicz, 2008). In terms of the nature of CDA, the students who completed more of the open-ended questions tended to provide better data for analysis (although there were some students who only answered one open-ended question but provided quite lengthy, insightful texts). The majority of my sample is drawn from students who made a larger contribution to the open-ended data within the survey. The final sample comprised 840 students' texts.

The data was analysed initially by identifying discursive types in the text using a semi grounded approach (Brown, 2011a). Having undertaken an extensive literature review of prevalent discourses about technology, potential themes were noted. Initially a small random group of texts was selected and read through to identify recurrent themes. The texts were sorted according to the initial categorization and re-read. Categories were adjusted and refined until a consistent theme begun to emerge. The corpus was then expanded (having learnt from the initial process that it was usually pointless reading "bits of text" less than 17 words in length) and the category definitions applied to the extended sample.

Once the sample had been categorised in these themes, a sub set of the texts from within each theme was selected in order to unpack how students viewed their ICT identity through the texts (Gee, 2005). This involved examining the texts to understand what this told us about how students thought, felt, valued, and used ICTs for their learning. A description was then formulated as to how students (who were operating from within a particular discourse) viewed the world of technology.

Gee's (2005) questions and framework were systematically used to analyse each theme in terms of building significance, activities, identities, relationships, politics, connections, sign systems and knowledge. In doing this the analysis moved between Fairclough's three levels of discourse practice as exemplified by others (Ng'ambi, 2008; Thompson, 2004; Tu & Kvasny, 2006).

Analysis at the text level concentrated on linguistic features (vocabulary, grammar, syntax). At the discursive level analysis focuses on how authors of the text drew on already existing discourses and genres in the consumption and interpretation of the text. Finally, analysis of the relationship between text and social practice, i.e. how the discursive practice reproduced or restructured existing order of discourse was examined to determine what consequences this had at a social level.

The object of analysis (text), the process by which the object was reproduced (practice) and the socio-historical conditions, which govern this process. In addition, each of these dimensions requires a different type of analysis namely "description of the text, interpretation of the relationship between text and interaction and explanation of the relationship between interaction and social context" (Fairclough, 2001).

This is encapsulated in the diagram below. As described the analysis started in the middle, at the level of what Fairclough calls the discourse practice. Here the focus was on Gee's concept of "Big C" Conversations and various worldview

Figure 1. Diagrammatic representation of the analytical framework

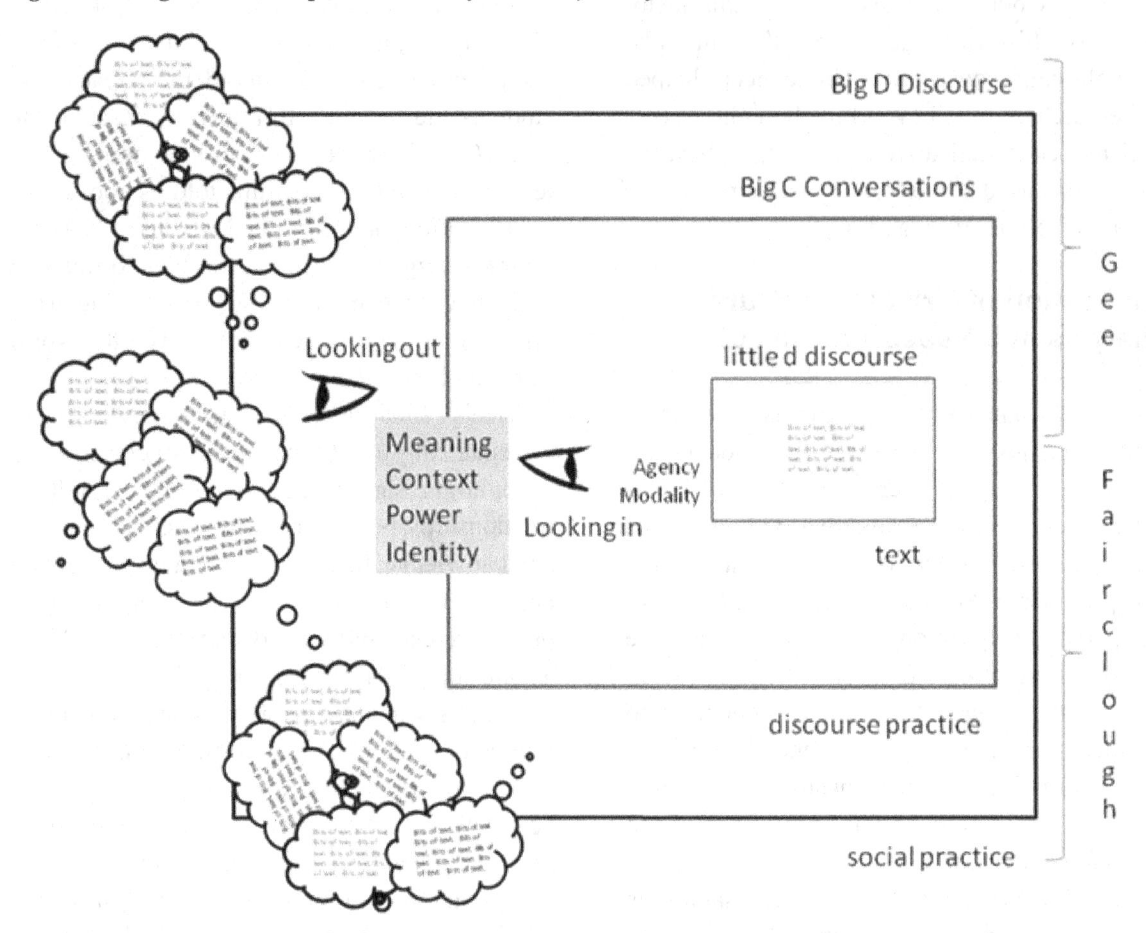

surrounding technology and its use in education were identified.

The analysis then proceeded to the level of social practice where these Big C Conversations were grouped into themes (represented on the outskirts of Figure 1 as clouds containing individual texts). Then went back to the level of the individual texts or as Gee calls it little d discourses where the analytical constructs of meaning, context, power, and identity were examined using Gee's guiding questions as tool. Finally we looked back out at the level of social practice to understand what Big D Discourses existed and what this could tell us about students' technological identity as a group.

Four key Discourses were identified amongst students within higher education contexts (Brown, 2011a, 2011b), namely:

- Globalisation Discourse with common elements of global opportunity, global citizenship, and the information society and notions of the digital divide, having and not having access, and disadvantage.
- Deterministic Discourse where the common themes which emerged were the strong, dominant (almost uncritical) view of technology as being essential, and by association the skills to use ICTs as being more highly valued and necessary than other skills.
- Learning Discourse with common elements include references to learning or studying activities in terms of their efficiency and effectiveness.

- Liberation Discourse, which involved a view of the use of ICTs primarily as a means of acquiring information and had a strong positive association with knowledge with a lesser awareness of the need to approach information with caution.

FINDINGS: UNDERSTANDING DISCOURSES AND CONVERSATION

The findings presented here focus on the Discourses of Globalisation and Determinism in order to examine who or what is constructing/ sustaining these Discourses at a macro level and how these relate to the Big C Conversations that exist about the information society and the digital divide. The role of these Discourses in maintaining students' social worlds and furthering interest of their social group is then examined as well as how Discourses empower or disempower students in terms of their use of technology.

As established earlier, context is crucial within a CDA framework. It includes a range of circumstances such as student's physical setting, the people present, the social relationship of people involved, and cultural and historical factors. The view between language and context in most contemporary discourse analyses is usually reflexive which means that the utterance (what students are writing) influences what they take to be the context and the context influences what they write (Gee, 2005).

It is therefore worth reflecting for a moment on the context in which South African universities students find themselves. "Infrastructure and access is obviously an issue for e-learning in South African HEIs. Issues related to resources (e.g. cost, bandwidth) are mentioned by many as a barrier and certainly limit e-learning practice" (Brown, Thomas, van der Merwe and van Dyk 2008, p. 75). Some key infrastructural issues which influence the way ICTs are used in higher education are that the number of Internet users in the country exceeds the availability of personal computers, Internet costs are the most expensive in Africa, limitations in bandwidth constrain what is possible in terms of teaching and learning, and access levels remain disparate between demographic groups (Brown & Czerniewicz, 2008).

Who/What is Constructing/ Sustaining Students' Discourses at a Macro Level?

Despite the infrastructural constraints in which students exist, their attitudes to ICTs are largely positive. This is not surprising as students are operating in a larger context where few people are negative about ICTs, and negativity with regards to ICTs is perceived to be associated with ignorance which is associated with backwardness.

This is one indication that students' are re-contextualising external Big "C" Conversations within their own discourses. Whilst it is unlikely that students are directly aware of government rhetoric, the ideas contained within government speeches, documents and policies do permeate society through more accessible forms of media such as radio and TV. Students with a Globalisation Discourse are also recontextualising concepts of the digital divide as espoused by the South African government. The ideas contained in government statements such as "most of the peoples of the world, especially from the developing countries, are confronted by the challenge of exclusion in the context of the global economy, in whose development modern Information and Communications Technologies (ICTs) play a vital role'" (Mbeki, 2005) and the education system as "severely stratified and dependent of personal wealth" (Department of Education and Department of Communication, 2001) are recontextualised by students as ICTs not being accessible for students who come from disadvantaged, especially poor, backgrounds. For example, students mention the digital divide directly "i think icts sometimes create digital divide, especially for students who find it hard to access Internet" [2922] and the issue of the personal cost of ICT access "i have to travel for 30 kilometers and pay ten rands for the taxi fare and after that i still have to pay 10 rands for 30 minutes using the Internet" [877].

Students within the Globalisation Discourses are also recontextualising concepts of global opportunity evidenced in various government and national policy documents and reports for example, "ICT education, training, learning, and competency are essential for the twenty first century" (Department of Training and Industry, 2000), and comment that "this is a computer dominated world, so it is essential for me to understand computers" [190].

ICTs can "create equal opportunity, promote social mobility and the well being of larger social constituencies (Council on Higher Education, 2004, p. 17) "learners will be equipped for full participation in the knowledge society" (Department of Education, 2003) point 2.5 and "At the centre of these changes is the notion that in the 21st century, knowledge and the processing of information will be the key driving forces for wealth creation and thus social and economic development." (Department of Education, 2001) This is another concept recontextualised in students texts "because every job requires a person to be computer literate, e.g. a person who will be able to type, do presentation on conferences using personal laptops" [3272].

Notions of ICTs improving chances of employability are also being repurposed in students texts, with ideas such ICTs being necessary "in a rapidly changing society, [for] the development of professional and knowledge workers with globally equivalent skills" (Department of Education, 2001) reiterated by students "i think ICTs should be a compulsory course because nowadays wherever a person is working, a computer is needed" [2494].

However the counter belief is also being recontextualised within the Deterministic Discourses as many believe if they don't have knowledge and ability about ICTS their chances of succeeding in life are less ""It's nothing more except or despite the fact that it's me because thus far I'm not computer competent due to the environment i hailed from. I'm still struggling to be one" [88].

Government rhetoric about the opposite side of this issue is equally strong in statements such as "Digital literacy is seen as a "life skill" in the same way as literacy and numeracy" (Department of Education, 2003) and "We must continue the fight for liberation against poverty, against underdevelopment, against marginalisation" and "... information and communication technology" (Department of Education, 2001). Students themselves used the opportunity within the survey to make a greater call for justice "i wish for greater information about developments in the field of ict's to be made to more people (vary the demographic)" [2252].

What is the Role of These Discourses in Maintaining the Students' Social Worlds or Furthering Interest of Their Social Group?

Students from a Globalisation Discourse are keenly aware of how their personal circumstances (their background and history) influence their opportunities. Their disadvantage in life is the reason they have lack of opportunities. Poverty denies them opportunity. Students with a deterministic discourse are not as explicit about the reason for their lack of ability but what is interesting is that they almost see ability as a social good. It's a currency, a ticket which they do not have.

Students with a Globalisation Discourse do have a sense of solidarity with others in their predicament and see themselves as part of a group of disadvantaged. For some they look back at where they came from and desire opportunities for learners at schools in rural areas because they feel they themselves are disadvantaged because their first exposure to ICTs was at university. They desire a greater exposure of ICTs across a wider range of people in South Africa or even a decrease in costs so ICTs are more accessible to more people. Even at the university level, they recommend changes in terms of faculty approaches to ensure equity of access. Their interest is in opening up opportunity and access to marginalized social groups. Yet only a few go beyond these demands for change and show evidence of agency themselves. When examining the transformative power of the texts, the two contrasts are those that see themselves denied opportunity and those that see ICTs as an opportunity. Interestingly, there is also a strong sense of competition to better themselves beyond and above others. There is a sense of themselves as individuals against the world.

Those students with a Deterministic Discourse who feel they have managed to acquire that ticket in to the world of ICTs are strongly in solidarity with those that have not gotten there yet. They are aware of the advantage that they now have and those that do not, referring generally to students as a group and staff as a group. Very few talk specifically about a social network.

How do Discourses Empower or Disempower Students' in Terms of Their Use of Technology?

Having access, and being seen to be ICT literate are big status symbols and students from Globalisation and Deterministic Discourses perceive this opportunity and use of ICTs to give them higher status and more power.

Interestingly there is a contradiction here as students within the Globalisation Discourse are both empowered and disempowered. Those with an information society theme seem to have a greater sense of agency and drive to achieve their goals fuelled by the rhetoric of the information society as being something that provides opportunities for all. Whilst these external Conversations might be merely a dream they do, in a sense give these students the impetus to give it a try. Students do talk in terms of upliftment, although sometimes at a personal cost. They feel they are operating in an environment with limited options and choices. Therefore, their sense of empowerment is limited.

Whereas, in contrast, the students with a digital divide theme feel as if they are already a product of their backgrounds and draw in external conversation such as economic disadvantage, racism, and elitism to explain their standing in life. These past disadvantages continue to be echoed in the present with students continuing to talk about how they are disadvantaged presently in terms of ICT access.

Students with a Deterministic Discourse also exhibit disempowerment. They feel personally inadequate, want to learn, are waiting for help, to be taught and feel that lecturers or the university needs to help students learning but the initiative for these

This response by students is very similar to the responses outlined by Kvasny and Trauth in examining how under-represented groups coped with IT related problems and what their responses were to the existing power structure. The conforming power response is similar to the first group of students described above whose discourses of the information society encouraged them to embrace technology but within its existing structures and processes, that is they did not challenge it in any way. The second group of students are similar to Kvasny and Trauth's (2002) group who conceded to power structures, as they are involuntarily excluded from IT practices. Kvasny and Trauth's (2002, p. 276) final group are those that challenge power structure through "commandeering IT to charts one's own usage and career course."

DISCUSSION

Discussion of the findings and the implication for how students' meanings and identities influence the way they use ICTs for learning is described in more detail elsewhere (Brown, 2011a, 2011b). In this section, we reflect on the value moving beyond the dominant IS Habermasian critical theory approach to explore the relevance and applicability of other critical discourse approaches to the field. In using and adapting existing theory to frame and structure an argument, we have shown how a unique combination of Gee and Fairclough can provide an alternative way of doing Critical Discourses Analysis within IS.

In doing this, we have addressed a number of criticisms leveled at critical theorists and critical discourse analysts, namely: making obvious the methodological process and analytical constructs, providing a pathway and process to move from the level of text to the level of social practice within a critical discourse analysis, and drawing on a large corpus of data to enable wider representations. In addition, we have sought not to obfuscate CDA as an insider but rather we hope through transparency to have made clear the reasons, values, and underlying assumption inherent in this approach.

This approach of drawing on more than one theorist in undertaken critical discourse analysis provides a fresh perspective on a problematic concept and has enabled new insights namely understanding Discourses of technology. However, the relationship of Discourses to technological identity warrants further research. It is also important not to fall into the trap of seeing these categories as fixed or static. Whilst naming the Discourses was an important aspect of acknowledging and identifying their existence, one must be cognizant that, in a different place and time, these may exist differently or not at all. Thus expanding the methodological toolkit to examine Discourses beyond text and talk and seeking new insight through a compatible yet different critical discourse lens will enable a better understanding of students' technological identity and how this influences and is in turn influenced by students' technological practices and the tools they use.

We have also demonstrated that it is possible to use this approach to analyze more deeply a source of data often overlooked in terms of meaning, namely the open-ended question; and thus have made a new contribution to empirical approaches in critical theory. The nature of the open-ended question was both a strength, and a limitation. As a data source they are limited in terms of the information they can provide. The language used is informal and not constructed in full sentences which means that many textual markers often used in a CDA are missing; they are written in a limited space so the text is short and the researcher does not have the opportunity to go back and question to clarify responses as, unlike an interview, there is no dialogue between researcher and participant. However as open-ended questions are often used in surveys to give people an opportunity to raise additional issues or clarify reasons for selection of particular scale item it is valuable to have a more multifaceted way to analyse this source of data. Most often they are drawn on in a high level content analysis or used to provide qualitative illustration of data. However, this study has shown that open-ended questions can be valuably analyzed using a Critical Discourse Approach and that it is possible to draw on this type of data not only to look at what people say but how they say it.

CONCLUSION

By using a critical lens in a systematic and rigorous manner we have demonstrated how students level of personal technological empowerment is closely related to their Discourses. This has enabled us to examine hidden meanings and imbalances of power and although in this process a range of discourses has been uncovered, as critical theorists we have on groups of students we felt were marginalized or under-represented.

We have questioned the ideas of progress that are associated with technological development within South Africa and believe it is important to acknowledge that despite huge efforts to position ICTs centrally within university education, the reality is that academic use is very limited in terms of teaching and learning.

These critical perspectives on students' experiences are important, as the various Discourse groups are all co-located within the university system. Teaching, technical staff and other students need to understand how students construct their ICT identities and position themselves in order to understand the feelings of powerlessness and isolation amongst some groups, so that we can better support them.

Whilst a dialectical relationship between Discourses and technology has been noted, it is important to foreground that students' Discourses are not singular. In fact they can be multiple and contradictory and may even cause conflict if, as Gee notes, there are contradictions between students' primary and secondary Discourses. In addition students' Discourses can and do open up various possibilities and options in terms of their use of technology, which in turn has the potential to transform that student's Discourse. Therefore, the view of Discourses adopted here is that they are a multifaceted, relational and a transformable concept, not a static, one-dimensional one.

In adopting a Foucauldian view of "power as choice," we have been able to foreground agency as a construct of power. As a recent report notes, the "first and most important empowering or disempowering factor for students' agency is their background"(Council on Higher Education, 2010). By understanding students' ICT identities we are better informed to work with (rather than against) the experiences and perceptions, and enable the freedom for them to set and pursue their own goals and interests.

REFERENCES

Alvarez, R. (2005). Taking a critical linguistic turn. In Howcroft, D., & Trauth, E. (Eds.), *Handbook of Critical Information Systems Research* (pp. 104–122). Cheltenham, UK: Edward Elgar Publishing Inc.

Avgerou, C., & Madon, S. (2004). Framing IS studies: Understanding the social context of IS innovation. In Avgerou, C., Ciborra, C., & Land, F. (Eds.), *The Social Study of Information and Communication Technology* (pp. 162–182). Oxford, UK: Oxford University Press.

Blommaert, J. (2005). *Discourse: A critical introduction*. Cambridge, UK: Cambridge University Press. doi:10.1017/CBO9780511610295

Broekman, I., Enslin, P., & Pendlebury, S. (2002). Distributive justice and information communication technologies in higher education in South Africa. *South African Journal of Higher Education, 16*(1), 29–35. doi:10.4314/sajhe.v16i1.25269

Brown, C. (2011a). *Excavating the meaning of information and communication technology use amongst South African university students: A critical discourse analysis*. (PhD Thesis). = University of Cape Town. Cape Town, South Africa.

Brown, C. (2011b). The influence of the recontextualisation of globalization: Discourses on higher education students' technological identities. In *Proceedings of the IADIS International Conference: ICT, Society and Human Beings 2011*, (pp. 73-80). IADIS Press.

Brown, C., & Czerniewicz, L. (2008). Trends in student use of ICTs in higher education in South Africa. In *Proceedings of the 10th Annual Conference of WWW Applications*. Cape Town, South Africa: Cape Peninsula University of Technology.

Brown, C., Thomas, H., van der Merwe, A., & van Dyk, L. (2008). The impact of South Africa's ICT infrastructure on higher education. In *Proceedings of the 3rd International Conference of E-Learning*, (pp. 69-76). Cape Town, South Africa: Academic Publishing Limited.

Cecez-Kecmanovic, D. (2005). Basic assumptions of the critical research perspectives in information systems. In Howcroft, D., & Trauth, E. (Eds.), *Handbook of Critical Information Systems Research* (pp. 19–46). Cheltenham, UK: Edward Elgar Publishing Inc.

Chigona, A., & Chigona, W. (2008). MXit up it up in the media: Media discourse analysis on a mobile instant messaging system. *Southern Africa Journal of Information and Communication, 9*, 42–57.

Chigona, W., Mjali, P., & Denzl, N. (2007). *Role of ICT in national development: A critical discourse analysis of South Africa's government statements*. Paper presented at QualIT 2007 Qualitative Research in IT. Wellington, New Zealand.

Cooper, D., & Subotzky, G. (2001). *The skewed revolution: Trends in South African higher education: 1988-1998*. Cape Town, South Africa: University of Western Cape.

Council on Higher Education. (2010). Access and throughput in South African higher education: Three case studies. *Higher Education Monitor, 9*. Retrieved from http://www.che.ac.za/documents/d000206/

Cross, M., & Adams, F. (2007). ICT policies and strategies in higher education in South Africa: National and institutional pathways. *Higher Education Policy, 20*, 73–95. doi:10.1057/palgrave.hep.8300144

Cukier, W., Bauer, R., & Middleton, C. (2004). Applying Habermas' validity claims as a stanadard for critical discourse analysis. In *Information Systems Research: Relevant Theory and Informed Practice* (pp. 233–258). London, UK: Kluwer Academic Publishers. doi:10.1007/1-4020-8095-6_14

Cukier, W., Ngwenyama, O., Bauer, R., & Middleton, C. (2009). A critical analysis of media discourse on information technology: Preliminary results of a proposed method for critical discourse analysis. *Information Systems Journal, 19*, 175–196. doi:10.1111/j.1365-2575.2008.00296.x

Czerniewicz, L., & Brown, C. (2007). Disciplinary differences in the use of educational technology. In *Proceedings of ICEL 2007: 2nd International Conference on e-Learning*, (pp. 117-130). New York, NY: ICEL.

Czerniewicz, L., & Brown, C. (2009). A virtual wheel of fortune? Enablers and constraints of ICTs in higher education in South Africa. In Marshall, S., Kinuthia, W., & Taylor, W. (Eds.), *Bridging the Knowledge Divide: Educational Technology for Development*. Denver, CO: Information Age Publishing.

Department of Education. (2001). *The national plan on higher education*. Pretoria, South Africa: Government Printers. Retrieved 12 June 2006, from http://www.polity.org.za/html/govdocs/misc/higheredu1.htm?rebookmark=1

Department of Education. (2003). *White paper on e-education: Transforming learning and teaching through ICT*. Pretoria, South Africa: Government Printers. Retrieved 1 June 2006, from http://www.info.gov.za/whitepapers/2003/e-education.pdf

Department of Education and Department of Communication. (2001). *A strategy for information and communication technology in education*. Retrieved from http://www.info.gov.za/otherdocs/2001/ict_doe.pdf

Department of Training and Industry. (2000). *South African ICT sector development framework*. Retrieved from http://www.dti.gov.za/saitis/docs/html/chap06.html

Doolin, B. (2009). Information systems and power: A Foucauldian perspective. In Brooke, C. (Ed.), *Critical Management Perspectives on Information Systems* (pp. 211–230). Oxford, UK: Butterworth Heineman.

Fairclough, N. (2001). *Language and power* (2nd ed.). London, UK: Longman.

Fairclough, N. (2006). *Language and globalisation*. Oxon, UK: Routledge.

Foucault, M. (1994a). The subject and power. In Faubion, J. (Ed.), *Power* (pp. 326–348). New York, NY: The New Press.

Foucault, M. (1994b). Truth and power. In Faubion, J. (Ed.), *Power* (pp. 111–133). New York, NY: The New Press.

Gee, J. (2005). *An introduction to discourse analysis*. New York, NY: Routledge.

Gras-Velazquez, A., Joyce, A., & Derby, M. (2009). *Women and ICT: Why are girls still not attracted to ICT studies and careers?* Brussels, Belgium: European Schoolnet (EUN Partnership AISBL).

Greenhalgh, T. (2006). Computer assisted learning in undergraduate medical education. *British Medical Journal, 322*, 40–44. doi:10.1136/bmj.322.7277.40

Hindess, B. (1996). *Discourses of power*. Oxford, UK: Blackwell Publishers Ltd.

Howcroft, D., & Trauth, E. (2005). Choosing critical IS research. In Howcroft, D., & Trauth, E. (Eds.), *Handbook of Critical Information Systems Research*. Chelteham, UK: Edward Elgar Publishing, Inc.

Johnson, S., & Aragon, S. (2003). An instructional strategy framework for online learning environments. *Bureau of Educational Research, 100*, 31–43.

Klecun, E. (2008). Bringing lost sheep into the fold: Questioning the discourse of the digital divide. *Information Technology & People, 21*(3), 267–282. doi:10.1108/09593840810896028

Kvasny, L., & Trauth, E. (2002). The digital divide at work and at home: Discourses about power and underrepresented groups in the information society. In Whitley, E., Wynn, E., & DeGross, J. (Eds.), *Global and Organizational Discourse About Information Technology* (pp. 273–294). New York, NY: Kluwer Academic Publishers.

Mbeki, T. (2005). *Address of the President of South Africa, Thabo Mbeki, at the world summit on the information society, Tunis, Tunisia.* Retrieved from http://www.thepresidency.gov.za/pebble.asp?relid=3168

McGrath, K. (2005). Doing critical research in information systems: A case of theory and practice not informing each other. *Information Systems, 15*, 85–101. doi:10.1111/j.1365-2575.2005.00187.x

Ng'ambi, D. (2008). A critical discourse analysis of students anonymous online postings. *International Journal of Information and Communication Technology Education, 4*(3), 31–39. doi:10.4018/jicte.2008070104

Pozzebon, M. (2004). Conducting and evaluating critical interpretive research: Examining criteria as a key component in building a research tradition. *International Federation for Information Processing, 143*, 275–292. doi:10.1007/1-4020-8095-6_16

Prensky, M. (2001). Digital natives, digital immigrants, part 2: Do they really think differently? *Horizon, 9*(6), 1–9. doi:10.1108/10748120110424843

Rogers, R. (2004b). Literate identities across contexts. In Rogers, R. (Ed.), *An Introduction to Critical Discourse Analysis* (pp. 51–78). Mahwah, NJ: Lawrence Erlbaum.

Roode, D., Speight, H., Pollock, M., & Webber, R. (2004). It's not the digital divide – It's the socio-techno divide. In *Proceedings of the 12th European Conference on Information Systems Turka*. IEEE.

Rowe, S. (2004). Discourse in activity and activity as discourse. In Rogers, R. (Ed.), *An Introduction to Critical Discourse Analysis in Education*. Mahwah, NJ: Lawrence Erlbaum Associates.

Selwyn, N. (2006). The use of computer technology in university teaching and learning: A critical perspective. *Journal of Computer Assisted Learning, 23*, 83–94. doi:10.1111/j.1365-2729.2006.00204.x

Seymour, L., & Fourie, R. (2010). *ICT literacy in higher education: Influences and inequality*. Paper presented at the Southern African Computer Lecturers' Association Conference Pretoria. Pretoria, South Africa.

Stahl, B. (2004). Whose discourse? A comparison of the Foucauldian and Habermasian concepts of discourse in critical IS research. In *Proceedings of the Tenth America's Conference on Information Systems*, (pp. 4329-4336). New York, NY: IEEE.

Steyn, A. G. W., & de Villiers, A. P. (2007). Public funding of higher education in South Africa by means of formula. *Review of Higher Education in South Africa*. Retrieved from http://www.che.ac.za/documents/d000146/5-Review_HE_SA_2007.pdf

Thompson, M. (2004). ICT, power, and developmental discourse: A critical view. *Electronic Journal on Information Systems in Developing Countries, 20*(4), 1–25.

Tu, L., & Kvasny, L. (2006). American discourses of the digital divide and economic development: A sisyphean order to catch up? In D. Howcroft, E. Trauth, & DeGross (Eds.), Social Inclusion: Societal & Organizational Implications for Information Systems, (pp. 51-66). London, UK: Kluwer Academic Publishers.

Van Dijk, T. (2001). Multidisciplinary CDA: A plea for diversity. In Wodak, R., & Myers, M. (Eds.), *Methods of Critical Discourse Analysis*. London, UK: Sage Publications. doi:10.4135/9780857028020.d7

Weiss, G., & Wodak, R. (2003). Introduction: Theory, interdisciplinarity and critical discourse analysis. In Weiss, G., & Wodak, R. (Eds.), *Critical Discourse Analysis: Theory and Interdisciplinarity* (pp. 1–34). New York, NY: Palgrave Macmillian.

Young, J. (2004). Cultural models and discourses of masculinity: Being a boy in a literacy classroom. In Rogers, R. (Ed.), *An Introduction to Critical Discourse Analysis in Education* (pp. 147–172). Mahwah, NJ: Lawrence Erlbaum Associates.

ADDITIONAL READING

Alvarez, R. (2008). Examining technology, structure and identity during an enterprise system implementation. *Information Systems Journal, 18*, 203–234. doi:10.1111/j.1365-2575.2007.00286.x

Budd, Y. (2005). Technological discourses in education. In *Proceedings of the International Conference on Critical Discourse Analysis: Theory into Research*, (pp. 75-83). IEEE.

Clegg, S., Hudson, A., & Steel, J. (2003). The emperor's new clothes: Globalisation and e-learning in higher education. *British Journal of Sociology of Education, 24*(1), 39–53. doi:10.1080/01425690301914

Doolin, B. (2009). Information systems and power: A Foucauldian perspective. In Brooke, C. (Ed.), *Critical Management Perspectives on Information Systems* (pp. 211–230). Oxford, UK: Butterworth Heineman.

Doolin, B., & McLeod, L. (2005). Towards critical interpretivism in IS research. In Howcroft, D., & Trauth, E. (Eds.), *Handbook of Critical Information Systems Research* (pp. 244–271). Cheltenham, UK: Edward Elgar Publishing, Inc.

Fairclough, N. (2003). *Analysing discourse: Textual analysis for social research*. London, UK: Routledge.

Fairclough, N. (2009). A dialectical-relational approach to critical discourse analysis in social research. In Wodak, R., & Myers, M. (Eds.), *Methods of Critical Discourse Analysis* (pp. 162–186). London, UK: Sage.

Gee, J. (2004). What is critical about critical discourse analysis? In Rogers, R. (Ed.), *An Introduction to Critical Discourse Analysis* (pp. 19–50). Mahwah, NJ: Lawrence Erlbaum. doi:10.1017/S0047404509990686

Gee, J. (2005). *An introduction to discourse analysis* (2nd ed.). New York, NY: Routledge.

Gee, J. (2008). *Social linguistics and literacies* (2nd ed.). London, UK: Falmer Press.

Goode, J. (2010). The digital identity divide: How technology knowledge impacts college students. *New Media & Society*, *12*, 497–513. doi:10.1177/1461444809343560

Howcroft, D., & Trauth, E. (2005). Choosing critical IS research. In Howcroft, D., & Trauth, E. (Eds.), *Handbook of Critical Information Systems Research*. Chelteham, UK: Edward Elgar Publishing, Inc.

Klein, H., & Huynh, M. (2004). The critical social theory of Jurgen Habermas and its implications for IS research. In Mingers, J., & Willcocks, L. (Eds.), *Social Theory and Philosophy for Information Systems*. Chichester, UK: John Wiley and Sons.

McGrath, K. (2005). Doing critical research in information systems: A case of theory and practice not informing each other. *Information Systems*, *15*, 85–101. doi:10.1111/j.1365-2575.2005.00187.x

Ndambuki, J., & Janks, H. (2010). Political discourses, women's voices: Mismatches in representation. *Critical Approaches to Discourse Analysis across Disciplines, 4*(1), 73-92.

Pozzebon, M. (2004). Conducting and evaluating critical interpretive research: Examining criteria as a key component in building a research tradition. *International Federation for Information Processing, 143*, 275–292. doi:10.1007/1-4020-8095-6_16

Rogers, R. (2004c). Setting an agenda for critical discourse analysis in education. In Rogers, R. (Ed.), *An Introduction to Critical Discourse Analysis* (pp. 237–254). Mahwah, NJ: Lawrence Erlbaum.

Rogers, R., Malancharuvil-Berkes, E., Mosley, M., Hui, D., & O'Garro Joseph, G. (2005). Critical discourse analysis in education: A review of the literature. *Review of Educational Research, 75*(3), 365–416. doi:10.3102/00346543075003365

van Dijk, T. (2009). Critical discourse studies: A sociocognitive approach. In Wodak, R., & Meyer, M. (Eds.), *Methods of Critical Discourse Analysis* (2nd ed., pp. 62–86). London, UK: Sage Publications.

Wilson, M. (2003). Understanding the international ICT and development discourse: Assumptions and implications. *The South African Journal of Information and Communication, 3*. Retrieved from http://link.wits.ac.za/journal/j0301-merridy-fin.pdf

Wodak, R., & Meyer, M. (2009). Critical discourse analysis: History, agenda, theory. In Wodak, R., & Meyer, M. (Eds.), *Methods of Critical Discourse Analysis* (pp. 1–33). London, UK: SAGE Publications.

KEY TERMS AND DEFINITIONS

Conversations: Big C Conversations are themes, debates or ideas that are familiar to particular social groups or within our society as a whole it (as used by Gee).

Critical Discourse Analysis: A perspective on doing linguistic, semiotic and discourse analysis that places the emphasis on critical standpoint, position, or attitude.

Critical Theory: A paradigm whose research purpose is to expose conflicts, contradictions and hidden structures, and to create knowledge as a catalyst for change.

Deterministic Discourse: The strong, dominant (almost uncritical) view of technology as being essential, and, by association, the skills to use ICTs as being more highly valued and necessary than other skills.

Dimensions of Discourse: The three dimensions of discourse are the object of analysis (text), the process by which the object is reproduced (practice) and the socio-historical conditions which govern this process (as used by Fairclough).

Discourse: Little d discourse refers to language in use. Words, phrases, sentences and the meaning or communicative purpose this form carries with it (as used by Gee). Big D Discourses refers to group identity – a way that an individual thinks, speaks and acts that is recognised by others in relation to the social world (as used by Gee).

Globalisation Discourse: Two themes, one of the information society and the other of the digital divide. On one side the view of Globalization of ICTs in terms of the information society and the opportunities that this offers and the other side the view of Globalization of ICTs as a disadvantage and the lack of opportunities it offers people who are on the underprivileged side of the divide.

Higher Education: Post school study at university.

Information Society: A term used to describe our current society where ICTs are perceived to be at the heart of our economy and our lives.

Recontextulization: The process of extracting text, signs or meaning from its original context (decontextualization) in order to introduce it into another context.

Technological Identity: Peoples world views on ICTs, how they think, feel, value, and use ICTs.

Section 3
Case Study Approaches

Chapter 11
The Role of Case-Based Research in Information Technology and Systems

Roger Blake
University of Massachusetts Boston, USA

Steven Gordon
Babson College, USA

G. Shankaranarayanan
Babson College, USA

ABSTRACT

Increasingly, academic research in Information Technologies and Systems (ITS) is emphasizing the application of research models and theories to practice. In this chapter, the authors posit that case-based research has a significant role to play in the future of research in ITS because of its ability to generate knowledge from practice and to study a problem in context. Understanding context—social, organizational, political, and cultural—is mandatory to learning and effectively adopting best practices. The authors describe some examples of case-based research to highlight this point of view. They further identify key topics and themes based on examining the abstracts from prominent case-based research over the past decade, analyzing their trends, and hypothesizing what the role of case-based research will be in the coming decades.

INTRODUCTION

More than ever before, academic research in Information Technology and Systems (ITS) is emphasizing the importance of its application to real-world problems. Even the best theoretical model fails to make an adequate contribution if its applicability cannot be demonstrated in real-world settings. To understand why, we opine that academicians (particularly in ITS) are increasingly relying on interactions with organizations to identify "interesting" problems that offer

DOI: 10.4018/978-1-4666-2491-7.ch011

Copyright © 2013, IGI Global. Copying or distributing in print or electronic forms without written permission of IGI Global is prohibited.

opportunities for academic research. Moreover, research solutions are being shared with organizations, especially those that researchers interact with. These organizations are looking to apply the research solutions to improve business processes, business strategies, and organizational performance. If research solutions are inapplicable, the contribution of the academic research appears marginalized. We believe that it is important to have a synergistic and symbiotic relationship with organizations, the entities that truly reflect the impact of research. We believe that case-based research has the potential to create and maintain such relationships.

One objective of this chapter is to highlight the importance of case-based studies for the future of ITS research. We posit that future research in ITS must offer more for practice. Benbasat, Goldstein, and Mead (1987) stated that case-based research is particularly well suited to capture the knowledge of practitioners and to develop theories from it. In the past, ITS researchers have learned from organizations. Organizations have innovated with technology and academics have learned how this is done and then extended research by proposing why this was done. In this process, academics have come up with prescriptive suggestions that have, in turn, helped organizations discern how to gain competitive advantage with ITS. Today and in the foreseeable future, in the face of global competition, organizations are and will be forced create and implement innovative ITS solutions. Success stories in Brazil, Russia, India, and China may have to be studied and duplicated in Europe, America, and other parts of the western world and vice versa. To achieve success, practical knowledge is important to understand. In this global economy, understanding practicality implies understanding specific differences and similarities across the globe in terms of culture, organizational structure and size, business practices, and societal influences.

We posit that case-based research offers the ability to provide this understanding. With its ability to study current phenomena in context, case-based research can highlight "why" a particular action or a set of actions were deemed appropriate, "how" these actions were executed, "why" this execution was successful (or not), and "how" culture, business practices, and societal factors contributed to the success (or failure) of the effort. This understanding allows us to transfer the knowledge to other situations while accounting for differences in culture, business practices, and societal influences. More importantly, it also offers the ability to inform decision-makers of the implications of these factors for practice.

Our specific objectives in this chapter are: (1) highlight the importance and benefits of case-based research and differentiate it from other traditional research methods that often substitute case-based research. (2) Present key examples of contributions offered by case-based research to show that such contributions are difficult to obtain from traditional research methods. (3) Examine the "identity" of case-based research by identifying the key themes and the core topics within each. (4) Reveal the trends in terms of how topics have evolved over time, based on the topics identified. This will help us understand the topics that are gaining importance and will offer directions for future research in case-based research.

The remainder of the chapter is organized as follows. The next section presents an overview of case-based research, describing the methodology and emphasizing the key differences between case-based research and related research methods. We then describe the importance of case-based research and the benefits that case-based research offers, along with an accompanying section describing some typical objectives and contributions of case-based research with illustrative examples. Next is a section that describes the methodology, Latent Semantic Analysis (LSA), we use in our analyses to identify themes and core topics.

Following that section, the preliminary results from our analyses are presented. We also discuss the implications of these findings and present a comparison of our core topics with the core IS topics identified by Sidorova, Evangelopoulos, Valacich, and Ramakrishnan (2008). We offer our conclusions and offer directions for further research and discuss the limitations of our research in the last section.

Overview of Case-Based Research

We start with a brief description of case-based research and the characteristics of good case-based research design. Simply stated, case-based research is an empirical inquiry into a contemporary phenomenon within its real-life context where the boundaries between the phenomenon and context are not explicit. Further, case-based research examines a "how" or a "why" question about contemporary phenomenon over which the researcher has little or no control (Yin, 2003). Yin further states that, using case-based research, a real-life event may be examined, as a whole, without losing its meaningful characteristics as organizational and managerial processes and individual life-cycles. Although case-based research is recommended for explanatory questions (the "how" and "why"), it can also be used in exploratory research. Case-based research can be used to describe events and uncover phenomena that generate hypotheses about relationships between events and outcomes. Case-based research may consist of a single-case study or multiple cases. While generalizability of single-case studies is low, multiple cases may support higher generalizability. Last, but, not least, case-based research can be a qualitative or quantitative study.

Management and organizational disciplines have used case studies as a way to collect data and for unstructured analysis. The unique advantage of case-based research is that it can examine research questions within an environment that is rich in contextual variables (Schell, 1992). The design of the case-based research and identification of the right application for case-based research are both critical – incorrect application or poor design may produce inconsistent and often useless findings. Hence, like any other research method, case-based research must be designed well. As stated by Yin (2003) and elaborated by Gordon (2008), good case-based research must address the following:

1. **Clarity of research question(s):** Ideally, the overall objective of the research must be stated and it must be divided into specific research questions examined by the case-based research. These questions, typically, translate into research propositions.
2. **Motivation for research:** Why is the problem interesting and how will solving the problem benefit research or practice. This helps define the contribution(s) of the case-based research.
3. **The context of the case must be described in as much detail as possible:** This helps readers understand how close their own context is with reference to the one described in the case. This, in turn, increases confidence in repeatability.
4. **Unit-of-analysis and format of data collection:** Must be thoroughly described to allow the readers to understand how the case-based research was conducted and to understand the applicability of the findings to their own context of application.
5. **The findings of the case must answer the research questions with appropriate caveats**: This highlights the value of the case-based research and allows the readers to understand the link between the analyses and the research propositions. As stated by Gordon (2008), no such link can be black and white. The research report must therefore include caveats, addressing other factors (besides the one stated in the link) that may have contributed to the observed effects/findings.

The choice of the case or cases for the study is also very critical in case-based research. As the case analysis provides insight into relationships/associations that may need generalization across a larger set of similar cases, the selection of the case becomes critical and poses a significant problem in case-based research. Gerring (2007) offers a classification scheme of nine case study types and provides a technique for choosing cases in each of the nine types.

Finally, this description will not be complete if we did not mention the validity requirements for case-based research. Developing criteria for evaluating case-based research requires logical tests of the validity and reliability of the research design. The important validity requirements are: construct validity, internal validity, external validity, and reliability. The first three are well explained by Yin (2003) for case-based research (and in general by Boudreau, Gefen, and Straub, 2001). Reliability, in case-based research, requires the investigator to follow the same process when repeating the same case process. Documentation of the research process is very important so that other researchers are to be able to repeat a research protocol. Hence, it is important to create and maintain a case database. Reliability is important during the data collection phase. Validity and reliability can thus be built into a well-designed case-based research, thus overcoming many of the key criticisms against case-based research.

The design science research paradigm, prescribed by Hevner, March, Park, and Ram (2004), emphasizes the need to validate artifacts produced by research to demonstrate their usefulness to practice. Van Aken (2005) offers interesting insights into the application of case-based research within the paradigm of design science research. The author also suggests that establishing relevance and usability must be a key requirement for management research, in addition to ensuring validity. Both design science and case-based research share a common goal—that of ensuring that the research artifact is relevant and useful for practice.

In this chapter, we first begin with a brief description of the literature on the importance of case-based research as a research methodology. We then present three examples of case-based research contributions from the academic world to emphasize how case-based research is different and why it can offer more for practice. We further describe the future of case-based research in ITS research and the steps we are taking to motivate case-based research as a research method for future research in ITS. We finally conclude by re-emphasizing the importance of case-based research and why we believe it has a significant role to play in the future of research in ITS.

BACKGROUND AND IMPORTANCE OF CASE-BASED RESEARCH

Benbasat et al. (1987) point out the benefits of case-based research in the context of information systems research. Citing some early proponents of case-based research, the authors note that the method is appropriate when the researcher (and research) are at the early stages of development (Roethlisberger, 1977) or when the researcher is examining practical problems where the participation of the actor and the context of action are both important (Bonoma, 1985). Benbasat et al. also describe the characteristics of case-based research and offer suggestions for conducting case-based research. They show that case-based research is typically, and appropriately, applied in the context of exploring and building theories. This is in sync with the demands of ITS research in the future—ITS theories will be in a state of flux and will be constantly developed and revised as learning continues. While it might be impossible to prove a connection between cause and effect using case-based research, it will be possible to link the effect to the cause through observations and through examination of the chain of events.

Besides the seminal article by Benbasat et al. (1987), a number of articles have highlighted the importance of case-based research (e.g., Stake, 1995; Yin, 2003). Baxter and Jack (1998) note that, in the eyes of both Stake and Yin, case-based research is a constructivist approach. This paradigm "recognizes the importance of the subjective human creation of meaning, but doesn't reject outright some notion of objectivity. One of the advantages of this approach is the close collaboration between the researcher and the participant. With action research methods (Lewin, 1946), the researcher can also be a participant. Whether neutral or participative through action research, the researcher is able to gain insights into participants' views of reality through interactions with them and thus able to better understand their actions. We believe such interactions are keys to learning and will have a significant role to play in the future of ITS research.

Not all case-based research follows a positivistic philosophy, which holds that laws of causation exist and can be deduced by following scientific principles. Information systems case researchers have also followed other philosophies, sometimes labeled "critical" or "interpretive" (see Klein & Myers, 1999). Critical research seeks to focus the reader's attention on social injustice or evil so as to improve social or working conditions, enhance productivity, or increase the value of information systems and technology. Interpretive research operates on the philosophy that reality is in the mind of the beholder. Phenomena are understood differently by different people, and reality is an interpretation of what has been seen or heard. The value of such research lies more in understanding complex social and organizational phenomena from multiple perspectives than in drawing or proving hypotheses. We acknowledge that case-based research, in general, may fall under different categories and need not be restricted to positivistic, interpretive, and critical studies. However, a majority of case-based research in ITS can be classified into these three categories.

Gordon (2008) observes that, in the past, academic researchers have not perceived case-based research as being sufficiently rigorous. He defines rigor as having two components: (a) the research demonstrates what it intended to demonstrate, unambiguously and (b) the research method/experiment is well-described so that it can be repeated. He argues that rigor is built into the research and is not really a characteristic of the research methodology. He compares statistical and case-based research from these two components of rigor and concludes that both research methodologies can be rigorous, but that they take different paths to achieving rigor. Further, as Yin (2003) emphasizes, for the research to be repeatable, it is critical to describe the unit-of-analysis, selection of subjects, techniques for observation, format for data collection, method(s) for data analysis, and it is critical to logically explain how conclusions were drawn from the analysis.

Gordon (2008) also compares statistical and case-based research from the perspective of relevance. He observes that while relevance of statistical studies depends on how well the independent variables explain the variance in the dependent variable, relevance of case-based research depends on how well the intent (what the research wants to demonstrate) of the study matches the chosen context (the case studied). Hence, it is difficult for practitioners to gauge the relevance of a statistical-study to their own application context. Alternately, in case-based research, if the context is described in sufficient detail, the practitioners can gauge the relevance of the study to their own environment.

It is hence argued that case-based and statistical research studies are different but complementary in the nature of their relevance (Gordon, 2008). Case-based research has the ability to increase the relevance of statistical research in the same domain and vice versa. A practitioner attempting to emulate a case-based research outcome gains confidence if statistical studies support that outcome, even if her situation is somewhat different from the case (gauged using the description of context). Likewise, a practitioner/researcher attempting to manipulate independent variables corresponding to a statistical study will gain confidence if a case study supporting the statistical research is similar to his or her context.

Case studies that provide counter-evidence to the results of statistical studies are equally valuable because they help identify situations for which theory based on statistical observation does not or may not apply. Researchers often fail to consider the case study method as a means for testing theory. Jans and Dittrich (2008) found only 23 theory-testing cases in a sample of 689 cases in five fields of business research published between 2000 and 2005 in top journals. Yet, the value of a case study in confirming or disconfirming theory cannot be denied.

Case-based research is one of many research methods applied to ITS. It is difficult, if not impossible to show that case-based research is the best choice for ITS research. In fact, we argue that such a demonstration would be incorrect. It is incorrect because the applicability of a particular research method is dependent on the context of the research. Some methods are more suited for some specific context. In fact, for a given context, there may be several methods that are suitable and the choice is up to the researcher. A second reason why such a comparison would be incorrect is that research methods are complementary. Benbasat et al. (1997), Yin (1993), Stake (1995), and the design science paradigm (Hevner, et al., 2004) all encourage researchers to use multiple methodologies in their research. For example, a researcher may develop a model and prove the correctness of the model mathematically, and next apply the model in one or more real-life settings to show the applicability and usefulness of the model. Alternately, a researcher could use a case study to understand the variables or factors in a research context and subsequently develop a model based on the understanding of those factors and the interactions between them. Similarly, as stated in the preceding paragraphs, statistical methods, and case-based methods may also be used to complement each other in the same research.

One of the key distinguishing features of case-based research is that it is multi-perspective. Case-based research can be exploratory, explanatory, or descriptive. Case-based research may be qualitative or quantitative. Case-based research may also be intrinsic (where the researcher is also part of the case study) or instrumental (understand more than what is observed). It is particularly important for examining social settings because the researcher not only understands the voices and perspectives of the actors but also the voices of the groups of actors and the interactions between them. As stated by Feagin et al. (1991), "it gives voice to the voiceless and powerless." In ITS research, given the number of stakeholders involved, case-based research can help the researcher understand the perspective of each different stakeholder.

Therefore, it is impossible to distinguish case-based research from other methods from any of these perspectives. Table 1 provides a summary of when and why case-based research may be considered a suitable method for ITS research.

As evidenced by the above, case-based research offers several unique benefits over other similar research methods. We next present some real evidence from research studies. These studies were chosen because each has employed case-based research to generate practical knowledge (in ITS) that may be difficult or impossible to generate with other research methods.

TYPES OF CASE-BASED RESEARCH CONTRIBUTIONS: SOME EXAMPLES

In this section, we briefly review three recently published articles that illustrate different ways in which case studies can contribute to ITS research.

Table 1. Summary of research contexts for case-based research

Research Context For Case-Based Research	Why is Case-based Research a Good Option for this Context?
When the research is in its early stages of development – good for understanding the research environment and research context (e.g., independent and dependent variables, causality, factors that affect research environment but are not easily observed) - Exploratory	Research context is not understood completely. Both statistical and quantitative models may be used – both presume an understanding of the research context during the initial modeling phase and have to be subsequently tested for usefulness and validation.
When testing theory that is built using statistical or mathematical or causal models. Typically the case when the researcher adopts the design science paradigm or chooses to increase rigor through multiple methods.	Testing requires the application of a model in a specific setting that conforms to the assumptions made in building the model. Such confirmatory processes are best done using a case-based study in a real-life setting.
When research context requires close collaboration between researcher and participants – both the participation of the actor and the context of action are important for the research	Neither statistical methods nor quantitative modeling methods can help the researcher understand the context of action. This is a unique advantage of case-based research that is not offered by other research methods, except action research. The researcher may choose to participate or stay neutral.
Research context requires multiple perspectives, specifically in social settings.	Allows the researcher to understand the voices and perspectives of the actors but also the voices of the groups of actors and the interactions between them.
Research context requires understanding unique/non-standard practices	No other method is as suitable for examining this context as case-based research.

The first, Silva and Hirschheim (2007), shows how a particular perspective or theoretical lens can draw from a case study to extend existing theory. The second, Koch (2010) illustrates how the differential analysis of two case studies can give rise to hypotheses that almost certainly could not be generated through other methodological means. The third, Chun, Sohn, Arling, and Granados (2009) offers an example of analyzing a unique practice to derive some best practice principles. Unique practices are, by definition, not amenable to study other than through case study methodologies.

Examination of a Process through a Theoretical Lens

Silva and Hirschheim's (2007) work was motivated by the desire to understand and generate theory to explain the success of Strategic Information Systems (SIS) implementations in the context of public organizations in developing countries. They observed that prior research on SIS implementation focused on private organizations in developed English-speaking countries, and that the implications of that research would likely not apply in other contexts. They drew upon prior research to link SIS implementation to an organization's "deep structure," a construct from the organization behavior discipline that includes such factors as "core beliefs and values" and "distribution of power," and whose transformation is believed to be critical for strategic reorientation. The authors examined, through this lens and through the lens of "punctuated equilibrium" (a theory of organizational change explaining the evolution of organizations as being characterized by long periods of stability and short periods of rapid change) attempts by the Ministry of Guatemala to implement an SIS between the years 1998 and 2000. They argued that a case study methodology was appropriate because prior theory for studying the relationship between deep structure and SIS implementation did not exist. But, it is also unlikely, because of sample size problems, that other methodologies could be directly applied to extend the theory of SIS implementation to public sector institutions in developing countries. Furthermore, a case study allowed the researchers to apply a theoretical lens to the SIS implementation process, which would be impossible with a statistical study.

The authors concluded that the case study contributed to the body of knowledge on SIS in public organizations in developing countries in six different areas: factors for the failure of SIS; impacts of SIS on the organization; reasons for SIS implementation uncertainty; whether implementation will entail radical change; when top management support is insufficient; and prerequisites for the acquisition and retention of necessary personnel. They observed that in many of these areas, their research contributed also to the overall understanding of SIS by identifying factors that might not otherwise appear in statistical studies since the samples of such studies were largely homogenous in context, providing insufficient variance on these factors.

Hypothesis Generation via Differential Analysis

Koch (2010) explores the dynamic capabilities that organizations must possess to successfully develop and sustain user participation in Business-to-Business (B2B) Electronic Marketplaces (EMPs). The researcher used an exploratory case study approach, selecting two cases based on their similarities and differences, and performing within-case and cross-case analyses over a three-year period. While it might have been possible to statistically analyze differences in dynamic capabilities between successful and unsuccessful EMPs over a larger sample, there are at least three advantages to the case study methodology in this context. First, it allowed Koch to select cases that were substantially alike in many ways, limiting the impact of factors other than dynamic capabilities that might have caused differences in success. Because many such factors could exist and because the sample size is necessarily small due to the small number of EMPs that succeeded or failed, it would be difficult to obtain statistically valid results in such a study. Second, the case study approach allowed Koch to explore in great depth the process by which EMPs developed, thereby identifying factors that might have been overlooked in a study designed to measure the impact of constructs predetermined from theory or prior research. For example, this study determined that pre-existing relationships and trust, which had not been considered in prior studies, were extremely important to creating successful B2B EMPs. Third, the case study approach required Koch to examine multiple actors in the conduct of the process over time, permitting a close examination of the roles they played, which, by comparing the two cases, Koch determined had an impact on success, and which likely could not be captured with any other methodological approach.

Assessment of Unique Practices

Chun et al. (2009) examine how one company applied "systems thinking" to implement a successful Knowledge Management (KM) practice. Many studies have explored the practice of KM, providing frameworks to guide those seeking to implement a KM initiative and identifying factors leading to the success or failure of KM initiatives. The authors cite many such studies, but also cite research showing that that despite this prior knowledge, many KM initiatives are "expensive, frustrate employees, and lack the focus needed to provide tangible value to the organization." They argue that these initiatives so often fail because the KM literature lacks a theoretical foundation and because it focuses on individual knowledge processes or technology-based systems rather than a holistic perspective on the organization and its KM and other processes. Their case study investigates a KM initiative whose success, they believe, is based on its foundation in systems thinking. The case study allows the reader to assess the extent to which this belief is supported.

It also offers four propositions, deduced from lessons learned in the case, which are amenable to further research. The case study methodology is necessary for this research because the authors know of no other cases where this approach to KM has been adopted and, even if there were others, the number is certainly small.

As illustrated by the above examples, case-based research allows us to examine phenomena that cannot be studied any other way. The examples illustrate that interpreting a practical implementation through a theoretical lens, superimposing theory on an event restricted by social/cultural/business factors to confirm/disaffirm the theory is difficult if not impossible without case-based research. Further, understanding the role of actors/processes on the outcome of an event and determining whether the actor/process impacted the outcome requires case-based research. Such complexities are difficult to comprehend using any other research methods. Finally, the above examples also illustrate that case-study findings can identify opportunities for further research. All of the above have significant implications for future research in ITS, underscoring the important role that case-based research will have to play in future research in ITS. In the competitive global economy of the future, successful organizations will be ones that adapt to the varying economic/political/social/business practices through innovative use of ITS. Case-based research appears to offer the best method to study such adaptation and innovation so as to inform organizations on the best practices from all over the globe.

LOOKING TO THE FUTURE

The Context

There are at least three reasons to believe that case-based methodologies will assume an increasingly important role and become more pervasive in future ITS research. First, ITS research has begun to focus more on process than relationships among theoretical constructs. Among the processes being studied are those of adoption, systems development, outsourcing, strategic planning and even business processes themselves. Cross-sectional statistical studies that take snapshots of many organizations at one or two points in time do not provide a very good understanding of the type of dynamic interactions that occur during the course of a process. Case-based research is ideal in this context.

Second, the theoretical frameworks that have seen increasing acceptance and use in ITS research, such as adaptive structuration theory (DeSanctis & Poole, 1994), favor complexity in the interaction between information technology and its users, the richness of which is hard if not impossible to capture in statistical studies. Adaptive structuration, for example, posits that rules and resources, which people use to accomplish goals, are both the medium and the outcome of the interaction between people and technology. Case-based research is ideally suited to study this type of complexity.

Third, information systems are becoming increasingly social. While it is possible to capture some of the social interaction of their adoption, use, and impact statistically, the social dimension is more amenable to case-based research methodologies, in which triangulation can help integrate multiple sources of data, such as documents, interviews, and direct observation or videos.

Case-based research has crossed several research disciplines, spanning social work, psychology, political science, planning, business and ITS. Even within ITS, case-based research that was once considered peripheral, has gained in stature and importance. It has been applied to several areas of ITS, such as database implementation, strategic planning of ITS, human-computer interaction, ITS project management/implementation and systems design. It is now becoming a unified body of knowledge. Journals such as JITCAR (*Journal of Information Technology Cases and Application Research*) have existed for over a decade. Given the growth and the potential for growth in case-based research, it is important for researchers to understand the key research themes addressed in case-based research and the popular research topics within each theme. This understanding is also termed as "establishing the identity" of case-based research. Establishing the identity of case-based research is an important motivation for us to embark on this next step.

Methodology for Analysis of Themes and Topics

We analyzed the abstracts of journal and conference articles proceedings where the central method is case-based research. Our analysis uses Latent Semantic Analysis (LSA), a statistical method for finding semantic relationships within a body of documents. LSA has been successfully used to predict the subjective ratings of essays made by human readers, to match human categorizations of terms (Laham, 1997), and measure textual coherence (Landauer, Foltz, & Laham, 1998). Our objective is to identify key themes and topics within each to help researchers gain a complete understanding on how and where case-based research has been applied with ITS. Further, we also hope to gain insights into how themes and topics have evolved over time to help researchers target the areas that have not been addressed and to identify the topic(s) that can garner the attention of the practitioners and academics in the future.

Studies in a range of fields have used LSA to analyze abstracts of published articles for the purpose of identifying the central topics and themes in an area of research as we do. This includes the Information Systems (IS) discipline as a whole. As a step towards defining the identity of the IS discipline Sidorova et al. (2008) used LSA to identify the core topics and themes of IS research. Their study was based on a corpus consisting of 1,615 abstracts from three top journals in the field, MIS Quarterly, Information Systems Research, and the Journal of Management Information Systems, published from 1987 through 2006. By using LSA, Sidorova et al. developed levels of research topics from an aggregate level with five topics (information technology (IT) and organizations, IT and individuals, IT and markets, IT and groups, and IS development) through a granular level of 100 topics.

Our methodology closely parallels their application of LSA to develop research topics from paper abstracts, and in using top journals acknowledged as at the core of the discipline. Accordingly, we used abstracts from papers in the six journals in the Association of Information Systems Senior Scholar's basket that were published from 2000 through 2010. These journals were the *European Journal of Information Systems* (EJIS), *Information Systems Journal* (ISJ), *Information Systems Research* (ISR), *Journal of the Association for Information Systems* (JAIS), *Journal of Management Information Systems* (JMIS), and *MIS Quarterly* (MISQ). Because none of those six journals exclusively focuses on case-based research we also included papers from the *Journal of Information Technology Case and Application Research* (JITCAR).

We reviewed the abstracts of the papers published in these journals, and as necessary the papers themselves, to select those that were case-based research. Studies utilizing only large-scale samples or quantitative analysis were excluded, studies that were either purely case-based or that used case-based methods in combination with others were included. This selection process yielded a corpus for analysis consisting of 229 abstracts from the seven journals as shown in Table 2.

In order to develop concepts from a corpus of documents, LSA uses the co-occurrences of words. LSA can determine that two documents have a semantic relationship if the same specific words are not used in each. For example, if different subsets of the words "site," "website," "url," and

Table 2. Counts of case-based abstracts for analysis

Journal	Abstracts
EJIS	67
ISJ	32
ISR	4
JAIS	15
JITCAR	66
JMIS	16
MISQ	29
Total	229

"www" appear in one set of documents but not elsewhere in a corpus, LSA can infer that those documents are representing the same underlying concept whether one uses the words "site" and another uses "website."

While LSA can handle this type of synonymy well, the ambiguities created by polysemy can pose difficulty. One such case could arise if the phrase "business intelligence" was to occur in one group of documents in the corpus and the phrase "business process" occurs in another; the word "business" is in common to both phrases. The degree to which LSA can distinguish their two different meanings depends on how frequently each phrase occurs and what other words appear in the same documents they do. As is often done, we developed a list of phrases, which have words that can be used in alternate contexts, and transformed these phrases into distinct words as a preprocessing step prior to applying LSA. For example, occurrences of "business intelligence" were transformed to "businessintelligence" and "enterprise resource planning" were transformed to "enterpriseresourceplanning."

Several other routine steps were used to preprocess the corpus prior to applying LSA. Commonly occurring words such as "the," "there," and "because" were removed; these are known as stop words and are removed because they are frequently used and do not contribute to distinguishing between contexts. For the same reason, a list of words specific to the abstracts being analyzed was developed and added to the list of stop words. These included words such as "paper," "discussion," and "conclusion" which are not found on generic lists of English stop words. For text processing applications, words appearing in the corpus that are semantically equivalent but have different suffixes or tenses are typically stemmed to terms. The stemming process designed by Porter, is a "process for removing the morphological and inflexional endings from words in English" used to normalize terms in information retrieval (http://tartarus.org/~martin/PorterStemmer/index.html). As an example, the words "analyze," "analyzed," and "analyzing" would be stemmed to the common root term "analy." We used the well-known Porter stemming algorithm for this purpose (Porter, 1980). These preprocessing steps are used to produce a raw term-document matrix with the frequency of each term in each document. With 390 terms from the preprocessing steps and 229 abstracts, for our analysis this was a 390 by 229 matrix. LSA uses a process similar to factor analysis to create factors based on the similarity of terms within documents. Landauer and Dumais (1997) provide a comprehensive explanation of LSA and Larsen and Monarchi (2004) provide a detailed explanation of how LSA can develop research topics from article abstracts.

The LSA methodology produces solutions with a varying number of factors. Solutions were analyzed for their consistency as research topics and coherence when considered as an aggregate of the topics from solutions with fewer factors. The process for identifying themes and topics within starts with the solution that has a large number of factors. If there is a consistent set of terms in each factor, we then attempt to what that factor is, i.e., what research topic in case-based research does that factor represent. If the terms are not consistent, we repeat the methodology with fewer factors in an attempt to create a more coherent set of terms under each factor. The number of factors is reduced in subsequent iterations of the LSA until the researchers can identify a consistent set of topics—consistent in that a coherent set of terms / documents load under each topic. We were able to obtain this consistency when the number of factors was 15. These were our topics. We then continued to reduce the factors to see how these 15 topics would "group and align" themselves, and we found that there are 8 large themes under which the 15 topics loaded. Both these steps were subjective and were determined based on the domain knowledge of the researchers. Our approach was consistent with how prior research had used LSA (e.g., see Sidorova, et al., 2008). We report here on the set of solutions with 8 themes covering 15 distinctive topics. In the next section, we present our preliminary findings from our analyses.

Table 3. Themes and topics in case-based research

Themes (From 8 Factor Solution)	Topics (From 15 Factor Solution)
Systems Development and Adoption	Systems Development Agile Development & IT Innovation Systems Adoption User Acceptance
Collaboration	Collaboration
Sourcing	Outsourcing Offshoring Control
ERP	ERP Implementation Control Critical Success Factors
Knowledge Management	Knowledge Management eGovernment
Group dynamics	Group dynamics
eCommerce	eBusiness Business Strategy
Process Improvement	Agile Development & IT Innovation Critical Success Factors

PRELIMINARY RESULTS

Our preliminary analyses presented in this chapter identified eight (8) core themes that were addressed by researchers using case-based research. This is based on articles that were published in years between 2000 and 2010. These core themes are listed in Table 3. The core themes were identified from the 8-factor solution generated by the LSA methodology. The topics from the 15-factor solution that are most closely related to each theme are also presented in Table 3.

Some themes such as "Collaboration" and "Group Dynamics" have just one topic while other themes have as many as 4 topics. Some topics appear in more than one theme. For example, control is a significant topic in both "ERP" and "Outsourcing." It should be noted that not all case-based publications fit neatly into these eight themes. Ironically, for example, Silva and Hirschheim's (2007) study, discussed in some depth earlier, does not significantly weigh into any of the themes (factors) shown in Table 3. This is both reasonable and instructive. Articles that do not weigh into themes help identify opportunities for new case-based research that will expand knowledge in our discipline on themes that have yet to be adequately explored.

Table 4 identifies the top ten terms weighing into each factor and Table 5 identifies the top 5 articles. In Table 5, we have chosen to include only the top 5 for brevity. Complete matrices of term and document weights on each factor for the 8 and 15 factor solutions are available from the authors by request.

For the most part, the fraction of case-based research that falls into each theme has remained relatively stable over the decade. Figures 1 and 2 illustrate trends in the themes that demonstrated significant changes over the period. Since 2005, the fraction of case-based research in the topics of Systems Development, Group Dynamics, and Agile Development and IT Innovation has been

Table 4. Terms in the 15-factor solution

Factor	Terms (Transformed & Stemmed)
Systems Development	Informationsystemsdevelop, project, role, practic, develop, social, emerg, require, feature, process
Business Strategy	Custom, industry, Internet, strategi, ecommerce, market, service, company, busi, competit
Outsourcing	Outsource, vendor, client, offshore, relationship, cost, firm, project, decis, offshoreoutsourc
ERP implementation	ERP, phase, function, implement, system, manag, structur, company, require, product
User Acceptance	User, resist, users, accept, implement, behavior, system, design, post, softwar
Group Dynamics	Inform, group, analysi, technique, interpret, data, coordin, standard, structur, emerg
eGovernment	Evalu, publicsector, invest, egovern, post, relat, practice, firm, framework, servic
Collaboration	Team, collabor, virtual, task, coordin, distribut, teammemb, offshore, geography, member
Knowledge Management	Knowledge, knowledgemanag, firm, strategi, align, busi, network, practice, technolog, process
Offshoring	Offshore, network, strategi, global, egovern, innov, model, govern, informationtechnolog, econom
Systems Adoption	Adopt, factor, ecommerce, decis, key, organiz, pressur, technolog, social, users
Critical Success Factors	Criticalsuccessfactor, public, implement, project, adapt, compar, identify, chang, manag, strategi
Control	Control, public, network, require, softwar, pattern, behavior, trust, produc, direct
Agile Development & IT Innovation	Agile, innov, softwar, adopt, methodology, agil, chang, network, diffuse, improv
eBusiness	Ebusi, methodology, element, manufactur, offshore, benefit, company, implement, large, chang

increasing, as shown in Figure 1. This figure plots the percentage of papers devoted to individual topics over distinct intervals of time; of the fifteen topics, these three have the highest rates of increase in recent years. Group dynamics typically refers to the understanding of the behavior of people in groups that are trying to solve a problem or make a decision. Given the growth in social networks and the ad hoc teams that emerge due to superior ability (offered by wireless networks and mobile devices) to connect with one another, the observed trend is not surprising. As seen in Figure 1, this topic trended downwards in the early years of the last decade but in the past few years has up-ticked significantly. We hypothesize that this trend is consistent with the emergence of the social network phenomenon. Further, this is an area that lends itself to case-based research and is difficult to examine using most other research methods.

The increasing trend in Systems Development is surprising. As seen from Figure 1, paralleling Group Dynamics, Systems Development shows a downtrend between 2000-2002 and 2003-2005, but from 2006-2008 those trends have reversed and recently these are the topics which have increased the most rapidly. During the early part of the last decade, this might have been expected for Systems Development because researchers have examined systems development extensively since the late 1980s and the topic has posed no new

Figure 1. Topics trending upwards

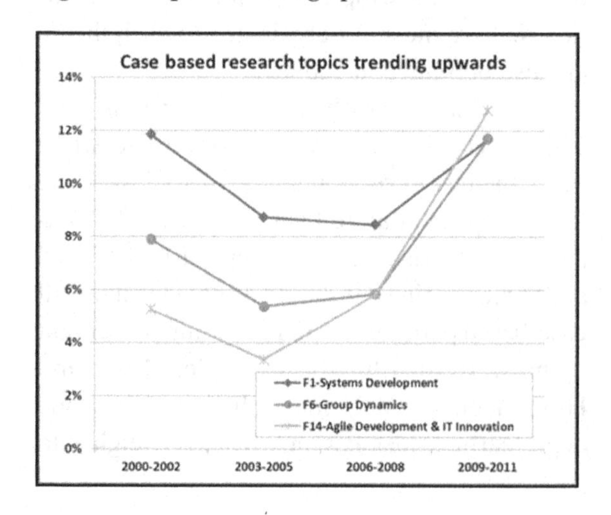

Table 5. Top 5 documents in the 15-factor solution

Factor	Documents
Systems Development	Ovaska and Stapleton-JITCAR-Vol. 12, No. 2 (2010) S Madsen et al.-EJIS-Vol. 15, No. 2 (2006) K Kautz-JITCAR-Vol. 11, No. 4 (2009) P Beynon Davies et al.-ISJ-Vol. 10, No. 3 (2000) K Kautz et al.-JITCAR-Vol. 6, No. 4 (2004)
Business Strategy	Y Yao et al.-JITCAR-Vol. 5, No. 3 (2003) B Shin and K Dick-JITCAR-Vol. 7, No. 2 (2005) TS Teo and P Kam Wong-JITCAR-Vol. 7, No. 2 (2005) S Poon-EJIS-Vol. 9, No. 2 (2000) BL Cooper et al.-MISQ-Vol. 24, No. 4 (2000)
Outsourcing	N Levina and JW Ross-MISQ-Vol. 27, No. 3 (2003) T Cata and V Raghavan-JITCAR-Vol. 8, No. 3 (2006) TF Madison et al.-JITCAR-Vol. 8, No. 4 (2006) P Haried and K Ramamurthy-JITCAR-Vol. 12, No. 1 (2010) O Petkova and D Petkov-JITCAR-Vol. 5, No. 1 (2003)
ERP Implementation	K Furumo and A Melcher-JITCAR-Vol. 8, No. 2 (2006) S Kumar and A Keshan-JITCAR-Vol. 11, No. 3 (2009) A Molla and A Bhalla-JITCAR-Vol. 8, No. 1 (2006) M Newman and C Westrup-EJIS-Vol. 14, No. 3 (2005) HL Wei et al.-EJIS-Vol. 14, No. 4 (2005)
User Acceptance	K Joshi-JITCAR-Vol. 7, No. 1 (2005) EH Ferneley and P Sobreperez-EJIS-Vol. 15, No. 4 (2006) L Lapointe and S Rivard –MISQ-Vol. 29, No. 3 (2005) MJ Gallivan and M Keil-ISJ-Vol. 13, No. 1 (2003) EL Wagner and S. Newell-JAIS-Vol. 8, No. 10 (2007)
Group Dynamics	EM Trauth and LM Jessup-MISQ-Vol. 24, No. 1 (2000) U Schultze-MISQ-Vol. 24, No. 1 (2000) G Rugg et al.-ISJ-Vol. 12, No. 3 (2002) A Powell and SE Yager-JITCAR-Vol. 6, No. 2 (2004) S Manwani et al.-JITCAR-Vol. 10, No. 3 (2008)

continued in following column

Table 5. Continued

Factor	Documents
eGovernment	Z Irani et al.-ISJ- Vol. 15, No. 1 (2005) S Jones and J Hughes-EJIS-Vol. 10, No. 4 (2001) V Serafeimidis and S Smithson-ISJ-Vol. 13, No. 3 (2003) D Gwillim et al.-ISJ-Vol. 15, No. 4 (2005) Z Irani, AM Sharif, and P Love-EJIS-Vol. 14, No. 3 (2005)
Collaboration	A Malhotra et al.-Vol. 25, No. 2-MISQ (2001) J Espinosa et al.-JMIS-Vol. 24, No. 1 (2007) G Piccoli and B Ives-MISQ-Vol. 27, No. 3 (2003) S Seshasai and A Gupta-JITCAR-Vol. 11, No. 4 (2009) J Kotlarsky and I Oshri-EJIS-Vol. 14, No. 1 (2005)
Knowledge Management	SR Nidumolu et al.-JMIS-Vol. 18, No. 1 (2001) MW Chun and C Griffy Brown-JITCAR-Vol. 10, No. 1 (2008) J Owen et al.-JITCAR-Vol. 6, No. 4 (2004) M Alavi et al.-JMIS-Vol. 22, No. 3 (2005) VA Martin et al.-JITCAR-Vol. 6, No. 2 (2004)
Offshoring	A Gokhale-JITCAR-Vol. 9, No. 3 (2007) H Olsson et al.-MISQ-Vol. 32, No. 2 (2008) C Canto and D Gorp-JITCAR-Vol. 9, No. 3 (2007) D van Gorp et al.-JITCAR-Vol. 9, No. 1 (2007) R Heeks and C Stanforth-EJIS-Vol. 16, No. 2 (2007)
Systems Adoption	N Al-Qirim-JITCAR-Vol. 9, No. 2 (2007) N MacKay et al.-EJIS-Vol. 13, No. 2 (2004) SC Chan and EW Ngai-ISJ-Vol. 17, No. 3 (2007) A Lin and L Silva-EJIS-Vol. 14, No. 1 (2005) C BenMoussa-JITCAR-Vol. 12, No. 1 (2010)
Critical Success Factors	VA Cooper-JITCAR-Vol. 11, No. 3 (2009) H Akkermans and K van Helden-EJIS-Vol. 11, No. 1 (2002) B Tan et al.-JMIS-Vol. 25, No. 3 (2008) W Lam-EJIS-Vol. 14, No. 2 (2005) K Pedersen et al.-JITCAR-Vol. 12, No. 2 (2010)
Control	ML Harris et al.-ISR-Vol. 20, No. 3 (2009) N Berente, U Gal, and Y Yoo-EJIS-Vol. 19, No. 1 (2010) V Choudhury and R Sabherwal-Vol. 14, No. 3-ISR (2003) V Kartseva et al.-EJIS-Vol. 20, No. 3 (2010) G Piccoli and B Ives-MISQ-Vol. 27, No. 3 (2003)

continued on following page

Table 5. Continued

Factor	Documents
Agile Development & IT Innovation	A Börjesson et al.-EJIS-Vol. 15, No. 2 (2006) DS Hovorka and KR Larsen-EJIS-Vol. 15, No. 2 (2006) R Vidgen and X Wang-ISR-Vol. 20, No. 3 (2009) K Lyytinen and GM Rose-ISJ-Vol. 13, No. 4 (2003) K Lyytinen and GM Rose-MISQ-Vol. 27, No. 4 (2003)
eBusiness	C Chu and S Smithson-ISJ-Vol. 17, No. 4 (2007) T Dotan-JITCAR-Vol. 4, No. 4 (2002) N Tang et al.-ISJ-Vol. 14, No. 2 (2004) J Versendaal and S Brinkkemper-JITCAR-Vol. 5, No. 4 (2003) B Clegg and D Shaw-ISJ-Vol. 18, No. 5 (2008)

Figure 2. Topics trending downwards

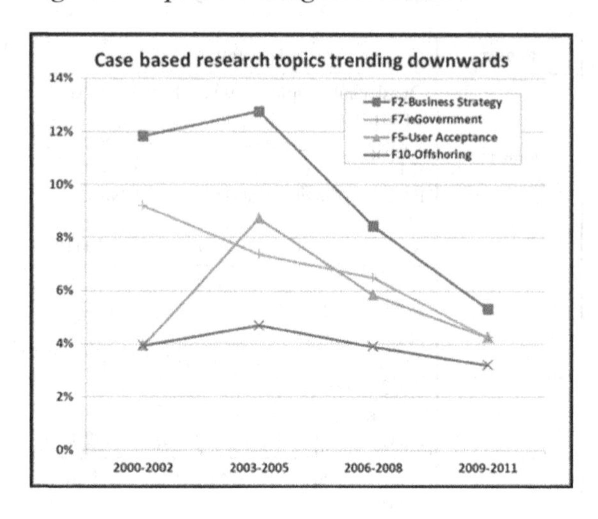

research challenges. However, one factor influencing the recent increase may be due to the downtrend in economy in 2008, resulting in significant cutbacks to IT spending. This motivated the quest to find and use quicker and cheaper methods for systems development, which in turn, motivated research examining the newer methods that were driven by the economic downturn. In particular, the fraction of research in Agile and IT innovation has increased from less than 4% in the 2003-2005 period to almost 13% in the 2009-2011 period. This may account for these two topics generating such interest in recent years.

In contrast, the fraction of case-based research themed as Business Strategy, eGovernment, User Acceptance, and Offshoring has been decreasing, the most rapidly of the fifteen research topics with the trends for these topics shown in Figure 2. The significant drop in research for the Business Strategy topic is surprising, since IT's contribution to setting business strategy has become increasingly important and Business Strategy is an ideal topic for case-based research. The downtick in offshoring is also surprising, especially when considering that ITS is the functional-unit tending to have the most offshoring. Looking at the number of research papers in this topic, a significant majority are based on research conducted in during 2004-2006 period, when offshoring activities were the most prevalent.

The downward trends in eGovernment and User Acceptance are not counter-intuitive. A majority of eGovernment initiatives were implemented in the very late 1990s and early 2000s. For instance, the US Internal Revenue Service started its e-filing in 1990, but after switching over to a superior filing system in 2003, has seen an enormous increase in the number of e-filers. Similarly, a majority of Web-based systems came into play after 1997 and between 2000 and 2005 there was a rapid movement towards M-Commerce and applications for mobile devices. This may be the reason for the upward trend in User Adoption between 2000-2002 and 2003-2005. Since then, research in User Acceptance has dropped off.

We have identified topics ripe for additional case-based research in two ways. First, we have examined articles that have not weighed heavily on any of the topics from our 15-factor solution. Second, we have compared our topics to those Sidorova et al. (2008) identified as "core" to the IS discipline, primarily by their 13-factor solution which identified 13 topics as constituting the "core" of the IS discipline. These are IS Development, IT Adoption and Use, Virtual Collaboration,

IT and Markets, IT for Group Support, Value of IT, IT Management, Project and Risk Management, Research Methods and Instruments, IS Discipline Development, HR Issues in IS, Decision Support Systems, and IT Use by Individuals. Table 6 presents our topics (from the 15-factor solution listed in the middle column) to Sidorova et al.'s topics (from the 13-factor solution listed in the rightmost column). The first column describes how we understand the comparative relationship between the two.

As shown in Table 6, a large number of topics have garnered the attention of both case-based and large-sample researchers. Key among these is IT Adoption and Use. That this topic is split into two topics in our case-based 15-factor solution may simply be a consequence of the need to combine and compress topics as the number of factors decreases (15 in our analysis and 13 in Sidorova et al.). The topic of IS Development, core to the IS discipline overall, is split into Systems Development and Agile Development and IT Innovation in case-based research. This split may be due in part to the fewer number of factors in Sidorova et al., but it could also indicate that case-based researchers examine focused areas within IS Development. Innovation within the IT function, including the use of agile methodologies, has been a common topic in the practitioner literature, so it is not surprising that it has captured the attention of case-based researchers. Collaboration is a central topic for both case-based and

Table 6. Comparing topics in case-based research with topics in IS core

Alignment	Case-Based Topics	Topics in IS Core
Tight	Systems Development	IS Development*
	Agile Development & IT Innovation	IS Development*
	Systems Adoption	IT Adoption and Use**
	User Acceptance	IT Adoption and Use**
	Collaboration	Virtual Collaboration
	eBusiness	IT and Markets
Loose	Control	Project and Risk Management
	ERP Implementation	IT Management
	Group Dynamics	IT for Group Support
No analogous topics in Sidorova et al. (2008)	Business Strategy	
	Critical Success Factors	
	eGovernment	
	Knowledge Management	
	Offshoring	
	Outsourcing	
No analogous topics in case-based research		Research Methodology/Measurement Instruments
		IS Field/Discipline Development
		HR Issues in IS
		Value of IT
		Decision Support Systems
		IT Use by Individuals

* IS Development is split among our factors into Systems Development and Agile Development
** IT Adoption and Use is split among our factors into User Acceptance and Systems Adoption

large-sample researchers. The use of the adjective "virtual" in column 3 seems to be artificially motivated by the inclusion of "virtual teams" as part of the Collaboration factor in Sidorova et al.'s cross loading between their 100- and 13-factor solutions. The difference between the term eBusiness in case-based topics and IT and Markets in core IS topics is largely one of labeling. The sub-topics in Sidorova et al.'s IT and Markets topic include electronic marketplaces, customer service, website design, electronic banking, online consumer behavior, and others that might easily be called eBusiness.

Three case-based research topics—Control, ERP Implementation, and Group Dynamics—have loose analogues to topics that have been identified as core to the IS discipline as a whole. One of these—ERP Implementation—is an area in which case-based research is more focused than the corresponding topic for the general core—IT Management. IT Management, for example includes not only ERP Implementation, but also such topics as information systems planning, executive information systems, and the role of top management. This comparison suggests opportunities for case-based research to extend into areas related to those that have been already explored in some depth. The other two case-based topics—Control and Group Dynamics—are broader in concept than their corresponding IS Core topics—Project and Risk Management and IT for Group Support. This comparison suggests that the core might better be more broadly defined by incorporation of case-based articles in a broader set of journals.

Of great significance, in our opinion, is the fact that six topics identified as core topics in case-based research have no analogous counterparts in the set of core IS topics identified by Sidorova et al. (2008). It is likely that these topics are subsumed under core IS topics such as IS Development and IT Management. For instance, a large body of research in Business Strategy does not examine IT/IS related issues and hence does not appear as a core IS topic. The few that examine IT/IS Strategy (such as business-IT alignment) are probably subsumed under IT Management or IS Development. It is very interesting that Knowledge Management has not been considered as a core IS topic. Additional investigation is required to understand its exclusion as a core IS topic while case-based research in IT includes it as part of its core. It is possible that for periods of time, one or more journals have focused on a particular topic of research, unduly influencing what appears to be central to case-based or general research. The possibility is greater for case-based research because of the more limited number of articles published. For example, it should be noted that four of the five articles most heavily weighted in the Outsourcing factor are from JITCAR.

Looking at the other side, topics considered "core IS topics" for the discipline as a whole, such as Human-Computer Interaction (HCI), Information storage/retrieval, and HR Issues in IS, have not received much attention in the case-based literature. These broad topics, and specific sub-topics, are potentially fruitful areas for investigation. Although HCI has been an important research area in IS research, it is interesting that the research in this area has not utilized case-based research. HCI lies at the intersection of several major fields of study including computer science, design, cognitive and behavioral sciences besides information systems. With HCI, it is critical to understand how and why the individual user derives benefit (satisfaction, quality of decision and/or acceptance), or not, from the interface and technology. Experimental settings that are typically used to derive this understanding, though controlling the manipulated variables effectively, do not examine the user in his/her natural work environment. We believe that case-based research in HCI has the ability to identify the features of an HCI that offer superior benefits to users. Given the proliferation of hand-held devices with different displays and the acceptance of visualization as an analytical tool, we believe that large-scale quantitative studies may become more difficult to conduct and that case-based research is well suited to play significant role in examining human-computer interactions.

Research has also extensively examined information storage/retrieval. Research topics in information retrieval do not easily lend themselves to case-based research. However, more recent research topics in this area such as managing data quality can be examined using case-based research. Research in this area deals with, among other things, the impact of data quality on decision performance and on organizational performance. In these evaluations, a key issue faced by researchers is that there are multiple factors that tend to mediate and moderate the effect of data quality on organizational and decision performance. Case-based research, with its ability to examine the action and process in context, appear to offer the ideal methodology for understanding the effect of data quality on individual and organizational performance.

CONCLUSION, LIMITATIONS, AND RESEARCH DIRECTIONS

In this chapter, we posit that case-based research has an important role to play in the future of research in ITS because of its ability to generate knowledge from practice and to study a problem in its context. To justify our statement, we first described some examples of case-based research to highlight the role of case-based research in ITS and to emphasize the contributions such studies offer. We then analyzed abstracts from over 200 articles that use case-based research in the context of ITS, all of which were published between 2000 and 2010. We used a repeatable methodology, Latent Semantic Analysis to analyze these abstracts.

From this analyses, we identified 8 core themes that have been examined by case-based research in ITS. We further identified 15 topics that fit into these 8 themes. These 15 topics are the key topics examined by case-based research in the period between 2000 and 2010. We further identified trends for how these topics have gained or lost importance over this time period. By comparing our 15 topics to the 13 topics identified as the core IS topics by Sidorova, we identified the mapping between our topics and the core IS topics. This mapping revealed that six of the case-based topics were well mapped to core IS topics. Some of the case-based research topics were relatively narrow in focus, which is characteristic of case-based research, and hence were subsumed by larger and more general topics in the IS core. Interestingly, we identified topics in case-based research that are not part of the IS core topics and hypothesized why this is so. Finally, we identified research topics that are part of the IS core but have not been examined by case-based research at all. We posited that these topics provide interesting research opportunities for case-based research and explained why case-based research may be suitable as a research method to examine them.

The results presented here are preliminary and limited by the extent of the analysis. In particular, a majority of abstracts analyzed came from two primary sources, EJIS and JITCAR. Our results may be skewed by the viewpoint of the editorial review board of these journals. Furthermore, the small size of our corpus may have restricted the factors we have identified in this research. We plan to expand our research to include articles that address case-based research from a larger sample of journals.

Our results may also be limited by our choice to use the abstracts for our analyses instead of the full text. However, Sidorova et al (2008) argue that the literature review and methodology sections of a full text document may introduce more noise into the analyses. Nevertheless, this is an important concern that should be addressed in the future versions of this research.

Our naming of themes and topics was subjective, based on the terms and documents that weighed most heavily into each factor identified by the LSA analysis. Future research should consider a more objective naming methodology, similar to that used by Larsen, Monarchi, Hovorka, and Bailey (2008).

One of the limitations stated by Sidorova et al. (2008), the bias towards North American authors, is not significant in our study. We have included a broad range of both European and North American authors and their research. Another limitation related to the methodology highlighted by Sidorova et al (2008) is that new/evolving research areas may not have a well-defined set of consistent terms, which, can interfere with how LSA works. We are attempting to mitigate the effect of such inconsistencies by carefully preprocessing the terms, manually examining terms for semantic similarities and combining phrases with equivalent meaning for consistency.

Case-based research can offer a number of benefits to IS research. This study has revealed areas of IS research that may not have fully utilized case-based research. Case-based research is a bridge between theoretically solving a problem and its implementation in practice. Given the rapidly evolving IS area and contextual/proprietary manner in which organizations leverage IS/IT, we believe that case-based research has a very significant role to play in the future of IS research. This research is a key step towards highlighting opportunities and needs for case-based research by identifying its trends and its gaps with research of the IS discipline as a whole.

ACKNOWLEDGMENT

Steven Gordon received funding from the Babson Faculty Research Fund for work on this project.

REFERENCES

Baxter, P., & Jack, S. (2008). Qualitative case study methodology: Study design and implementation for novice researchers. *Qualitative Report*, *11*(4), 544–559.

Benbasat, I., Goldstein, D. K., & Mead, M. (1987). The case research strategy in studies of information systems. *Management Information Systems Quarterly*, *11*(3), 369–386. doi:10.2307/248684

Bonoma, T. V. (1985). Case research in marketing: Opportunities, problems, and a process. *JMR, Journal of Marketing Research*, *22*(2), 199–208. doi:10.2307/3151365

Boudreau, M. C., Gefen, D., & Straub, D. W. (2001). Validation in information systems research: A state-of-the-art assessment. *Management Information Systems Quarterly*, *25*(1), 1–16. doi:10.2307/3250956

Chun, M., Sohn, K., Arling, P., & Granados, N. (2009). Applying systems thinking to knowledge management systems: The case of Pratt-Whitney Rocketdyne. *Journal of Information Technology Case and Application Research*, *11*(3), 43–67.

DeSanctis, G., & Poole, M. S. (1994). Capturing the complexity in advanced technology use: Adaptive structuration theory. *Organization Science*, *5*(2), 121–147. doi:10.1287/orsc.5.2.121

Feagin, J., Orum, A., & Sjoberg, G. (Eds.). (1991). *A case for case study*. Chapel Hill, NC: University of North Carolina Press.

Gerring, J. (2007). *Case study research: Principles and practice*. Cambridge, UK: Cambridge University Press.

Gordon, S. (2008). A case for case-based research: Editorial preface. *Journal of Information Technology Case and Application Research*, *10*(1), 1–6.

Hevner, A. R., March, S. T., Park, J., & Ram, S. (2004). Design science in information systems research. *Management Information Systems Quarterly*, *28*(1), 75–105.

Jans, R., & Dittrich, K. (2008). A review of case studies in business research. In Dul, J., & Hak, T. (Eds.), *Case Study Methodology in Business Research*. Oxford, UK: Butterworth-Heinemann.

Klein, H., & Myers, M. (1999). A set of principles for conducting and evaluating interpretive field studies in information systems. *Management Information Systems Quarterly*, *23*(1), 67–97. doi:10.2307/249410

Koch, H. (2010). Developing dynamic capabilities in electronic marketplaces: A cross-case study. *The Journal of Strategic Information Systems*, *19*(1), 28–38. doi:10.1016/j.jsis.2010.02.001

Laham, D. (1997). Latent semantic analysis approaches to categorization. In *Proceedings of the 19th Annual Conference of the Cognitive Science Society*, (pp. 412-417). Mahwah, NJ: Lawrence Erlbaum.

Landauer, T. K., & Dumais, S. T. (1997). A solution to Plato's problem: The latent semantic analysis theory of acquisition, induction, and representation of knowledge. *Psychological Review*, *104*, 211–240. doi:10.1037/0033-295X.104.2.211

Landauer, T. K., Foltz, P. W., & Laham, D. (1998). Introduction to latent semantic analysis. *Discourse Processes*, *25*, 259–284. doi:10.1080/01638539809545028

Larsen, K. R., & Monarchi, D. E. (2004). A mathematical approach to categorization and labeling of qualitative data: The latent categorization method. *Sociological Methodology*, *34*(1), 349–392. doi:10.1111/j.0081-1750.2004.00156.x

Larsen, K. R., Monarchi, D. E., Hovorka, D., & Bailey, C. (2008). Analyzing unstructured text data: Using latent categorization to identify intellectual communities in information systems. *Decision Support Systems*, *45*(4), 884–896. doi:10.1016/j.dss.2008.02.009

Lewin, K. (1946). Action research and minority problems. *The Journal of Social Issues*, *2*, 34–46. doi:10.1111/j.1540-4560.1946.tb02295.x

Porter, M. F. (1980). An algorithm for suffix stripping. *Program*, *14*, 130–137. doi:10.1108/eb046814

Roethlisberger, F. J. (1977). *The elusive phenomena*. Boston, MA: Harvard University Press.

Schell, C. (1992). *The value of case-study as a research methodology. Research Report*. Manchester, UK: Manchester Business School.

Sidorova, A., Evangelopoulos, N., Valacich, J. S., & Ramakrishnan, T. (2008). Uncovering the intellectual core of the information systems discipline. *Management Information Systems Quarterly*, *32*, 467–482.

Silva, L., & Hirschheim, R. (2007). Fighting against windmills: Strategic information systems and organizational deep structures. *Management Information Systems Quarterly*, *31*(2), 327–354.

Stake, R. E. (1995). *The art of case study research*. Thousand Oaks, CA: Sage Publications.

Van Aken, J. E. (2005). Management research as a design science: Articulating the research products of mode 2 knowledge production in management. *British Journal of Management*, *16*, 19–36. doi:10.1111/j.1467-8551.2005.00437.x

Yin, R. (2003). *Case study research: Design and methods* (3rd ed.). Thousand Oaks, CA: Sage.

KEY TERMS AND DEFINITIONS

Case-Based Research: An empirical inquiry into a contemporary phenomenon within its real-life context where the boundaries between the phenomenon and context are not explicit, and the number of entities from which data are drawn is small.

Information Systems and Technology: The application of computer and communication technologies to capture, store, process, and apply data for use in business and other endeavors.

Latent Semantic Analysis: A statistical method for finding semantic relationships within a body of documents.

Research Method: The set of processes and tools that helps conduct research. This can include methods to examine the artifact of interest, collect and analyze data and draw conclusions.

Research Theme: A broad area of research typically encompassing several research topics.

Research Topic: A category of a research interest, typically encompassing many studies and articles.

Research Trend: A change over time in the extent of interest and number of studies of a particular research topic or theme.

Chapter 12
Mapping Participatory Design Methods to the Cognitive Process of Creativity to Facilitate Requirements Engineering

Nicky Sulmon
Centre for User Experience Research (CUO), Belgium

Jan Derboven
Centre for User Experience Research (CUO), Belgium

Maribel Montero Perez
ITEC – Interdisciplinary Research on Technology, Education, and Communication, Belgium

Bieke Zaman
Centre for User Experience Research (CUO), Belgium

ABSTRACT

This chapter describes the User-Driven Creativity Framework: a framework that links several Participatory Design (PD) activities into one combined method. This framework, designed to be accordant with the mental process model of creativity, aims to integrate user involvement and creativity in the early stages of application requirements, gathering, and concept development. This chapter aims to contribute to recent discussions on how user-centered or participatory design methods can contribute to information systems development methodologies. The authors describe a mobile language learning case study that demonstrates how an application of the framework resulted in system (paper) prototypes and unveiled perceptions of learners and teachers, effectively yielding the necessary in-depth user knowledge and involvement to establish a strong foundation for further agile development activities. This chapter provides engineers or end-user representatives with a hands-on guide to elicit user requirements and envision possible future application information architectures.

DOI: 10.4018/978-1-4666-2491-7.ch012

Copyright © 2013, IGI Global. Copying or distributing in print or electronic forms without written permission of IGI Global is prohibited.

INTRODUCTION

Creativity

Creativity is a very important component in the process of innovative product and application design. This is true especially in the early stages of the design process (Snider, Dekoninck, & Culley, 2011), where creativity is often considered an essential prerequisite for designers to come up with innovative ideas for new products or software. However, in some accounts, the role of creativity in design is somewhat elusive, up to the point of becoming the 'mystical' element in design (Fallman, 2003), valuing the designer's values and taste over methodology and control. Adding this somewhat 'mystical' perception of creativity to the often 'ill-structured' early design phases (Guindon, 2012), designers can use some helpful tools facilitating creative thinking. For this reason, several aids have been developed to facilitate and stimulate creative thinking in product design cycles (Lucero & Arrasvuori, 2012; Shneiderman, 2007; Vass, Carroll, & Clifford, 2002); in addition, several academic (e.g. the DESIRE conferences on creativity and innovation in design, ACM Creativity and Cognition) and non-academic (e.g. HOW Design Conference) conferences focus on the role of creativity in design.

The concept of creativity has been approached in literature from a variety of different angles, ranging from a psychological point of view describing (Findlay & Lumsden, 1988) and assessing (Kim, 2006) creative processes, to more practical approaches, such as the development of both physical (Lucero & Arrasvuori, 2012) and digital (Vass, et al., 2002) tools to support creativity and ideation. Specifically in PD, most literature describes strategies on how to create new designs based on the users' current practices (Blomberg, Giacomi, Mosher, & Swenton-Wall, 1933), though some publications have already stressed and explored the role of creativity in participatory design (Bodker, Nielsen, & Graves Petersen, 2000; Dalsgaard & Halskov, 2010; Muller, 2003; Steen, 2001).

When considering IS Requirements Engineering (RE), the impact of creative thinking has long been underestimated as a decisive factor for building competitive and imaginative products that avoid re-implementing the obvious (Maiden & Gizikis, 2001), existing solutions with little added value. However, the requirements analysis has been described as a task of discovery (Robertson & Heitmeyer, 2005) where the requirements engineer has an overview of the business problem, stakeholder requirements and available technology, and is thus ideally placed to innovate (Robertson, 2005). The importance of creativity, both from a philosophical and a more pragmatic point of view, has recently been echoed by requirements engineers in a focus group study (Cybulski, Nguyen, Thanasankit, & Lichtenstein, 2003). The study unveils a conceptual framework for understanding creativity in RE and argues that further research needs to investigate how this creativity in RE may be facilitated.

User Involvement

It is a well-known practice in user-centered design to involve end users at the earliest stages in the design process (Cooper, 2007; Courage & Baxter, 2004; Hackos & Redish, 1998). In early design stages, it is essential for designers to gather as much insights into the users and their contexts of use, in order to create effective designs that address the users' needs and concerns.

The concept of user involvement is not restricted to user-centered or User Experience (UX) design practices and also receives attention in several Information System (IS) development methodologies. The latest generation of these methodologies is characterized by short iterations and continuous feedback from stakeholders to validate the incremental development process, and is often referred to as Agile Development (Highsmith & Fowler, 2001). Nevertheless, one of the limitations of agile methods is that they tend to overlook how this stakeholder and especially the actual end users' feedback can be elicited. It is here that UX approaches can make a valuable contribution to IS development (Patton, 2002). However, there is still little guidance on how the two perspectives can be combined (Ferreira, Sharp, & Robinson, 2011).

Stimulating End-User Creativity

Participatory Design (PD) represents an approach towards computer systems design in which the people destined to use the system play a critical role in designing it (Schuler & Namioka, 1993). Over the last decade, design researchers have shown a growing interest in participatory design, shifting away from traditional user-centered design methods characterized by an expert mind-set (Sanders, 2008). Whereas the latter methods focus on designing for the user, PD implies designing with the user (co-designing). Numerous examples of different co-design methods can be given (Botero, Kommonen, Koskijoki, & Ollinki, 2003; Gaver, Dunne, & Pacenti, 1999; Laurel & Lunenfeld, 2003) and many focus on generating creative ideas. However, we found little research on how a combination of these individual methods can be mapped onto the cognitive process of creativity.

Combining creativity and user involvement as two early-stage requirements reveals exciting opportunities. The question at stake is how we can stimulate the creativity of the users while they participate in the idea generation and elicitation of user requirements on new design ideas. Methods to involve users usually focus either on gathering requirements, or on a more creative approach, using brainstorm techniques to come up with innovative ideas. In order to tap the end users' creativity in a more effective way, a structured methodology that combines and integrates the mental creativity processes and user requirements gathering is needed—such an integrated approach is still largely lacking in the participatory design and information systems literature. Hence, this chapter introduces the User-Driven Creativity Framework (UDC), a participatory design methodology that aims to address the end users' creativity in better ways by structuring and pacing the framework based on creativity theories. Moreover, the framework provides a means to gather initial requirements and develop an information architectural vision to serve as the foundation for further IS development. A mobile language learning case study is used as a practical example to illustrate the framework. The detailed report of this case study should give those agile team members interested in guarding the stakeholders' needs and wishes a hands-on guide to effectively apply the framework in future projects.

BACKGROUND

There has been an increasing interest in the means to involve stakeholders during the initial requirements gathering phase of IS development methodologies. Agile Development comprises a number of incremental and iterative IS methodologies that all share a common vision and core. One of the core beliefs implies continuous feedback and interaction with stakeholders. Some methodologies that fit within the Agile Development framework include Scrum, Extreme Programming (XP), RAD, JAD, DSDM, and Crystal. Although the need for stakeholder involvement is generally accepted by the aforementioned methodologies, the assumptions regarding these stakeholders often vary. Most agile methodologies acknowledge that customer affinity (Fowler, 2006) and a good collaboration between the customer and development team can significantly increase the success of a project (Schuh, 2004), but few actually define how people playing the customer role on the team can learn what the real end user needs and how they can accurately represent those needs. Moreover, agile methods tend to assume that users can say what they want if asked; that users can articulate their tasks, motives and goals (Beyer, 2010). Some believe that customer teams just come up with user stories and throw them at developers, which also relates to the misconceived notion that requirements are just lying out there, waiting to be gathered (Fowler, 2006). Others have a limited view on who the stakeholders are in a IS development project (Augustine, 2005); or tend to neglect the requirements of the actual end user in favour of the project sponsor's preferences.

For instance, the notion of the on-site customer in XP expects the end user to be available at all time as an integral part of the team to ensure continuous interaction during development iterations. In practice, this type of involvement is difficult to reach, especially in those instances where the actual end users are most likely not co-located with the rest of the development team (e.g. commercial products). Consequently, the customer is usually a role played out by one or more XP team members. Likewise, in Scrum projects, it is the product owner who represents the actual end user. It is his responsibility to find out the end users' needs and attitudes, reconcile these with other stakeholder requirements, and communicate this to the rest of the team.

Moreover, while XP calls for real customer involvement, it is not specified what XP customers should do, nor how they should do it (Martin, Biddle, & Noble, 2009). A typical task of the requirements engineer or customer representative is the creation of user stories that describe the basic behaviour of an application, or the writing of a Product Requirements Document (PRD); but it is seldom specified how they should be generated.

JAD processes tend to expect that all stakeholders have the competency to express themselves using one or more modelling techniques; or that different stakeholders, ranging from actual end users to marketing and management staff, can jointly attend a JAD session to successfully elicit user requirements.

These are the cases where user-centered and participatory design methods can offer a valuable contribution to IS development methodologies. Beyer (2010) proposes that effective agile processes make room for a phase 0 or sprint 0 where high-level user research and design can be conducted to reach a rich understanding of real user work practice, culture, goals, strategies and issues. The author's ideal organization of this preliminary phase consists of an 8-week process with a strong focus on contextual design (Beyer & Holtzblatt, 1997) techniques. This looks to be in direct violation of the "Big Requirements Up Front Approach" (BRUF). BRUF claims that too much planning or requirements research in the beginning can lead to significant waste during the course of the project, since most requirements are expected to alter anyway. However, successful projects that have user interfaces with significant impact on how users work have found that some level of up-front design is necessary (Martin, et al., 2009), also referred to as "Big Picture Up Front" (BPUF). Fundamental user requirements are usually stable. The fact that it might look like requirements change after each development iteration can often be attributed to end users not being able to articulate their real needs in the first place (Beyer, 2010). Although we approve of the BPUF philosophy and confirm that contextual inquiries and field research provide excellent means to gain profound insight in the end user's needs and desires, we also acknowledge the reality that most companies are not inclined to budget the time required to adequately apply these means. Therefore, we propose the User-Driven Creativity (UDC) framework as a viable alternative to enable requirements engineers or customer representatives to efficiently attain active stakeholder participation for initial requirements gathering within a limited amount of time. The approach takes into account all stakeholder groups with a specific focus on the actual end users, who might not be able to think in terms of diagrams and workflows, but whose participation is critical for a project's success.

The framework matches with the envisioning iteration 0 in Agile Model Driven Development (AMDD), which is divided into two sub-activities, namely initial requirements envisioning and initial architecture envisioning (Ries, 2011). The UDC framework provides a suitable work process to execute these activities and reach their respective goals, namely building a shared understanding and an initial information architecture that has a good chance of working.

THE USER-DRIVEN CREATIVITY FRAMEWORK

Motivation

It is not self-evident in early-stage requirements gathering to encourage users to be as creative as possible and to elicit the most important ideas and knowledge from end users. The observation made in BRUF that requirements tend to change in the course of the project can, at least in part, be attributed to the fact that users have difficulties in expressing their tasks, motives, and goals. System requirements are not readily available inside end users' heads, waiting to be gathered. As these requirements are not readily available, gathering them is an activity that takes time, and effort. In order to make this phase as successful as possible, it is important that end users are prepared in the best possible way to comment on their experiences, goals, and tasks.

This idea of preparing end users for requirement gathering triggered the authors to create the user-driven creativity framework. The framework was created to allow researchers to better capture the participants' knowledge and inspiration during the actual requirements gathering/PD sessions by taking theories about creativity into account when planning and executing the workshops.

Creativity as a Process: Theoretical Foundations and Method Description

In this section, we will discuss the theoretical foundation of the user-driven creativity framework, and its various stages. We also point out benefits and drawbacks of the method.

To devise the blueprint of the general methodology, we set out to take the mental mechanisms that steer the process of creativity into account. Creativity as a cognitive process is typically divided into four or five different stages (Boden, 2003; Bullinger, Müller-Spahn, & Rössler, 1996; Csikszentmihalyi, 1997; Weisberg, 1986).

- The first stage, preparation, is characterized by a deep immersion into the problem. One analyses the situation and collects all relevant information to reach a problem definition.
- The second stage is described as an incubation period during which the problem is being processed passively. The problem and the related information that have been collected during the preparation phase, now have the time to ripe in one's subconsciousness.
- The third stage, inspiration, is that brief moment where a possible solution or new insight suddenly occurs. A famous example of this stage is the Eureka effect attributed to Archimedes.
- The fourth stage is the transformation stage, in which the insight or idea needs to be carried through. Some authors discern two separate stages within the transformation stage: evaluation, i.e. deciding whether the idea is valuable, and elaboration, i.e. validating and communication the idea (Csikszentmihalyi, 1997).

Figure 1. User-driven creativity framework linked to the cognitive process of creativity

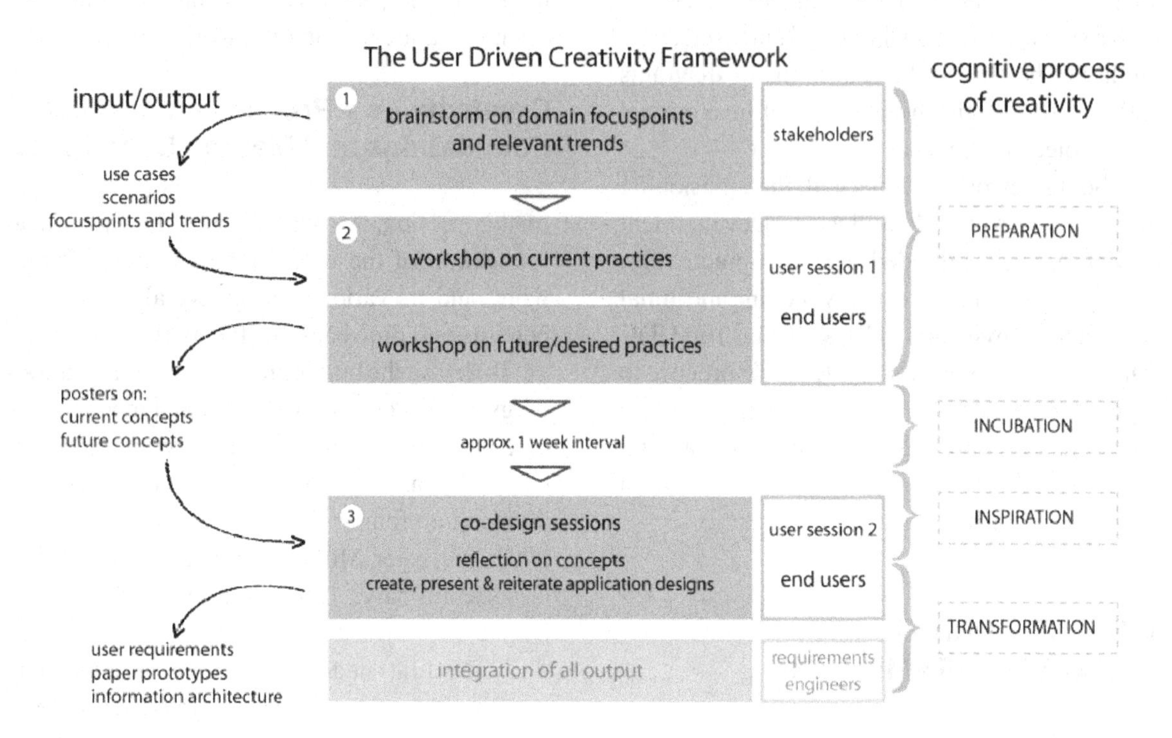

The procedure of the User-Driven Creativity Framework has been mapped to the creative process as described above. Figure 1 shows how the creative stages correspond to the different phases of the methodology.

Preparation Stage

In the user-driven creativity framework, the preparation stage consists of two activities. An initial brainstorm with subject matter experts and other stakeholders (e.g. marketing or management staff, developers, product managers, system installers, etc.) serves to collect as much background information as possible. From this information, the main focus points of the following end user sessions will be determined. The initial preparatory brainstorm aims to arrive at a joint understanding of the main complexities and issues surrounding the topic at hand. This joint understanding is used as input in the first end user session. During this first, three-hour user session, actual end users are encouraged to actively engage with each other in small group discussions and general presentations to analyse the relevant information based on the focus points as determined at the initial brainstorm. Specifically, the first user session is divided into two parts, the first part being a reflection on current practices in the domain. The second part, in its turn, focuses on the desired future practices, and the technologies facilitating that future situation. The division between these two parts can help the requirements engineer to reach a problem definition (Gause & Weinberg, 1989). For both parts, the results of the discussions are summarized on a poster and presented to the other groups, with a short discussion afterwards. Having these short presentations and discussions has a double advantage. First, the groups can learn from each other, swap ideas, and maybe even inspire each other for later assignments. Second, the presentations are very informative for the researchers or requirements engineers, since the participants' presentations already start to point out user requirements.

Together, the preparatory session and the first user session constitute a typical preparation stage, in which all participants are deeply immersed in the problem.

Incubation Stage

At the end of the first user session, participants are informed that, during the second user session, they will be asked to build a concrete application, i.e. a low fidelity prototype, based on the discussions on current and future practices from the first user session. This notification aims to trigger the incubation stage during the time between the two user sessions. Often, this incubation stage is neglected in participatory design research: participatory design sessions often jump straight to the inspiration phase after preparation. However, the incubation stage has been described as 'the most creative part of the entire process' (Csikszentmihalyi, 1997) and it has been noted that inspiration and ideas emerge especially in stages of relaxation (Abra, 1989). Figure 2 shows the performance of creativity in relation to the level of activity.

At best, typical participatory design or brainstorm sessions report on merely offering a coffee or lunch break between briefing participants on the intent and the actual working part of the session. In the user-driven creativity framework, leaving a week between the brainstorm session and the participatory design session has specific advantages. As both sessions take approximately three hours, combining both sessions into one day creates a full-day programme. Apart from availability of participants being problematic, the results of the (critical) second half of a full-day session probably suffer from the participants' fatigue. Instead, programming two half-day sessions—with a week in between—enables participants to start the second session with renewed energy and enthusiasm. Typically, when meeting again at the start of the second session, the atmosphere is prominently more familiar than during the first user session; participants heartily greet each other and engage in informal conversations, which we believe in turn leads to a more fruitful and fun design session. Leaving a few days between the two phases also offers the participants some 'room for thought,' effectuating the desirable incubation stage. On the one hand, participants remember the outcome of the first session in a

Figure 2. Performance of creativity (according to Deutsche Gesellschaft für Gedächtnistraining e. V.)

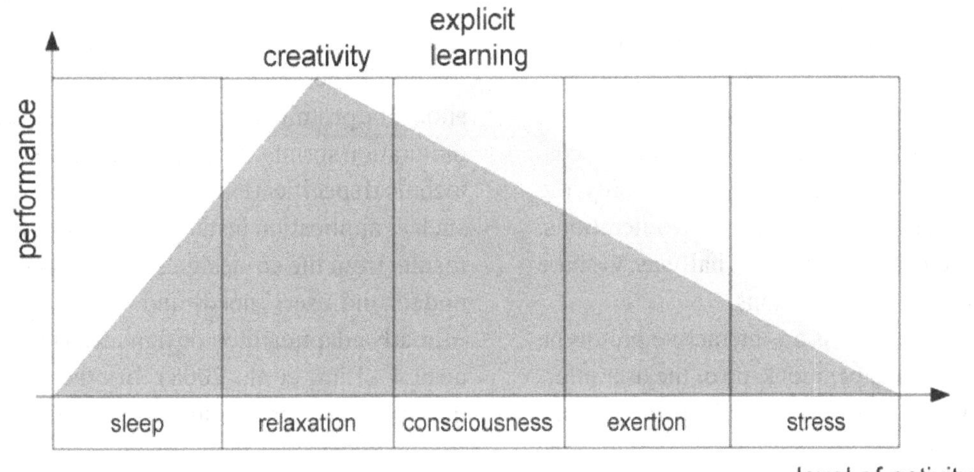

lively way. On the other hand, it offers the participants time to process that outcome (at least on a subconscious level), making the second session more productive. In this light, we would recommend a break of at least 3 days to a maximum of two weeks.

Based on the theoretical foundation presented above, we believe that separating the first user session and the second, participatory design session generates more diverse results than workshops that are conducted on a single day. In our own experience, participants in single-day workshops often only remember what was said during the preparation, trying to copy exactly that. We argue that the break of multiple days between sessions allows participants to truly contemplate new ideas or new paths, and expand their minds. In this perspective, the break is sufficient to let participants remember what was said during the previous session, while idea generation is not hindered by a too-vivid recollection of the previous session.

Inspiration Stage

'Inspirationalists' point out that luck favours the prepared mind, thereby turning to the study of how preparation and incubation lead to moments of illumination (Shneiderman, 2000). The methodology presented here aims to have these prepared minds geared up for entering the next creative stage and maximize inspiration in the second user session. In itself, this is a co-design session, serving as a way to make the outcome of the first user session more concrete using participatory paper prototyping, as prototypes can 'help users articulate their needs' (Beaudouin-Lafon & Mackay, 2003). Ideally, the second user session is split up in two iterations, with intermediate presentations halfway, to share some first ideas and intentions.

A paper prototype is an interactive prototype that consists of a paper mock-up of the user interface (Arnowitz, Arent, & Berger, 2006), often used in participatory design sessions. In participatory design, paper prototypes are especially useful to 'actively engage users in creating and exploring design ideas' (Lim, Stolterman, & Tenenberg, 2008). Muller et al. lists some of the benefits of this way of prototyping, the most important for our purposes being:

- It enhances communication and understanding through creating and discussing concrete artefacts.
- It allows participants to express their ideas directly, ensuring an improved incorporation of new ideas.

In participatory prototyping, it is important to know what end users can and cannot do (Beaudouin-Lafon & Mackay, 2003): end users are no designers. It is preferable to let the end users freely 'design' their own prototypes, with little guidance from interaction designers that moderated the sessions, to get more pure design ideas from the users, unmediated by too much designer's input. In creating the prototypes, participants themselves start making the shift from the inspirational stage to the final creative stage of transformation using participatory design techniques: participants create a concrete prototype to communicate and, ultimately, evaluate their ideas. Researchers or requirements engineers should wait for these sessions to be finished before they start integrating the input and translating these into a prototype that takes into account other critical aspects. Usability, user experience, and other design guidelines should be prominent in the ensuing design process or iteration sprints, as well as a translation to more technical specifications, taking into account issues such as application installation and updating. The results from the co-design session can be used to understand users' needs and values, rather than 'directly adapting their design ideas into the □nal design' (Lim, et al., 2008). In other words, the designers and professional researchers finalise the design process themselves.

Transformation Stage

To aid this transformation process, a clear role division between facilitators during the user sessions is advisable. While an active moderator is essential during the user sessions, an extra researcher can take notes during those sessions, especially during the 'plenary' presentations of posters and prototypes. As mentioned before, the intermediate presentations already hint at several user requirements—these requirements, and other ideas from the presentations can be captured immediately. This basic information can provide a solid starting point for later in-depth analysis of the results.

Separation of (End-User) Roles during Sessions

During the end user sessions, it is important to decide whether to keep different types of end users together, or to split them up in different user sessions. In the UDC Framework, we argue that it is advisable to keep end user roles separated by organizing separate sessions.

An important reason for keeping end user roles separated is to make sure requirements engineers can pay attention to everyone's opinion: in mixed group discussions and co-design sessions, it can be difficult to ensure that everyone participates (Paulus, Dugosh, Dzindolet, Coskun, & Putman, 2012), making the outcome biased towards the opinion of the most dominant person in the group. To avoid session outcomes of mixed groups that represent mainly the view of one end user group, end user groups can be divided into separate sessions. In this light, one distinct benefit of separating user groups is the clean 'separation' of the sometimes quite different points of view of the groups: mixing groups would make it difficult for requirements engineers to pinpoint exactly which parts were considered important by which group.

Separating groups can also be a consideration for the initial brainstorm involving other stakeholders, especially when factors like hierarchy might influence the discussion outcomes. A JAD session might even be appropriate for conducting the initial brainstorm under the circumstance all stakeholders share a similar level of requirements modelling skills. In any case, we advise against a joint session for actual end users and other stakeholders; having an overview of uncompromised end user requirements offers unmatched value.

Although separating the end user groups allows for a more clear-cut distinction between their points of views, it does leave an extra interpretive step to the design team. Mixed group participatory design results already represent a consensus of sorts between the groups. In contrast, separating end user roles allows requirements engineers to get a clear view of the distinct end user groups' requirements, and leaves the trade-offs between potentially conflicting viewpoints to be made by the requirements engineers. These trade-offs can be validated afterwards, for instance in user tests of application prototypes. Separating end user groups does, however, prevent a creative merging of ideas from different sides of the spectrum. It should be noted that while this might be problematic for user groups that do not normally interact with each other, this can be considered less of a problem for user groups that do collaborate closely on a regular basis and therefore are familiar with each other's situations.

Case Study: A Mobile Language Learning Platform

We chose mobile language learning as a concrete case for our research. While some authors have used PD for developing language learning applications (Cardenas-Claros & Gruba, 2010; Zaphiris & Constantinou, 2007) and others for mobile applications (Massimi, Baecker, & Wu, 2007; Spikol, 2009), mobile language learning is a relatively novel application domain, especially suitable for the User-Driven Creativity Framework. Recent studies point out that mobile learning is not simply a variant of e-learning (Sharples, 2009); further research on design research methods is required (Kramer, 2009), making the application of the UDC framework very appropriate.

Mobile learning is not simply a variant of e-learning; it's a field that has recently seen a new wave emerging with the true arrival of the mobile Web and a significant market penetration of smart phones. Especially in Europe, this happened only very recently, with 2010 often being quoted as the year of the mobile Web, since many of the barriers blocking access to it finally broke down. Device and bandwidth costs have decreased, while the device's capabilities in terms of multimedia processing, connection bandwidth, storage, etc. have increased considerably. These mobile devices are radically changing the learning landscape: they allow learning to become a truly mobile, interactive experience (m-learning). Users can learn anywhere, at any time.

Profound changes in computer usage brought about by social networking and User-Generated Content (UGC) are challenging the idea that educators are completely in charge of designing learning. With social networking and UGC, learners all over the world can communicate and exchange learning material without educators having to intervene. The combination of social networking and mobility create a changing learning landscape, offering entirely new ways for people to interact with information. This new way of interacting and learning creates a previously unexplored field of possibilities and challenges.

In this context, the aim of our research was to design an adaptive electronic learning environment for personal mobile devices. There are a number of challenges in this, both on a technical and an educational level. On a technical level, the platform had to be compatible with multiple devices: the learning platform had to be optimized taking into account the possibilities and limitations of the device used. On the educational and design levels, the platform had to offer a coherent, pleasant, and meaningful user experience, keeping the pedagogical focus points in mind.

In order to cope with these challenges, we were faced with many design issues. For instance, how should we design for effective and pleasurable mobile learning? Which (combination) of methods is most appropriate? What are the user requirements for new mobile and adaptive learning system? These and other questions will be addressed in the description of the participatory design process using the UDC framework. In this discussion, the main focus will be on the methodological framework. By giving a detailed description of the case study procedure, we hope to offer requirements engineers or end user representatives a hands-on guide on how to elicit user requirements and envision possible future application information architectures during the requirements gathering phase.

Participants

A total of 30 individuals participated in the study. We recruited language teachers (n = 10, Mage = 38 years, all female) as well as currently enrolled language learners (n = 12, Mage = 27 years, 8 male and 4 female). In addition, other participants included interaction designers (n = 4, Mage= 29 years, 3 male and 1 female) and pedagogical researchers (n = 4, Mage = 36 years, all female). Both teachers and learners were recruited through two different channels: a call for participation was launched in a newsletter from a company offering language courses for professionals and from a language institute targeting a broader audience. In this way, we were able to include teachers and learners from both professional and more recreational courses.

Apparatus

For the first user sessions, all groups received blank posters and a variety of markers to create their mindmap or scheme of concepts. For the second user sessions, we provided cardboard mock-ups of enlarged smartphones that could hold a number of A4 sized sheets representing different UI screens. In addition, participants had access to various materials such as sticky notes, a printout of interface icons, stickers, markers, etc. For inspirational purposes, we projected the focus points, user scenarios, use cases, and user aspects in a presentation loop on the wall. To record the user sessions, we set up one digital video camera and one audio recorder as backup. One of the researchers present took notes and made a mindmap of the participant's ideas, comments, and discussions during the participants' presentations.

Procedure

In order to address the design issues relevant to mobile language learning, we used the UDC framework. This methodology was used to elicit m-learning user requirements from both language learners and teachers. The participatory design process especially targeted 'learning experience' requirements (i.e., how can the product be made enjoyable, satisfying and motivating) and the mobile context, rather than focusing on learning objectives and didactic approaches (Parsons, Hokyoung, & Cranshaw, 2006).

Preparation Stage

As described previously, the process started with a brainstorm session with pedagogical researchers and interaction designers. This preparatory brainstorm session was organized to identify important themes concerning (e-)learning. The session, lasting two hours, actively engaged researchers and designers with each other in small group discussions and general presentations to analyze and collect as much relevant information as possible. During this brainstorm session, the main focus points for the sessions with teachers and learners were set. These focus points include themes such as motivation, collaboration, interaction, preferred media, (shared) content, etc.

In a second step a few days later, the first user session was conducted with end users. In the case study, this amounted to brainstorm sessions with teachers and learners. Teachers were divided into three groups, learners into four groups, each group containing three to four members. The groups consisted of a mix of participants recruited through the two different channels to ensure a representation of divergent motivations to learn languages, ranging from professional (promotions) to more personal (cultural interest, holidays, relationships, etc.). Each session took approximately three hours. It started by a short presentation on the project and its goals, making the aim of the project more tangible using a concrete user scenario.

After this short introduction, participants were asked to think about their current language courses in a first assignment: which aspects did they like, which aspects could be improved? The main focus points during this assignment were the themes identified during the initial brainstorm with pedagogical experts. The results of the discussions were summarized on a poster and presented to the other groups (see Figure 3), with a short discussion afterwards, in which groups could learn from each other, swap ideas, and inspire each other.

In the second part of the user session, the participants were asked to think about future language learning and how advanced mobile devices might fit in this context: they were asked which elements from the posters could be interesting, would be rather inappropriate for mobile language learning, etc. To facilitate this discussion, three user scenarios in the form of short texts explaining a potential learning situation in a specific context were presented. These scenarios served to remind the participants of the project goal, and to help them envision appropriate contexts for future mobile language learning. Additionally, a quick overview of trends in mobile environments was presented, referring to aspects such as social networks, user generated content,

Figure 3. User presentations

ubiquitous technologies, nomadism, and personalization. Offering this overview was especially important in this case because for most people, learning has a very static connotation, with a classroom and a teacher up front and students responding when appointed. The trend overview aimed to encourage participants to break out of the traditional, rather rigid learning paradigm, and think more in terms of the possibilities offered by mobile new media. Again, all groups created a poster with ideas around these topics and presented it to the other groups. Especially during these plenary presentations of posters (and, in the second user session, of prototypes), one facilitating researcher took notes. In this way, requirements and other design ideas from the presentations or emerging from the ensuing discussions were immediately captured in mind maps. These mind maps, initially structured per poster (in the second user session per prototype), were afterwards enriched with links between similar ideas and requirements, providing a solid starting point for in-depth analysis of the results by affinity diagramming.

After a final discussion involving all groups, the first user sessions were concluded.

Incubation and Inspiration Stage

The third part of user-driven creativity framework, i.e. the second user session, followed after a one-week break. The one-week break served as the essential incubation stage, in which participants could process the outcome of the first user session (on a subconscious level). The same teachers and learners were engaged in participatory design sessions of two to three hours to create concrete representations in the form of paper prototypes (based on their ideas on mobile learning applications from the first user session). To facilitate these participatory design sessions, the process and importance of paper prototyping was explained to the participants using a video in which paper prototyping was demonstrated. In addition, the results of the brainstorm sessions recapitulated and the posters and user scenarios materials hung up on the wall. The participants were asked to create their ideal mobile language learning application and received no further constraints. In order to create a paper learning application prototype, participants received a variety of creative materials, ranging from a giant smart phone in cardboard to post-its, coloured pens, and printed versions of common user interface elements. The session was divided into two iterations, with intermediate presentations in the middle of the workshop, to share some first ideas and intentions. At the end of the session, concluding group discussions were held.

Transformation Stage

The transformation stage took place after the last user session and allowed the researchers or requirements engineers to process the outcomes and translate these into communicable forms for the rest of the development team. The brainstorm and co-design sessions resulted in seven prototypes of language learning applications (see Figure 4). Based on these prototypes, usable application designs still had to be made by the interaction designers. Some prototypes created by the participants did not have a clear application concept, incorporated too much functionality to be usable on a mobile device, or sketched a general application framework while others elaborated only on one specific application part. However, since the main application goal was to get a varicty of application ideas and feedback from end users, and not to end up with a specific, instantly deployable application, our approach to participatory design worked quite well. A lot of different ideas and viewpoints were discussed during the sessions; the most important issues in mobile language learning emerged in different forms in various prototypes. This provided the designers with a wealth of design ideas and enough insight to draft an initial information architecture to complete the architectural envisioning task of the AMMD lifecycle.

The next paragraphs present an extract of some insights from the user sessions preceding the prototyping sessions. By giving a short summary of these insights, we aim to demonstrate how we were able to derive high-level user requirements that are likely to remain stable during the course of the project, but nevertheless provide an important foundation for further development activities.

Figure 4. Paper prototypes

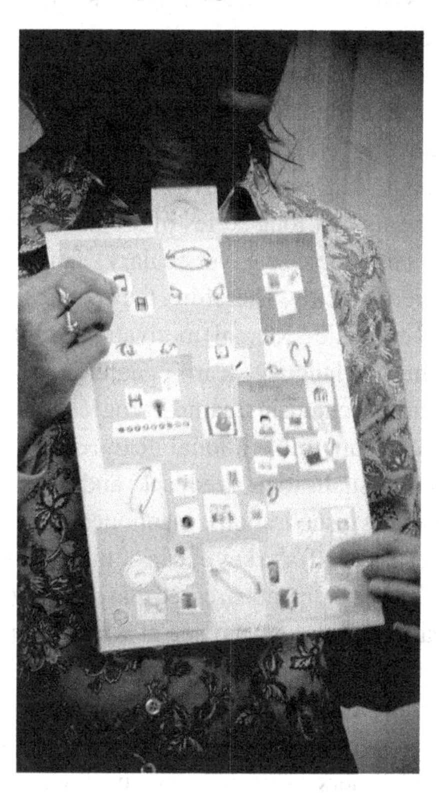

Progress Visualisation and Feedback, Rewarding, and Motivation

The initial workshop among pedagogic researchers and interaction designers identified the themes of progress visualisation and feedback, rewarding and motivation as being very important. However, the sessions with language teachers and learners not only provided us with ideas about how to design a mobile language learning application that motivates users and provides good feedback, but also pointed out that these themes are very interrelated. This deep relation between these aspects that emerged from the user sessions resulted in an integrated strategy in incorporating motivational aspects, rewarding, and feedback in the resulting application.

Motivation, a topic frequently returning during discussions, is an important, multi-faceted topic in mobile learning (Jones, Issroff, & Scanlon, 2007). Co-design participants (teachers as well as students) looked for ways in which working on a mobile device can be rewarding and interesting, keeping the students' attention levels high. While mobile learning can initially be motivating due to a novelty effect (Montero Perez, Senecaut, Clarebout, & Desmet, 2010), it was evident to participants that long series of similar exercises on the same topic (e.g. vocabulary) can become tiring very quickly. Therefore, they came up with several extrinsic (external) motivational factors to make mobile language learning interesting, such as reward mechanisms, games, progress tracking, etc. Important motivational features included progress visualisation, feedback, and rewarding.

Progress visualisation is mentioned in Montero Perez et al. (2010) as a key extrinsic motivator. The workshop participants stressed that it is important for learners to know and see at what point they have arrived, both at a high level within the entire course, and at a low level within specific exercises. At a high level, the progress indication is important to keep learners motivated, and give them a sense of goal, instead of having them wade through a pool of seemingly endless exercises. Reaching regular in-course milestones and goals can serve well as a mechanism to keep motivation levels high. This mechanism offers the learner a sense of achievement after reaching particular goals.

Progress tracking can work as a motivator in another way, as well. Learning pace is at least as important. The pace set by the teacher or the learning application must suit the learner's needs. When establishing the pace of learning, it is also important to keep in account the medium: pace will be quite different in a classroom, compared to a mobile application. When using a mobile application, user concentration generally is quite low: often, the environment provides additional, distracting stimuli. This means that exercises and learning tasks on mobile devices should be short, simple tasks (e.g. picture tagging).

Evaluation and progress are tightly linked potential motivators, although progress in itself is a more intrinsic motivator, and evaluation (results) a more extrinsic motivator. Ideally, progress and evaluation are linked to the extent that they run almost completely in parallel. As already mentioned above, most learners prefer a constant evaluation, rather than the classic model of one test/evaluation at the end of a course. Teacher guidance is very useful for constantly tracking and evaluating the progress made, and identifying and tackling areas the learner is struggling with. Moreover, a competitive element can also be built into the evaluation. By allowing learners to share the results they have achieved on social network sites, learners can compare their progress, and make sure they stay on the same level as their peers.

Feedback, when used right, can also be seen as a means of personalisation. Immediate feedback, completed with additional explanations and examples at the appropriate time can make learners understand the material better. In addition, it gives the learner the impression that the application really 'helps': leaving out unnecessary explanations when all goes well, but jumping in with an extra helping hand when things get difficult, without trying to force the learner to go through all additional explanations if they are redundant.

Especially rewards received a lot of emphasis from both learners and teachers during the brainstorm session. Teachers pointed out the necessity of giving rewards; they referred to physical rewards like candy or fake medals. Learners took this idea to the digital world and referred to achievements, which are currently very popular in the video game industry. The introduction of achievements and experience points (giving players a higher status, etc.) or goals makes games such as MMORPGs virtually endless, never becoming repetitive or boring (Ng & Wiemer-Hastings, 2005).

Rewarding in mobile language learning can be tightly linked with the learner's progress: the progress can be represented visually in graphs. The further a learner progresses in the application (and thus, the better the learner masters the language) the more credits the learner can achieve. A similar mechanism could be introduced in language learning, to prevent learning and courses from becoming boring.

The insights in the relation between progress visualisation and feedback, rewarding and motivation resulting from the two user sessions proved to be of invaluable importance in the later development of the language learning application. This observation suggests that it is important that the focus points determined during the initial brainstorm with subject matter experts should not be treated too rigorously. In other words, researchers moderating the end user discussions should not be too rigid in imposing the discussion subjects. While the focus points should serve as input for the user sessions, it is important to let the end users decide for themselves which elements are foregrounded, and which are considered less important. Unexpected insights and interesting links between topics can emerge when the end users are given the freedom to select the most interesting and important discussion topics.

FUTURE RESEARCH DIRECTIONS

The mobile language learning case study presented above is a case study in which the UDC framework is used to unite participatory design principles with a more theoretical foundation of creative processes. Future work includes a refinement, and an in-depth evaluation of the method, the latter aiming to confirm the strengths of the UDC framework. In addition, the framework will be applied in design processes in other, novel application domains, in which a creativity-focused, user-oriented approach is appropriate.

Future sessions will help in specifying the method in more detail, e.g. making clear whether participants should collaborate in pairs or in small groups during the sessions, and determining the optimal number of participants per session. In addition, future experimentation with the framework can provide extra information on whether end user group separation during the sessions generates important differences in the outcome or not.

In order to assess the framework in detail, a comparative study will be executed, in which a control group of participants works in a more traditional participatory design or JAD session, while another group of participants takes part in a series of brainstorm sessions and participatory design sessions such as prescribed by the UDC framework. A preliminary creativity assessment of the individuals will have to determine which participants can take part in the sessions. By selecting equally creative participants in this way, personal creativity levels can be ruled out to some extent in comparing the outcomes of the parallel trajectories.

CONCLUSION

This chapter describes the User-Driven Creativity Framework as a contribution to Participatory and Information System Design by integrating several PD methods into one combined method that structures and supports the cognitive process of creativity. The framework aims to offer requirements engineers or stakeholder representatives in an IS development team the necessary guidance to extract user requirements and derive an informed information architecture in the early stages of a system development process.

The UDC framework is illustrated by a case study based on a mobile language learning application design process.

Methodologically, expanding the participatory design method to the UDC framework supporting the mental processes that steer creativity proves to be valuable: case study results were rich and elaborate, and obtained in a limited amount of time. By mapping specific PD methods to each stage of the creativity process, participants are given the occasion to immerse themselves into the subject matter, to mature their ideas during an incubation period and to transform their inspiration into communicable designs.

Since the main application goal was to get a variety of application ideas and feedback from end users, and not to end up with a specific, workable application framework, the use of the UDC framework paid off. Besides the hands-on design ideas resulting from the prototypes, the study unveiled stable high-level requirements of learners and teachers covering issues like motivation, evaluation, course content and collaboration, effectively yielding the necessary in-depth user knowledge and involvement to establish a strong foundation for further agile development activities. This research will serve as a base for future studies on stimulating end user creativity in the early design process of innovative applications to elicit user requirements.

ACKNOWLEDGMENT

The IBBT MAPLE project is a project cofunded by IBBT (Interdisciplinary Institute for Technology), a research institute founded by the Flemish Government, and with project support of IWT.

REFERENCES

Abra, J. (1989). Changes in creativity with age: Data, explanations, and further predictions. *International Journal of Aging & Human Development*, *28*(2), 105–126. doi:10.2190/E0YT-K1YQ-3T2T-Y3EQ

Arnowitz, J., Arent, M., & Berger, N. (2006). *Effective prototyping for software makers*. San Francisco, CA: Morgan Kaufmann.

Augustine, S. (2005). *Managing agile projects*. Upper Saddle River, NJ: Prentice Hall PTR.

Beaudouin-Lafon, M., & Mackay, W. (2003). Prototyping tools and techniques. In *The Human-Computer Interaction Handbook: Fundamentals, Evolving Technologies and Emerging Applications* (2nd ed., pp. 1006–1031). Hillsdale, NJ: Lawrence Erlbaum Associates.

Beyer, H. (2010). *User-centered agile methods*. New York, NY: Morgan & Claypool Publishers.

Beyer, H., & Holtzblatt, K. (1997). *Contextual design: Defining customer-centered systems*. San Fransisco, CA: Morgan Kaufmann Publishers Inc.

Blomberg, J., Giacomi, J., Mosher, A., & Swenton-Wall, P. (1933). Ethnographic field methods and their relation to design. In *Participatory Design: Principles and Practices* (pp. 123–155). Hillsdale, NJ: Lawrence Erlbaum Associates.

Boden, M. A. (2003). *The creative mind: Myths and mechanisms* (2nd ed.). London, UK: Routledge.

Bodker, S., Nielsen, C., & Graves Petersen, M. (2000). Creativity, coopeartion and interactive design. In *Proceedings of the 3rd Conference on Designing Interactive Systems: Processes, Practices, Methods, and Techniques,* (pp. 252-261). ACM.

Botero, A., Kommonen, K., Koskijoki, M., & Ollinki, I. (2003). *Codesigning visions, uses and applications*. Retrieved from http://arki.uiah.fi/arkipapers/codesigning_ead.pdf

Bullinger, H.-J., Müller-Spahn, F., & Rössler, A. (1996). *Encouraging creativity - Support of mental processes by virtual experience. Virtual Reality World.* IDG.

Cardenas-Claros, M. S., & Gruba, P. A. (2010). *Bridging CALL & HCI: Input from participatory design*. Retrieved from http://www.postgradolinguistica.ucv.cl/pr_curriculum_pub_doc.php?pid=430

Cooper, A. (2007). *About face 3.0: The essentials of interaction design*. New York, NY: John Wiley & Sons, Inc.

Courage, C., & Baxter, K. (2004). *Understanding your users: A practical guide to user requirements methods, tools, and techniques*. San Francisco, CA: Morgan Kaufmann Publishers Inc.

Csikszentmihalyi, M. (1997). *Creativity: Flow and the psychology of discovery and invention*. New York, NY: Harper Perennial.

Cybulski, J., Nguyen, L., Thanasankit, T., & Lichtenstein, S. (2003). Understanding problem solving in requirements engineering: Debating creativity with IS practitioners. *PACIS 2003 Proceedings*. Retrieved from http://aisel.aisnet.org/pacis2003/32

Dalsgaard, P., & Halskov, K. (2010). Innovation in participatory desing. In *Proceedings of the 11th Biennial Participatory Design Conference,* (pp. 281-282). ACM.

Fallman, D. (2003). Design-oriented human-computer interaction. In *Proceedings of the SIGCHI Conference on Human Factors in Computing Systems,* (pp. 225-232). ACM.

Ferreira, J., Sharp, H., & Robinson, H. (2011). User experience design and agile development: Managing cooperation through articulation work. *Software, Practice & Experience, 41*(9), 963–974. doi:10.1002/spe.1012

Findlay, C. S., & Lumsden, C. J. (1988). The creative mind: Toward an evolutionary theory of discovery and innovation. *Journal of Social and Biological Systems, 11*(1), 3–55. doi:10.1016/0140-1750(88)90025-5

Fowler, M. (2006, July 28). *Customer affinity*. Retrieved from http://martinfowler.com/bliki/CustomerAffinity.html

Gause, D. C., & Weinberg, G. M. (1989). *Exploring requirements: Quality before design*. New York, NY: Dorset House Pub.

Gaver, B., Dunne, T., & Pacenti, E. (1999). Design: Cultural probes. *Interaction, 6*, 21–29. doi:10.1145/291224.291235

Guindon, R. (1998). Designing the design process: Exploiting opportunistic thoughts. *Human-Computer Interaction, 5*(2), 305–344. doi:10.1207/s15327051hci0502&3_6

Hackos, J. A., & Redish, J. C. (1998). *User and task analysis for interface design*. New Your, NY: John Wiley & Sons, Inc.

Highsmith, J., & Fowler, M. (2001). The agile manifesto. *Software Development Magazine, 9*(8), 29–30.

Jones, A., Issroff, K., & Scanlon, E. (2007). Affective factors in learning with mobile devices. In *Big Issues in Mobile Learning* (pp. 17–22). Nottingham, UK: University of Nottingham.

Kim, K. H. (2006). Can we trust creativity tests? A review of the Torrance tests of creative thinking. *Creativity Research Journal, 18*(1), 3–14. doi:10.1207/s15326934crj1801_2

Kramer, M. A. M. (2009). *The case for MobileHCI and mobile design research methods in mobile and informal learning contexts. Researching Mobile Learning: Frameworks*. Methods, and Research Designs.

Laurel, B., & Lunenfeld, P. (2003). *Design research: Methods and perspectives*. Cambridge, MA: The MIT Press.

Lim, Y.-K., Stolterman, E., & Tenenberg, J. (2008). The anatomy of prototypes: Prototypes as filters, prototypes as manifestations of design ideas. *ACM Transactions on Computer-Human Interaction, 15*(7), 1–7, 27. doi:10.1145/1375761.1375762

Lucero, A., & Arrasvuori, J. (2010). PLEX cards: A source of inspiration when designing for playfulness. In *Proceedings of the 3rd International Conference on Fun and Games*, (pp. 28-37). New York, NY: ACM.

Maiden, N., & Gizikis, A. (2001). Where do requirements come from? *IEEE Software*, *18*(5), 10–12. doi:10.1109/52.951486

Martin, A., Biddle, R., & Noble, J. (2009). *XP customer practices: A grounded theory*. New York, NY: IEEE.

Massimi, M., Baecker, R. M., & Wu, M. (2007). Using participatory activities with seniors to critique, build, and evaluate mobile phones. In *Proceedings of the 9th International ACM SIGACCESS Conference on Computers and Accessibility*, (pp. 155–162). New York, NY: ACM.

Montero Perez, M., Senecaut, M., Clarebout, G., & Desmet, P. (2010). Designing for motivation: the case of mobile language learning. In *Proceedings of the XIVth International CALL Conference*. CALL.

Muller, M. J. (2003). Participatory design: The third space in HCI. In *The Human-Computer Interaction Handbook: Fundamentals, Evolving Technologies and Emerging Applications* (2nd ed., pp. 1051–1068). Hillsdale, NJ: Lawrence Erlbaum Associates.

Ng, B. D., & Wiemer-Hastings, P. (2005). Addiction to the internet and online gaming. *Cyberpsychology & Behavior*, *8*(2), 100–113. doi:10.1089/cpb.2005.8.110

Parsons, D., Hokyoung, R., & Cranshaw, M. (2006). A study of design requirements for mobile learning environments. In *Proceedings of the Sixth International Conference on Advanced Learning Technologies*, (pp. 96-100). IEEE.

Patton, J. (2002). Hitting the target: Adding interaction design to agile software development. *OOPSLA 2002 Practitioners Reports*, (pp. 1–ff). New York, NY: ACM.

Ries, E. (2011). *The lean startup: How today's entrepreneurs use continuous innovation to create radically successful businesses*. New York, NY: Crown Business.

Robertson, J. (2005). Requirements analysts must also be inventors. *IEEE Software, 22*(1), 48, 50.

Robertson, J., & Heitmeyer, C. (2005). Point/counterpoint. *IEEE Software*, *22*(1), 48–51.

Sanders, L. (2008). On modeling: An evolving map of design practice and design research. *Interaction*, *15*, 13–17. doi:10.1145/1409040.1409043

Schuh, P. (2004). *Integrating agile development in the real world*. New York, NY: Cengage Learning.

Schuler, D., & Namioka, A. (1993). *Participatory design: Principles and practices*. Mahwah, NJ: Lawrence Erlbaum Associates.

Sharples, M. (2009). Methods for evaluating mobile learning. In *Researching Mobile Learning: Frameworks, Tools and Research Designs*. Retrieved from http://oro.open.ac.uk/31417/

Shneiderman, B. (2000). Creating creativity: User interfaces for supporting innovation. *ACM Transactions on Computer-Human Interaction*, *7*(1), 114–138. doi:10.1145/344949.345077

Shneiderman, B. (2007). Creativity support tools: Accelerating discovery and innovation. *Communications of the ACM*, *50*(12), 20–32. doi:10.1145/1323688.1323689

Snider, C., Dekoninck, E., & Culley, S. (2011). Studying the appearance and effect of creativity within the latter stages of the product development process. In *Proceedings of the DESIRE 2011 Conference on Creativity and Innovation in Design*, (pp. 317-328). ACM.

Spikol, D. (2009). Exploring novel learning practices through co-designing mobile games. *Researching Mobile Learning: Frameworks, Methods, and Research Designs*. Retrieved from http://www.ungkommunikation.se/documents/ungkommunikation/documents/ungkommunikation/rapporter/spikol,%20david.pdf

Steen, M. (2001). Cooperation, curiosity and creativity as virtues in participatory design. In *Proceedings of the DESIRE 2011 Conference on Creativity and Innovation in Design,* (pp. 171-174). ACM.

Vass, M., Carroll, J. M., & Clifford, A. (2002). Supporting creativity in problem solving environments. In *Proceedings of the 4th Conference on Creativity & Cognition,* (pp. 31-37). New York, NY: ACM.

Weisberg, R. W. (1986). *Creativity: Genius and other myths*. New York, NY: W H Freeman & Co.

Zaphiris, P., & Constantinou, P. (2007). Using participatory design in the development of a language learning tool. *Interactive Technology and Smart Education, 4*(2), 79–90. doi:10.1108/17415650780000305

ADDITIONAL READING

Agars, M. D., Kaufman, J. C., & Locke, T. R. (2007). Social influence and creativity in organizations: A multi-level lens for theory, research, and practice. *Research in Multi Level Issues*, 7, 3–61. doi:10.1016/S1475-9144(07)00001-X

Alexander, B. (2004). Going nomadic: Mobile learning in higher education. *EDUCAUSE Review, 39*(5), 28–35.

Ally, M., Schafer, S., Cheung, B., McGreal, R., & Tin, T. (2007). *Mobile learning: Transforming the delivery of education and training*. Athabasca, Canada: AU Press.

Amabile, T. M. (1996). *Creativity in context: Update to the social psychology of creativity*. Boulder, CO: Westview Press.

Arieti, S. (1976). *Creativity: The magic synthesis*. New York, NY: Basic Books.

Attewell, J., & Savill-Smith, C. (2004). *Mobile learning anytime everywhere*. MLEARN.

Baer, M., & Oldham, G. R. (2006). The curvilinear relation between experienced creative time pressure and creativity: Moderating effects of openness to experience and support for creativity. *The Journal of Applied Psychology, 91*(4), 963–970. doi:10.1037/0021-9010.91.4.963

Battarbee, K., & Kurvinen, E. (2003). Supporting creativity – Co-experience in MMS. In *Proceedings of COST269*. COST.

Biskjaer, M. M., Dalsgaard, P., & Halskov, K. (2010). Creativity methods in interaction design. In *Proceedings of the 1st DESIRE Network Conference on Creativity and Innovation in Design,* (pp. 12–21). Lancaster, UK: Desire Network. Retrieved from http://portal.acm.org/citation.cfm?id=1854969.1854976

Boden, M. (1991). *Dimensions of creativity*. Cambridge, MA: MIT Press.

Bodker, S., & Gronbæk, K. (1991). *Cooperative prototyping: Users and designers in mutual activity*. London, UK: Academic Press Ltd. Retrieved from http://portal.acm.org/citation.cfm?id=122825.122806

Bowers, J., & Pycock, J. (2004). *Talking through design: Requirements and resistance in cooperative prototyping*. Paper presented at the CHI 2004. Vienna, Austria.

Camacho, L. M., & Paulus, P. B. (1995). The role of social anxiousness in group brainstorming. *Journal of Personality and Social Psychology, 68*(6), 1071–1080. doi:10.1037/0022-3514.68.6.1071

Chirumbolo, A., Mannetti, L., Pierro, A., Areni, A., & Kruglanski, A. W. (2005). Motivated closed-mindedness and creativity in small groups. *Small Group Research, 36*(1), 59–82. doi:10.1177/1046496404268535

De Roeck, D., Rutten, C., & Godon, M. (2008). *Co-design and the city*. Mobile HCI.

Diehl, M., & Stroebe, W. (1987). Productivity loss in brainstorming groups: Toward the solution of a riddle. *Journal of Personality and Social Psychology, 53*(3), 497–509. doi:10.1037/0022-3514.53.3.497

Farmer, R., & Gruba, P. (1992). Towards model-driven end user development in CALL. *CALL, 19*(2), 149–191. doi:10.1080/09588220600821529

Kelly, J. R., & Karau, S. J. (1993). Entrainment of creativity in small groups. *Small Group Research, 24*(2), 179–198. doi:10.1177/1046496493242002

Kujala, S., & Kauppinen, M. (2004). Identifying and selecting users for user-centered design. [ACM Press.]. *Proceedings of NordiCHI, 2004*, 297–303. doi:10.1145/1028014.1028060

Laine, T. H., Vinni, M., Sedano, C. I., & Joy, M. (2010). On designing a pervasive mobile learning platform. *ALT-J. Research in Learning Technology, 18*(1), 3–17. doi:10.3402/rlt.v18i1.10672

Lee, I., Yamada, T., Shimizu, Y., Shinohara, M., & Hada, Y. (2005). In search of the mobile learning paradigm as we are going nomadic. In *Proceedings of World Conference on Educational Multimedia, Hypermedia and Telecommunications 2005*. ACM.

Lim, Y.-K., Stolterman, E., & Tenenberg, J. (2008). The anatomy of prototypes: Prototypes as filters, prototypes as manifestations of design ideas. *ACM Transactions on Computer-Human Interaction, 15*(7), 1–7, 27. doi:10.1145/1375761.1375762

Massimi, M., Baecker, R. M., & Wu, M. (2007). Using participatory activities with seniors to critique, build, and evaluate mobile phones. In *Proceedings of the 9th International ACM SIGACCESS Conference on Computers and Accessibility*, (pp. 155–162). New York, NY: ACM.

Moggridge, B. (2007). *Designing interactions*. Cambridge, MA: The MIT Press.

Morch, A. I., Engen, B. K., & Asand, H.-R. H. (2004). The workplace as a learning laboratory: The winding road to e-learning in a Norwegian service company. In *Proceedings of the Conference on Participatory Design*, (pp. 142-151). ACM Press.

Mumford, M. D., Hunter, S. T., & Bedell-Avers, K. E. (2008). Constraints on innovation: Planning as a context for creativity. *Research in Multi Level Issues, 7*, 191-200. Retrieved from http://www.emeraldinsight.com/books.htm?chapterid=1757182&show=pdf

Nash, S. S. (2007). Mobile learning, cognitive architecture and the study of literature. *Issues in Informing Science and Information Technology, 4*.

Nijstad, B. A., & Paulus, P. B. (2003). Group creativity: Common themes and future dorections. In *Group Creativity Innovation through Collaboration*, (pp. 326–339). Oxford, UK: Oxford University Press. Retrieved from http://ezproxy.library.arizona.edu/login?url=http://search.ebscohost.com/login.aspx?direct=true&db=psyh&AN=2003-88061-014&site=ehost-live

Nyíri, K. (2002). *Towards a philosophy of m-learning*. Paper presented at the International Workshop on Wireless and Mobile Technologies in Education (WMTE 2002). Växjö, Sweden.

Osman, A., Baharin, H., Ismail, M. H., & Jusoff, K. (2009). Paper prototyping as a rapid participatory design technique. *Computer and Information Science, 2*(3).

Parsons, D., Hokyoung, R., & Cranshaw, M. (2006). A study of design requirements for mobile learning environments. In *Proceedings of the Sixth International Conference on Advanced Learning Technologies*, (pp. 96-100). IEEE.

Parsons, D., Ryu, H., Lal, R., & Ford, S. (n.d.). *Paper prototyping in a design framework for professional mobile learning*. Retrieved from http://www.massey.ac.nz/~dpparson/Paper%20 Prototyping%20in%20a%20Design%20Framework%20for%20Professional%20Mobile%20 Learning.pdf

Paulus, P. B., Dugosh, K. L., Dzindolet, M. T., Coskun, H., & Putman, V. (2002). Social and cognitive influences in group brainstorming: Predicting production gains and losses. *European Review of Social Psychology*, *12*(1), 299–325. doi:10.1080/14792772143000094

Paulus, P. B., Dzindolet, M. T., Poletes, G., & Camacho, L. M. (1993). Perception of performance in group brainstorming: The illusion of group productivity. *Personality and Social Psychology Bulletin*, *19*(1), 78–89. doi:10.1177/0146167293191009

Polson, D., & Morgan, C. (2007). *MILK: The mobile informal learning kit*. Retrieved from http://eprints.qut.edu.au/26094/2/26094.pdf

Rietzschel, E. F., De Dreu, C. K. W., & Nijstad, B. A. (2009). What are we talking about, when we're talking about creativity? Group creativity as a multifaceted, multistage phenomenon. In Mannix, E. A., Neale, M. A., & Goncalo, J. A. (Eds.), *Research on Managing Groups and Teams: Creativity in Groups* (*Vol. 12*, pp. 1–28). Bingley, UK: Emerald Press.

Rietzschel, E. F., Nijstad, B. A., & Stroebe, W. (2006). Productivity is not enough: A comparison of interactive and nominal brainstorming groups on idea generation and selection. *Journal of Experimental Social Psychology*, *42*(2), 244–251. doi:10.1016/j.jesp.2005.04.005

Sanders, E. B. N., & Stappers, P. J. (2008). *Co-creation and the new landscapes of design*. Retrieved from http://www.maketools.com/articles-papers/CoCreation_Sanders_Stappers_08_preprint.pdf

Sawyer, R. K. (2006). *Explaining creativity: The science of human innovation*. Oxford, UK: Oxford University Press.

Sharples, M. (2000). The design of personal mobile technologies for lifelong learning. *Computers & Education*, *34*, 177–193. doi:10.1016/S0360-1315(99)00044-5

Smith, G. F. (1998). Idea-generation techniques: A formulary of active ingredients. *The Journal of Creative Behavior*, *32*(2), 107–133. doi:10.1002/j.2162-6057.1998.tb00810.x

Steiner, I. D. (1972). *Group process and productivity*. New York, NY: Academic Press Inc.

Trigg, R. H., Bodker, S., & Gronbaek, K. (1991). *Open-ended interaction in cooperative prototyping: A video-based analysis*. Retrieved from http://aisel.aisnet.org/sjis/vol3/iss1/5/

Vavoula, G., Pachler, N., & Kukulska-Hulme, A. (Eds.). (2009). *Researching mobile learning*. Oxford, UK: Peter Lang Publishers.

Section 4
Mixed-Methods Approaches

Chapter 13
Combining Research Paradigms to Improve Poor Student Performance

Roelien Goede
North-West University – Vaal Campus, South Africa

Estelle Taylor
North-West University – Potchefstroom Campus, South Africa

Christoffel van Aardt
Vaal University of Technology, South Africa

ABSTRACT

The aim of this chapter is to demonstrate the advantages of combining methods from different research paradigms. Positivism, interpretivism, and critical social theory are presented as major paradigms in Information Systems research. The chapter demonstrates the use of methods representative of these three research paradigms in a single research setting. The main problem in the research setting is the poor performance of students in a specific module of their academic programme. This problem is addressed by initiating an action research project using methods representing different research paradigms in the different phases of the project. The argument for using mixed methods is presented by providing information on research paradigms, discussing the problem environment, describing the research process, and finally, reflecting on research paradigms and their application in this environment.

INTRODUCTION

The chapter seeks to contribute to the discussion on the use of mixed methods in Information Systems (IS) education research. Various authors, such as Mingers (2001), Bryman (2006), and Symonds and Gorard (2010) describe the combination of methods related to traditional research paradigms in a single research project. The objective of researchers from different schools of thought differs in terms of research outcome and in terms of generation of knowledge. Some researchers have

DOI: 10.4018/978-1-4666-2491-7.ch013

Copyright © 2013, IGI Global. Copying or distributing in print or electronic forms without written permission of IGI Global is prohibited.

objective measuring as their main objective where understanding might be the motivation for others. On the far end of the spectrum some researchers has participative change and emancipation as their main objective. This chapter begins with an explanation of research paradigms to illustrate different perspectives of researchers in order to create sensitivity towards the complexity of mixing methods from different paradigms.

This chapter describes a problem situation where methods from all three typical research paradigms, namely Positivism, Interpretivism, and Critical Social Theory, are used to address a problem. The problem under investigation is the poor performance of university engineering students in a module on computer networking. The research team used a combination of methods in an action research project to effect changes to the teaching environment in order to improve the performance of these students.

The argument for using mixed methods is presented by providing information on research paradigms, discussing the specific problem environment, describing the research process and finally reflecting on research paradigms and their application in this environment.

BACKGROUND

Brief information on research methodology terminology is provided in order to clarify the position of the research team. Research approaches are then discussed in terms of paradigms. A short discussion on research paradigms, mixed methods, and research quality follows.

The usage of the terms qualitative and quantitative research is sometimes confusing. Some authors such as Bryman (2006) use it to distinguish between research approaches while others such as Myers (1997) use it to distinguish between the natures of the data used. In terms of the first group, one can distinguish between qualitative and quantitative research as:

Quantitative purists believe that the observer should be objective and separate from the entities that are observed. They should remain emotionally detached and uninvolved. This is in accordance to the positivism paradigm described in the following section.

Qualitative purists reject this and argue for idealism, humanism, and hermeneutics. They are characterised by a dislike of detached writing and prefer detailed and rich description (Johnson & Onwuegbuzie, 2004). This is in accordance with the interpretive paradigm described in the following section.

In terms of the group focusing on the nature of the data (Denzin & Lincoln, 2003; Myers, 1997; Straus & Corbin, 1998; Oates, 2006) the following distinction is given:

Quantitative research is used to study natural phenomena and include experiments and numerical methods. It is data based on numbers / values. This is the main type of data generated by experiments and are primarily used by positivists, but can also be used by interpretivists or critical researchers.

Qualitative research, on the other hand, uses quantitative data, for example interviews, documents, observations and stories, usually to understand and explain social occurrences. The emphasis is on processes and meaning that cannot always be measured in terms of quantity, amount or frequency. It includes words, images, and sound.

In this chapter the term qualitative refer to textual data and quantitative refer to numerical data. The assumptions regarding the validity of data are handled in terms of research paradigms rather than type of data. A short discussion on research paradigms, mixed methods, and research quality follows.

Research Paradigms

Our philosophical assumptions guide the way in which we do our work, also in research. Hughes (1990) wrote: "Every research tool or procedure is inextricably embedded to particular versions of the world and to knowing that world." This implies that our research methods are linked to philosophical assumptions we make about the world we live in. A research paradigm is a description of research methodology according to a specific set of assumptions. The term paradigm is used here in accordance with the discussion of Mingers (2001) who describes it as "a construct that specifies a general set of philosophical assumptions covering, for example, ontology (what is assumed to exist), epistemology (the nature of valid knowledge), ethics or axiology (what is valued or considered right), and methodology."

Kuhn made the concept of a paradigm popular. According to Kuhn (1977) a paradigm is a general concept including a group of researchers having a common education and an agreement on examples of high quality research or thinking. Researchers whose research is based on shared paradigms are committed to the same rules and standards for scientific practice. Paradigms guide research by direct modelling as well as through abstracted rules. A paradigm governs not a subject matter but rather a group of practitioners (Kuhn, 1970).

Three research paradigms are widely described in Information Systems (IS) research, namely Positivism, Interpretivism, and Critical Social Theory (CST). These paradigms can be investigated on three levels of understanding, namely philosophy, methodology, and practice as indicated in Figure 1.

Philosophical Level

On the philosophical level, the ontological and epistemological assumptions can be investigated. Positivism is grounded in the ontological stance of realism, and fact rather than value is investigated. The aim of positivistic research is often to measure the influence of specific variables in a situation. Although there are many differences among positivist philosophers, such as Comte and Durkheim, Giddens (1974) identifies the following claims or perspectives that make up positivistic philosophy:

- Reality consists in what is available to the senses.
- Science constitutes a framework by which any form of knowledge can be determined.
- The natural and human sciences share common logical and methodological foundations, and methods of natural sciences can be applied in social sciences.
- There is a fundamental distinction between fact and value. Science deals with facts, while values belong to an entirely different order of discourse beyond the remit of science.

These philosophical assumptions led to the development of empiricism as research methodology for social research.

In contrast, interpretivists argue that people create and attach their own meanings to the world around them and to the behaviour they manifest in that world (Schutz, 1962). Interpretivism is grounded in the ontological stance of relativism, where the aim of research is to understand the phenomenon. Lee (1999) argues that the study of the subjective meaning human subjects attach to behaviour, requires procedures that have no counterparts among those of the natural sciences. Hughes (1990, p. 90) states that "The method recognises the actions, events and artefacts from within human life; not as the observation of some external reality."

Finally, critical social research is underpinned by a critical-dialectical perspective, which attempts to dig beneath the surface of historically specific, oppressive, social structures (Harvey, 1990). Critical social theorists see knowledge as being structured by existing sets of social relations

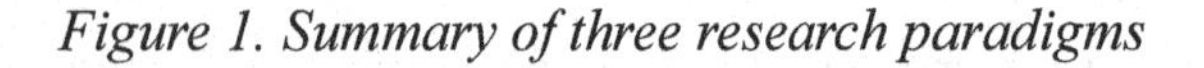
Figure 1. Summary of three research paradigms

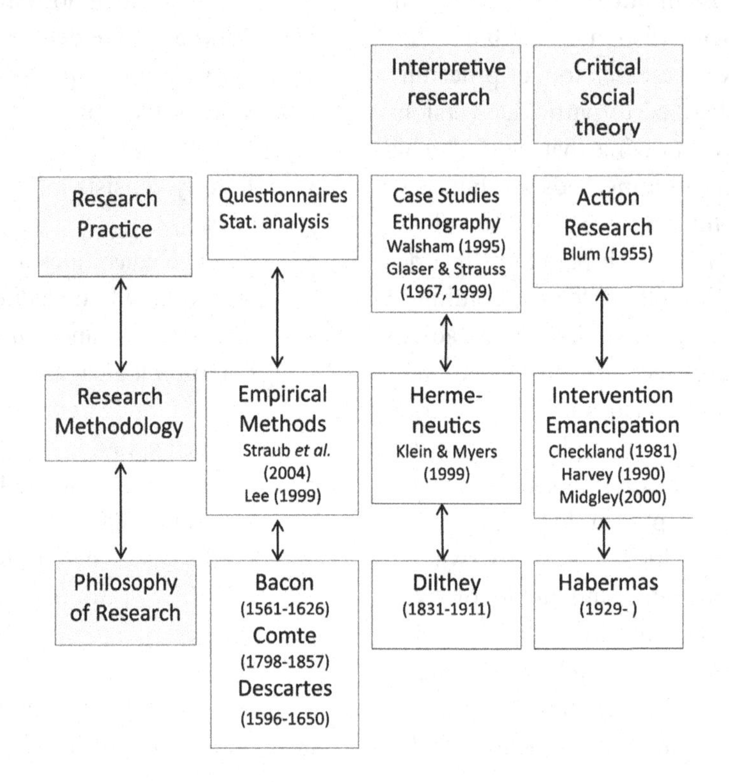

that are oppressive. This can be class, gender or race oppression. "Knowledge is critique… It is a dynamic process not a static entity…It is the process of moving towards the understanding of the world and of the knowledge which structures our perceptions of the world" (Harvey, 1990). In critical social research, the aim of the researcher is to emancipate or change suppressive behaviour.

Methodological Level

On the methodological level, one finds principles guiding the practice of these methods. For positivism, these principles describe the typical scientific method and are focused on objectivity, repeatability, and relationships between variables.

The following principles for interpretive IS research can be found in the work of Klein and Myers (1999):

1. **The fundamental principle of the hermeneutic circle:** This principle suggests that all human understanding is achieved by iteration between the interdependent meaning of parts and the whole they form. This principle of human understanding is fundamental to all the other principles.
2. **The principle of contextualisation:** Requires critical reflection on the social and historical background of the research setting, so that the intended audience can see how the current situation under investigation emerged.

3. **The principle of interaction between the researchers and the subjects:** Requires a critical reflection on how the research materials (or "data") were socially constructed through the interaction between the researchers and the participants.
4. **The principle of abstraction and generalisation:** Requires relating the idiographic detail revealed by the data interpretation through the application of principles one and two to the theoretical general concepts that describe the nature of human understanding and social action.
5. **The principle of dialogical reasoning:** Requires sensitivity to possible contradictions between the theoretical preconceptions guiding the research design and actual findings ("the story which the data tells") with subsequent cycles of revision.
6. **The principle of multiple interpretations:** Requires sensitivity to possible differences in interpretations among the participants as are typically expressed in multiple narratives or stories of the sequence of events under study. They are similar to multiple witness account, even if all tell it as they saw it.
7. **The principle of suspicion:** Requires sensitivity to possible "biases" and systematic "distortions" in the narratives collected from the participants.

These principles are guided by the hermeneutic circle, which is the process of switching between words and whole sentences to understand the meaning of the words in the context of the sentence. This idea is generalised beyond sentences and words to for example understand specific answers of a participant in context of the whole interview and, conversely, to understand the interview in the context of the individual answers.

Critical social theory research principles are guided by emancipation. The researcher should equip him-/ herself to identify and resolve oppressing structures in a problem situation. Harvey (1990) provides the following principles for critical social research projects:

1. Through abstraction, critical social research aims to reveal underlying structures that are otherwise taken for granted. These structures specify the nature of the abstract concepts, which have themselves been assimilated uncritically into the prevailing conceptualisation.
2. Totality refers to the view that social phenomena are interrelated to form a total whole. Social phenomena should not be investigated in isolation but always as part of a larger context.
3. Essence refers to the fundamental element of the analytical process. Critical social researchers view essence as a fundamental concept that can be used as the key to unlocking the deconstructive process.
4. According to Harvey (1990, p. 22), praxis means practical reflective activity. It is activity that changes the world. The critical social researcher is not only interested in understanding the world; he/she aims to change the world. It is not the actions of an individual that is of interest but rather the actions that change the social formations.
5. Ideology: The positive view of ideology sees it as false consciousness which hides the interests of dominant groups from themselves. According to the negative view of ideology, it cannot be detached from the material conditions of their production; it is constantly reaffirmed through everyday practice. The nature of the ideology needs to be revealed by the researcher through the identification of the essence of social relations and the separation of this essence from structural forms through a process of dialectical deconstruction and reconstruction.
6. Structure is seen by the critical social researcher as more than the sum of the elements. It is viewed holistically as a com-

plex set of interrelated elements which are interdependent and which can be conceived adequately only in terms of the complete structure.

7. Critical research history is not so much interested in the historical facts as in the circumstances within which it occurred. It investigates the social and political contexts, addresses the economic constraints, and engages the taken-for-granted ideological factors. It also takes the situation of the researcher into account.
8. The critical researcher aims to deconstruct the situation into abstract concepts in order to study the interrelations between the concepts with the purpose of discovering the key to the structure of the situation. The core concept is used to reconstruct the situation. This is an ongoing process to expose the ideology underpinning the situation in order to identify the oppressive mechanism, which requires change.

Practice Level

On the practice research level, the paradigms are used to describe practical methods for research. Positivistic research practice entails methods such as questionnaires and experiments to gather data and statistical methods such as mean, t-tests, regression, and others to analyse the data.

Interpretive research practice entails methods such as ethnography and case studies to gather data. Since the aim of data gathering is understanding, the researcher fulfils the role of a student of the participants' world who aims to learn as much as possible. Therefore, follow-up discussions are used to clarify information. Content analysis methods such as grounded theory are used to analyse the data. Interviews are transcribed and coded to develop a theory grounded in the data. Saturation occurs when all information deducted from new data is already represented in the data. Glaser and Strauss (1967) provide a clear description of this process.

Critical social theory researchers use Action Research (AR) as a method to intervene in a problem environment. AR researchers often make use of methods from other paradigms to gather and analyse data. Lewin (1948) argued that one could only understand the inner structure of a social system by trying to change it. Action research is a cyclic research method where a problem situation is diagnosed in terms of identifying oppressive structures and subsequently a solution is designed and implemented. After implementation, the success of the solution is investigated and the cycle is repeated until the identified oppression is relieved. Baskerville (1999) provides a very good tutorial on AR in IS.

There are two major differences between positivistic research practices on the one hand and interpretive and critical social research practices on the other hand. The first major difference is the objectivity of the researcher. Positivistic research practices require and are designed to ensure that the researcher is objective in his research activity. This implies that the researcher does not influence the research environment in the data collection activity. Interpretive methods allow for the personal interpretations of the researcher, and the researcher is encouraged to learn as much as possible from the research environment in order to give a reliable interpretation of the environment. Critical researchers are not only interpreting the data in the environment but are also designing and affecting change in the problem environment (typically an organisation). The second major difference between positivistic research practice and the other two approaches is the reduction of the problem situation through sampling. Positivistic methods assume that a sample, if carefully selected, represents the population, while interpretive and critical methods study the problem situation as a whole.

Mixed Methods

According to Kuhn (1970) rational debate about competing paradigms is almost impossible, one that can never be settled by logic and experimentation alone or by proofs, and many scientists believe that one science can accommodate only one paradigm. Kuhn describes the debate about paradigm choice as a circular debate, because each group uses its own paradigm to argue in that paradigm's defence. Johnson and Onwuegbuzie (2004) also refer to followers of different paradigms advocating the incompatibility thesis—that qualitative and quantitative research paradigms cannot and should not be mixed. These authors agree with (among others) Johnson and Onwuegbuzie (2004) and Niaz (2008) that recent research has shown that research paradigms are not displaced (as suggested by Kuhn) but can rather lead to integration.

Mixed methods refer to the combination of research methods from different research paradigms in a single research project (Symonds & Gorard, 2010) rather than a mixing of paradigms in terms of trying to change the underlying assumptions of specific paradigms. Mingers (2001) gives comprehensive information on mixed method research arguing that we need methods to understand all aspects of a problem situation as different methods applied from different perspectives provide us with an understanding of different aspects of reality. He further argues that research projects are often divisible in phases displaying different characteristics, which require different research strategies. Teddlie and Tashakkori (2009) discuss the following advantages of using methods from different paradigms in the same project: Triangulation (validating research results by combining different data sources, research methods and research practitioners); creativity (discovering new or paradoxical aspects that requires more research); and expansion (highlighting influential aspects previously regarded as outside the scope of the project).

Methods can be combined on different levels. Mingers (2001) classifies mixed research as sequential, parallel, dominant, multi methodology, or multilevel.

Creswell (2012) describes the following designs for a mixed method project:

1. **Convergent Parallel Design:** Where methods from different paradigms are used in parallel and the results are compared and related.
2. **Explanatory Sequential Design:** Where one method is followed up with another method to provide an explanation of the results.
3. **Exploratory Sequential Design:** Where on method is used to do exploration work before another method is used as the dominant method.
4. **Embedded Design:** Where some of the data gathered during the main data collection is of such a nature that different methods are required to collect analyse the data.
5. **Transformative Design:** Where transformation is part of the research in order to address social issues for "marginalised or underrepresented population (Creswell, 2012, p. 546).
6. **Multiphase Design:** Where the study is viewed as smaller studies each with its own research design.

A dominant mixed method project is reported on in this research where one method (action research) is dominant and interpretive case studies and positivistic quantitative data analysis is used to reach the (critical) research objectives. In terms of classification of Creswell (2012) the study can overall be classified as a transformative design, using both explanatory and exploratory designs in different phases of in different phases of the research process. This will be discussed in more detail later in the chapter.

Quality Criteria for Research

One of the most important advantages for the categorisation of specific research projects in terms of research paradigms is the evaluation of the research project from a quality perspective. Traditionally, the concepts of validity and reliability have been the key tools for research quality assessment. Many writers, however, claim that these tools are no longer meaningful because of their lack of semantic rigour in non-positivistic research. Methodologists have introduced tens of alternative concepts and views without coherence, which may cause confusion about the diversity of alternatives (Heikkinen, et al., 2007). The issue of quality and quality criteria has become increasingly prominent in methodological discussions. The rise of qualitative (interpretive and CST) research over the last 30 years represents one of the reasons for the growing interest in research quality criteria. It is assumed that criteria for positivistic research are well known and agreed, but that this is not the case for interpretive and CST research (Bryman, et al., 2008).

The quality of a research project categorised as positivistic may be determined by way of specific measures. Reviewers are able to evaluate a set of aspects of positivistic methods in order to justify their decision on the suitability of a project for academic publication. Conventional research criteria apply to positivistic research, such as the following (Guba & Lincoln, 1989):

1. **Internal Validity:** The extent to which variations in a dependable variable can be attributed to controlled variation in an independent variable.
2. **External Validity:** Inference that the presumed casual relationship can be generalized across alternative measures of cause and effect.
3. **Reliability:** Consistency of a given inquiry involving predictability, dependability and / or accuracy and replicability.
4. **Objectivity:** Neutrality and freedom from bias.

Klein and Myers (1999) published a paper on principles of interpretive case studies in IS which became the standard criteria for good interpretive IS case studies. These principles were presented in section 2.1. According to Guba and Lincoln (1989) the conventional positivistic criteria can be adjusted to be meaningful for the constructivist (interpretive) inquiry:

1. **Credibility:** This is parallel to internal validity. Instead of focusing on a presumed reality, the focus is on establishing a match between realities of respondents and realities represented by the evaluator. It is concerned with how believable findings are.
2. **Transferability:** This is parallel to external validity. The positivistic paradigm requires both sending and receiving contexts to be random samples of the same population. In the interpretive paradigm this is replaced by an empirical process for checking the degree of similarities between sending and receiving contexts. The main technique for establishing transferability is thick description.
3. **Dependability:** Dependability is parallel to reliability. It is concerned with the stability of data over time.
4. **Confirmability:** This may be thought of as parallel to objectivity. It is concerned with assuring that data, interpretations, and outcomes are rooted in context. The interpretive paradigm's assurances of integrity of data are rooted in the data themselves. This means the data can be tracked to their sources and that bias have been kept in check.

Although the nature of critical social research projects is more diverse than other research projects, Harvey (1990) described principles for critical social research that may be used to review such projects. These were presented in section 2.1.

Heikkinen et al. (2007) present the following five principles for the evaluation of action research as narrative:

1. **Principle of historical continuity:** This emphasises awareness of the socio-historical frame of the research project.
2. **Principal of reflexivity:** This includes analysis of the ontological presumptions (the presumptions concerning reality), as well as the epistemological analysis (concerning knowledge). The researcher must consciously reflect on insights, and the research must be transparent (the material and methods should be described).
3. **Principal of dialectics:** The research must combine different interpretations and voices as authentically as possible.
4. **Principle of workability:** The evaluation of action research must pay attention to whether it has given rise to changes in action.
5. **Principle of evocativeness**: Good research should awaken and provokes a person to think in a new and different way.

When combining methods in one project as is the case in mixed method research, Bryman (2006) argues that one might use one of three approaches to evaluate quality:

1. **Convergent Criteria:** Where the same criteria are used for the interpretive and the positivistic phases of the research.
2. **Separate Criteria:** Where separate criteria are used for the interpretive and the positivistic phases of the research.
3. **Bespoke Criteria:** Where new criteria are devised for mixed method research projects.

Bryman (2006) argues that as the outcomes of mixed method research are often unplanned, it is difficult to develop criteria that may be used generally. Instead, researchers in mixed method research community such as Bryman (2006) and Feilzer (2010) advocate the use of a more pragmatic approach to determine the value of research. Bryman (2006) stresses the importance of the research question. How could there be any other measure than the answering of the research question? Each project should be reviewed on its own merits. Most researchers feel that a combination of traditional and alternative criteria should be employed in relation to mixed method research (Bryman, et al., 2008). Researchers employ different criteria for different components of mixed method research. In a study of Bryman, Becker, and Sempik (2008) interviews brought out a number of criteria. The most important were:

1. Relevance to research question.
2. Transparency about the nature and content of procedures.
3. Integration of mixed method components and findings.
4. Rationale for using mixed method research.

Reviewers should use applicable criteria such as internal validity or trustworthiness that is suitable in the specific project to evaluate the quality of the research project. Comments on this approach are provided in the conclusion.

This chapter reports on a research project in higher education to demonstrate the use of mixed methods in a practical environment.

EMPIRICAL STUDY: NETWORKING STUDENTS

Computer Systems engineering students at the Vaal University of Technology in South Africa struggled to pass the module in Networking Systems. In the year prior to the start of this project the pass rate of this module was 61%. The aim of this research project was to facilitate improvement in the academic performance of these students. Until

Figure 2. The action research cycle

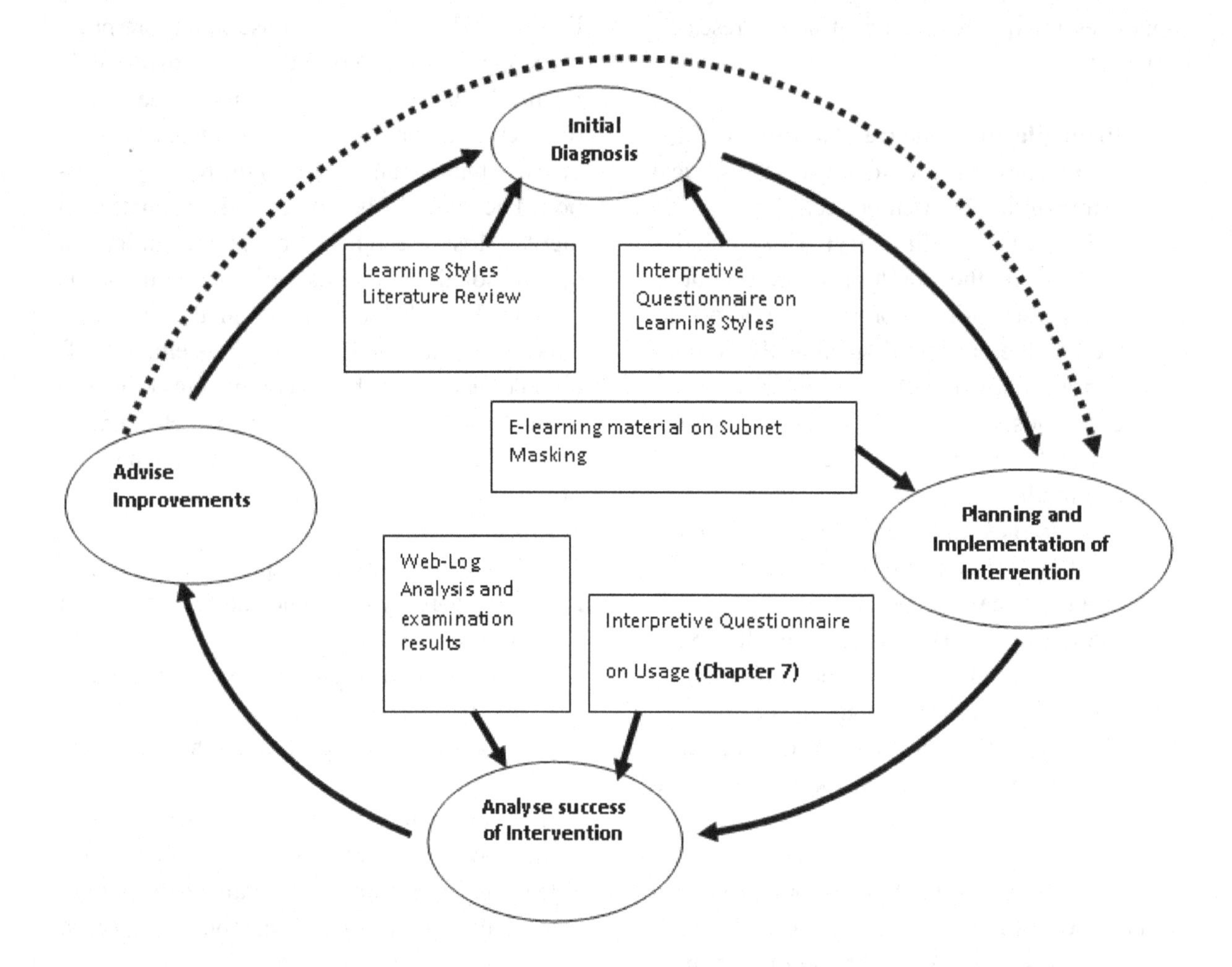

recently, this module was taught in a face-to-face environment using traditional methods of black board explanations and textbook referencing.

Research Strategy

This problem description is clearly in line with the CST paradigm. Action Research (AR) is an iterative research methodology for guiding change in a situation. It is participative in nature and the research team consisted of faculty members and researchers. Students who completed the module left the campus, since this module is presented in their final semester. Stakeholders, however, were sensitive towards the interests of the students involved. An iterative AR project was designed, depicted in Figure 2, according to the general phases of a typical participative AR project, discussed by Baskerville (1999):

- Initial diagnosis.
- Planning and implementation of the intervention.
- Analyse success of the intervention.
- Advise improvements.

From a research methodology perspective, the project could be seen as an action research project using mixed methods. This research method is pluralistic in the sense that all three paradigms, namely positivism, interpretivism, and constructivism are being used individually in different phases of the research. Each phase will be discussed in more detail.

Initial Diagnosis

After discussions with stakeholders (lecturers, head of department, and research team), it became clear that the ways in which students learn (learning styles) are central in understanding their difficulties. It was obvious that the learning styles of these students needed to be understood. Understanding is the central principle in the interpretive research paradigm, thus an interpretive questionnaire was developed to test the learning styles of the students. As it is difficult to cater for different learning styles in a face-to-face classroom learning environment, it was decided to extend the learning environment to a blended learning environment where face-to-face lectures are complemented by electronic (e)-learning material.

According to Felder and Silverman (1988), a learning style model classifies students according to the way in which they receive and process information. Although different models on learning styles exist, the model of Felder and Silverman was chosen for this project, as it was developed for use by engineering students. The Felder and Silverman model incorporates four dimensions, namely perception, processing, input, and understanding:

- Perception addresses intuitive learners and sensing learners. Intuitive learners look at possibilities and relationships within the learning environment, while sensing learners prefer to learn facts.
- Processing addresses active and reflective learners. Active learners like to participate in lectures and class discussions, while reflective learners like to work in a quiet environment on their own.
- Input addresses visual and verbal learners. Visual learners learn what they see and verbal learners learn written and spoken words.
- Understanding addresses sequential and global learners. Sequential learners like to study in clear linear steps, while global learners prefer seeing the bigger picture before learning the smaller sections.

Addressing multiple learning styles within a group of students will ensure better retention of module specific information.

It was decided to develop a blended learning environment. *Blended learning* can be defined as the mix of different didactic methods and delivery formats (Kerres & De Witt, 2003), or as a blend of different approaches, e.g. face to face and e-learning, the use of different technology-based tools, or the blending of classroom-based and work-based learning (Allan, 2007). *Blended e-learning* is defined as a parallel learning medium which combines traditional learning and e-learning (Littlejohn & Pegler, 2007; Carliner, 2004). In this study, the term *blended learning* will be used.

An interpretive study to investigate the students' readiness for e-learning, as well as the learning styles and preferences of the students, was done. Questionnaires were handed out during class and collected immediately thereafter. Being interpretive, the questionnaires gave the opportunity for the students to motivate their answers. A shortcoming in most educational settings is that students believe that there is a correct answer to the question. The purpose of the questionnaire was carefully explained to the students. All questionnaires were returned. Analysis indicated that all the learning styles in the Learning Styles Index of Felder and Soloman, as reported in Felder (2009) and developed by these two authors to evaluate the Felder and Silverman model, were present in this group.

From analysing questions about the students' preferences, the following conclusions could be made:

- Learners still preferred to make use of the traditional teaching-learning methods (74%). This motivates the choice of a blended learning environment with traditional lectures as the main form of tuition.
- Most learners preferring traditional teaching and learning methods also showed a tendency towards an active learning style.
- Some students preferred to be alone when studying, which is an indication of a reflective learning style.
- Students showed a strong tendency towards sequential study methods.
- Many students stressed the importance of looking at the complete picture encompassed by the study material before starting to study. This learning style is referred to as a global learning style.
- Most students preferred written text. This is an indication of a verbal learning style.
- The motivation for learning was very important to some students. They were more practical in their approach, which is indicative of a sensing learning style.
- Many students indicated their preference for a combination of learning styles.

It also transpired that these students were not used to e-learning, but were willing to experiment with it. After interpretive content analysis of the participants' answers was done, it became clear that, although students' learning styles differed substantially, the group as a whole was representative of the learning styles described by Felder and Silverman (1988).

Planning and Implementation of Intervention

The intervention phase of the AR project consisted of the design and implementation of e-learning material on a subset of the module called subnet masking. Lecturers identified subnet masking as the most problematic section of the module content and new e-learning material was developed on this topic. Multimedia material was developed specifically to accommodate the different learning styles of the Felder and Silverman (1988) model. The material could, for instance, be viewed globally or sequentially, or explanations could be read or listened to. Examples and exercises were included. Pictures as well as text were used. The new e-learning material was used in conjunction with face-to-face lecturing to create a blended learning environment, made available through the Internet.

The e-learning material was implemented on the Moodle Learning Management System (LMS). The Moodle LMS keeps an audit log of the usage of the e-learning material. Students signed into the system with a username and password and Moodle kept track of all their activities. This was an important data source in the analysis of the intervention.

Analysing the Success of the Intervention

In any critical social research project, the analysis of the success of the intervention is most important. Critical social research projects are only completed and successful when the oppressed party is emancipated. In this case, the oppressed party is the computer networking student and the project will only be successful when the pass rate of the module has increased substantially.

The implementation phase was done over 2 years (referred here as year 1 and year 2) until the pass rate of the students was satisfactory. This discussion will report on these two years simultaneously. Analysis of the intervention was done in three phases: The Web-logs (containing usage statistics) created by the learning management system were analysed; the correlation between time spent on the e-learning material and the marks obtained by the students summative evaluations were tested statistically; and an interpretive questionnaire was used to better understand the attitude of the students towards the e-learning material.

Phase 1: Analysis of the Web-Logs

The first part of the analysis of the intervention was done according to positivistic methods, owing to the quantitative nature of the data. Data available were the Weblogs and test and examination results of students. A log is a list of records showing how many users visited a page, when the user visited the page and the time he/she spent on the page. An assumption is made that students use the page actively while the page is open in their Internet browser. It is very difficult to establish the level of attention a student actually pays to an open page. This fact also motivated the use of interpretive interviews in the third analysis phase. These logs were analysed. The analysis of the Web usage in the first iteration (year 1) indicated that only 52% of the students used the newly developed e-learning material. In discussions on the success of the intervention with all the stakeholders, it was decided that the usage in year 1 was unsatisfactory. As the results of the t-tests were very encouraging (the students who used the e-learning material did significantly better in the question related to the e-learning material than the students who did not use the e-learning material), it was decided to use the e-learning material unchanged in year 2. From the action research project's perspective, this constitutes a second iteration. An important change was decided upon by all stakeholders for this second iteration. Lecturers were advised to refer to the available e-learning material frequently during classes, as students needed to be encouraged to use the e-learning material. The lecturers would point out to students exactly which explanations, as well as additional examples with exercises, were available on the e-learning material.

At the end of year 2, logs were again analysed and the intervention of lecturers were found successful as many more students used the e- learning material. The increase in usage was from 52% (year 1) to 88% (year 2) of all students enrolled for this module. The pass rate of students in the examination was also notably higher. It appeared as if the increased usage of e-learning material had a positive impact on the students' results. The pass rate percentage of learners increased from 60.6% in the year prior to the intervention to 91.9% in year 2. This initial comparison did not prove causality between usage of the e-learning material and examination success, but was encouraging in terms of the pragmatic success of the intervention.

Phase 2: Analysis of the Correlation between Time Spent on the e-Learning Material and Marks Obtained by the Students

Lecturers kept track of performance of students in specific questions in the examination. This implies that lecturers had marks for students for questions on module content not supported by the e-learning material as well as marks for questions on module content (specifically subnet masking) covered by the e-learning material. All this data is quantitative by nature and were captured according to the standards for good positivistic research data collection strategies. Statistical analysis could be used to gain information from this data.

In order to determine whether a meaningful correlation between the usage of e-learning material and the student's performance in the examination exists, the Pearson correlation was used. Usage

Table 1. Pearson correlations (year 1)

Variable	Time in minutes
QE	**0.2801**
	p=.017
E	-0.0604
	p=.614

Table 2. Pearson correlations (year 2)

Variable	Time (minutes)
QE	.0674
	p=.594
E	-0.1245
	p=.323

was studied in terms of number of minutes spent on using the e-learning material. The results of this analysis are presented in Table 1 (year 1) and Table 2 (year 2). In Table 1, the only meaningful correlation ($p< .05$) was the correlation between the mark obtained for the specific examination question on work related to the content of the e-learning material (QE) and time spent on this material. No meaningful correlation was found between the time spent on the e-learning material and the total mark obtained in the examination (E). It may be concluded that in year 1, time spent on e-learning material had a positive influence on the students' examination results with regard to work covered by the e-learning material.

Table 2 shows that in year 2 no meaningful correlations were found between the mark obtained for the examination question (QE) on work related to the content of e-learning material and time spent on the e-learning material. These results differ from the results in year 1.

In order to determine whether users of the e-learning material performed better in tests and the examination than non-users, t-tests were used. Results of this analysis for year 1 and year 2 are presented in Table 3 and 4 respectively. Non-users are represented in columns with a 0-heading, whereas users are represented in columns with a 1-heading. The mean for the examination question related to the content of the e-learning material is 47.8 for non-users, while for the same question, the mean for users is 65. The value of p is $< .0001$, showing a very good correlation. The mean value for the examination (all the questions) is 44.5 for non-users and the mean value is 52.7 for users. The value of $p < .05$ indicates a correlation between these values. From Table 3, it can be seen that students in year 1 who used the e-learning material performed significantly better in the examination question concerning the e-learning material than those not using the e-learning material.

In year 2, unlike year 1, comparing results of the group using the e-learning material to those not using it, revealed no significant difference in marks obtained for the examination (Table 4). This may be attributed to the fact that there was a much higher usage of e-learning material, resulting in a very small group not using it. The mean, however, of the users are higher for the question on subnet masking in the examination than that of the non-uses. The standard deviations need to be investigated further in order to ascertain causality.

Table 3. Student t-test for unequal variance (year 1)

	Valid N Non-users	Valid N Users	Mean Non-users	Mean Users	Standard deviation Non-users	Standard deviation Users	P
QE	70	72	47.82609	65.06410	19.15465	20.89495	**0.000001**
E	70	72	44.50000	52.77778	13.20600	13.84239	**0.000377**

Table 4. Student t-test for unequal variance (year 2)

Variable	Valid N Non-users	Valid N Users	Mean Non-users	Mean Users	Standard deviation Non-users	Standard deviation Users	P 2-sided
QE	9	65	50.82540	67.90350	29.87404	25.50145	0.069126
E	9	65	58.11111	59.38462	10.25237	11.00950	0.744120

Phase 3: Interpretive Questionnaire used to Understand the Attitude of Students towards e-Learning Material

The results of the statistical analysis in the first iteration were very encouraging, but the percentage of students using the e-learning material was unsatisfactory. After specific intervention from the lecturers, the usage improved during the second iteration, but the correlation between student marks and time spent on the e-learning material was statistically not as significant as the previous iteration. However, as the pass rate on the module increased substantially during the second iteration, the issue of causality, however, came to the front. Is the increased pass rate of the students a result of the availability of the e-learning material, or not? From a pragmatic point of view, it did not really matter, because their performance improved. However, a decision needed to be taken whether the e-learning material should be extended to cover other sections of the syllabus as well.

The researchers developed an interpretive questionnaire to better understand the attitude of the students towards the e-material. The questionnaire was distributed to the students who used the e-learning material in year 1 and year 2, and who could still be reached.

All of the participating students used the e-learning material. The analysis of the interpretive questionnaire revealed a positive attitude towards this material:

- 94% of the students indicated that the e-learning material was helpful, and that they would want other subjects to use blended learning.
- 87.5% of the students felt that the e-learning material helped them to score higher marks.

Several suggestions for improvement were made by the students.

- Promote shorter lessons with less contact time.
- Promote interaction among students, as well as between students and lecturer.

Advising Improvements

After the first year iteration, it was advised to keep the material unchanged and encourage usage (as described earlier). After the second year, results proved satisfactory and recommended improvements included the following:

- Improve synchronisation of e-learning material with class lectures.
- Include more examples and exercises.
- In order to be able to better analyse the usage of the e-learning material through Web-logs more active controls can be added to the designed e-material forcing the students to perform more recordable actions.
- Action research differs from consultation in that it uses some theoretical framework to guide the proposed intervention. In this

instance, learning styles theory was used. However, very little contribution was made to learning style knowledge from this research. In future, a learning style investigation, such as the one done during the diagnosis phase of this project, can be linked to the Web-logs of specific students. Interesting information can be gained from comparing the questionnaire answers of a specific student to actions of that same student, when confronted with choices of study material catering for different learning styles. In practical terms: Does the learner whose answers indicated that he is a global learner use the e-material aimed at global learners?

- Extend the material to cover course content as a whole and not just part of it. This process can be started by identifying other problematic aspects, and developing e-learning material for those.
- Although these students were not used to blended learning environment the outcome of this research indicated that they are willing to use electronic media to enhance their learning experience. This observation might be used as motivation to dramatically extend the learning environment towards the use of more electronic media. Negative perceptions around specific module content sections can be addressed by using popular media such as social networks. In the South African context, one should however never make the assumption that the students have access to all social media networks without verifying it.

REFLECTION ON RESEARCH PARADIGMS

The chapter began with a discussion of research paradigms in terms of a three level model presented in Figure 1. The model consisting of research practice, research methodology, and research philosophy (paradigm) levels was used to reflect on this research project.

Reflection on the Practice Level

When reflecting on the methods used, it is best to do so according to the phases of the project. Overall, action research was used from a critical social research perspective. The success of the method is measured pragmatically in terms of the research problem. Did the student performance improve? As the answer on the question is positive, one may argue that the project was successful.

In the diagnosis phase, interpretive interviews were used. Data were gathered using an interpretive questionnaire with many open-ended questions, which were analysed using content analysis. The principles for interpretive case studies apply for this phase and will be discussed in the following section on research methodology. Similar methods were used in the analysis of intervention where interpretive questionnaires were used to clarify the positivistic analysis.

During the analysis of intervention phase, methods were used from a positivistic perspective. Web logs were analysed by using descriptive statistics and correlations were investigated between student performance and Web usage information. This analysis proved to be inconclusive and we could not prove that it was the additional e-learning material that caused improved performance of students. This might be due to the fact that it is impossible to know if students actually make use of the learning material while their Internet browser is open on the page containing the e-learning material. The method for usage measuring is unreliable from this point of view, but from a pragmatic perspective, the time spent on the material is very valuable information for the research team. In future blended learning material more required actions will be embedded on the e-learning material in order to better understand the actual usage of the students when analysing the Web-logs. However, pragmatically (student performance) and according to the students (from interpretive questionnaire data) the blended-learning environment proved successful.

Typical positivistic research design would require two similar groups to be used, one group to undergo the intervention and a control group. In this case such an approach would have ethical implications as the initial diagnoses indicated that a blended learning environment sensitive to learning styles should have a positive impact. All students enrolled for the module have a right to the best tuition possible.

Reflection on Methodology Level

As this is overall a critical social theory project, the principles of Harvey (1990) may be used to reflect upon the quality of the research done:

1. **Abstraction and Structures**: Many lecturers stand uncritical towards the teaching methods they use. During the diagnosis phase, it was discovered that the different learning preferences of students may cause our traditional methods to hinder rather than enhance learning.
2. **Totality:** We needed to include as many as possible stakeholders in the discussions. This could have been done better since students were not involved due to practical reasons discussed. In asking open ended questions during the diagnosis phase the students were granted the opportunity to tell us more than what we expected them to tell us.
3. **Essence:** The essence that came to the fore is that students struggle mostly with a specific section of the work. The focus of the research project should be on learning. This implied that we should do whatever different methods required to facilitate their learning. In this case, we realised that they have different learning styles and by accommodating these in the e-learning material we can relief the inherent problems of face-to-face only teaching.
4. **Praxis (Means practical reflective activity):** We developed new material and acted when not enough students used it. We were not only interested in measuring or understanding we want to improve student performance through intervention.
5. **Ideology:** Learning was viewed separately from face-to-face traditional teaching. The traditional methods only catering for specific students' learning preferences were identified as oppressing and they were enhance by the development of e-learning content sensitive towards wider learning preferences.
6. **Complete Structure:** This principle is evident in the analysis of the intervention phase where more than one method was used to gain a better understanding of the success of the intervention. The fact that students requested the e-learning material to be expanded to cover the entire module content inspired the research team even if the statistical analysis in year 2 did not show a significant correlation between the usage of the material and the performance of the students.
7. **Critical Research History:** The context of students today is vastly different from before as students live in an information consumer society where information is always available. The specific political factors in South Africa could also play a role—discussion of this did take place in the research project but falls outside the scope of this chapter.
8. **Deconstruction / Reconstruction:** The learning experience of the students was deconstructed in terms of learning styles to understand how individual students learn best. A new learning environment was constructed from this knowledge to better suit the learning preferences of this students.

Care was taken to analyse the success of the intervention to ensure that no new oppressive structures were created.

In order to validate the quality of the research one can reflect on the interpretive phases from the Klein and Myers (1999) perspective:

1. **The fundamental principle of the hermeneutic circle:** During both phases where interpretive questionnaires were used, the analysis of each answer was done in terms of all the answers from a specific participant as well as in terms of all the participants' answers for that question. This means that individual answers were better understood by looking at other answers from the same participant. Each answer was also compared with the other participants' answers to that question.
2. **The principle of contextualisation:** During the initial diagnosis care was taken to discuss the historic background of the students. It also came to fore that students were not used to e-learning as none of their other modules were presented electronically.
3. **The principle of interaction between the researchers and the subjects:** The research team was sensitive to the fact that students may regard the team member who did the data collection as a superior figure. Care was taken to explain to students that there are no correct or incorrect answers to the questions as the motive of the research team is learning from the students.
4. **The principle of abstraction and generalisation:** Content analysis was used for the coding of the questionnaires for both phases. The data of each phase were analysed twice to ensure that consistency was achieved. The understanding of the learning styles of these students can be used in the design of their other modules to enhance their learning experiences.
5. **The principle of dialogical reasoning:** The research team expected that individual learners would strongly demonstrate learning preferences according to the framework of Felder and Silverman (1988). The data indicated that individual students indicated preferences for styles such as visual and verbal learning, which are regarded as opposing strategies.
6. **The principle of multiple interpretations:** Very few instances of multiple interpretations were identified.
7. **The principle of suspicion:** The answers of both questionnaires were treated with caution as the participants might have answered what they believed the research team wanted to hear in order to achieve some form of academic gain.

In terms of reflection from a positivistic perspective the criteria of Guba and Lincoln (1989) may be used. Positivistic methods were used to investigate the cause-effect relationship between the usage of the e-learning material and the performance of the students. The positivistic research may be evaluated as follows:

1. **Internal Validity:** After analysis of the first year's usage statistics and the performance of the students a correlation between time students spent working on the e-learning material and their performance in the examination question on the corresponding topic could be established. A relative small number of students used the material. In the second year when many students used the material, this correlation could not be achieved.
2. **External Validity:** External validity could not be achieved.

3. **Reliability:** Care was taken to record all quantitative values accurately. There is nothing that prevents the same data to be collected and analysed for future groups in this module.
4. **Objectivity:** All student work was evaluated in the examination according to accepted model answers. Accepted repeatable methods were also used. Examination questions and student answers were moderated by external parties to ensure fairness.

Reflection on Paradigm Level

In this section, the problem is briefly reflected upon from the viewpoint of individual research paradigms. The aim of the section is to illustrate the advantage of using a combination of research methods by highlighting problems while applying a single paradigm.

From a purely positivist perspective, the initial diagnosis would have been done differently. A standardised questionnaire would have to be completed by the lecturers. In pure positivist research, very few open-ended questions are used. It is possible that the issue of learning styles would not have been raised. The analysis of the data also proved complex from a positivistic perspective, as no meaningful correlations were found between the performance of students and their time spent on e-learning material after the second iteration. Causality between the usage of e-material and the performance of students could not be proven.

From a purely interpretive viewpoint, the overall drive would not have been to improve the students' performance. Understanding would have been the central theme, such as understanding of the problems experienced by lecturers, which would necessitate an understanding of the learning styles of students. The abundance of information stored in the Web logs and test and examination marks would not have been used, as typical interpretive methods do not focus on analysing large amounts of quantitative data.

One might argue that this research is purely CTS. It is true that pragmatism plays an important role in CTS, and any methods required (in this case positivistic and interpretive methods) to solve the problem would be utilised. From a purely CST evaluation of the research, the omission of students from the decision making group is troublesome, as they are the affected party in the problem situation which needs to be emancipated. Except for the fact that successful students in the module are no longer present on campus, faculty members are also not used to involving students in decision making.

The research strategy of this project can be viewed as a true mixed method application. Papers on mixed methods are currently more frequently published, and the main point of discussion in this field is the evaluation of mixed research projects. The ongoing debate is whether specific guidelines for such projects should be different from guidelines for evaluating purist research projects, or whether phases in mixed methods research projects should be evaluated individually according to the specific paradigm used. The latter method was followed in this research environment. The omission of students from the research team can be seen as a major shortcoming from a CST perspective.

REFLECTIONS ON MIXED METHODS

In terms of the mixed method classification of Creswell (2012) this study can overall be classified as transformative as the object was to improve the performance of networking students. Pragmatically, this objective was the driving force behind the chosen methods. The diagnosis phase can be views as an exploratory sequential design as the aim was to better understand the learning styles of the individual students and the group as a whole in order to understand what type of learning environment to design. This was followed by the design and implementation of a blended learning environment designed to assist the students in a problematic area of the learning content. After the quantitative Web-logs were analysed, the research team were not satisfied that they had enough information to understand the improvement of the student results. An explanatory sequential design was used in terms of interpretive questionnaires to better understand the experience from the perspective of the students.

The possibility of different methods in different phases of the project allowed better understanding of the problem environment in order to solve the problems of the students. As indicated in the previous section the application of a pure positivistic or interpretive design would not have had the same effect on the performance of the students. In dividing the project in clear phases, it was possible to combine methods from different paradigms while using each method true to the research paradigm underpinning the method. Each phase could then be evaluated from the methodological quality perspective of the representative paradigm as done in the previous section.

CONCLUSION

In choosing a research strategy for a problem situation, the underpinning paradigms guiding the application of specific research methods should be clearly understood. Once understood, these paradigms and their grounding assumptions should guide the researcher to divide a problem into phases, each with a specific nature. It is simpler to apply mixed methods to different phases of a project than in a single phase.

In the planning phase of the project, decisions need to be taken on the evaluation of the validity of the research results. Research methods should be true to the evaluation criteria used by practitioners in the specific paradigm.

The case study represents a complicated mixed method design where most of the types of mixed methods designs described by Creswell (2012) could be demonstrated. However, from the perspective of the research team the design was motivated by the overall objective of improving student performance rather than that of demonstrating the advantages of a complex mixed method design. It incorporated different methods to achieve each phase in an action research project. The diagnosis was done interpretively in order to better understand the students and their learning style preferences. The outcome of the analysis guided the development of e-learning material that accommodated different learning style preferences. The analysis of intervention phase is always of great importance in any action research project as it proves the success of the intervention. In this case, a large amount of quantitative data was available in terms of student marks and the usage logs of the Web pages. However, causality could not be proved satisfactory from the positivistic perspective. Too many uncertainties led to questions such as: 'Are the students reading the material while it is displayed on their computer screens?' It was decided to test the attitude of the students towards the material by means of an interpretive questionnaire. The outcome was encouraging and motivated the research team to extend the scope of the e-learning material. The students also gave helpful insights towards better coordinating the material with existing lectures.

As students demonstrated willingness to extend their learning activities to include electronic learning, the use of e-learning can be, in future, extended to other forms of media such as social networking. The impact of such an extension can be investigated in a similar study.

Another prospect for future research is a comparison between Human-Centred Design (HCD) and action research. These methods can be evaluated from the framework of philosophy, methodology, and practice used in this chapter. References on HCD are provided in the additional reading list at the end of the chapter.

A problem with mixed methods research is the publication of such work. If the reviewer applies a fixed paradigm set of quality criteria there are bound to be shortcomings in the work. If the responsibility lies with the researcher to provide criteria for the specific project, the burden of proof of good research may overshadow the actual research case. The authors believe that the reviewer community is not yet ready to accept pragmatism as quality criteria and that self-evaluation is the best course of action.

Mixing methods is not a way of hiding poor research strategy, but rather to enrich the research outcomes by viewing the problem environment through different paradigmatic lenses.

REFERENCES

Agar, M. H. (1980). *The professional stranger: An informal introduction to ethnography*. New York, NY: Academic Press.

Allan, B. (2007). *Blended learning: Tools for teaching and training*. London, UK: Facet.

Baskerville, R. L. (1999). Investigating information systems with action research. *Communications of the Association for Information Systems*, *2*(19), 2–31.

Behr, A. L. (1983). *Empirical research methods for the human sciences: An introductory text for students of education, psychology and the social sciences*. Pretoria, South Africa: Butterworths.

Blum, F. H. (1955). Action research: A scientific approach? *Philosophy of Science*, *22*(1), 1–7. doi:10.1086/287381

Bryman, A. (2006). Paradigm peace and the implications for quality. *International Journal of Social Research Methodology*, *9*(2), 111–126. doi:10.1080/13645570600595280

Bryman, A., Becker, S., & Sempik, J. (2008). Quality criteria for quantitative, qualitative and mixed methods research: A view from social policy. *International Journal of Social Research Methodology*, *11*(4), 261–276. doi:10.1080/13645570701401644

Carliner, S. (2004). *An overview of online learning*. Amherst, MA: HRD Press.

Checkland, P. (1981). *Systems thinking, systems practice*. Chichester, UK: Wiley.

Creswell, J. W. (2012). *Educational research: Planning, concucting, and evaluating quantitiative and qualitative research* (4th ed.). Boston, MA: Pearson.

Denzin, N. K., & Lincoln, Y. S. (2003). Introduction: The discipline and practice of qualitative research. In Denzin, N. K., & Lincoln, Y. S. (Eds.), *The Landscape of Qualitative Research* (pp. 1–45). New Delhi, India: SAGE Publications.

Feilzer, M. Y. (2010). Doing mixed methods research pragmatically: Implications for the rediscovery of pragmatism as a research paradigm. *Journal of Mixed Methods Research*, *4*(1), 6–16. doi:10.1177/1558689809349691

Felder, R. M. (2009). *Index of learning styles*. Retrieved September, 19, 2009, from http://www4.ncsu.edu/unity/lockers/users/f/felder/public/ILSpage.html

Felder, R. M., & Silverman, L. K. (1988). Learning and teaching styles in engineering education. *Journal of Engineering Education*, *7*(7), 674–681.

Giddens, A. (Ed.). (1974). *Positivism and sociology*. London, UK: Heinemann.

Glaser, B. G., & Strauss, A. L. (1967). *The discovery of grounded theory: Strategies for qualitative research*. New York, NY: Aldine de Gruyter. doi:10.1097/00006199-196807000-00014

Glaser, B. G., & Strauss, A. L. (1999). *The discovery of grounded theory: Strategies for qualitative research*. New York, NY: Aldine de Gruyter. doi:10.1097/00006199-196807000-00014

Guba, E. G., & Lincoln, Y. S. (1989). *Fourth generation evaluation*. Thousand Oaks, CA: Sage.

Harvey, L. (1990). *Critical social research*. London, UK: Unwin Hyman.

Heikkinen, H., Huttunen, R., & Syrjälä, L. (2007). Action research as narrative: Five principles for validation. *Educational Action Research, 15*(1), 5–19. doi:10.1080/09650790601150709

Hughes, J. A. (1990). *The philosophy of social research* (2nd ed.). London, UK: Longman.

Johnson, R. B., & Onwuegbuzie, A. J. (2004). Mixed method research: A research paradigm whose time has come. *Educational Researcher, 33*(7), 14–26. doi:10.3102/0013189X033007014

Kerres, M., & De Witt, C. (2003). A didactical framework for the design of blended learning arrangements. *Journal of Educational Media, 28*(2/3), 101–113. doi:10.1080/1358165032000165653

Klein, H. K., & Myers, M. D. (1999). A set of principles for conducting and evaluating interpretive field studies in information systems. *Management Information Systems Quarterly, 23*(1), 67–94. doi:10.2307/249410

Kuhn, T. S. (1970). *The structure of scientific revolutions* (2nd ed.). Chicago, IL: University of Chicago Press.

Kuhn, T. S. (1977). *The essential tension: Selected studies in scientific tradition and change*. Chicago, IL: University of Chicago Press.

Lee, A. S. (1999). Researching MIS. In Currie, W. L., & Galliers, R. (Eds.), *Rethinking Management Information Systems* (pp. 7–27). Oxford, UK: Oxford University Press.

Lewin, K. (1947). Frontiers in group dynamics. *Human Relations, 1*(1), 5–41. doi:10.1177/001872674700100103

Littlejohn, A., & Pegler, C. (2007). *Preparing for blended e-learning: Connection with e-learning*. Florence, KY: Routledge.

Lubbe, S. (2003). The development of a case study methodology in the information technology field: A step by step approach. *Ubiquity: A Web-Based Publication of the ACM, 4*(27). Retrieved October 5, 2003 from http://www.acm.org/ubiquity/views/v4i27_lubbe.pdf

Midgley, G. (2000). *Systemic intervention: Philosophy, methodology and practice*. New York, NY: Kluwer Academic/Plenum.

Mingers, J. (2001). Combining IS research methods: Towards a pluralist methodology. *Information Systems Research, 12*(3), 240–259. doi:10.1287/isre.12.3.240.9709

Myers, M. D. (1997). Qualitative research in information systems. *Management Information Systems Quarterly, 21*(2), 241–242. doi:10.2307/249422

Niaz, M. (2008). A rationale for mixed method (integrative) research programmes in education. *Journal of Philosophy of Education, 42*(2), 287–305. doi:10.1111/j.1467-9752.2008.00625.x

Oates, B. J. (2006). *Researching information systems and computing*. London, UK: SAGE Publications.

Schutz, A. (1962). *Collected papers I: The problem of social reality*. The Hague, The Netherlands: Martinus Nijhoff.

Straub, D. W., Boudreau, M., & Gefen, D. (2004). Validation guidelines for IS positivist research. *Communications of the Association for Information Systems, 13*, 380–427.

Strauss, A., & Corbin, J. (1998). *Basics of qualitative research*. New Delhi, India: SAGE Publications.

Symonds, J. E., & Gorard, S. (2010). Death of mixed methods? Or the rebirth of research as a craft. *Evaluation and Research in Education*, *23*(2), 121–136. doi:10.1080/09500790.2010.483514

Teddlie, C. B., & Tashakkori, A. (2009). *Foundations of mixed methods research: Integrating quantitative and qualitative approaches in the social and behavioral sciences*. Thousand Oaks, CA: SAGE Publications.

Walsham, G. (1995). The emergence of interpretivism in IS research. *Information Systems Research*, *6*(4), 376–394. doi:10.1287/isre.6.4.376

ADDITIONAL READING

Baskerville, R. L., & Wood-Harper, A. T. (1996). A critical perspective on action research as a method for information systems research. *Journal of Information Technology*, *11*, 235–246. doi:10.1080/026839696345289

Bernstein, R. J. (Ed.). (1985). *Habermas and modernity*. Cambridge, UK: Polity.

Clark, P. A. (1972). *Action research and organisational change*. London, UK: Harper and Row.

Dilthey, W. (1989). The relationship of the human sciences to the natural sciences. In Makkreel, R. A., & Frithjof, R. (Eds.), *Wilhelm Dilthey: Selected Works* (*Vol. 1*, pp. 66–71). Princeton, NJ: Princeton University Press.

Elden, M., & Chisholm, R. F. (1993). Features of emerging action research. *Human Relations*, *46*(2), 275–298. doi:10.1177/001872679304600207

Hanington, B., & Martin, B. (2012). *Universal methods of design: 100 ways to research complex problems, develop innovative ideas, and design effective solutions*. Beverly, MA: Rockport.

KEY TERMS AND DEFINITIONS

Action Research: An iterative research methodology for guiding change in a situation.

Blended Learning: The mix of different didactic methods and delivery formats.

Critical Social Research: Research paradigm underpinned by a critical–dialectical perspective, which attempts to dig beneath the surface of historically specific, oppressive social structures.

Mixed Methods: The combination of research methods from different research paradigms in a single research project.

Research Criteria: Criteria to evaluate research quality.

Research Paradigm: A construct that specifies a general set of philosophical assumptions regarding research.

Chapter 14
Experiences in Applying Mixed-Methods Approach in Information Systems Research

Guo Chao Peng
University of Sheffield, UK

Fenio Annansingh
University of Plymouth, UK

ABSTRACT

Mixed-methods research, which comprises both quantitative and qualitative components, is widely perceived as a means to resolve the inherent limitations of traditional single method designs and is thus expected to yield richer and more holistic findings. Despite such distinctive benefits and continuous advocacy from Information Systems (IS) researchers, the use of mixed-methods approaches in the IS field has not been high. This chapter discusses some of the key reasons that led to this low application rate of mixed-methods design in the IS field, ranging from misunderstanding the term with multiple-methods research to practical difficulties for design and implementation. Two previous IS studies are used as examples to illustrate the discussion. The chapter concludes by recommending that in order to apply mixed-methods design successfully, IS researchers need to plan and consider thoroughly how the quantitative and qualitative components (i.e. from data collection to data analysis to reporting of findings) can be genuinely integrated together and supplement one another, in relation to the predefined research questions and the specific research contexts.

INTRODUCTION

Research designs and methods adopted by Social Sciences researchers in general, and in the Information Systems (IS) field in particular, can be broadly classified into two main categories, namely quantitative and qualitative (Jick, 1979; Orlikowski & Baroudi, 1991; Mingers, 2001; Creswell, 2003; Saunders, et al., 2003). However, it is widely understood and recognized that both quantitative and qualitative designs have their own advantages and limitations. For example,

DOI: 10.4018/978-1-4666-2491-7.ch014

Copyright © 2013, IGI Global. Copying or distributing in print or electronic forms without written permission of IGI Global is prohibited.

questionnaire survey, as a typical quantitative method, is a very efficient and economical way for collecting data from a large sample in a wide geographical area at the same time (Bryman, 2004, pp. 133-134). Nonetheless, a questionnaire is arguably a less efficient method to be used in exploratory studies that aim to investigate and explore sophisticated social contexts (Robson, 2002, p. 234). On the other hand, interview as a typical qualitative tool is very useful and efficient in gathering and exploring in-depth human insights and perceptions on complex social phenomena (Saunders, et al., 2003, p. 246; Bryman, 2004, p. 321). Nevertheless, as interviews can very often last more than one hour, it is very time-consuming to carry out interviews with a large group of respondents.

The realization of the inherent limitations of quantitative and qualitative approaches results in the emergence and use of an alternative research design, namely mixed-methods research. Mixed-methods research integrates and combines both quantitative and qualitative methods to investigate the same underlying phenomenon in one single study (Leech & Onwuegbuzie, 2009). This approach is deemed to be efficient in supplementing the weaknesses of single method designs and thus leading to richer findings and higher quality research (Jick, 1979; Mingers, 2001; Creswell, 2003; Fidel, 2008; Leech & Onwuegbuzie, 2009). It has been used as a distinct approach in social sciences research for more than five decades (Campbell & Fiske, 1959; Jick, 1979; Rocco, et al., 2003). In the IS field, a considerable number of researchers have advocated the use of mixed-methods approach since the early 1990s (Galliers, 1991; Lee, 1991; Robey, 1996; Mingers, 2001; Petter & Gallivan, 2004; Fidel, 2008). Curiously, despite strong and continuous support from IS researchers, the actual use of the mixed-methods approach in IS research has not been prevalent. In particular, Mingers (2003) reviewed the IS literature between 1993 and 2000, and found that only 20% of articles published in this period of time adopted multiple research methods. In a more recent study, Fidel (2008) reviewed 465 articles published in four major journals in Library and Information Science (LIS) during 2005-2006. The study found that only 17% of these LIS articles adopted multiple methods, and only 5% could be considered as 'truly' mixed-methods research (Fidel, 2008). This low application rate of mixed-methods research in the IS field might be caused by a number of issues such as:

- There may exist some misunderstandings among IS researchers about the actual meaning of mixed-methods research (i.e. is 'mixed-methods' the same as 'multiple-methods' design?).
- IS researchers may find it difficult to decide which mixed-methods design would be suitable for a particular study (e.g. what priority or weight should be given to the quantitative and qualitative methods? In what sequence these methods should be conducted?).
- IS researchers may have difficulties in integrating and making sense of quantitative and qualitative components effectively across the entire study.

This chapter provides an in-depth discussion and some practical guidelines related to the above issues. It aims to help IS researchers make more appropriate decisions when designing and implementing mixed-methods research, and thus leading to more rigorous and meaningful findings. We also use two previous IS studies as examples to illustrate the discussion. These two studies adopted mixed-methods designs to investigate respectively ERP post-implementation risks and knowledge leakage risks associated with the design and use of 3D modelling.

The chapter is structured as follows. The next section provides a discussion and clarification on the concept of mixed-methods research, followed by a discussion on various commonly used mixed-methods designs and their associated benefits and practical difficulties. Subsequently, the chapter provides two examples (respectively adopting QUAN + qual and QUAL + quan approaches) to illustrate how mixed-methods design can be applied in IS studies. The chapter concludes with a recommendation that in order to apply the mixed-methods approach effectively, IS researchers need to plan and consider thoroughly how the quantitative and qualitative components (i.e. from data collection to data analysis to findings) can be integrated together to support one another, in relation to the predefined research questions and the specific research contexts.

MIXED-METHODS RESEARCH

Definition of Mixed-Methods Research

Although mixed-methods research is not a new concept, its definition has been relatively misrepresented by a number of IS researchers over time. In particular, some advocators of mixed-methods approach in the IS field, such as Gable (1994) and Mingers (2001, 2003), simply named the concept as 'multiple-methods' research in their articles. However, a further review of the literature on Social Sciences, where mixed-methods approach originated from and has been widely used, suggests that multi-methods is not equivalent to mixed-methods research, due to at least two main reasons:

- When adopting multiple-methods, researchers can select methods from just a single approach (e.g. two quantitative methods or two qualitative methods), rather than combining the use of both quantitative and qualitative methods as required in mixed-methods design (Jick, 1979; Rocco, et al., 2003; Petter & Gallivan, 2004; Fidel, 2008).
- A study can be considered as a multi-methods research whenever more than one research method is employed (Petter & Gallivan, 2004). However, a mixed-methods research requires not only the use of multiple methods, but also that quantitative and qualitative approaches and findings need to be properly integrated and actually complement each other (Rocco, et al., 2003; Bryman, 2007; Fidel, 2008).

Thus, for the purposes of this chapter, we would adopt a more rigorous definition for mixed-methods research as proposed by Tashakkori and Creswell (2007). That is, mixed-methods research is defined as "research in which the investigator collects and analyzes data, integrates the findings, and draws inferences using both qualitative and quantitative approaches or methods in a single study or program of inquiry" (Tashakkori & Creswell, 2007). Furthermore, Fidel (2008) reinforces that 'mixing' is the core of a mixed-methods research and can occur in different stages of the study:

- In the design stage, features of all selected research methods need to be taken into consideration to establish the research design of the study.
- In the data collection stage, one approach should provide insights that improve the process of data collection of the other approach.
- In the analysis stage, data collected and results derived from both approaches need to be integrated and support each other.

These fundamental needs for integrating quantitative and qualitative elements across the entire study lead to both strengths and practical difficulties of mixed-methods research, as discussed later in this chapter.

Types of Mixed-Methods Designs

In relation to the above rigorous definition, a mixed-methods research can actually be designed very flexibly. In particular, an extensive literature review conducted by Tashakkori and Teddlie (2003) identified that, researchers in Social and Behavioral Sciences developed around fourty different mixed-methods designs that employed at least two quantitative and qualitative methods concurrently or in sequence. Nonetheless, the six (i.e. including three sequential and three concurrent) designs proposed by Creswell (2003) are some of the most commonly used ones in mixed-methods research (Ivankova, et al., 2006). These six mixed-methods designs are therefore described in detail in Table 1.

When there are so many different combinations to establish a mixed-methods design, it should be highlighted that one is not necessarily better than the other. In fact, the actual design that researchers should use, depends entirely on the research question and research context being studied. In order to illustrate how a mixed-methods design can be selected and applied in IS research, two practical examples are given in section 3.

Strengths of Mixed-Methods Design

Mixed-methods research is particularly useful and suitable for research projects where no single approach can fully explain or explore the phenomenon being investigated, especially when this phenomenon is complex and multifaceted (Fidel, 2008). Moreover, the combination of the use of quantitative and qualitative methods in a single study can help researchers to supplement the limitations of each single method, as well as to achieve triangulation (Creswell, 2003; Fidel, 2008).

The term 'triangulation' is broadly defined by Denzin (quoted by Jick, 1979) as "the combination of methodologies in the study of the same phenomenon." In practice, the concept of triangulation can often be applied in four ways (Denzin, 1978; Jick, 1979; Patton, 1999):

Table 1. Six types of mixed-methods designs (Creswell, 2003)

Mixed-methods design	Characteristics
Sequential designs:	
Sequential explanatory Design	This design contains two phases and is characterized by the collection and analysis of quantitative data followed by the collection and analysis of qualitative data. The priority is given to the quantitative part. The purpose of this design is to use qualitative results to further explore and explain the findings of a primarily quantitative study.
Sequential exploratory Design	This design has an initial phase of qualitative data collection and analysis. This qualitative component is then followed by a phase of quantitative data collection and analysis with the aim of increasing generalisability of the findings. The priority is given to the qualitative aspect.
Sequential transformative Design	This design contains two distinct data collection phases. However, either method may be used first when collecting data, and the priority can be given to either the quantitative or the qualitative phase.
Concurrent designs:	
Concurrent triangulation Design	In this design, both quantitative and qualitative methods are used simultaneously in one phase, with the aim to confirm and cross-validate findings. Both components are equally important.
Concurrent nested design	This design contains one data collection phase, during which both quantitative and qualitative data are collected simultaneously. However, one method (either quantitative or qualitative) must take the predominant position.
Concurrent transformative design	This design combines the features of both concurrent triangulation and concurrent nested designs. Specifically, it may involve a triangulation of quantitative and qualitative components that are equally important. It is also embedded with a supplement method to further explore the issue.

- **Method Triangulation:** Multiple methods are used to collect different sets of data to study the same concept/phenomenon.
- **Triangulation of Sources:** Use the same method to collect data from different samples or data sources at different times, locations and/or contexts.
- **Analyst Triangulation:** Multiple techniques are used to analyze and interpret the same set of data from different dimensions.
- **Theory/Perspective Triangulation:** Multiple theories or perspectives are used to interpret the data.

It is evident that, when multiple approaches are used to collect data in mixed-methods research, method triangulation would be achieved. According to Jick (1979) and Creswell (2003, pp. 15-16), method triangulation can provide researchers with a range of benefits:

- Allows researchers to be more confident on their results.
- Helps to neutralize or cancel the biases that may exist in single method approach.
- Stimulates the creation of new ways to combine different approaches, strategies and methods to answer a specific research question.
- It helps to uncover the unique or deviant dimension of a sophisticated phenomenon that might be otherwise overlooked when using a single method, and thus amplifies the richness of the research findings.

Overall, in contrast with the single method approach, mixed-methods research can allow researchers to explore and investigate sophisticated issues more holistically and widely (Fidel, 2008). This approach is thus deemed to be particularly suitable to study IS issues, which are always multi-dimensional involving a wide range of socio-political, socio-technical, regional, cultural, and organizational factors.

Practical Difficulties of Mixed-Methods Design

Despite the above crucial benefits, mixed-methods research is however not easy to design and implement in actual practices (Ivankova, et al., 2006; Fidel, 2008). In fact, while mixed-methods research opens a variety of opportunities and possibilities for research design, it also raises a wide range of questions and issues that need to be considered cautiously by researchers.

Specifically, at the design stage, researchers need to carefully consider which quantitative and qualitative methods should be combined in the study, in what order (e.g. sequentially or concurrently) and what priority (e.g. equally important or one predominates the other), how these methods can be used, and what objectives each component should attempt to achieve (Fidel, 2008; Ivankova, et al., 2006; Creswell, et al., 2003). It is evident that this complex set of decisions needs to be made in response to the predefined research questions and research context. Inappropriate decisions made at the early stage will not just affect the rigorousness and reliability of the mixed-methods research design, but will also impact the richness and significance of the subsequent research findings.

Moreover, researchers' attitudes and abilities to implement the research design represent further challenges for adopting mixed-methods approach (Fidel, 2008). In fact, it is clear that quantitative and qualitative approaches are very different in terms of underlying epistemologies, data collection procedures, nature of data collected, and data analysis techniques. As a result, it will be difficult for inexperienced researchers (e.g. PhD students) to mix these different approaches in one single study. On the other hand, senior researchers having traditional preference in using one approach (i.e. either quantitative or qualitative approach) may very often find it difficult to accept and in fact, may also not be equipped with adequate skills, to use the other (Fidel, 2008; Bryman, 2007). The situation can become particularly complicated when a mixed-methods study is undertaken by a team of researchers and each of them has a very strong stance on their own single approach and considers other approaches as secondary (Patton, 2002; Bryman, 2007; Fidel, 2008).

Consequently, these practical issues and difficulties can hinder efficient integration of quantitative and qualitative findings in mixed-methods research (Bryman, 2007). In particular, Bryman (2007) highlights that in many mixed-methods studies, quantitative and qualitative results may either not be genuinely integrated or be integrated to only a limited degree. The reasons for this may be related to a wide set of barriers, e.g. inefficient research design, methodological preferences and skills of researchers, and even publication tendency (i.e. only accept quantitative or qualitative findings) of some journals (Bryman, 2007). In such circumstance, the quantitative and qualitative components may neither be related nor complemented with each other. As a result, the mixed-methods design may not result in more holistic and significant findings than a single method approach, and in fact will also lose its essential meaning and original value.

APPLICATION OF MIXED-METHODS DESIGN IN IS RESEARCH

In order to provide further insights and guidance to address the above practical challenges, this chapter presents two studies as examples to illustrate how mixed-methods research can be designed and implemented effectively in the IS field. These two studies respectively adopted a sequential explanatory (QUAN + qual) and a sequential exploratory (QUAL + quan) design, as proposed by Creswell (2003) and outlined earlier. The presentation of both studies follows a common structure, namely research aims and objectives; choice and rationale of the research design; quantitative or qualitative data collection and analysis; and finally the integration of findings.

Identification and Assessment of Risks associated with the Post-Implementation of ERP Systems in China

Research Aims and Objectives

Enterprise Resource Planning (ERP) systems have nowadays been widely adopted by Chinese companies in order to improve operational efficiency and enhance core competencies. However, successful implementation of the system is not the end of the ERP journey. Very often, the system post-implementation or exploitation stage is where the real challenges will begin and more critical risks may occur. Thus, the research project reported in this section aimed to identify, assess and explore potential risks that Chinese firms may encounter when using, maintaining and enhancing ERPs in the post-implementation phase. It also attempts to explore the causes, impacts, probability of occurrence and frequency of occurrence of identified risk events, as well as to investigate the relationships between them.

Research Design: QUAN + Qual Approach

At the initial stage of the study, the first temptation of the researchers was to undertake a national study of the whole of China. However, this soon proved to be virtually impossible, given the size of China and the fact that the current economic situation and context in the country are very complicated and fluid. As a consequence, after studying the national and business environment of China by conducting a Political, Economic, Social and Technological (PEST) analysis, the researchers decided to focus and base the study on a specific set of Chinese companies, namely

Figure 1. Theoretical ontology of ERP post-implementation risks

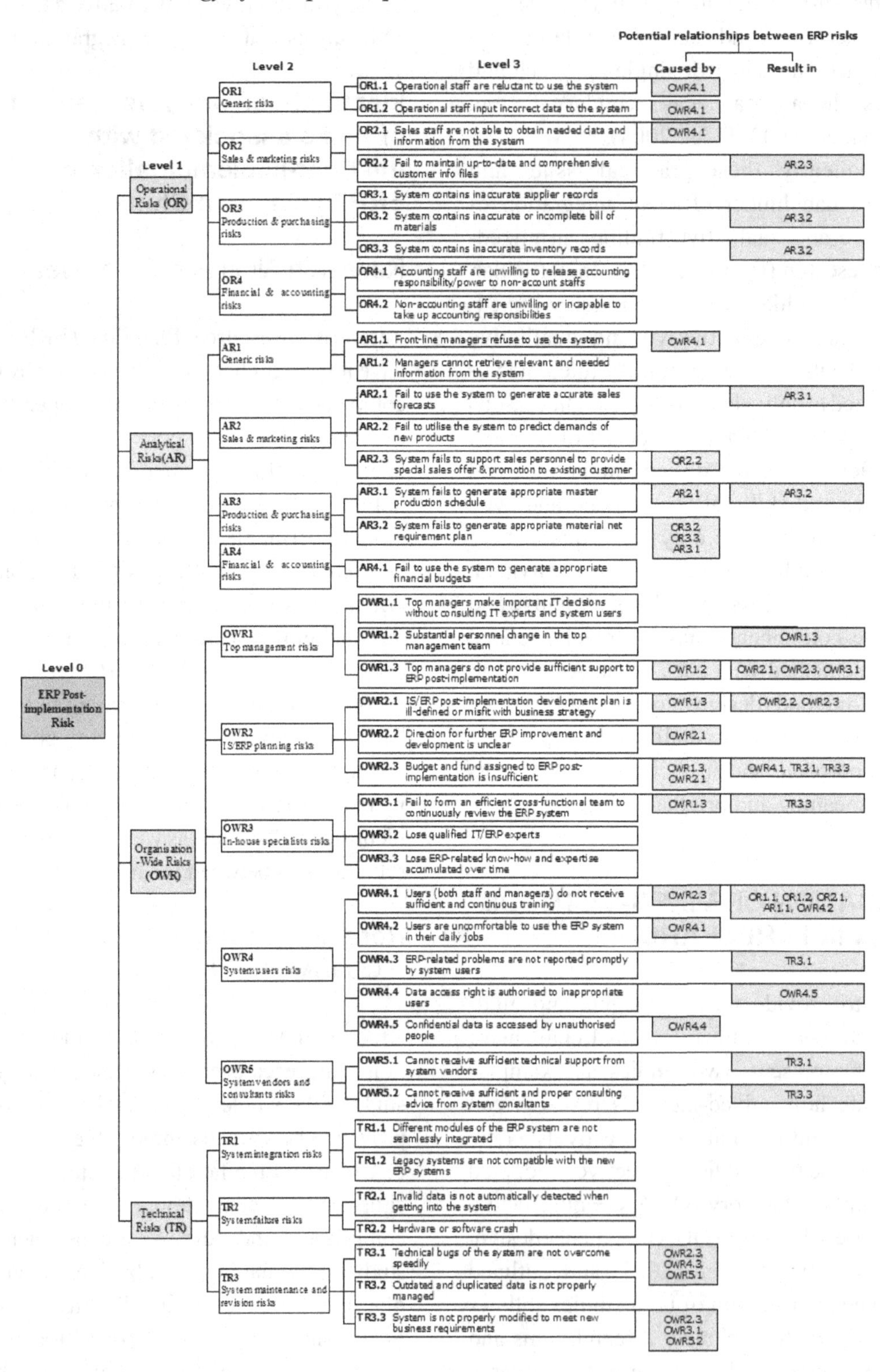

State-Owned Enterprises (SOEs) in the electronic and telecommunication manufacturing sector in Guangdong province in China.

After a feasible set of companies was selected, the next step of the study was to establish explicit IS lens, in order to frame the study and generate meaningful and significant findings. As a consequence, the researchers carried out a desktop study based on a critical literature review. The process of this extensive literature review did not return any specific studies on ERP post-implementation risks. Nevertheless, the researchers identified and retrieved, through the critical literature review, a large amount of IS and general business research studies, case studies and theoretical papers in both English and Chinese. By analysing, comparing, and synthesising these articles and materials, the researchers established and proposed a total of 40 ERP exploitation risks, as well as explored and analysed their potential causes and impacts. These 40 ERP exploitation risks were concentrated around operational, analytical, organisation-wide, and technical areas. A risk ontology was subsequently developed to organise and present these identified ERP risks and their relationships, as shown in Figure 1.

Moreover, when the study focused on such a specific context as discussed above, the researchers considered that findings that were generalisable to this particular context would not just be essential but in fact also highly valuable and meaningful. It therefore became apparent that a deductive quantitative questionnaire-based study was needed to produce generalisable statements, as well as to test and examine the suitability of the theoretical ERP risk ontology in the context of Chinese SOEs.

Moreover, due to a lack of study in ERP post-implementation in general and in the Chinese context in particular, IS and ERP literature used to ground the theoretical basis of the quantitative study were published mainly in the West. However, it was then anticipated that some findings derived from Western contexts may not be entirely applicable to the Chinese one. Therefore, the early quantitative study might yield findings that would differ from the original theory. It was thus considered that a follow-up qualitative study should be carried out. This second study aimed at using a process of interviews to explore any unexpected findings derived from the quantitative component.

Consequently, these considerations clearly pointed to the selection and adoption of a two-phase sequential explanatory (QUAN + Qual) design for this research. In particular, a quantitative questionnaire survey was carried out as the first phase of this mixed-methods design and took the predominant position of the entire research. Subsequently, a follow-up qualitative case study component was conducted to explore further the quantitative findings and thus achieve triangulation and theory extension.

Figure 2 provides a summary and a more visual view on the whole research design being discussed so far.

Quantitative Data Collection

The questionnaire used in the quantitative phase was designed based on the theoretical ERP risk ontology. It aimed to seek Chinese managers' perceptions on the 40 predefined ERP risks (e.g. which of the 40 proposed events would be perceived as risks to ERP exploitation in Chinese SOEs, and which risks Chinese managers would consider as the most important ones), as well as to explore the correlations between these risks. In order to achieve these objectives, each of the 40 predefined risk events was examined in the questionnaire through four questions:

1. Whether this event could be perceived as a risk to ERP exploitation (1 = yes, 2= no).
2. What the probability of occurrence of this risk event could be (measured on a 3-point Likert scale, ranging from high [3] to low [1]).

Figure 2. Overall research design of the ERP project in China

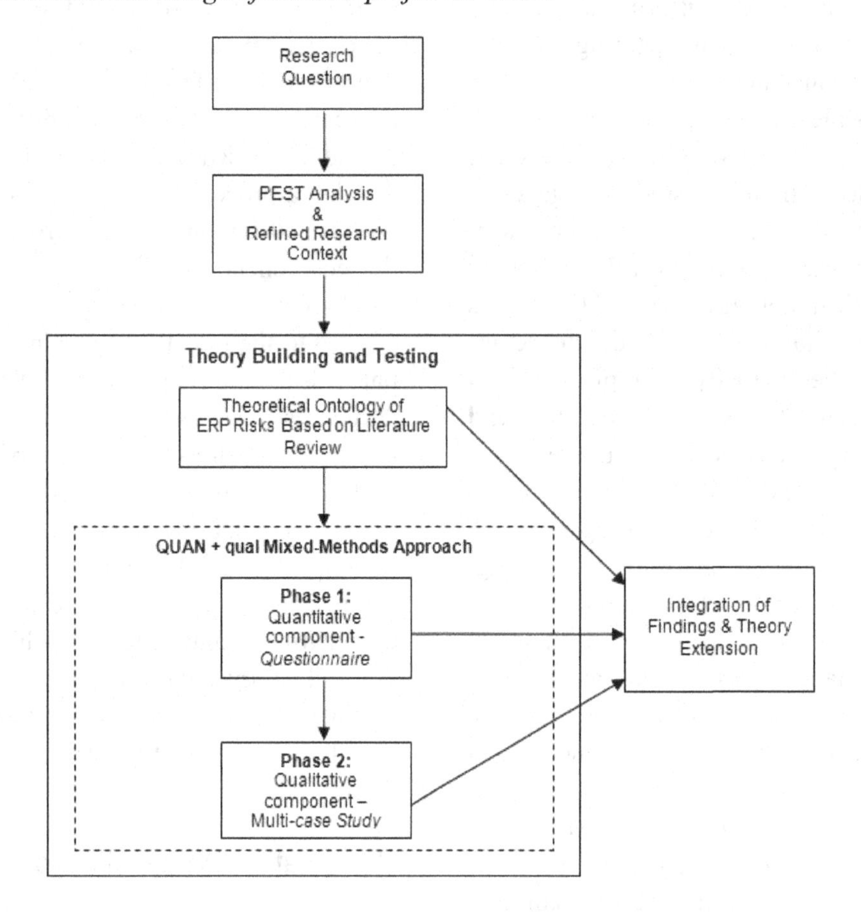

3. What level of impact this risk could result in (measured on a 3-point Likert scale, ranging from high [3] to low [1]).
4. What the frequency of occurrence of this risk event could be (measured on a 5-point Likert scale, ranging from very often [5] to very rarely [1]).

Moreover, the questionnaire was originally developed in English and then translated into Chinese. Substantial attention had been paid during the translation process in order to ensure that both the English and Chinese versions of the questionnaire were conceptually equivalent, and thereby ensure high internal validity. In order to further improve its validity, the Chinese version of questionnaire was pilot tested with a group of Chinese postgraduate students and researchers in the authors' department as well as 5 Chinese managers working in a SOE. This pilot test resulted in a number of minor corrections to the description of some risk items given in the questionnaire.

According to statistical data provided by the local statistical bureau, there are 118 SOEs operating in the Electronic and Telecommunication Manufacturing Sector in Guangdong. A complete contact list of these companies was retrieved from the Guangdong Statistical Bureau. Consequently, the designed questionnaire was sent to IT managers and operational managers of the 118 target Chinese SOEs, from which 42 valid and usable responses were received and analyzed. This survey thus achieved a response rate of 35.6%.

Quantitative Data Analysis and Findings

The survey findings showed that all of the 40 events proposed in the risk ontology were confirmed by the majority (80% or more) of respondents as risk events to ERP post-implementation in Chinese SOEs. Moreover, and as discussed above, the questionnaire also asked respondents to assess the importance of each risk item from three aspects, namely probability of occurrence, impact and frequency of occurrence. The need for all this information lies in the fact that from a risk management perspective, a risk event that has a high probability of occurrence may not have a high impact, and vice versa. As a typical example, system crash is a risk event that often has high impact but low probability of occurrence. Moreover, while probability refers to 'how likely' a risk event may occur, frequency refers to 'how often' this event may happen. Therefore, when evaluating the importance of a risk event, it was considered necessary and vital to take into account all these three risk aspects. Consequently, and in order to facilitate risk assessment, the following formula was developed:

*Risk score of each ERP risk = Σ [W *(Probability + Impact + Frequency)]*

The structure of this formula is consistent with and clearly reflects the design of the questionnaire. Based on this formula, the calculation of the risk score for each identified ERP risk event should go through the following 3 steps:

Step 1. (Probability + Impact + Frequency): sum up the values given by each respondent for the three independent dimensions of a risk event, namely probability of occurrence (i.e. 3, 2, or 1), level of impact (i.e. 3, 2, or 1) and frequency of occurrence (i.e. 5 to 1).

Step 2. W*(Probability + Impact + Frequency): 'W' refers to whether or not the respondent perceived this risk event as an ERP risk, with '1' stands for 'yes' and '0' means 'no.' In case that the respondent did not perceive the given risk event as an ERP risk, the formula will turn the value generated from Step 1 into 0: W*(Probability + Impact + Frequency) = 0*(Probability + Impact + Frequency) = 0.

Step 1 and 2 thus generate the individual score that each respondent gave for a specific risk event.

Step 3. Σ [W*(Probability + Impact + Frequency)]: sum up the individual score that each of the 42 respondents of the survey gave for a particular risk event, and thus generate the total risk score that this risk event received.

By using this formula, the researchers calculated the risk scores for all of the 40 ERP risk events examined, and then prioritised these risks based on their risk scores. The top 10 ERP risks ranked by their risk scores are shown in Table 2.

It is clear from Table 2 that the top ten ERP exploitation risks did not cluster around a specific subset of the main categories. This means that critical risks seem to be found across the organizational processes and not conveniently localised around one category, namely not around the technical category. In fact, since only 2 of the top ten risks are related to technical aspects, these findings seem to suggest that potential ERP post-implementation failure in Chinese SOEs may be owing to business-oriented risks rather than technical problems.

In order to seek further statistical evidence to support these findings, a bivariate analysis of the data was carried out. A bivariate analysis is a statistical technique that aims at identifying the correlation between two variables. This study used bivariate analysis to examine potential correlations of risks in the context of Chinese SOEs. In particular, it explored if the probability of

Table 2. Top 10 ERP post-implementation risks in Chinese SOEs

Category	The top 10 ERP exploitation risks		Rank	Risk Score
Operational risks	OR2.2	Customer files contained in ERP are out-of-date or incomplete	6	246
	OR3.2	ERP system contains inaccurate or incomplete bill of materials	8	243
	OR3.3	ERP system contains inaccurate inventory records	1	263
Analytical risks	AR1.2	Managers cannot retrieve needed information from ERP	4	247
	AR2.1	Sales forecast generated by ERP is inaccurate or inappropriate	3	250
	AR4.1	Fail to use ERP to generate appropriate financial budgets	4	247
Organization - wide risks	OWR1.3	Support from top managers to ERP exploitation is insufficient	10	242
	OWR3.3	Lose ERP-related know-how accumulated over time	2	252
Technical risks	TR1.1	Seamless integration is not achieved between modules of ERP	8	243
	TR1.2	ERP is not able to seamlessly integrate with other IS application	7	233

occurrence of a particular risk event was related to the increase of the probability of occurrence of other risks. As illustrated earlier, Likert scales were used in the survey to examine the likelihood of each identified risk, data variables generated were therefore ordinal data sets. According to Field (2005, pp. 130-131) and Bryman and Cramer (2005, p. 225), Spearman's rho (r_s) is the most commonly used approach to measure bivariate correlations between ordinal variables. As a consequence, Spearman's rho was adopted for this study. Moreover, one-tailed test was used to test the statistical significance (P value) of each directional correlation identified. By following this approach, the researchers identified and confirmed 10 statistically significant correlations between all of the 40 identified ERP risks, as shown in the conceptual map in Figure 3.

The findings of the bivariate analysis were very illuminating. In particular, by investigating the correlations highlighted in Figure 3, it becomes apparent that the majority of the correlations occurred between analytical and organisation-wide risks. On the other hand, technical risks that are very often seen as the main perpetrators in ERP failure seem to be important but not strictly related to other risks. Together with the top ten risks presented in Table 2, this survey study seems to confirm that failure of ERP systems may not just be conveniently related to the technical infrastructures and software packages. Actually, what this study confirms is that it is in operational, management and strategic thinking areas that the majority of critical risks were identified. Moreover, these business-oriented and organisation-wide risks seemed to be interwoven and closely related with other similar risks. Consequently, the occurrence of these risks is much more difficult to manage, mitigate, and contain in Chinese SOEs.

Nevertheless, although the questionnaire resulted in a set of interesting and meaningful findings, it was not able to gather in-depth human insights to explore fully the critical ERP risks being identified. This is in fact a typical and inherent weakness of questionnaires, as discussed above. Therefore, in order to collect richer human perceptions to verify and further explore the questionnaire findings, a follow-up case study was carried out at the second stage of the research.

Figure 3. Conceptual map of correlations between ERP risks in Chinese SOEs

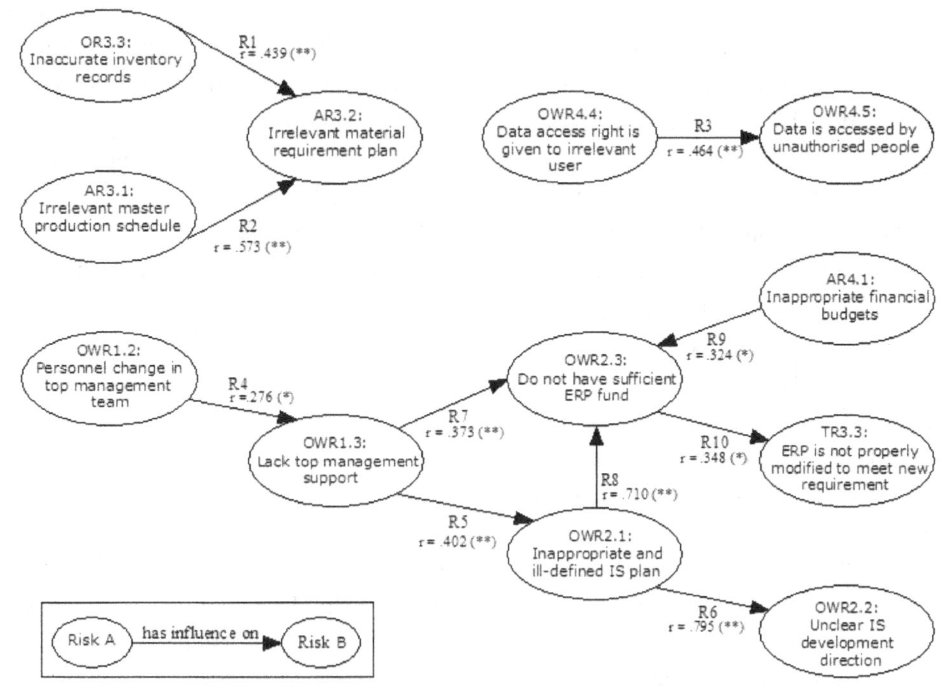

OR = Operational risks; AR = Analytical risk; OWR = Organisation-wide risks; TR = Technical risks

Qualitative Data Collection

At the end of the questionnaire, respondents were asked whether or not they would be willing to participate in the multi-case study stage to discuss further ERP-related risks and issues in their companies. Two volunteer companies were thus identified to participate in the second phase of the study.

As discussed by Saunders et al. (2003, p. 246), semi-structured interview is a very efficient tool to collect in-depth human insights to explore a list of themes and questions that are predefined prior to the collection of data (e.g. the questionnaire findings derived from the first phase). Moreover, researchers can extend and change the predetermined questions flexibly during semi-structured interviews, in order to fully explore the views of interviewees (Saunders, et al., 2003, p. 246). Given these features, semi-structured interview was used as the most suitable data collection method of the case study component.

The interview instrument was designed based on a set of refined and selected questionnaire findings (i.e. the top 10 risks, correlations between risks, and unexpected outcomes related to certain risks). Consequently, 25 semi-structured interviews were carried out with the CEOs, IT managers, and departmental managers and system users in diverse business areas (i.e. sales, financial, production, and purchasing department) of the two case companies. Moreover, all interviews were digitally recorded with prior permission, and lasted for 40 minutes to 1 hour. In order to enhance the trustworthiness of the data, written transcription was done on the same day that the interview had taken place. In addition, the transcription of each interview was sent to the interviewee to read through, and thus allowing the researcher to identify and remove any potential bias or inappropriate interpretation. Consequently, a very rich set of qualitative data was collected from the two case companies, and was then systematically analysed.

Qualitative Data Analysis and Findings

The interview data was analysed by using a thematic analysis approach with *a priori* coding. Thematic analysis is a process of searching, identifying and exploring codes and themes that emerged from the data as "important to the description of the phenomenon" (Daly, et al., 1997). This data-driven inductive approach can often be used together with a deductive priori coding (Fereday & Muir-Cochrane, 2006). In this study, the set of questionnaire findings used to construct the interview questions were also used as a set of priori codes, while a wide range of codes were also identified from data. Following guidelines given by prior researchers (Braun & Clarke, 2006; Rice & Ezzy, 1999), the thematic analysis conducted in this study consisted of the following five stages:

Step 1: Getting familiar with the interview data (get known the data by reading and re-reading the data set).

Step 2: Coding the data (develop the coding scheme, and code the textual data in a systematic fashion across the entire data set by using NVivo).

Step 3: Connecting codes with themes (identify a number of central themes, collate codes into the identified themes, and gather all data relevant to each theme).

Step 4: Reviewing themes (check and verify if the themes work in relation to the coded quotes and the entire data set).

Step 5: Reporting findings (final analysis of selected quotes, relate results back to the research question and literature, and then present the findings).

This thematic analysis process resulted in a very rich and interesting set of qualitative findings. In conjunction with the results of the literature review and the questionnaire, the case study provided further insights on the causes and consequences of the top 10 risks identified. For instance, one of the top 10 risks highlighted in Table 2 is related to the issue that different modules of the ERP system may not be seamlessly integrated. In fact, an integrated solution from one single ERP vendor may not satisfy all business needs of the company. Therefore, it is common for modern companies to procure suitable software modules from different system vendors to form their own version of ERP (Pan, et al., 2011). This approach however may increase complexity and difficulty in harmonizing integration issues. In other words, companies may face a risk that seamless integration may not be achieved between current modules, or between current and new modules, of the ERP system. The questionnaire findings showed that, a vast majority (87.8%) of respondents considered this risk event had a high to medium probability of occurrence. The analysis of the interview data identified that the occurrence of this risk event can often lead to system fragmentation in user companies, through the creation of technological islands which are very often totally isolated and non-communicant. This type of system fragmentation can significantly affect ERP data quality. The Financial Manager of company A concluded that, "since ERP requires very high data accuracy in order to work efficiently, poor data quality can make the implemented ERP fail substantially in exploitation."

Moreover, the 10 statistical correlations identified in the questionnaire findings (in Figure 3) were further validated in the case study, which also indicated and confirmed a further set of casual relationships between the identified ERP risks in Chinese SOEs. Consequently, an extended conceptual map (Figure 4) was developed to highlight all risk relationships identified in this mixed-methods study. As clearly emerged in the extended conceptual map, the majority of risk relationships occurred between the identified organisation-wide risks. This type of risk was also found to be the key triggers of many operational, analytical, and technical risks. In contrast, technical risks were neither found to be interwoven with each other nor to be the main causes of other risks. These qualitative findings thus reinforced the questionnaire results as discussed in section 3.1.4. In particular, the case study findings supported the quantitative results by confirming that business and human-related risks are the main triggers of the other ERP exploitation risks, including the technical ones. This type of risk is thus not just more dangerous but also more difficult to contain and manage in Chinese SOEs.

Overall Integration of Findings

As discussed above, the critical literature review at the early stage of the study resulted in the establishment of a risk ontology, which contained 40 potential risks associated with ERP post-implementation. These 40 risk items were organised into four categories, namely operational risks, analytical risks, organisation-wide risks, and technical risks.

Subsequently, the questionnaire examined the suitability of this theoretical ontology in the context of Chinese SOEs. The survey findings identified that all of the 40 predefined risk events were perceived by the majority of respondents as risks to ERP exploitation. The 10 top prioritised risks were distributed across organisational processes and operation. Moreover, the findings also identified a set of statistically significant correlations between the 40 identified risks. The majority of these correlations occurred between analytical and organisation-wide risks. Since these types of risks seemed to be interwoven and closely related with each other, it was concluded that the occurrence of these risks is much more difficult to manage and

Figure 4. Extended conceptual map of relationships between ERP risks in Chinese SOEs

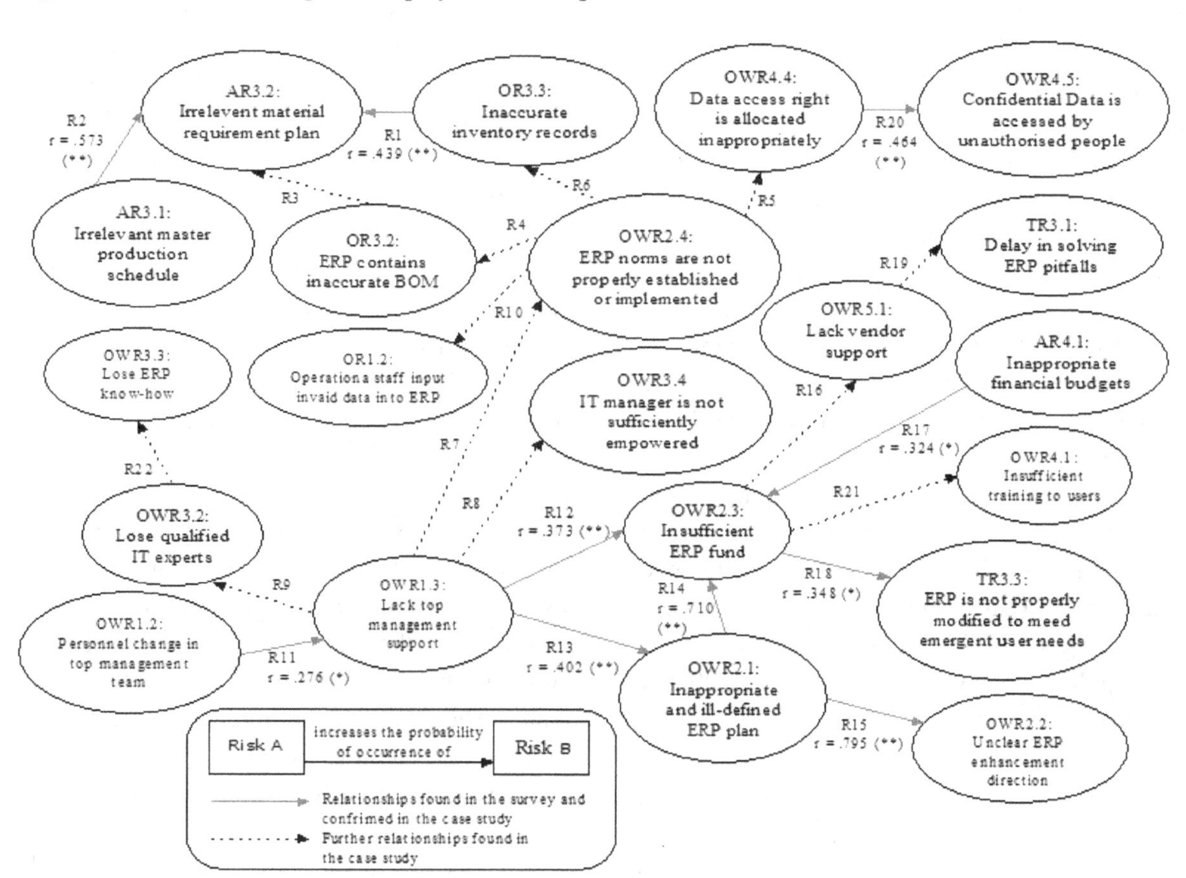

mitigate in the SOEs studied. In contrast, technical risks that are very often expected as the main perpetrators in ERP failure seem to be important but not strictly related to other risks.

These quantitative findings were then further explored, validated and confirmed in the follow-up case study. In particular, the interview findings supported that crucial ERP exploitation risks did not conveniently localise around technical aspects. In fact, organisation-wide risks are often the direct triggers for the operational and technical risks and indirect triggers for the analytical risks.

Moreover, by analysing, triangulating and synthesising the quantitative and qualitative findings, it was identified that the most critical ERP risks, which originated the entire complicated risk network, were associated with 3 main human factors, namely top management, in-house IT experts, and system users.

Overall, the study concluded that potential failure of ERP systems cannot be conveniently attributed to technical aspects, such as the software package and the ICT infrastructure, in the context of Chinese SOEs studied. In fact, the integrated findings of the study suggest that it is in organisation processes and human-oriented aspects that the more dangerous and difficult-to-manage risks can be found in these companies. Therefore, Chinese managers need to become fully aware of the importance and critical impacts of these organisational risks, as well as to take proper risk mitigation actions to manage them. Effectively managing the identified organisational risks can substantially contribute to the mitigation of other types of risks, including the technical ones, and thus helping Chinese SOEs to achieve long-term ERP success. Further discussion about the findings of this ERP study can be found in our other publications (e.g. Peng & Nunes, 2009a, 2009b, 2012).

Reflection of the QUAN + Qual Design

The selected QUAN (i.e. questionnaire) and qual (i.e. case study) mixed-methods approach proves to be a suitable, rigorous and effective approach for generating both generalisable and in-depth findings for this study. In particular, the questionnaire survey conducted in the first stage of the research was an efficient method to examine and validate the large number of ERP risks, as proposed in the theoretical risk ontology, in the context of Chinese SOEs. By analysing the questionnaire data, the researchers successfully prioritised the risks and identified a set of most critical ERP exploitation risks for SOEs. Subsequently, the follow-up case study allows the researchers to gather in-depth human insights to further explore, validate, and complement the questionnaire findings. Such breadth and depth of the findings could not have been simultaneously achieved by using either a single quantitative or qualitative approach.

Moreover, it is important to note that the very comprehensive ERP ontology derived from the critical literature review provides not just the theoretical foundation but also the prerequisite for implementing this QUAN and qual design. In fact, the nature of this critical review is very different from a traditional literature review (Peng & Nunes, 2009). That is, it did not just simply aim to provide a descriptive summary of the literature for the topic being studied. Instead, it required the researchers to systematically and critically search, analyse and synthesise a wide range of IS and ERP literature (e.g. books, conference papers, journal articles, industrial papers, and technical reports), in order to construct and propose a very extensive list of ERP exploitation risks. This extensive risk ontology was then used as the theoretical basis to construct the questionnaire in the QUAN component. It should also be highlighted that, since the original risk ontology contains a wide range of risk items, it will be very time-consuming, and in fact infeasible, to adopt a qualitative method (e.g. interview) at the first stage to explore all the proposed risks. It is therefore fundamental to carry out a quantitative study first to refine and prioritise the list of ERP risks, before the qualitative component is conducted.

Nonetheless, and in another scene, if the context and matters being investigated in a specific research are relatively ambiguous, and/or if a comprehensive research framework could not be built at the initial stage of the study, the QUAN and qual design will not be suitable. In this case, a qualitative approach should be considered to use at the first phase in order to clarify and explore further the phenomenon being studied. Subsequently, a quantitative survey may be carried out at the second stage in order to generalise the findings. In other words, a QUAL (e.g. case study) and quan (e.g. questionnaire) design can be used, as further illustrated in the example below.

Risk Assessment of Knowledge Exposure Risks associated with 3D VRE

Research Aims and Objectives

High-tech companies (e.g. Aerospace sector) have increasingly invested in 3D and virtual reality applications to support customers in understanding and using their products, as well as to provide complex and specialised training to employees. These applications take the form of 3D Virtual Reality Environments (VREs) where users can navigate, browse and learn in an authentic and close to realistic contexts. However, given the fact that VRE applications are realistic, easy to navigate and contain holistic specialised and technical explicit knowledge, the use of VREs can lead to possible risks of explicit knowledge exposure and/or leakage, and thus representing a threat to internal knowledge management.

Thus, the study reported in this section aimed to explore and identify these risks. Specifically, it attempted to identify the base-events that may trigger the risks, define and characterise the risks, and finally propose solutions and recommendations on how to minimise and remediate their occurrence.

Research Design: QUAL and Quan Approach

In fact, the original idea of the research was triggered by the Technical Director (TD) of Rainmaker 3D. Rainmaker 3D is an engineering company that provides a wide range of specialised services, including sales of metrology equipments, system integration, measurement services, and training. One of its important areas of expertise is the creation of virtual environments and 3D models of plants, machinery, aircraft and ships that contain detailed and holistic information that can be easily navigated and queried by users or employees. However, despite the success of Rainmaker's VREs, the nature of this technology poses both potential and apparent knowledge exposure risks. The challenge for the researchers was to identify and assess these Knowledge Leakage (KL) risks and make recommendations on how the company could address them.

In order to explore the situation of Rainmaker and conduct in-depth investigation of knowledge leakage risks, the research developed and used the Practice-Based IS Research (PB-ISR) framework seen in Figure 5. This PB-ISR framework is particularly useful for projects where research questions emerge from real-life organizational processes and functions (Annansingh & Nunes, 2005).

By using the framework, theory building and theory testing were preformed via triangulation of literature review, case study analysis and questionnaire survey. In order to investigate the situation and current knowledge leakage risks faced by Rainmaker 3D, an exploratory case study approach was employed at the first phase of the research design. However, the results of any case

Figure 5. The PB-ISR framework (Annansingh & Nunes, 2005)

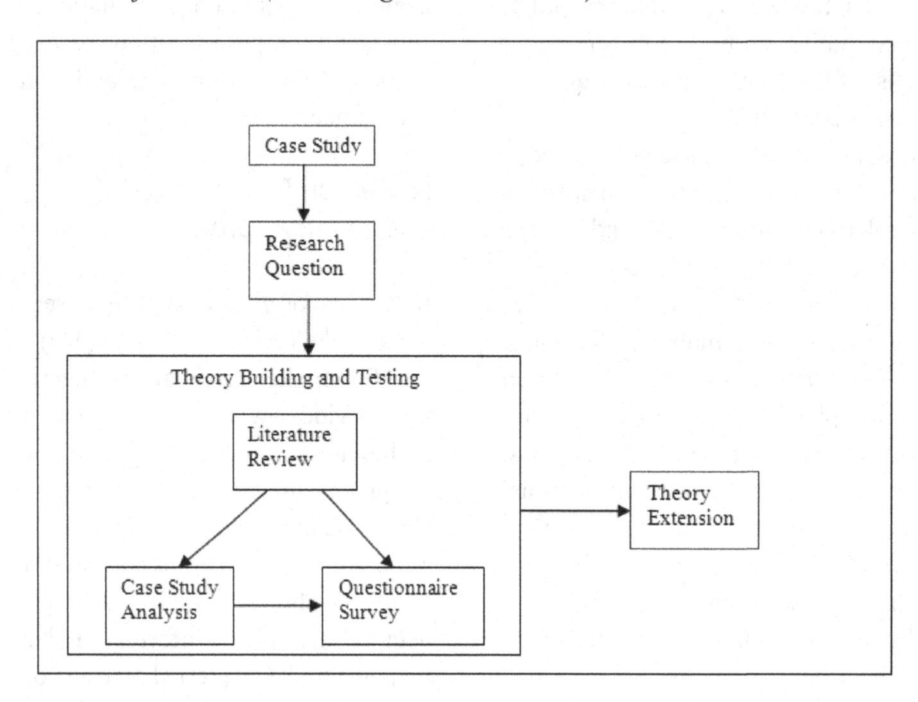

study are always difficult to generalise to a wider context. In order to address this inherent limitation of the case study approach and produce generalisable findings beyond the immediate case study, it was considered meaningful and necessary to carry out a further cross-sectional questionnaire survey. This survey queried a meaningful sample of companies in order to generalise and test acceptance of the findings emerging from the case study analysis.

Consequently, these considerations and the specific context of the research led to the selection and adoption of a two-phase sequential exploratory (QUAL + quan) design for this study. Specifically, an exploratory case study was conducted for discovery and identification purposes. Subsequently, a cross-sectional questionnaire survey was used to validate and generalise the findings of the first phase.

Qualitative Data Collection

In order to establish appropriate understandings for the concepts being studied and support theoretical sensitisation, an initial literature review was undertaken at the early stage of the study. As a result of the literature review, the researchers established a high level conceptual framework that was adapted from the Yeates and Cadle (2001) Risks Identification Framework. This framework allows multi-characterisation of risks from three categories, namely business/ organizational, project, and technical categories. Supported by the risk typology and conceptual understandings drawn from the literature review, initial exploratory interviews were undertaken with the TD of Rainmaker, as well as with with key management and technical personnel in the firm. Semi structured interviews were used in addition to the predetermined questions, as it provides the interviewer the freedom to probe for answers (Berg, 1995; May, 2003). For this type of interview the interviewer is free to seek both elaboration and clarification on the phenomenon being investigated. General open-ended questions were chosen as they gave respondent complete freedom to reply to the questions and also allowed the interviewer the chance to probe for more in depth responses. Moreover, an amalgamation of specific, reflective, hypothetical and leading questions were used to further explore the responses (Keats, 2000; Warwick, 1984). These interviews allowed for an early identification and assessment of risks as well as to provide a more technical description of the software development process.

Further to this, axial coding was used to identify the relationships between the categories of data that emerged from the open coding process. As the relationship between categories were identified they were rearranged based on a hierarchal system with sub-categories emerging (Saunders, et al., 2003). Axial coding was used to determine the risks as well as the consequences and impact to the organization. From the categories emerging from the open codes, selective coding was used to categorize the risks into key concepts around KL, thus resulting in the identification a number of key concepts (Saunders, et al., 2003). Concepts were used as they provide useful mental images or perceptions. The results were then discussed in depth with the Rainmaker's TD and the concept maps revisited and fine-tuned. These concept maps allowed a comparative analysis between the results collected from the case study with the theoretical stance, as well as a constant re-interpretation of new findings and already accepted causal chains. These first stage findings produced a number of interesting and surprising results that, according to the argumentation above now require validation in order to be generalizable.

Qualitative Data Analysis and Findings

Being an interpretivist research, perceptions were obtained across the management and development team of Rainmaker. These varied perspective, were considered relevant to develop an understanding of the strategic and operation decisions made and their effects on the project. Embedded in this discussion was the unearthing of any trace of a risk management strategy exploited by Rainmaker. This was achieved by examining a number of key concepts: KL, security, training, testing, and maintenance, each of which was examined in lieu of risk identification and assessment procedures. Consequently, several risks were identified not only as a result of the technology used, but as a result of the project environment, requirements capture, documentation and testing during design and development. In terms of the risks inherent to the nature of the software, risks classification fell under security and KL risks.

The risks identified from the developers were then sorted and grouped together based on the Cadle and Yeates (2001) risks identification framework. This framework was used to establish clear categories under which the risks identified from the case study could be categorized, thus removing any ambiguities and extraneous factors. These risks categories selected are:

Business/Commercial Risk:

1. Commercial Background;
2. Contract;
3. Functional Requirements;
4. Stakeholders/Users;
5. Knowledge Leakage.

Technical Risks:

1. Technical Requirements;
2. Maintainability;
3. Reliability;
4. Security;
5. Third Party Consideration;
6. System Architecture.

Project Risks:

1. Development Environment;
2. Tools and Methods;
3. Acceptance;
4. Developers Skills;
5. Project Plan.

Based on the risks identified from the case, testing of these phenomena was done via the use of a questionnaire survey.

Quantitative Data Collection

The in-depth data gathered through exploratory interviews however are limited to the knowledge, experience, and perspectives of the respondents within the case company. As discussed above, in order to validate and generalise the findings obtained from the case study, as well as to gain a more comprehensive understanding of the leakage risks identified, a cross-sectional questionnaire survey was carried out. The aim was to determine whether the risks identified from the case study were a true representation of perceptions in the sector. The sample was selected from the Kompass UK Portal and the FAME database. The selection was done via the identification of a number of product codes, which are relevant to the design, development, and use of 3D models/VREs. These codes were then used to query the databases. A sample of 500 companies was then randomly selected from a list of UK companies that are involved in the design, development, and use of 3D models/VR environment. This selection tried to produce a geographically well distributed and size balanced sample (Annansingh & Nunes, 2005). Four sets of pilot test were conducted and a number of changes were made. Questionnaires were sent to small and medium size companies as well as large corporations and targeted a wide target group within the organization with similar job functions from the case study. Likewise, the questionnaire followed a similar format to the case study questions where unstructured open-ended questions were used to encourage participants to construct their own meaning of the phenomena—meanings that have been forged through discussions and interactions with other individuals within the organization. Thus, participants were able to express their views. Closed questions were also used and these were employed primarily to validate the risk identified in the case study by Rainmaker as well as the risks mitigation strategies and/or security technologies employed to protect the organization from the KL risks. Subsequently, the quantitative data collected were analysed by using SPSS. The questionnaire findings were then integrated and cross-referenced with the findings of the case study.

By testing and establishing relationships between the variables identified from the case study, sequential triangulation is achieved. This is due to the fact that the first phase of the research—the case study—is used to inform the second phase—the survey (Creswell, 1994). From the PB-ISR framework, this cross sectional survey forms the third and final vertex for the triangulation of methods outlined in the theory building and testing phase.

Quantitative Data Analysis and Findings

Descriptive data analysis was used to analyzed the data from the cross sectional survey. For which, descriptive univariate and bivariate analysis are conducted and made use of percentages, frequencies, diagrams, cross tabulation and correlations. Data representation for the questionnaire took the form of frequency or distribution tables, graphs, and charts. These univariate tools were used to reduce the bulk of the data as well as to explore each variable in the data set.

In order to assess the strength of the relationship between two variables, a non- parametric test was used, namely the Spearman's rank correlation coefficient which was employed to determine the significance of the cross tabulations (Somekh & Lewin, 2005; Saunders, et al., 2000). The coefficient of this rank correlation varies between -1 and +1 therefore it provides information on both the strength and direction of the relationship. Owing to the fact that this rank is a non-parametric test it can be used in a wide variety of context as it makes fewer assumptions about the variables (Bryman & Cramer, 1992). Spearman's rank correlation coefficient is especially useful with small numbers or where the items have unique ranks and categorical data (Blaikie, 2004).

The quan results identified a number of variables as well as established clear relationships between them. However, in addition to preparing the foundation for the relationships identified, the questionnaire findings also provided a detailed discussion in relation to risk management, KL and the security strategy in an organization. Under each category, security and risks management issues were addressed as well as the relationships based on these concepts. These risks were then categorized according to the Cadle and Yeates (1996) risk identification technique, which provides a detailed listing of these risks under a number of sub headings. Hence, an exhaustive check list of risks and KL risks associated with VREs were produced. Some of the risks identified are presented in Table 3.

Overall Integration of Findings

As mentioned above, the conceptual framework derived from the literature review highlighted three categories for risk identification, including business/organizational, project and technical categories. By integrating the case study and questionnaire findings, the study identified, explored and assessed a comprehensive set of risks associated with the design and use of 3D models. In particular, the integrated findings suggested that critical knowledge leakage risks could occur in relation to the organization's operational and management characteristics (business risks), during the design and development cycle of the 3D models (project risks), as well as associated with the use of and the inherent nature of these models (technical risks). Moreover, the findings also identified that companies participated in the study had neither proper knowledge management nor holistic risk management and information security approach to handle the identified knowledge leakage risks. This poses a significant threat to both 3D software development companies and their direct customers, as well as to third party companies (i.e. secondary customers). Therefore risk management thinking should be an integral part of the overall strategic decision making process of the organization. By systematically and consistently adopting an organizational wide approach to risk management, an organization facilitates the management of both the negative and positive consequences of risk. A thorough and continuous risk management approach should be adopted in order to exploit the opportunities often associated with these risks. The results highlight the importance that the organization maintains written policies and procedures that clearly outline the risk management policies for derivatives activities, since regardless of the organization and its functions or activities, risks will be associated with its practices. The task of management therefore, is to prepare for these risks, thus maximizing or enhancing the likelihood of achieving organizations' objectives (Cadle & Yeates, 2001; Pressman, 2000) and exploiting opportunities. Consequently, everyone in the organization should be given risk management responsibility.

Based on the concepts identified throughout the research technologies can, and have been implemented to prevent or mitigate risk occurrences. For knowledge leakage this is particularly challenging, as there is no knowledge leakage technology per se, which can prevent this leakage, since once information is leaked it cannot be retrieved. Technology therefore, has to be combined with strategy—whether KM or risk management—in order to prevent knowledge leakage. In addition, since knowledge leakage can be isolated as a unique concept, as well as exist within another,

Table 3. Knowledge leakage risks associated with VREs

Business/Commercial Risk	Technical Risks	Project Risks
Commercial Background • Absence of risk management strategy; • Insufficient management control; • Politics place over substance; • Lack of senior management commitment to the project; • Inappropriate decision making; • Sector specific risks; • Absent KM strategy; • Intellectual property; • Absence of strategic security measure; • Loss of competitive advantage; • Knowledge sharing encouraged without consideration to KL or other KM issues; • Substitution of security technologies for risk management strategies/thinking; • Over reliance on security technologies	**Technical Requirements** • System built inaccurate; • Technical information not readily available or inaccurate; • Inadequate functionality	**Development Environment** • Unauthorized access to development site • Inefficiencies; • Simultaneous update of data/ issues; • Bugs; • Wasted resources; • Multiple copies of system documentation posing KL risks
Contract • Accountability; • Opposing interest; • Failure to highlight confidentiality clause; • Breach of confidentiality	**Maintainability** • Contactability of developers; • Inadequate maintenance and updating of the networks defences; • Poor maintenance of the system; • Stop the progress of the system; • Inconsistencies.	**Tools and Methods** • Unaware of hardware software conflicts; • Compatibility with existing systems; • Unfamiliar software; • Users need to be aware that it is based on supposition; • Information easily outdated; • Need constant update; • Scalability; • Granularity ; • Verisimilitude nature of software
Functional Requirements • Functional overlaps; • Incorrect/Incomplete requirement; • Unrealistic expectations/assumption • Usability functional requirement not captured; • Lack of /inadequate user requirement; • Inaccurate requirement capture; • System does not match the requirement or the expectations of the users.	**Reliability** • Faulty product; • Poor software quality; • Error prone; • Debugging; • Delay response time	**Project Plan** • Insufficient planning; • Budget cuts; • Schedule constraint; • Time constraint; • Optimistic assumption; • Unrealistic assumption
Security • Developers former employees work for clients main competitors; • Prone to malicious attacks—viruses, etc from competitors; • Breach in security; • Open to theft and mismanagement of the information; • Typical risks associated with Internet and WWW; • Misinterpreting of data; • Information leakage;	**Knowledge Leakage** • Employees deliberately/unintentionally leak information to clients competitors; • Loss of sensitive information to competitors or rogue traders; • Knowledge loss from employee leave the organization; • Causal information exchange; • Whistle blowing; • Unwanted customer leverage; • Information not updated/inaccurate • Misleading information; • Information untimely or outdated; • Incomplete information;	

continued on following page

Table 3. Continued

Business/Commercial Risk	Technical Risks	Project Risks
• Theft; • Carelessness • Won't happen here mentality; • Deletion of data; • Unauthorized use; • Increased risks of theft; • Password breach; • Copies not updated simultaneously; • Remote attack of the system; • Inadequate physical precautionary infrastructure; • System penetration; • Lack of security skills; • Data interception;	• Disgruntled Employees; • Theft of transaction information; • No counter balance of signature; • Unauthorised access/use; • Theft—back up disk or tape; • Loss of sensitive information; • Employees don't buy into the idea of confidentiality or see the need; • Maliciously copying or deleting the data; • Malicious intruders; • Multiple copies of system documentation posing KL risks; • Disgruntled employee; • Redundancies; • Loss of intellectual property; • Theft of transactional information; • Unauthorised modification of information;	
• Unauthorised software installations; • Internet attacks—hackers, viruses; • Easy breach of different access level; • Easy propagation of viruses; • Insider abuse of network access; • Industrial espionage; • Easy accessibility; • Password cracking/guessing; • Invalid access attempts; • Complex network; • Dynamic configurations; • Multiple access points; • Difficult to know who is accessing the system; • User error; • Equipment failure; • Unauthorised access by disgruntled employee;	• Information disclosure; • Hacking; • Unauthorized access to confidential information; • Repudiation of data; • Security breach; • Inaccurate information; • Untimely information; • Incorrect estimate of security needs; • Obscuring information; • Withhold vital information; • Relying on employees discretion: to keep data secret; • Unauthorized transfer of information; • Malicious copying and deletion of files; • Data readily interpreted/analyzed; • User neglect and carelessness;	
• Industrial spies; • Inadequate security; • Loss of sensitive information; • Inadequate access control; • Denial of service; • Unauthorized inspection of software by former employee; • Inadequate physical security; • Malicious users; • KL—intra and external; • Lack of risk management consideration; • Failure to highlight confidentiality clause; • Former employees; • Periodic download; • Failure to consider KL; • Lack of security action when employees leave; • Remote connections/access; • Remote workers; • Security risks; • Losing a copy;	• Intra organization KL; • High staff turnover; • Lack of risk management consideration; • Breach of confidentiality; • Former employees; • High staff turnover; • Periodic download; • Failure to consider KL; • Lack of security action when employees leave; • Failure to modify right of access; • Free distribution of information/ knowledge to employees; • Employees not given information on a need to know basis; • Failure to consider disgruntled employees in RM plan; • Remote connections/access;	

continued on following page

Table 3. Continued

Business/Commercial Risk	Technical Risks	Project Risks
• Copies not updated simultaneously; • Loss of proprietary information ; • Unauthorized access or use; • Denial –it won't happen here syndrome; • Absence of strategic security measures; • Theft of proprietary information; • Identity theft; • Email; • Over reliance on security technology; • Inadequate security control; • Unauthorised personnel; • Industrial espionage; • Third party maintenance; • Default password; • Default account; • Third party companies; • Ready interpretation and analysis of data; • Visualization of data; • Information easily outdated	• Remote workers; • Increased risks of theft; • Losing a copy; • Copies not updated simultaneously; • Password breach; • Loss of proprietary information; • Training activities; • Unauthorized access or use; • Knowledge loss; • Theft of proprietary information; • Identity theft; • Inadequate security control; • Unauthorised personnel; • Industrial espionage; • Third party maintenance; • Third party companies; • Ready interpretation and analysis of data; • Loss of employees;	
Stakeholders/Users • Inadequate training; • Employee not understanding software; • Employee not using software; • Misuse of software—lack of knowledge about the limitation; • Help file insufficient; • Frustrated end users; • Input inconsistent data; • Passing the buck; • Misuse of system; • Barriers to change	**System Architecture** • Unrealistic/optimistic assumption about the capabilities of the system; • Unrealistic expectations; • Feature creep; • Problem may occur --- patch needs patching; • Real time responses; • Asynchronous event handling; • Multi user interaction; • Recovery time; • Interruption handling; • Deficiency/Incorrect modelling of data	
	Third Party Consideration Competitors	

strategist should not fail to incorporate knowledge leakage or consider its impact on these concepts during risk management thinking.

Besides protecting their information system and more specifically the 3D models, an organization needs to implement physical security measures in order to protect its other assets. The various physical security measures were considered sufficient to adequately protect the physical environment of the organization. The results indicate that respondents believe the security policy for any VRE should depend significantly on its applications. Nonetheless, access control mechanisms as well as group and member authentication schemes were the most likely choice of protection used in these environments. However, despite these beliefs the results show management attitude and perception to the security precautions in the organization as a whole and more specifically in the 3D VREs were very lax if not too optimistic.

Reflection of the QUAL and Quan Design

This study aims to investigate and explore KL risks resulting from the adoption and use of VREs, by obtaining both the perceptions and perspectives of various stakeholders in a case company as well as reaching a consensus within a population assisted with the generalization of the findings. It should be highlighted that very little previous research has been done on risk identification and assessment on VREs. This points to a clear need of the use of the qualitative component (i.e. exploratory case study) at the first phase in order to explore further the phenomenon being studied. Subsequently, the cross-sectional questionnaire survey helps to generalize the in-depth qualitative findings in a wider context. By using the dominant qualitative (QUAL) and the less dominant quantitative (quan) method, the researcher was able to address different areas of the same research question. This extended the breadth of the study and essentially increased the quality of the research. Therefore, conclusions drawn would be more likely to be correct and accepted.

FURTHER GUIDANCE FOR APPLYING MIXED-METHODS DESIGNS

A set of key practical challenges for designing and implementing mixed-methods research have been discussed in section 2.4 (e.g. in what order and priority the quantitative and qualitative methods should be used; researchers' skills and abilities in implementing the design; and difficulties in integrating quantitative and qualitative findings). By comparing and drawing on the experience of the two IS projects above, this section provides some further guidance to address these practical difficulties.

In particular, and as discussed above, quantitative and qualitative methods can be combined and used very flexibly in mixed-methods research (that is, either the quantitative or qualitative component can be carried out first and take the predominant position). However, there is no best combination or order of the use of these methods. As clearly emerged from the above IS exemplifications, decisions about whether the quantitative or qualitative component should be carried out first and take a higher weighting, need to be made based on the nature of the research and its specific context. For instance, the use of a dominant quantitative method (i.e. the questionnaire survey) in the first stage of the above ERP study was due to the fact that a very comprehensive ERP risk ontology was developed from the critical literature review, and that quantitative tools would be more efficient than qualitative methods to examine the large amount of risk items in the target Chinese companies. In contrast, the VRE project discussed above was initiated by the Technical Director of a case company, of which the situation was complicated and relatively ambiguous. Moreover, not much research has been done on knowledge leakage risks in the 3D VRE environment. Therefore, it makes sense for this project to adopt a qualitative (rather than a quantitative) approach at the first stage in order to explore the context and phenomenon under studied.

Moreover, when designing mixed-methods studies, researchers should also take into consideration their skills and abilities in implementing the selected quantitative/qualitative methods. This is not to suggest researchers to only choose methods that they are good at and avoid methods that they are less capable of. Instead, and as stressed above, any decisions about the selection and combination of methods, should be made based on the actual needs of the study. If researchers have less knowledge and experience on one or some of the selected quantitative/qualitative methods, they should treat this as a learning opportunity to develop new research skills. In fact, in order to carry out the above IS projects effectively and successfully, the research team had undertaken additional research training on quantitative/qualitative data collection and analysis methods at the initial stage of the study. Nonetheless, it should be acknowledged that learning new research skills can be time consuming and costly. Researchers thus need to bear in mind the time limitation of the project and availability of resources when designing a mixed-methods study.

Finally, in order for the quantitative and qualitative findings to be genuinely integrated, appropriate links should be established in the series of data collection and analysis activities of the mixed-methods study. Specifically, findings of early stages need to shed light on the design of the data collection tool of later stages of the design. As in the above ERP study, interview instruments used in the follow-up case study were designed based on the findings derived from the early questionnaire. Similarly, in the VRE project, findings generated from the exploratory case study were used to guide the design of the cross-sectional questionnaire survey. On the other hand, when analysing the quantitative and qualitative data, findings derived from later stages need to complement with and be triangulated with results generated from early stages. For example, in the ERP project, the case study results were analysed, interpreted, and reported by cross-referencing with findings of the questionnaire and the literature review. In the VRE project, the questionnaire findings were also interpreted and discussed in conjunction with the early qualitative results. With these efforts, the quantitative and qualitative findings of both studies were integrated very naturally. Consequently, more significant and meaningful conclusions were drawn based on the integrated findings.

CONCLUSION

This chapter discussed the use of mixed-methods approach in IS research, in order to resolve the limitations of single quantitative or qualitative approach and thus lead to more comprehensive, rigorous, and significant findings. As emerged from our discussion, the key determinant of the success of mixed-methods design is the researcher's ability to combine genuinely the quantitative and qualitative elements throughout the project, from design and implementation of the research to integration and reporting of findings. In fact, and as illustrated from the above IS exemplifications, a mixed-methods study can be designed in a very flexible manner. Nonetheless, any decisions made for the mixed-methods design need to be in accordance to the nature of the research question and actual needs of the study as well as the specific context of the research. Moreover, a clear and justifiable rationale must be embedded in the design to allow that quantitative and qualitative data collection and analysis can complement each other. As demonstrated in the above examples, this can more likely result in the subsequent findings to be related and supplement with each other. To conclude, when these principles are kept rigorously and applied effectively, mixed-methods approach can be a fundamental tool in the IS researcher's arsenal.

REFERENCES

Annansingh, F., & Baptista Nunes, J. M. (2005). Validating interpretivist research: Using a cross-sectional survey to validate case study elicitation of knowledge leakage risks associated with the use of virtual reality models. In *Proceedings of the 4th European Conference on Research Methodology for Business and Management Studies*. Paris, France: Université Paris Dauphine.

Berg, B. L. (1995). *Qualitative research methods for the social scientist*. London, UK: Allyn and Bacon.

Blaikie, N. W. H. (2003). *Analyzing quantitative data: From description to explanation*. London, UK: Sage Publications Ltd.

Braun, V., & Clarke, V. (2006). Using thematic analysis in psychology. *Qualitative Research in Psychology, 3*, 77–101. doi:10.1191/1478088706qp063oa

Bryman, A. (2004). *Social research methods* (2nd ed.). Oxford, UK: Oxford University Press.

Bryman, A. (2007). Barriers to integrating quantitative and qualitative research. *Journal of Mixed Methods Research, 1*(8), 8–22. doi:10.1177/2345678906290531

Bryman, A., & Cramer, D. (2005). *Quantitative data analysis with SPSS 12 and 13: A guide for social scientists*. East Sussex, UK: Routledge.

Campbell, D. T., & Fiske, D. W. (1959). Convergent and discriminate validation by the multitrait-multimethod matrix. *Psychological Bulletin, 56*, 81–105. doi:10.1037/h0046016

Creswell, J. W. (2003). *Research design: Qualitative, quantitative, and mixed methods approaches* (2nd ed.). Thousand Oaks, CA: SAGE Publications.

Daly, J., Kellehear, A., & Gliksman, M. (1997). *The public health researcher: A methodological approach*. Oxford, UK: Oxford University Press.

Denzin, N. K. (1978). *Sociological methods*. New York, NY: McGraw-Hill.

Fereday, J., & Muir-Cochrane, E. (2006). Demonstrating rigor using thematic analysis: A hybrid approach of inductive and deductive coding and theme development. *International Journal of Qualitative Methods, 5*(1), 1–11.

Fidel, R. (2008). Are we there yet? Mixed methods research in library and information science. *Library & Information Science Research, 30*, 265–272. doi:10.1016/j.lisr.2008.04.001

Field, A. (2005). *Discovering statistics using SPSS: And sex, drugs and rock'n'roll* (2nd ed.). London, UK: SAGE Publications.

Gable, G. (1994). Integrating case study and survey research methods: An example in information systems. *European Journal of Information Systems, 3*(2), 112–126. doi:10.1057/ejis.1994.12

Galliers, R. (1991). Choosing appropriate information systems research approaches: A revised taxonomy. In Galliers, R. (Ed.), *Information Systems Research: Issues, Methods and Practical Guidelines* (pp. 144–162). Oxford, UK: Blackwell.

Ivankova, N. V., Creswell, J. W., & Stick, S. L. (2006). Using mixed-methods sequential explanatory design: From theory to practice. *Field Methods, 18*(1), 3–20. doi:10.1177/1525822X05282260

Jick, T. D. (1979). Mixing qualitative and quantitative methods: Triangulation in action. *Administrative Science Quarterly, 24*(4), 602–611. doi:10.2307/2392366

Keats, D. M. (2000). *Interviewing: A practical guide for students and professionals*. Buckingham, UK: Open University Press.

Lee, A. (1991). Integrating positivist and interpretivist approaches to organizational research. *Organization Science, 2*, 342–365. doi:10.1287/orsc.2.4.342

Leech, N. L., & Onwuegbuzie, A. J. (2009). A typology of mixed-methods research designs. *Quality & Quantity, 43*(2), 265–275. doi:10.1007/s11135-007-9105-3

May, T. (2003). *Social research: Issues, methods and process*. Maidenhead, UK: Open University Press.

Mingers, J. (2001). Combining IS research methods: Towards a pluralist methodology. *Information Systems Research, 12*(3), 240–259. doi:10.1287/isre.12.3.240.9709

Mingers, J. (2003). The paucity of multimethod research: A review of the information systems literature. *Information Systems Journal, 13*, 233–249. doi:10.1046/j.1365-2575.2003.00143.x

Novak, J. D., & Cañas, A. J. (2006). *The theory underlying concept maps and how to construct them*. Technical Report IHMC CmapTools. Tallahassee, FL: Florida Institute for Human and Machine Cognition. Retrieved from http://cmap.ihmc. us/Publications/ResearchPapers/TheoryUnderlyingConcept Maps.pdf

Orlikowski, W., & Baroudi, J. (1991). Studying information technology in organizations: Research approaches and assumptions. *Information Systems Research*, *2*(1), 1–28. doi:10.1287/isre.2.1.1

Pan, K., Nunes, J. M. B., & Peng, G. C. (2011). Risks affecting ERP post-implementation: Insights from a large Chinese manufacturing group. *Journal of Manufacturing Technology Management*, *22*(1), 107–130. doi:10.1108/17410381111099833

Patton, M. Q. (1999). Enhancing the quality and credibility of qualitative analysis. *Health Services Research*, *34*(5), 1189–1208.

Patton, M. Q. (2002). *Qualitative research and evaluation methods* (3rd ed.). Thousand Oaks, CA: Sage.

Peng, G. C., & Nunes, J. M. B. (2009). Surfacing ERP exploitation risks through a risk ontology. *Industrial Management & Data Systems*, *109*(7), 926–942. doi:10.1108/02635570910982283

Peng, G. C., & Nunes, J. M. B. (2009b). Identification and assessment of risks associated with ERP post-implementation in China. *Journal of Enterprise Information Management*, *22*(5), 587–614. doi:10.1108/17410390910993554

Peng, G. C., & Nunes, J. M. B. (2012). Establishing and verifying a risk ontology for ERP post-implementation. In Ahmad, M., Colomb, R. M., & Abdullah, M. S. (Eds.), *Ontology-Based Applications for Enterprise Systems and Knowledge Management*. Hershey, PA: IGI Global. doi:10.4018/978-1-4666-1993-7.ch003

Petter, S. C., & Gallivan, M. J. (2004). Toward a framework for classifying and guiding mixed method research in information systems. In *Proceedings of the 37 Hawaii International Conference on Systems Sciences*. Hawaii, HI: IEEE.

Pressman, R. S. (2000). *Software engineering: A practitioner's approach*. Berkshire, UK: McGraw-Hill.

Rice, P., & Ezzy, D. (1999). *Qualitative research methods: A health focus*. Oxford, UK: Oxford University Press.

Robey, D. (1996). Diversity in information systems research: Threat, promise and responsibility. *Information Systems Research*, *7*, 400–408. doi:10.1287/isre.7.4.400

Robson, C. (2002). *Real world research: A resource for social scientists and practitioner-researchers* (2nd ed.). Oxford, UK: Blackwell.

Rocco, T. S., Bliss, L. A., Gallagher, S., & Perez-Prado, A. (2003). Taking the next step: Mixed methods research in organisational systems. *Information Technology, Learning and Performance Journal*, *21*(1), 19–29.

Saunders, M., Lewis, P., & Thornhill, A. (2003). *Research methods for business students* (3rd ed.). Essex, UK: Pearson Education.

Somekh, B., & Lewin, C. (2005). *Research methods in the social sciences*. London, UK: Sage Publications.

Tashakkori, A., & Creswell, J. W. (2007). Editorial: The new era of mixed methods. *Journal of Mixed Methods Research*, *1*, 3–7. doi:10.1177/2345678906293042

Teddlie, C., & Tashakkori, A. (2003). Major issues and controversies in the use of mixed methods in the social and behavioral sciences. In Tashakkori, A., & Teddlie, C. (Eds.), *Handbook on Mixed Methods in the Behavioural and Social Sciences* (pp. 3–50). Thousand Oaks, CA: Sage.

Warwick, D. (1984). *Interviews and interviewing*. London, UK: Education for Industrial Society-Management in Schools.

Yeates, D., & Cadle, J. (2001). *Project management for information systems* (3rd ed.). Harlow, UK: Prentice Hall.

Chapter 15
Examining Web 2.0 E-Learning Tools:
Mixed Method Classroom Pilot

Janet Holland
Emporia State University, USA

Dusti Howell
Emporia State University, USA

ABSTRACT

With so many fields using new technologies in e-learning, we are all challenged with selecting and effectively implementing new Web 2.0 tools. This chapter provides a mixed method research approach to quickly evaluate available Web 2.0 tools and instructional implementation. Class observations and pilot study surveys were used to determine students' levels of satisfaction after using various numbers of Web 2.0 tools and varying student work group sizes. The pilot studies were designed to model initial classroom examinations when integrating emerging Web 2.0 technologies. Use of this type of pilot study approach is necessitated as many individual class sizes are too small for a full research study, and the time needed to conduct a full study using multiple classes could cause the results to quickly be out of date, thus not providing the needed immediate classroom data for just in time learning. Fast emerging technologies pose a unique challenge to traditional research methodology. Where immediate specific classroom data is needed, a needs analysis with a pilot study is the best option. Note, with emerging technologies, it is difficult to find appropriate literature to determine its effectiveness in the classroom. If desired, compiling the results from many small pilot studies offers an additional benefit of fleshing out key issues to be examined later in greater detail using a full research study for extending theory or scientific practices.

DOI: 10.4018/978-1-4666-2491-7.ch015

Copyright © 2013, IGI Global. Copying or distributing in print or electronic forms without written permission of IGI Global is prohibited.

INTRODUCTION

Web 2.0 e-learning tools are often referred to as "those interactive Web sites where we, the public, supply the material" (Pogue, 2008, p. 1). Pogue goes on to define Web 2.0 as offering "a direct, more trusted line of communications than anything that came before it" (Pogue, 2008, p. 2). Pogue indicates it is more work to facilitate, but the positive gains are worth it. In essence, Web 2.0 e-learning tools allow for a window to get to know individuals' needs on a much deeper level. When we know more about the learners' needs, instructors can better facilitate knowledge acquisition. The biggest challenge is working to overcome fear from all parties involved. Even though the author is referring to business, the same concepts hold true in the educational arena, as well.

At Emporia State University, in the department of Instructional Design and Technology, our students have a wide range of professional interests, with all looking to us, the faculty, to help them find innovative ways to teach their content areas to others. In addition, our University administration has our IDT faculty training our colleagues, from all discipline areas across campus, to integrate Web 2.0 e-learning technologies, as a way to invigorate their classroom and online teaching. As a result, we have students in our classes expressing their excitement not just about our teaching and our tools, but about the new technologies our colleagues are now using. It is rewarding, especially knowing we have a hand in training them to make this happen.

We are finding Web 2.0 e-learning tools can be used to embrace new digital learning environments by having learners actively research, collaborate, innovate, and share their ideas. Many of the Web 2.0 e-learning collaborative communication tools can be used to increase knowledge acquisition quickly and efficiently while making global connections for broader perspectives. Providing meaningful integration of new technologies through the use of quality instructional practices can alter how learners and instructors engage with concepts and each other to achieve powerful learning and meet workplace challenges.

With the vast amount of technology used in the modern work environment, it was just a matter of time until educational variants filtered down to the public school systems in an effort to prepare students for eventual workplace realities. One example is reflected in the new United States National Standardized Test on Technology and Engineering Literacy Assessment. Assessment data collection is anticipated in 2014 by the National Assessment of Educational Progress (NAEP) for K-12 public schools. This highlights the increasing need to get up to speed on the knowledge and skills needed to work with new and emerging e-learning technologies for teaching, learning, innovating, and collaborative endeavors.

EVALUATION OF WEB 2.0 E-LEARNING ISSUES

One of the driving objectives of the pilot study is to begin an examination of the overall information systems used for instruction to make wise educational decisions directed towards the effective and efficient integration of Web 2.0 e-learning technologies. The following six questions were the basis for the immediate action research pilots designed to optimize student learning when working with emerging Web 2.0 e-learning tools.

The guiding pilot study research questions included:

1. What Web 2.0 e-learning tools can be found through an extensive online search for instructional purposes?
2. How can the Web 2.0 e-learning tools be implemented into teaching by aligning instructional tools to the curriculum goals and objectives based on classroom observations?
3. What are the optimal number of Web 2.0 e-learning tools to use based on students'

quantitative level of satisfaction ranging from either one tool, a small group of six tools, or open-ended student selections?

4. What data would students' qualitative open-ended questions provide and how would it compare to the quantitative data in regards to students level of satisfaction when working with either one Web 2.0 tool, six tools, or open-ended tool choice?
5. What is the optimal student group configurations when working with Web 2.0 e-learning tools based on students qualitative level of satisfaction ranging from one individual, two students, or larger group of four?
6. What data would students' qualitative open-ended questions provide and how would it compare to the quantitative data in regards to students' level of satisfaction when working with one, two, or a group of four students when working with Web 2.0 e-learning tools?

The questions listed have become the basis for the pilot "design most appropriate for the question[s] being studied" (Shavelson, 1981, p. 1) as seen in Table 1. The pilot studies were conducted by instructional practitioners with the goal of improving just in time classroom teaching and learning, specifically when using emerging technologies. Applied research, such as found in this pilot study, can be defined as "research designed to develop and test predictions and interventions that can be used directly to improve practice" (Gall, Gall, & Borg, 1999, p. 525). "Action research enables teachers, administrators, school counselors, and other education practitioners to investigate and improve their performance in systematic, personally meaningful ways" with "the purpose of improving local practice rather than producing theory or scientific generalizations" (Gall, Gall, & Borg, 1999, p. 467). The quality of the research depends on how well the project serves immediate, local needs. Table 1 provides a quick overview of the mixed method approach used in the pilot studies.

Table 1. Summary of mixed method pilot studies

Mixed Methods	Data Collection For Six Key Questions
Tool Research Class Observations Quantitative Survey Qualitative Feedback Quantitative Survey Qualitative Feedback	Web 2.0 E-Learning Tool Options Web 2.0 Tool Implementation Number of Tools Satisfaction Number of Tools Open-Ended Feedback Student Group Configuration Size Satisfaction Student Group Configuration Open-Ended Feedback

One problem instructors, students, and trainers face when selecting Web 2.0 tools is, finding quality sites to assist in making appropriate selections to align with desired goals and objectives. Learners need assistance learning how to navigate Web resources, especially with so many options to pick from. As Jenson stated, "we are witnessing a radical shift in how we establish authority, significance, and even scholarly validity" (Jensen, 2007, p. 6). The paradigm shift from independent learners to the opposite end of the spectrum, crowd sourcing with the co-construction of knowledge requires a very different approach. With Web 2.0 e-learning, it is about a "change in focus to participation, user control, sharing, openness, and networking" (Eisenberg, 2008, p. 22). Social learning networks are more open, participatory, conversational, and democratic (Cohen, 2007). With learners and workers using new media content resources with questionable validity, it is even more important now to seek out multiple means of comparison for cross checking the sources, content, and analysis of each resource.

It is easy and comfortable to stay with what is familiar, but the payoff is much greater and worth the time invested to expand tool options for fostering creative thinking and expanded perspectives. The good news is many of these Web 2.0 e-learning tools are free to use so the cost becomes less of an issue. The bad news, many instructors have not grown up using the new technologies, thus pointing to the need for professional training and support with some hands on experience to help embrace the path of continued lifelong learning. Let us face it, instructors and trainers are very busy and need all the assistance they can get, to create quality technology integrated lessons or workplace challenges. With the myriad numbers of Web 2.0 e-learning tool choices available, it can be overwhelming just trying to decide where to start. Having a good organizational plan will help make the needed digital tool selections to align to the desired goals and objectives. In addition, instructors need to be aware of the impact with regards to the number of tools used at one time and the optimal learner grouping configurations for designing successful technology infused instructional experiences.

When selecting Web 2.0 e-learning media tools for student learning activities such as research, collaborative wikis, discussion blogs, micro-blogs, video, slide presentations, apps, conferencing, cloud-based sharing, and other emerging technologies, it is important to teach learners how to select and implement appropriate Web 2.0 e-learning tools. By initially assisting learners in selecting appropriate tools, they can be taught to develop the knowledge and skills needed to be responsible for verifying their own digital tool selections.

As a part of that assistance, learners need to be educated on issues surrounding the use of creative commons shared content resources. In addition, learners need to be aware of the freedoms allowed through educational copyright rather than focus on what they cannot do. "We need to stop fighting against Wikipedia and Twitter...Demonizing any particular information source that the world values makes us look clueless" (Valenza & Johnson, 2009, p. 32).

Internet filtering found in many school districts and work environments can become an issue in regards to having access to Web 2.0 e-learning tools. While realizing the need for security it is important to find ways to keep the filters from blocking access to important educational opportunities. It is critical to find ways to resolve these issues in order to be technologically literate while being prepared to compete and succeed in college and in the workplace. It will be important to work on administrative buy in and understanding of the needs of instructors, learners, trainers, and different disciplines. Knowledge and understanding are key to gaining access so learners are better prepared for their digital futures. Most of the current tools have options for controlling secured private groups, if needed or desired.

One additional consideration with the use of Web 2.0 tools surrounds having the needed infrastructure to support the technology, such as enough bandwidth for streaming media. Advanced planning will help to ensure long-term stability and continued use.

We begin with extensive research of viable Web 2.0 e-learning tools able to infuse greater learning into classroom lessons based on targeted learning goals and objectives. In the following section, classroom observations are used to further refine and better align appropriate tools with the learning tasks. The sections below are devoted to classroom pilot studies used to determine the optimal number of tools and student group configurations for working effectively with Web 2.0 e-learning tools for immediate classroom implementation.

INSTRUCTIONAL PRACTICES

A good starting place for instructional planning is to begin with an analysis of learner needs, goals and objectives, to align the content and Web 2.0 e-learning tools selected. Quality instructional practices often include an intellectual challenge to assist learners in thinking critically about the content. Intellectual rigor and engagement can be accomplished by allowing learners to build their own understandings, promoting meaningful collaborations, encouraging critical and creative thinking, and showcasing through authentic audiences (Dockter, Haug, & Lewis, 2010). Many of the current Web 2.0 e-learning tools align perfectly for creating compelling learning and workplace opportunities. Teaching learners to "effectively and creatively find, evaluate, analyze, use, and communicate information" are important lifetime skills (Valenza & Johnson, 2009, p. 31). One of the important hallmarks of the use of Web 2.0 tools is the use of collaboration to build a richer expanded perspective within a social environment. By being socially and globally connected, learners can develop academic relationships with individuals and groups, thereby increasing the pace and knowledge base at the same time.

The author Weinberger has a wonderful analogy describing knowledge as no longer being organized as trees, but as a pile of leaves (Weinberger, 2005). It will then be up to the student with support from instructors, classmates, and experts to create meaning in virtual spaces that otherwise would be a pile of leaves (Hedberg & Brudvik, 2008, p. 141).

By targeting learning goals and objectives, instruction can be directed towards a wide variety of Web-based tools to "increase student achievement, help meet state and national standards, and capitalize on existing investments yields a excellent return on technology investments, whether measured by use or products or learner outcomes" (Baumbach, 2009, p. 16).

It may feel pretty overwhelming, with the number of new technologies being released. However, by trying just one new Web 2.0 tool at a time, one can build up a nice repertoire over time. Start by looking at the needs of the learners, and matching the best tool to the learning goals. In the end, it is about what learners need to know and be able to do. "It is important for educators to find the appropriate tool to unlock learning possibilities" (Fredrick, 2010, p. 34). Or, in the true Web 2.0 e-learning style, mix and mash tools to more fully meet the needs of the learning community (Hyatt & Craig, 2009). To keep pace with the rapidly shifting landscape of digital tools, the best tip is "recognizing that change is no longer an option" and "we need to be willing to change to be open to new ideas" (Brooks-Young, 2008, p. 56). "Find one tool that you think has use in your classroom or work environment and use it. Start small. Start with a tool that compliments what you currently do either as a teacher or with your students. Start with something you are comfortable with" (O'Brien & Scharber, 2010, p. 602). Integrating technology effectively will have far-reaching effects for preparing learners and workers for the realities of a globally connected world. Foster an inquiry-based approach to building a technology rich vision for teaching, learning, and working. Tap into students' desires to work together, expand ideas, and work efficiently.

Level One: Getting Started

Recently during a break at a Web 2.0 e-learning conference, a couple of tech savvy colleagues confided they were completely overwhelmed. They were teaching online classes but had not yet integrated any Web 2.0 tools into their curriculum. In their efforts to correct that deficiency, they had just spent the entire morning watching over a hundred great Web 2.0 tools being demonstrated. "Where do we start?" The easy answer is to just pick one tool and get started. A simple place to start is to identify a need that could help one or two students in your class. A student that is a terrible speller could benefit from practice as spellingcity.com. A student needing to improve their speed with basic multiplication math facts can practice at studystack.com. Struggling math students and gifted students wanting to be challenged can go to Khanacademy.org to develop their proficiency in a growing hierarchy of skill exercises, beginning with elementary levels of math then moving through calculus. The Khan Academy uses short video clips to teach new skills followed by practice exercises. Students getting ten correct answers in a row are given merit badges. Badges are built-in incentives to motivate students to continue to improve while extending their growing levels of math skills. Even better, students and teachers can see a visual map of their growth as well as the new areas they can work on.

Level Two: Improving Teaching

One great way to improve your teaching with Web 2.0 e-learning tools is to give pretests and posttests on material you are covering in class. Surveys are a great way to do this and they give immediate feedback. One tool is a Google Doc tool called Google Forms (http://www.google.com/google-d-s/createforms.html). Students do not need to sign in to take this assessment. They just go to the link you've provided, take the assessment and submit. Teachers get immediate feedback along with a visual graphic of results if they want to share the results in real time with students. This allows teachers to focus on teaching the things students are struggling with the most, instead of covering all topics at the same level.

Another great Web 2.0 tool teachers can use is the free and extremely easy to use Jing.com This screen capture program is a great way to introduce a new website or new piece of software. For example, if you want to introduce Khanacademy.org to students; first, start the Jing recorder. Then, open your browser and go to the Khanacademy website. Talk out loud, explain what you may be clicking on and where you're going in the website. Jing.com records your mouse movements on the screen, it records your voice and allows students to see and hear exactly what they need to do.

Level Three: Class Improvements

Most schools in the USA have to meet Annual Yearly Progress (AYP) goals. These goals are built around the biggest academic problems each school needs to address for their students. As a classroom instructor, you will most likely be required to address these specific needs in your class. Web 2.0 tools can be a great resource to address these needs. Table 2 presents a summary of recommendations based on classroom observations.

Table 2. Summary of classroom level 1, 2, 3 web 2.0 observation recommendations

Level 1	Level 2	Level 3
Select one tool	Improve existing lesson	Align to standards

WEB 2.0 E-LEARNING TOOLS

Tool selection options can vary widely. It will be up to the instructor, learners, or workers, to determine the primary objective by which to narrow the Web 2.0 e-learning tool search to make the final technology selections. Invigorating teaching, learning, and workplace challenges with Web 2.0 tools requires a mixture of vision and excitement to challenge learners. "The new generation of Web 2.0 solutions are easier to use, more engaging and are making a larger impact upon collaboration and communication" (Yan, 2008, p. 30). Part of the challenge is finding innovative ways to bring together instructors, students, mentors, and experts for social scholarship to create the "cyberinfrastructure needed for increased innovation, globalization, and knowledge networking" (Greenhow, 2009, p. 46). Web 2.0 learning environments providing opportunities for collaboration are game changers for the "way learners can retrieve, share and evaluate information, and create knowledge" (Benson & Brack, 2009, p. 74). By designing learning challenges within the social environment, one can promote cognitive engagement through the co-construction of knowledge and exchange of ideas. By making learning interesting, fun, and challenging it is a great way to capture learners' imaginations and passions for the topic while continuing the learning process. The author Walling has a great analogy about working with the new media, that of standing on a "railroad track facing a speeding high-tech train. They can stand pat and get run down. They can step aside and get passed by. Or, if they recognize the promise of tech-savvy teaching, they can swing aboard and join their students on a fascinating journey of discovery" (Walling, 2009, p. 22).

Bloom's Taxonomy of Web 2.0 E-Learning Tools

Bloom's Taxonomy of Web 2.0 e-learning tools are great for teachers, elementary and on up. With the goal of integrating new technologies into teaching, learning, and the work environment, many Web 2.0 resources were examined to assist in sorting through the vast number of digital tools available to meet targeted objectives. Using models, including Bloom's Taxonomy and others to be discussed below, was designed to assist in selecting tools by the type of learning or work activity targeted. The first resource selected is a derivation of the Bloom's Taxonomy Model for higher-level thinking skills created by author Samantha Penney, from the University of Southern Indiana (2010). This model includes the main categories of 1) creating, 2) evaluating, 3) analyzing, 4) applying, 5) understanding, and 6) remembering. Below is an example of the model in text form with the Web 2.0 tools applied. Penney added the Web 2.0 tools to the online digital pyramid from Krathwohl and Anderson (2001), which is an adaptation of Blooms Taxonomy (1956). Using this as a guide is one way to begin selecting appropriate Web 2.0 tools for targeted goals. Penny's Model is a great beginning point for aligning the tool options to the desired outcomes. Many additional tools could be added to this list and it is important to keep in mind, some tools can be placed into more than one area.

1. **Creating:** Prezi, Voicethread, Protagonize, Glogster Edu, Wikispaces.
2. **Evaluating:** RubiStar, YouTube, PollDaddy, iRubric, Protagonize, Rcampus, E Portfolio, Survey Monkey, Nota.
3. **Analyzing:** iExploratree, Google Analytics, Google Trends 10x10, Google Finance, Create a Graph, Pipes, Google Earth.
4. **Applying:** WolframAlpha, Google Sketchup, Go2Web20.net, Scribble Maps, Gliffy, Evernote, Pipes.
5. **Understanding:** The Periodic Table of Videos, Wikipedia, Google News Timeline Labs, Footnote, Webspiration, JeopardyLabs, JohnLocker.com, SlideShare, Bubbl.us.
6. **Remembering:** Technorati, Wordnik, FlashcardExchange, Creately, NinjaWords, CoboCards, Visuwords online graphical dictionary, CarrotSticks, Zoho Work Online, Linolt, Delicious, Flickr.

One of our graduate students in Instructional Design and Technology at Emporia State University took the concept of Bloom's Digital Taxonomy one step further. As an instructional technologist working in the K-12 setting, they used it as a basis for linking to related Web 2.0 tutorials for teacher training. The categories used included 1) creating, 2) evaluating, 3) analyzing, 4) applying, 5) understanding, and 6) remembering. The digital tool tutorials are listed below and the site can be located at http://lleimbach.glogster.com/blooms-resources.

1. **Creating:** Animoto, Google Sites, Voki, VoiceThread.
2. **Evaluating:** EduBlogs, Wordpress Blogs, Bubbl.us Collaborative Mapping, Google Earth, Twiddla Collaborate Wkspace.
3. **Analyzing:** Wordle, Survey Monkey, OmniGraffe, Facebook, Delicious Bookmarking, Digital Graphic Organizers.

4. **Applying:** Games, Flickr, SchoolTube, Glogster, Go2web20.net.
5. **Understanding:** Google WonderWheel Search, Diigo, iTunes, Lucid Outliner, Online Summarizing Tool.
6. **Remembering:** Visual Thesaurus, Stixy, Wallwisher, Internet Flashcard Database, MindMap.

Stripling Inquiry Model of Web 2.0 E-Learning Tools

The Stripling Inquiry Model would be great to use for secondary schools. The Inquiry Model was selected as another way to narrow the search for appropriate Web 2.0 e-learning tools, since it includes active engagement through questioning and critical thinking for a deeper retention of knowledge acquisition. The Stripling Inquiry-Based Model (2010) was then designed as a student centered framework for learning. It includes six main phases of student generated critical thinking, including 1) connect, 2) wonder, 3) investigate, 4) construct, 5) express, and 6) reflect. Below is a condensed example of the approach with Web 2.0 tool examples added to assist in making tool selections. Again, many additional tools can be added to this list or be included in more than one category. Below is a copy of the Stripling Inquiry and Web 2.0 Tools Integration Guide.

1. **Connect:** Observe, experience, connect a subject to self and previous knowledge.
   a. **Teaching Strategy:** Dialogue, research, journal, log, chart, organizer, engage, explore.
   b. **Web 2.0 Tools:** EduBlogs, Ning, Wikispaces, Skype, Google Docs, Zoho Suite, Mindmeister, Bubbl.us, Mind42, LooseStitch, Google Earth, TeacherTube, Flickr.
2. **Wonder:** Predict, develop questions and hypotheses.
   a. **Teaching Strategy:** Brainstrom, questioning, anticipation guides.
   b. **Web 2.0 Tools:** Google Docs, templates, Mindmeister, Bubbl.us.
3. **Investigate:** Find and evaluate information to answer questions, test hypotheses.
   a. **Teaching Strategy:** Research, notes, guided practice, organize, evaluate.
   b. **Web 2.0 Tools:** Google, Clusty, Ask, Kartoo, Exalead, Intute, Google Docs, Zoho Notebook, iOutliners, SpringNote, Wikispaces (pathfinders), Jing, Voicethread, Google Reader, Diigo, Delicious, SimplyBox, Netvibes, Pageflakes, 30 Boxes, TaDaList, Mindmeister, Bubbl.us.
4. **Construct:** Conclusions, arrive at new understandings.
   a. **Teaching Strategy:** Chart, map, compose, question.
   b. **Web 2.0 Tools:** Edublogs, Wikispaces, PBWorks, GoogleDocs, Zoho Suite, Polleverywhere, Google Docs, Zoho Suite, Edublogs, E-mail, Instant Messenger, Skype, Twitter.
5. **Express:** Apply understanding to a new context, share learning with others.
   a. **Teaching Strategy:** Rubric, conferencing.
   b. **Web 2.0 Tools:** Google Docs, Zoho Suite, Voicethread, Glogster, Podcast, Animoto, Flickr, TeacherTube, Skype, Blogs, Nings.
6. **Reflect:** Examine one's own learning and ask new questions.
   a. **Teaching Strategy:** Feedback, reflection log: I use to think, now I know.
   b. **Web 2.0 Tools:** EduBlogs, Wikispaces, E-mail, Ning, GoogleDocs, Voicethread, Podcast.

K-12 Web 2.0 Website Recommendations

The K-12 Web 2.0 website recommendations are ideal for getting advice from an expert like your local librarian. As a result of their research efforts of examining quality resources for including Web 2.0 websites in the K-12 learning environment, The American Association of School Librarians listed their top 25 sites for teaching and learning. School librarians are often on the forefront of working with new media in the public school sector and serve as a great resource for Web 2.0 site integration into the classroom within all discipline areas. Many additional websites can be added to the list as it grows in importance to providing a good starting place for selecting sites to meet specific digital learning needs at the K-12 setting.

1. **Media Sharing Websites:** Glogster, Masher, Prezi, Professor Garfield, SchoolTube, Scratch, WatchKnow.org.
2. **Digital Storytelling Websites:** International Children's Digital Library, Jing, Storybird.
3. **Manage and Organize Websites**Evernote, Jogtheweb, Live Binders, MuseumBox, Pageflakes, Weblist.
4. **Social Networking and Communication:** Creative Commons, Learn Central, TED.
5. **Content Collaboration Websites:** Debategraph.
6. **Curriculum Sharing Websites:** Exploratree, The Jason Project, National Science Digital Library.
7. **Content Resources: Lesson Plans and More Websites:** Edsitement, National Archives Digital Classroom.

Popular Business Web 2.0 E-Learning Sites

Popular business related Web 2.0 e-learning sites are great for marketing and business programs; also, users looking for the most popular tools can benefit from these resources. When searching for the most popular Web 2.0 websites on the Internet with the highest number of hits, eBizMBA offers some great resources at http://www.ebizmba.com/articles/web-2.0-websites. The idea behind it is, by using current digital tools and sites with the highest popularity, they can provide the services people need to reach people where they search, create, communicate, collaborate, share, and purchase on the Internet. This site lists the top 15 most popular Web 2.0 websites, as of the frequently updated list on the site. On the site, users will find the eBizMBA ranking composed of an average of each website's Alexa global traffic rank with a U.S. traffic rank generated from both Compete and Quantcast.

UNDERSTANDING THE WEB 2.0 E-LEARNING GOAL

The primary goal for getting students to participate and collaborate with Web 2.0 e-learning tools is to foster an environment in which each individual helps add value to a group initiative. Some schools in their rush to use Web 2.0 e-learning tools are requiring every student, from kindergarten on up, to add material to their individual Blog and Wiki, every day. Although students making comments to other students' blogs can be a great incentive for students to post quality work, collaboration is not a focus and, therefore, the final results are not as strong as they could be.

A group project that could benefit the entire class would be related to academic areas that are growing in information. Whether it is an area of science like biology or an area of technology, new information is discovered daily. How does a teacher keep up with the new information? One idea is to create an evaluation tool allowing students to participate. Here at Emporia State University, we are always evaluating new Web

2.0 tools. This type of undertaking can take time and effort to get going and maintain but it is well worth it.

The first step is to create an evaluation form. This form should be a self-explanatory easy guide for students to complete. The form we created states "The goal of this form is to help teachers gain confidence with emerging Web 2.0 e-learning technologies." These technologies tend to allow for more collaboration to take place in education. "You will research, analyze and review one Web 2.0 technology. You will demonstrate to the class how this technology could be used in the classroom. This presentation must include a handout sharing with their fellow students how to use this technology in the classroom." After a section for further comments, students needed to share links to a project they created with the Web 2.0 resources including other related and useful links. Students can then give a three or four minute demonstration using slides, their voice and their face using a Web 2.0 tool like http://present.me. However, at the end of a couple years of using this process, with students filling out evaluation forms and turning them in for a grade, the power of group collaboration was lost. The teacher may have a stack of forms at the end of the year, and links to online presentations, but students don't benefit as much as they could if this material was moved to a shared Wiki.

The second step is to create a Wiki allowing students to share this information online. Many schools have Wikis built into their Learning Management System. At Emporia State University, a couple of options are available. Campus Pack is a plugin tying in nicely with Blackboard. Google Apps is also available. In the online Wiki a number of key areas were designated and a basic spreadsheet was created (https://sites.google.com/a/emporia.edu/webwiki/web-2-0). Web 2.0 categories were selected for being useful to teachers. These categories were social bookmarking, social networks, Blogs, Wikis, main ideas from text, project sites, photo editing sites and miscellaneous sites. Because the Wiki we selected to use was supported by the university, students log onto the Wiki just as they would if they were logging into their Blackboard class. Only one student can edit the information on the site at any time. If a second student tries to log on, the site will open up a message indicating who is currently editing the page.

The third step, nearly two years after beginning this project, and after dozens of sites have been evaluated, is to rank the sites in each category. Four-star rating will indicate the best tool in each category. This rating system will be useful for teachers that do not have time to experiment with each site in a category and need to pick a winning program quickly. In this way, the students are participating and collaborating to make this database a useful tool for future use. When done successfully, this type of value added initiative could contribute to creating an end project that is extremely useful for all involved. The teacher alone can't possible keep up with all the new information that continues to grow in this area. Also, students need to see how they can work together to create a project that improves their abilities to use Web 2.0 tools along with their abilities to choose the best tools to teach with.

WEB 2.0 CLASSROOM APPLICATION EXAMPLES

For pre-service teachers looking to get started, the following Web 2.0 tools are some great resources, though constantly changing, constantly improving, take advantage of them. The tools, presented below, are selected for their potential to help learners increase their knowledge in reading, writing, content area projects and basic study skills. The recommendations are based on naturalistic classroom observations. Sax defines the observations as "a method of data collection in which an unobtrusive observer records uncontrived and typical behavior" (Sax, 1997, p. 607).

Teaching Reading

One area of reading, critical to comprehension, is the ability to identify the main idea from the text. There are some neat Web 2.0 e-learning tools that will help do just that. Main ideas and key words from reading material can be extracted and summarized with Wordle, Word Sift, and Great Summary.

Wordle (http://www.wordle.net) creates "word clouds" from the most frequently used words in the text. The site provides dozens of ways to format the word cloud to change the look and layout. One can customize the word cloud or just choose the version the site originally created. Some great uses for Wordle are to paste a president's speech into Wordle to see the main ideas he spoke about. Another idea is to have students write down descriptive words of their peers and make a Wordle of what the kids think of them.

Word Sift (http://www.wordsift.com/) is similar to word cloud but has more tools for comprehension built into the site. After placing your text into the site window, the fifty most frequently used words in the text will appear in a tag cloud. The bigger the word in the cloud, the more frequently it is used in the text. Our favorite vocabulary enhancement tool, built into the page, is the visual thesaurus. The most frequently used word in the text is automatically added to the visual thesaurus to give the reader a quick overview of related words.

Great Summary (http://www.greatsummary.com/) allows you to type in the Uniform Resource Locator (URL) address of any website then it delivers a quick summary of the page.

Teaching Writing

Tools motivating students to write can be great resources. There are some fantastic Web 2.0 e-learning tools to help in this area. The resources available can help younger students to write stories and can assist older students to get their work online in the form of Blogs and Wikis.

Story Jumper (http://www.storyjumper.com/) is a tool students can use to build stories online. With Story Jumper, teachers and students write, illustrate, and publish their stories. If the story turns out to be really great, they can order a hard copy, as well. This is an excellent resource for reluctant writers because it generates prompts and cues to keep students moving through their writing projects.

Blogs are a great way for individual students to showcase their works online. Blogs can be private or public or just open to those invited to view their materials. Blogger ([REMOVED HYPERLINK FIELD]http://www.blogger.com) and EduBlogs (http://edublogs.org) are great places to start. Additionally, a Blog is a great place for a teacher to invite parents and students to view a class newsletter, to post assignments or videos, and to showcase student work online. Instructors may allow learners to be authors on the Blog so they can post entries, too. The comment area allows parents and students to ask questions or carry on a classroom discussion. When students are required to post their writing materials on their own blogs, they often will put more time and effort into their writing because it is posted to a public forum for the world to see. Even if the blog is made semi-public, just for parents and fellow students to see, they are more apt to put more effort into their work.

Wikis are an excellent place for your entire class, or several classes, to work on year round projects. Try taking a project you do on an annual basis and consider creating a Wiki of what students are learning. Students add material, edit, refine, and become part of a group project. Year after year, students add value to what their peers did in previous years while continuing to add value to the selected topics. This gets teachers out of a redundant style, of teaching the same things every year, and allows students to contribute to a growing legacy of knowledge improvement. Wikispaces for Educators (http://www.wikispaces.com/site/for/teachers) is a wonderful place to get started.

Project-Based Learning

Having students create projects about what they are learning is a great constructivist approach to learning. Presented below are a number of excellent Web 2.0 e-learning tools for creating multimedia-enhanced projects.

Voice Thread (http://voicethread.com/) is a tremendous resource allowing users to upload pictures, images, video, then put voice descriptions over their creations. You can also draw on pictures while you are talking, pause and draw on movie clips, and write comments in a voice bubble. Teachers can easily make tutorials for students. Students can create projects to be assessed by the instructor or peer reviewed or to communicate with other classrooms around the world.

Glogster (http://www.glogster.com/) is an easy way to create an online poster. The poster can include videos, songs, and photos. Teachers can create a poster to introduce a new topic or a new book to be covered in class. Students can create posters showcasing what they've learned in a unit.

Museum Box (http://museumbox.e2bn.org/) is a creative way for students to publish persuasive and informational reports. Learners can create boxes full of video, pictures, sound, and text, as well as hyperlinked Webpages. This could be an excellent alternative to PowerPoint and requires little instruction for high quality reports. This is a nice way to build an argument or description of an event, person, or historical period.

Video, Animation, Comics

Animoto (http://animoto.com/) is great at making videos from images.

Xtranormal (http://www.xtranormal.com) is ideal for creating 3D cartoons. This site is used to help students learn in a visual method. Instructors or students can put together an animated movie to make education more appealing while using technology. It is very easy, just type, drag and drop. Monitor students while they are on the site to make sure they do not view other movies. Many are not appropriate for young students.

GoAnimate (http://goanimate.com/) is also excellent for creating a 3D cartoon. Create a cartoon about whatever you would like. You can create your own characters, scenery, word bubbles, and whatever animations you want! This is a great way to switch up a lesson plan and add some variety. In addition, kids love cartoons! It is easy, fast, and free.

Pixton (http://www.pixton.com/) is perfect at creating online Web comics. Create a comic strip with pre-made characters. Just add bubbles for words, change the scene, and make it whatever you want. It is free and simple, you just have to monitor the pre-made comics on the site as some are not appropriate, again, for younger students.

Teacher Tools and Study Skill Enhancers

Gliffy (http://www.gliffy.com/) is for making concept and mind maps.

Remember the Milk (http://www.rememberthemilk.com/) is for keeping online To Do lists, and is effective as a task manager.

Study Stack (http://www.studystack.com/) is good for creating flash cards. Upload vocabulary words and definitions, then use them in a fun interactive way through game choices like matching, crosswords and, of course, flash cards. What makes it even better is that you can print them off!

Survey Monkey http://www.surveymonkey.com/ is for creating a quick survey of ten questions or less at no charge. It has an excellent data analysis section to help identify students' needs.

WEB 2.0 E-LEARNING QUALITATIVE AND QUANTITATIVE PILOT STUDIES

Three classroom pilots were conducted to try to find the most effective way to work with new Web 2.0 e-learning sites and tools. All three studies were conducted separately within one pre-service teacher course during one semester. The majority of pre-service teachers in the course were planning to work in elementary level classrooms upon completion of their degree programs. In the first classroom test intervention, one Web tool called VoiceThread was selected by the instructor for the classroom lesson because it provided the opportunity to create presentations at high cognitive levels using Bloom's Digital Taxonomy while incorporating images, text, voice, drawing board, and interactive two-way communications directed towards the desired learning goals. Based on the initial pilot test, it was easy to conclude, any content area could be integrated. When comparing past lessons to the new Web 2.0 enhanced lesson, the outcome was a much richer learning experience with better quality instructional materials being created. Through the development of the lesson projects the online interactions became more authentic, interactive, and immersive for the student participants through the use of written words, drawings, images, and personal audio narratives. The online communications became more natural and accessible to student learners while supporting collaborative teaching and learning practices.

In the second classroom pilot, the college students researched a new Web 2.0 tool to present to the class using a traditional slide presentation. The instructor provided students with six topic choices placed in categories to help guide the pre-service teachers into covering the desired Web 2.0 tools. After seeing the results of the presentations, it was modified further to have students use their Web 2.0 tool or site researched in their classroom presentations rather than using traditional slides. The additional practice with the Web 2.0 tools and accountability for demonstrating each increased the level of student engagement and the resulting quality of the projects presented. Placing students into groups increased the number of creative ideas generated and shared.

In the third classroom pilot, the students were allowed to select their own tool or tools based on their needs, content, goals, and objectives. For this round, the students were free to make their own choices without any instructor intervention. Students were given a satisfaction survey to determine whether they preferred having the instructor select one Web 2.0 tool, preferred to have the instructor limit the selection to a small group of tools such as six to select from, or if they would like to have it completely open for students to make the selections depending on the project goal. Even though the pilot studies were directed towards elementary classroom instruction, they can easily be modified and adapted for any grade level classroom or working professionals.

RESULTS

Optimal Number of Web 2.0 E-Learning Tools

The three pilot survey results on the optimal number of Web 2.0 e-learning tools included 20 pre-service teachers from a small Midwestern University. The students were provided a link to participate in an online survey. The participants included 20 undergraduate students in one course. The ages ranged from 19 to 22 years old. The participants included 1 male and 19 female students.

The pilot surveys consisted of 3 items using a five-point Likert scale, with items rated as *Very Low* (1), *Low* (2), *Average* (3), *High* (4), *Very High* (5). The survey concluded with an open-ended question to gather additional information on students' satisfaction about using Web 2.0 tools for teaching and learning. Table 3 presents the students' responses to the 3 survey items. Table

Table 3. Summary of student quantitative survey results (N=20)

Question Items	Very Low	Low	Average	High	Very High
1. Rate your level of satisfaction when you are able to select any Web 2.0 tool or tool to accomplish the project goals.	0.0%	5.0%	20.0%	50.0%	25.0%
2. Rate your level of satisfaction when the instructor selects six Web 2.0 tools for you to use.	0.0%	0.0%	30.0%	45.0%	25.0%
3. Rate your level of satisfaction when the instructor selects one Web 2.0 tool for you to use.	0.0%	15.0%	40.0%	40.0%	10.0%

4 presents all of the students' responses to the open-ended question item on the survey.

For the quantitative assessment, the overall preliminary data indicated the pre-service teachers were fairly evenly split. Some preferred having the instructor limit the number of Web 2.0 tool choices to a manageable number while about an equal number preferred leaving it open for students to choose depending on the project goals.

For the qualitative assessment student feedback provided additional insights to consider when implementing Web 2.0 tools. One important issue raised was to consider the additional time required for open-ended tool choice exploration. In addition, for younger learners, the tools selections could be reduced to keep it to a more manageable level.

Optimal Web 2.0 E-Learning Group Configurations

One aspect to consider when working with Web 2.0 tools for teaching and learning deals with the optimal learner configurations. Is it better to have learners work individually, in pairs, with a group of four, or a combination of groupings? Below are the results of a survey given to a class after implementing the various configurations when working with Web 2.0 tools to examine the students' levels of satisfaction.

A separate or fourth pilot survey was conducted on optimal learner configurations. Once completed, it included responses from 20 pre-service teachers from a small Midwestern University. The students were provided a link to participate in an online survey. The participants included 20 undergraduate students in one course. The ages ranged from 19 to 32 years old. The participants included 1 male and 19 female students.

The pilot survey consisted of 4 items using a five-point Likert scale, with items rated as *Very Low* (1), *Low* (2), *Average* (3), *High* (4), *Very High* (5). The survey concluded with an open-ended question to gather additional information on students' satisfaction with regards to working with Web 2.0 tools individually, in a group of

Table 4. Summary of student qualitative survey results (N=20)

Student Excerpts
1. It is nice to work in groups and individually using the Web 2.0 tools for teaching and learning.
2. Maybe if the teacher were to pick one Web 2.0 tool for the whole class to use, have different groups use different variations of the Web 2.0 tool to pick out the advantages and disadvantages of each program.
3. I think that if students are given a small amount of tools to use then the students aren't going to spend the whole class period deciding what tool to use. I think it's important for the students to be able to explore different options but to narrow it down for specific projects.
4. It just depends on the project and the class you are teaching. For upper levels, giving them the option would be all right but for lower levels you might want to select the tool you want them to use so they do not take more time choosing then actually doing the project. For the most part, it depends on the class, the grade level and what you want to accomplish.
5. It is easier when teacher explains it.

Table 5. Summary of student quantitative survey results (N=20)

Question Items	Very Low	Low	Average	High	Very High
1. Rate your level of satisfaction when you are able to work alone on class projects.	0.0%	25.0%	30.0%	25.0%	20.0%
2. Rate your level of satisfaction when you are able to work with two students to complete a class project.	0.0%	5.0%	25.0%	50.0%	20.0%
3. Rate your level of satisfaction when you are able to work in a group with four students to complete a class project.	5.0%	10.0%	0.0%	50.0%	35.0%
4. Rate your level of satisfaction when the instructor varies the projects from individual, to 2 students, to groups of 4 students.	0.0%	5.0%	45.0%	30.0%	20.0%

two, group of four, or a mixture of configurations. Table 5 presents the students' responses to the 4 survey items. Table 6 presents all of the students' responses to the open-ended question item on the survey.

CONCLUSION

For the quantitative assessment, the overall preliminary data indicated the pre-service teachers had a variety of preferences on whether to work on projects alone, with pairs, groups of four, or a combination of student groupings. Observations of the data trends indicated a general satisfaction when working alone while there was a very high satisfaction with working in groups of four students. Working in pairs just edges out varying the size of the work groups consisting of one, two, or four. Therefore, in order of satisfaction, students prefer working in groups of four, then groups of two, mixing up the groups, while working alone is least preferred.

For the qualitative assessment, student feedback provided additional insights to consider when implementing student group configurations when working with Web 2.0 e-learning tools while working alone, in groups of two, four, or a mixture. One important issue raised was to consider the diversity of student preferences and levels of satisfaction based on the configurations. Based on student comments, important issues surfaced such as - be sure to match the appropriate Web 2.0 tool with the best student group configurations as some projects lend themselves better to different sizes of groups. For example, working on a video is a great project for group work since all the students can be actors, there is someone to run the camcorder, do the video editing, place the results into a website to share, etc. Students liked the idea of sharing to increase creativity, thus ideas, and increase productivity time. Learners indicated a desire to have more "hands on" time for technology work, except that working in groups can detract from this unless an equitable division of tasks is in place. Changing the students within the groups helped to keep the learners more engaged in the learning tasks. It also had the side benefit of distributing the different levels of achievement and motivation to keep group work more balanced. The results indicated when placing students into larger groups with clearly defined goals and roles, it will help to ensure more even workloads. This study was conducted in a synchronous course for the explicit reason, it allows for additional observations and fine-tuning of the instruction. We really like one student's comment about how working with others makes the class seem more like the "real world," as justification for the potential value of trying out different student configurations when working with Web 2.0 tools.

To move group configurations to the online environment when working with Web 2.0 tools, it will be important to have a technology interface al-

Table 6. Summary of student qualitative survey results (N=20)

Student Excerpts
1. If you prefer to work alone to complete class projects, state why.
I think I get more done in this class when working alone. Perhaps if we could maybe switch up the tables for the groups so we are not working with the same people every project that would help.
Working alone gives us more hands on time.
I usually seem to be one of the first ones done with the individual projects, which I don't mind, but I really enjoy and prefer to work with my group.
I like to work alone on most projects, while still sitting with others so we can ask each other questions if we need to.
2. If you prefer to work in a group with two students to complete class projects, state why.
Two students make the workload not so intense. It gives you someone else to talk to about ideas and concepts when making a project. It just seems handy to have another person helping along the way.
I think ideas are more creative when working with a group.
Working with one other person is fine. I just do not like working by myself because I enjoy having someone else to discuss the assignment with and work together to get the assignment done accurately and creatively.
I think when working on computers it is hard to work in groups but working with partners is not so bad.
Working with others gives us more experiences. It teaches us to learn to share the project responsibilities.
It is nice to just have one partner because sharing a computer is easiest with two or less people.
3. If you prefer to work in a group with four students to complete class projects, state why.
I like working in a group because we are able to split up the work and it goes much quicker.
It is nice to combine all the ideas and be able to work together.
I prefer to work in large groups on projects. But I have been lucky our group is cooperative and wants to work together. Some groups have issues with attendance making it difficult to work as a large group. If I had a second choice, it would be working in groups of two.
I like working as a group with my table because I feel like we can all contribute to the project. I feel like the more people, the more ideas that can be thrown around and I feel like the project will turn out better because we can all work on it.
I think the more people we get involved on a project the more ideas get thrown around and the better the outcome is.
I think it is easier to complete the requirements because sometimes the other computers don't always work for each program.
We are able to help each other figure out the different programs for projects.
I prefer to work in a group because if I do not understand how to do something someone else can help and we can work together to be creative and have different ideas besides just one.
I loved working as a group for the video project. We got to make it our own and we all had a certain part in the making of the movie.
I really enjoy working with my group. We collaborate well and everyone pulls their own weight on the projects and we all seem to get a lot out of it.
This has been kind of hard because usually most of the work is done on one computer, and that person ends up doing most of the work. If we do split up the work the others have to email it to one person and then when the final project is complete they have to email it to all the group members again
4. If you prefer to have a combination of individual, 2 student, and 4 student groupings for completing class projects, state why.
I think it is very important to work with a variety of groupings. It makes class seem more like the "real world."
Sometimes it is better to be able to work with your own idea instead of having to collaborate.
I think a combination of groupings is very appropriate for this class because it really depends on the programs we are using as to what grouping works best.
I like to have someone else's opinion on my ideas, and it is nice to only have to rely on one other person.
When I work in a group of more than two it seems the two or three extra people take a vacation while the one person does all of the work.
I really prefer to work alone, but I do not mind working with a partner.
Some projects work better on an individual level- if it was in a group, some people wouldn't have anything to work on. It is better to decide if it should be individual or group on case by case.
I don't like mixing it up because then it becomes too confusing and I lose focus on what is really needs to be done.
I really enjoyed working in a combination because it showed me I can do these projects on my own with little problems.
I think it is best to have a combination of groupings and individual work. This way the students are able to have all of these experiences. The students who do not do much during the group projects need the individual so they can truly learn something.
Don't really mind a combination of group and individual work. I like working with my group the best but I haven't had any issues working on my own either.
I would prefer mostly individual projects, some partner projects, and a few four-student projects because certain projects are better with a certain number of people.

lowing for easy visual and verbal communications with a shared digital space such as a combined Whiteboard, Wiki, and Skype type of learning environment. In addition, it would be important to consider issues surrounding student groupings when students may well be in different time zones, have jobs, and family responsibilities, in order to effectively coordinate group work. One possible alternative would be to offer students more choices in regards to student groups and the distribution of responsibility.

When the last pilot class began, learners worked individually on the first project to get use to working with the computer and class software. From that point on, students worked in groups of four students. At the semester mid-term, we noticed the students' grades were higher than they had ever been and all students turned in all assignments with no missing grades. It seemed like the group work was increasing the individual students' feelings of responsibility to their classmates. There was some concern as to whether individual learners were carrying the main body of the load and whether all students were acquiring the needed knowledge and skills, so some individual work was implemented. We began seeing some assignments not turned in and a slight drop in scores. Next, we implemented a project working in pairs with students selecting a new partner they had not worked with before and observed a renewed spark in motivation. We realize there is no magic solution but rather it is an ongoing process and could vary depending on a variety of needs, goals, tools, learner differences, and the level of prior knowledge. As there is no real end, we can only continue experimenting and seeking feedback from the learners themselves, in order to continue improving our teaching.

Since all four pilot tests were limited in numbers and restricted to one discipline area, it would be good to conduct a full research study with a larger sample size to see if the trends hold true across additional discipline areas. It would be good to identify the differences in students' levels of satisfaction when changing the number of Web 2.0 e-learning digital tools implemented and using different student group configurations.

The pilot studies consisted of both qualitative and quantitative data about available Web 2.0 e-learning tool resources, classroom observations, and pilot study surveys of students' levels of satisfaction when using various numbers of tools and different sizes of student group configurations. This type of pilot study was needed since the class sizes were too small for a full research study, and the time needed to conduct a full study using multiple classes would have caused the data to be out of date due to the pace of changes in emerging technologies. In addition, full research studies would not provide the needed immediate classroom feedback for just in time learning. Therefore, it is recommended, in those instances where immediate specific classroom data is needed, a needs analysis with pilot studies is the best approach.

REFERENCES

American Association of School Librarians. (2010). *Best websites for teaching and learning: Top 25 websites for teaching and learning.* Retrieved April 4, 2011 from: http://ala.org/ala/mgrps/divs/aasl/guidelinesandstandards/bestlist/bestwebsitestop25.cfm

Baumbach, D. (2009, March/April). Web 2.0 and you. *Knowledge Quest.*

Benson, R., & Brack, C. (2009). Developing the scholarship of teaching: What is the role of e-teaching and learning? *Teaching in Higher Education, 14*(1), 71–80. doi:10.1080/13562510802602590

Berger, P., & Stripling, B. (2010). Student inquiry and web 2.0. *School Library Monthly, 26*(5). Retrieved April 4, 2011 from http://www.schoollibrarymonthly.com/articles/Berger2010-v26n5p14.html

Brooks-Young, S. (2008, May). Web tools: The second generation. *District Administration.*

Cohen, L. (2007, October). Information literacy in the age of social scholarship. *Library 2.0: An Academic's Perspective.*

Dockter, J., Haug, D., & Lewis, C. (2010, February). Redefining rigor: Critical engagement, digital media, and the new English/language arts. *Journal of Adolescent & Adult Literacy*. doi:10.1598/JAAL.53.5.7

Eisenberg, M. (2008). The parallel information universe. *Library Journal*, *133*(8).

Fredrick, K. (2010). In the driver's seat: Learning and library 2.0 tools. *School Library Monthly, 26*(6).

Gall, J., Gall, M., & Borg, W. (1999). *Applying educational research: A practical guide* (4th ed.). New York, NY: Addison Wesley Longman, Inc.

Greenhow, C. (2009). *Social scholarship: Applying social networking technologies to research practices*. Knowledge Quest.

Hedberg, J. G., & Brudvik, O. C. (2008). Supporting dialogic literacy through mashing and moddling of places and spaces. *Theory into Practice*, *47*, 138–149. doi:10.1080/00405840801992363

Hyatt, J., & Craig, A. (2009, October). Adapt for outreach: Taking technology on the road. *Computers in Libraries*.

Jensen, M. (2007). The new metrics of scholarly authority. *The Chronicle*, *53*(41).

Leimbach, L. (2010). Bloom's mashup: Bloom's taxonomy tools and tutorials. *Teacher Tech: Changing Education One Byte at a Time*. Retrieved April 4, 2011 from http://rsu2teachertech.wordpress.com/2010/12/04/blooms-mashup/

O'Brien, D., & Scharber, C. (2010). Teaching old dogs new tricks: The luxury of digital abundance. *Journal of Adolescent & Adult Literacy*, *53*(7).

Penney, S. (2010). *Bloom's taxonomy of web 2.0 tools*. Retrieved April 4, 2011 from http://www.usi.edu/distance/bdt.htm

Pogue, D. (2008). Art you taking advantage of web 2.0? *New York Times: Pogue's Posts: The latest in Technology from David Pogue*. Retrieved January 19, 2012 from http://pogue.blogs.nytimes.com/2008/03/27/are-you-taking-advantage-of-web-20/?scp=1&sq=Are%20you%20taking%20advantage%20of%20Web%202.0&st=cse

Sax, G. (1997). *Principles of educational psychological measurement and evaluation* (4th ed.). New York, NY: Wadsworth Publishing Company.

Shavelson, R. (1981). *Statistical reasoning for the behavioral sciences* (3rd ed.). Needham Heights, MA: Allyn and Bacon.

Valenza, J. K., & Johnson, D. (2009, October). Things that keep us up at night. *School Library Journal*.

Walling, D. R. (2009). Tech-savvy teaching and student-produced media: Idea networking and creative sharing. *Tech Trends: Linking Research and Practice to Improve Learning*. Retrieved April 4, 2011 from http://webcast.oii.ox.ac.uk/?view=Webcast&ID=20051130_109

Yan, J. (2008, Winter). Social technology as a new medium in the classroom. *New England Journal of Higher Education*.

KEY TERMS AND DEFINITIONS

Action Research: Allows testing for the purpose of improving local practices.

Applied Research: Designed to test interventions used to directly improve instruction.

Collaboration: Working together to increase knowledge and skills.

E-Learning: Internet based instruction.

Instructional Design: Teaching effectively through the use of best practices.

Instructional Technology: Imparting knowledge or skills through the effective use of technology.

Levels: The various amounts of technology experience requiring more scaffolding support.

Mixed Methods: Quantitative and qualitative data is collected to produce better research results.

Open Source: Source code available to the public without a charge.

Web 2.0: User generated content.

APPENDIX

Additional Resources

Table 7 provides some additional Web 2.0 online supplemental resource sites for finding new Web 2.0 tools for integrating into online teaching and learning.

Table 7. Recommended resources

Source	Website
Centre for Learning & Performance: Top 100 Tools for Learning 2010 List (2011).	http://www.c4lpt.co.uk/recommended/top100-2010.html
Discovery Education: Web 2.0 Tools (2011).	http://school.discoveryeducation.com/schrockguide/edtools.html
Go2Web20 (2011).	http://www.go2web20.net/
Internet 4 Classrooms: Web 2.0 Tools (2011).	http://www.internet4classrooms.com/web2.htm
ISTE Wikispaces: Favorite Web 2.0 Tools (2011).	http://sigilt.iste.wikispaces.net/Favorite+Web+2.0+Tools
SlideShare, Top 20 Web 2.0 Tools for Teachers and Librarians (2011).	http://www.slideshare.net/scyuen/top-20-web-20-tools-for-teachers-and-librarians
Teaching with Technology Tools for Creating, Editing, and Sharing (2011).	http://ipt286.pbworks.com/w/page/10618966/Index
Web 2.0 Cool Tools for Schools (2011).	http://cooltoolsforschools.wikispaces.com/
Web 2.0 Guru: Web 2.0 Resources for 21st Century Instruction (2011).	http://web20guru.wikispaces.com/Web+2.0+Resources

Chapter 16
Evaluation of Web Accessibility:
A Combined Method

Sergio Luján-Mora
University of Alicante, Spain

Firas Masri
University of Alicante, Spain

ABSTRACT

The Web is present in all fields of our life, from information and service Web pages to electronic public administration (e-government). Users of the Web are a heterogeneous and multicultural public, with different abilities and disabilities (visual, hearing, cognitive, and motor impairments). Web accessibility is about making websites accessible to all Internet users (both disabled and non-disabled). To assure and certify the fulfillment of Web accessibility guidelines, various accessibility evaluation methods have been proposed, and are classified in two types: qualitative methods (analytical and empirical) and quantitative methods (metric-based methods). As no method by itself is enough to guarantee full accessibility, many studies combine these qualitative and quantitative methods in order to guarantee better results. Some recent studies have presented combined evaluation methods between qualitative methods only, thus leaving behind the great power of metrics that guarantee objective results. In this chapter, a combined accessibility evaluation method based both on qualitative and quantitative evaluation methods is proposed. This proposal presents an evaluation method combining essential analytical evaluation methods and empirical test methods.

INTRODUCTION

Nowadays, the Web is present in all fields of our life, from access to information and service Web pages to electronic public administration (e-government). The social and economic impact of the Internet cannot be denied. Many people cannot imagine their lives without the Internet these days. However, many users of the Web can encounter various problems if websites do not accomplish a minimum level of Web accessibility. Therefore, Web accessibility is becoming increasingly criti-

DOI: 10.4018/978-1-4666-2491-7.ch016

Copyright © 2013, IGI Global. Copying or distributing in print or electronic forms without written permission of IGI Global is prohibited.

cal to the Internet experience. Tim Berners-Lee, inventor of the World Wide Web, once noted, "*The power of the Web is in its universality. Access by everyone regardless of disability is an essential aspect*" (W3C, 2011).

Traditionally, accessibility is a term more associated with architectural thought, rather than website development. With websites, the term traditionally refers to the development of websites accessible to all users who may want to access them, independent of the abilities or disabilities of the users. When websites are correctly designed and developed, all users can have equal access to information and functionality. A simple definition of Web accessibility is "*the property of a site to support the same level of effectiveness for people with disabilities as it does for non-disabled people*" (Slatin & Rush, 2003). An alternative definition of accessibility id "making Web content available to all individuals, regardless of any disabilities or environmental constraints they experience" (Mankoff, Fait, & Tran, 2005).

Web accessibility primarily benefits people with disabilities. However, as an accessible website is designed to meet different user needs, preferences, technical knowledge, and situations, this flexibility can also benefit people without disabilities in certain situations, "*such as people using a slow Internet connection, people with temporary disabilities such as a broken arm, and people with changing abilities due to aging*" (W3C, 2011). Moreover, an accessible website can also help people who have limited access to certain technology, such as slow computers or slow Internet connections.

To provide access to all possible users represents a huge challenge. Web accessibility aims to address the needs of heterogeneous users with different impairments, such as visual impairments, mobility impairments, hearing impairments, cognitive impairments, and learning impairments.

In 1999, the Web Accessibility Initiative (WAI), a project by the World Wide Web Consortium (W3C), published the Web Content Accessibility Guidelines (WCAG) version 1.0 (W3C, 1999a). These guidelines have been widely accepted as the definitive guidelines on how to create accessible websites. On 11 December 2008, the WAI released the WCAG version 2.0 (W3C, 2008) to be up to date while being more technology neutral. Conformance to the WCAG is based on four ordinal levels of conformance (none, A, AA, and AAA).

Nevertheless, verifying a website's accessibility can be a time-consuming task and needs expert evaluators to validate. If the intention is to fulfill Web accessibility guidelines WCAG 1.0 and WCAG 2.0 or other national and international guidelines and laws (Jefatura del Estado de España, 2002; Ministro per l'Innovazione e le Tecnologie de la Repubblica Italiana, 2005; US Government, 1998) regulating and protecting the rights of disabled users to access information, the task can be very complex and time consuming.

To assure Web accessibility, several studies have suggested numerous evaluation methods (Brajnik, 2006; Bühler, Heck, Perlick, Nietzio, & Ulltveit-Moe, 2006; Vigo, Arrue, Brajnik, Lomuscio, & Abascal, 2007) as a means to verify, measure and certify the fulfillment of the accessibility guidelines and therefore to supply full accessibility to disabled people. Currently, there are two types of evaluation methods: analytical and empirical qualitative methods and quantitative methods.

The qualitative methods have been the most used until now, specifically the analytical ones, which are characterized by their low cost and ease of use. Automatic evaluation tools such as AChecker (ATutor, 2011), A-Prompt (University of Toronto, 2011), Cynthia Says (HiSoftware, 2003), EvalAccess 2.0 (Universidad del País Vasco, 2011), eXaminator (2005), TAW (Fundación CTIC, 2011), and WAVE 4.0 (Web Accessibility in Mind, 2011b) have been the pioneers and are the most well-known, due to their usability, ease of use and its quick results, although they are not the final and complete solutions, since a comparison between them can show quite contradictory results (Thatcher, et al., 2006; Diaz & Cachero, 2009). It is clear no one tool, alone, can determine if a website meets Web accessibility guidelines. Knowledgeable human evaluation is required to determine if a site is accessible.

However, other analytical evaluation methods, based on the manual heuristic inspection of code, do not guarantee full accessibility (Brajnik, 2008) and depend largely on the evaluator's experience and the adopted guidelines. On the other hand, empirical methods are generally more expensive, but more accurate, because they clearly show the most catastrophic accessibility faults. User test is the most reliable and complete one (Masri & Lujan, 2010a), despite the inconveniences experienced while using it.

The quantitative methods help to understand, control and improve the final product (Fenton & Pfleeger, 1998), thus its main goal is to assure the quality and monitor the accessibility level by establishing values and summarizing results. These methods, due to their nature, are not sufficient to assess accessibility and evaluators cannot depend on them only.

As a consequence of the above-mentioned situation related to evaluation methods and because no method on its own will guarantee the detection of all accessibility barriers, some studies (Lopez, 2010; Villegas, Pifarré, & Fonseca, 2010), considering W3C and researchers recommendations, start to apply in their evaluations process a combined evaluation method based completely on qualitative methods (analytical and empirical methods). However, these works ignore totally the inclusion of the important quantitative methods to their methodology.

Again, due to the above situation, and addressing the growing need to provide combined evaluation methods, arises the idea to present both kinds of methods in a clearly documented evaluating method, with a clear user-centered orientation, specifically no homogeneous and unified practice in this field (Masri & Lujan, 2010b), those already existing are not complete enough to guarantee the coverage of all kinds of evaluations tasks from one side and do not provide clear and objective results from the other side.

The method presented in this chapter combines the essential analytical methods with the empirical user test methods and concludes with the Web Accessibility Barrier metric (Parmanto & Zeng, 2005) to objectively summarize the results. This method explicitly shows the way of systematically carrying out an accessibility evaluation.

This book chapter is divided into six parts. After this introduction, the following section provides a brief background discussion about Web accessibility. Then, the section "Accessibility evaluation methods" is devoted to examining the most popular accessibility evaluation methods. After that, a combined method for evaluating Web accessibility is introduced with the whole evaluation process. Finally, this book chapter ends with the main conclusions and the references.

A BACKGROUND ABOUT WEB ACCESSIBILITY

Concerns about websites accessibility has been steadily growing since the mid-1990s. This cause has been supported by researchers, community organizations, Web standards bodies, and governmental agencies. This broad community has attempted to raise awareness of accessibility issues and to encourage accessible design practices through a number of different methods, laws, and techniques. Web standards organizations have developed technical guidelines attempting to codify accessible design techniques. In brief, the main goal of accessibility is to ensure that all users, with or without disabilities, are able to use websites.

Impaired or disabled people use special devices called assistive technology to enable and assist Web browsing. Basically, the following disabilities and assistive technologies exist:

- Severe vision impaired user uses screen reader, such as Jaws, Window-Eyes, or NVDA, which can read out, using synthesized speech, what is happening or being displayed on the monitor of a computer.
- Medium vision impaired user uses magnification software, such as Zoomtext, which enlarges what is displayed on the computer monitor, making it easier to read.
- Mild vision impaired user uses large text in browser with personal color preferences.
- Severe motor difficulties user uses switch/head operated mouse access and on-screen keyboard.
- Medium motor difficulties user only uses keyboard or special input devices.

Making a website accessible can be complex or simple, depending on many different factors, such as the size of the website, the complexity of the website, and the type of content. In order to achieve this goal, many different techniques must be used:

- Use standard HyperText Markup Language (HTML) and Cascading Style Sheet (CSS) code.
- Provide alternative content for screen readers through alternative texts, image long descriptions, and appropriate form labels.
- Ensure an acceptable default color contrast.
- Use acceptable typography and providing the ability to alter text size.
- Provide appropriate link texts and explanatory information for acronyms and abbreviations.
- Place important content higher up the Web page.
- Use headings and other structural and semantic markup correctly.
- Use skip navigation links at the top of the Web page.

The World Wide Web Consortium (W3C), the main international standards organization for the World Wide Web, is engaged in promoting the creation of accessible websites. The Web Accessibility Initiative (WAI) is a special working group established by the W3C in April 1997. The WAI provides ten "quick tips" that summarize key concepts of accessible Web development. These tips are not complete guidelines; however, they are a good starting point to understand and to achieve Web accessibility:

1. **Images and Animations:** Use the alt attribute to describe the function of each visual.
2. **Image Maps:** Use the client-side map and text for hotspots.
3. **Multimedia:** Provide captioning and transcripts of audio, and descriptions of video.
4. **Hypertext Links:** Use text that makes sense when read out of context. For example, avoid "click here."
5. **Page Organization:** Use headings, lists, and consistent structure. Use CSS for layout and style where possible.
6. **Graphs and Charts:** Summarize or use the longdesc attribute.
7. **Scripts, Applets, and Plug-ins:** Provide alternative content in case active features are inaccessible or unsupported.
8. **Frames:** Use the noframes element and meaningful titles.
9. **Tables:** Make line-by-line reading sensible. Summarize.
10. **Check your work:** Validate, use tools, checklist, and guidelines at http://www.w3.org/TR/WCAG.

The W3C develops a series of accessibility standards and guidelines. Web accessibility depends on several components working together, mainly:

- Authoring tools, software used by Web developers to create the content of websites.
- Web content, the information in a Web page, including text, images, sounds, videos, etc.
- User agent, software used by people to get and interact with the content, such as Web browsers, media players, and assistive technologies.

The W3C develops Web accessibility guidelines for the three main components:

- Authoring Tool Accessibility Guidelines (ATAG) addresses authoring tools.
- Web Content Accessibility Guidelines (WCAG) addresses Web content, and is used by developers, authoring tools, and accessibility evaluation tools.
- User Agent Accessibility Guidelines (UAAG) addresses Web browsers and media players, including some aspects of assistive technologies.

Regarding the Web content, the W3C has developed the most important guidelines concerning Web accessibility, the WCAG versions 1.0 and 2.0 (W3C, 1999a, 2008). In recent years, these guidelines have been widely accepted as the definitive guidelines on how to create accessible websites.

WCAG 1.0 (W3C, 1999a) is one element in a comprehensive accessibility strategy; other WAI recommendations address the authoring tools used to create Web content (W3C, 2000) and the user agents that display that Web content (W3C, 2002a).

WCAG 1.0 is composed of fourteen guidelines, or general principles of accessible design:

1. Provide equivalent alternatives to auditory and visual content.
2. Do not rely on color alone.
3. Use markup and style sheets and do so properly.
4. Clarify natural language usage
5. Create tables that transform gracefully.
6. Ensure that pages featuring new technologies transform gracefully.
7. Ensure user control of time-sensitive content changes.
8. Ensure direct accessibility of embedded user interfaces.
9. Design for device-independence.
10. Use interim solutions.
11. Use W3C technologies and guidelines.
12. Provide context and orientation information.
13. Provide clear navigation mechanisms.
14. Ensure that documents are clear and simple.

Besides, each guideline includes a list of checkpoints. Each checkpoint is intended to be specific enough so that someone reviewing a Web page or website may verify that the checkpoint has been satisfied. Each checkpoint has a priority level based on the checkpoint's impact on accessibility:

- **Priority 1:** A Web content developer must satisfy this checkpoint. Otherwise, one or more groups will find it impossible to access information in the document. Satisfying this checkpoint is a basic requirement for some groups to be able to use Web documents.
- **Priority 2:** A Web content developer should satisfy this checkpoint. Otherwise, one or more groups will find it difficult to access information in the document. Satisfying this checkpoint will remove significant barriers to accessing Web documents.
- **Priority 3:** A Web content developer may address this checkpoint. Otherwise, one or more groups will find it somewhat diffi-

cult to access information in the document. Satisfying this checkpoint will improve access to Web documents.

WCAG 1.0 and 2.0 (W3C, 1999a, 2008) define four ordinal levels of accessibility (none, A, AA, and AAA) and provide a set of checkpoints or success criteria for each level. A Web page must satisfy all priority A checkpoints or criteria to be considered minimally accessible. Web developers may implement priority AA and priority AAA checkpoints or criteria to provide increased accessibility for users. Unfortunately, this system of evaluation does not reflect the real accessibility of a website: if a website satisfies many checkpoints in addition to all level A checkpoints, the website will only conform to level A of WCAG, but the additional efforts to achieve a better level of accessibility will not be visible.

Although Web accessibility guidelines, such as WCAG, are designed to be easy to follow, verifying a website's accessibility can be a time-consuming task and needs expert evaluators to validate. For that reason, several researchers (Brajnik, 2006; Vigo, et al., 2007) have proposed methods to support evaluators.

These methods are generally divided into two types: analytical and empirical methods.

Researchers and organizations have created automatic evaluation tools for checking adherence to these guidelines (W3C, 2006), such as AChecker, A-Prompt, Cynthia Says, EvalAccess 2.0, eXaminator, TAW, and WAVE 4.0 to assist Web developers in evaluating accessibility of websites. These tools differ in several ways, manners, and capabilities, ranging from functionalities (testing vs. fixing) to supported interaction form (online service vs. desktop application integrated in authoring tools), effectiveness, reliability, cost, etc.

Automatic tools are the most useful analytical methods due to their quick results and ease of use. These tools may assist developers in the creation of accessible websites, but may not be able to identify all accessibility issues. Several comparisons between various analytic methods show quite contradictory results (Díaz & Cachero, 2009; Thatcher, et al., 2006). On the other hand, other analytical methods mostly used like manual revisions also do not guarantee full accessibility (Brajnik, 2008a) while depending largely on the evaluators' experiences and the adopted guidelines to achieve satisfactory results.

For this reason, automated tools are often used in combination with some type of manual evaluation and empirical methods like the user test method (Masri, 2010b), a combination strongly recommended to be applied in order to guarantee full accessibility to all disabled people.

Automatic tools generally verify the presence of a valid element or attribute, such as the alt attribute (alternative text) or the label element (description of a form control). However, human judgment is also needed, because some questions are very relevant, such as whether or not the value of the alt attribute clearly and effectively conveys the function of the image. For example, there is a big difference between the alternative text that an active or inactive image needs.

LEGAL ASPECTS CONCERNING WEB ACCESSIBILITY

Governments worldwide have also begun to consider the application of Web accessibility to electronic and digital products (Web Accessibility in Mind, 2011a).

In the United States, Section 508 of the Rehabilitation Act of 1973 (US Government, 1998) stipulates that all electronic information produced by federal agencies must be accessible to people with disabilities. Section 508 also provides a set of mandatory accessibility checkpoints for federal government websites.

The United Kingdom also adopted laws and legislations concerning Web accessibility, the Disability Discrimination Act (DDA) (Ministry of

Justice, 1995) were introduced to end discrimination against disabled people.

The European Union in its term, adopted many important legislations concerning Web accessibility especially that such laws comply with the principle of non-discrimination set up in the Treaty on the European Union, eEurope 2002, eEurope 2005, i2010, and Digital agenda for Europe 2010-2020 action plans were launched in 2000, 2002, 2005, and 2010, respectively (European Union, 2012, 2010).

Recently, the EU adopts WCAG 1.0 guidelines and level AA as mandatory for all public sites, they also boosts and pressures all member countries to have a continuing development policies related to people with disabilities and Web accessibility.

ACCESSIBILITY EVALUATION METHODS

When developing a new website or redesigning an existing website, evaluating accessibility early and throughout the development process can identify accessibility problems early when it is easier and most effective to address them. Simple techniques such as changing some settings in a browser can determine if a website meets some accessibility guidelines. However, a complete evaluation to determine if a website meets all accessibility guidelines is much more complex.

Different methods for finding accessibility problems in Web pages have been developed during the last ten years. At the same time, different comparisons of these different methods have been carried out (Mankoff, Fait, & Tran, 2005).

Qualitative methods include two essential types of evaluation methods: analytical and empirical. The analytical methods include: standard expert's revisions (conformity revisions), automatic tools, and the barrier walkthrough method. The empirical methods include: user tests, subjective revisions, and screen techniques (Masri & Lujan, 2010a).

Qualitative evaluation methods are used to perform formative evaluations (identification of the list of problems) during the development phase; and summative evaluations (validation and comparison) in the final phase of the product and after the final users have used it. These methods estimate the accessibility of an interface so as to validate it. The results will always show the descriptions of the failure modes, defects or even solutions and recommendations for the developers.

Quantitative methods include: Failure Rate Metric, Web Quality Evaluation Metric (WEBQEM), Web Accessibility Barrier (WAB) Metric, Unified Web Evaluation Methodology (UWEM), A3 - Aggregation Metric, Web Accessibility Quality Metric (WAQM), and T1 Metric. These methods help evaluators to monitor and improve accessibility levels, providing objective summarized results, where these results can be used to compare quality among Web pages, or to track quality improvement in the quality assurance process (Freire, Fortes, Turine, & Paiva, 2008; Sirithumgul, Suchato, & Punyabukkana, 2009; Vigo, et al., 2007).

Comparison between Web Accessibility Evaluation Methods

Evaluation methods, whether qualitative (analytical and empirical) or quantitative methods, have advantages and disadvantages when used alone to evaluate Web accessibility.

Analytical methods are characterized by their great capacity for identifying a wide range of diverse problems for diverse audiences, in addition to their ability to clearly highlight the exact violations of the adopted guidelines. They are criticized for requiring skilful evaluators and for not distinguishing between the important from unimportant Web accessibility problems (Brajnik, 2008). Analytical methods have also proved methodologically weakness when qualifying the gravity of identified problems (Petrie & Kheir,

2007). A comparison between automatic tools has also shown quite contradictory results (Diaz & Cachero, 2009; Thatcher, et al., 2006).

Empirical methods are more exact when qualifying a website; they discover the more catastrophic accessibility mistakes in real time, especially if they are applied in the correct context taking into account the specific characteristics for Web accessibility (Masri & Lujan, 2010a). Many studies (Dey, 2004; DRC, 2004) have shown empirical methods and specific user tests are the best ones to detect and qualify the severity of real accessibility mistakes faced by users. They have also been credited with identifying 45% of accessibility difficulties emerge with the application of user tests directly with disabled people after being evaluated with software and other methods. However, user tests are criticized for their elevated costs due to the necessity to have well equipped labs and real users.

Quantitative metrics are being used in large-scale evaluation processes; only the metric for WEBQEM did not. WAB, UWEM, A3, WAQM, and T1 metrics have shown a great correlation between their results (Freire, et al., 2008; Sirithumgul, et al., 2009), for that reason, it is not possible to state which metric could be more effective in general cases. Each metric may be more suitable for different projects, according to their needs. In this sense, in order to help the definition of good metrics, it is needed to identify important characteristics of useful software metrics. According to Daskalantonakis (1992), software metrics must be:

1. Simple to understand and precisely defined.
2. Objective.
3. Cost effective.
4. Informative (ensure that changes to metric values have meaningful interpretations).

From all the above mentioned characteristics, we determine a combined accessibility evaluation method must contain both a user test method, due to its reliability and effectiveness, and a quantitative metric, to control and monitor accessibility results and cover all task types. Particularly, in our proposal, we consider the WAB metric presented by Parmanto and Zeng (2005) as the best metric to fit our approach due to its simplicity, objectivity and task coverage. Besides, it has a fixed defined barrier weight corresponding to WCAG levels (A, AA, and AAA). In addition, it is considered inclusive because it includes all user groups.

Why is the User Test the Ideal One to Combine with Analytical Methods?

The user test is the only empirical evaluation method that detects and qualifies the severity of the real mistakes faced by users while evaluating in real time. The user test validates the compliance of WCAG applied by developers during implementation or adaptation of accessibility guidelines in any website.

This is the most effective and reliable of all empirical methods. The disadvantages of this method are minimal, especially if the evaluation objective is well defined (Masri & Lujan, 2010b). Unfortunately, the main disadvantage of the user test is the need to set different groups of users. Fortunately, according to Jakob Nielsen (2000), "*the best results come from testing no more than 5 users and running as many small tests.*" As more users are included in a user test, less knowledge is learnt because extra users keep showing the same results as previous users. It is a waste of time and resources to keep observing the same findings multiple times. It is better to distribute users across many small user tests instead of a single, elaborated, and huge study.

As noted earlier, many researchers and reports (Dey, 2004) have shown that 45% of accessibility difficulties emerge with the application of the user test to disabled people after being evaluated with software and other methods based on direct inspection of the code. In addition, some studies also assure that just following the accessibility guidelines and good practices is not enough to get accessible applications and therefore user tests should be applied with disabled people as an empirical method to demonstrate and prove accessibility. For all the above mentioned reasons, we consider the user test a more reliable method to guarantee access to all disabled people and, thus, its presence is essential in any evaluation method.

THE COMBINED METHOD FOR EVALUATING WEB ACCESSIBILITY

When we talk about accessibility evaluation, first of all, we must think directly of what our principal objective is and what we know about it. When assessing accessibility, we can face three different situations:

- The first is evaluating a website during development in order to see if it fits all accessibility guidelines.
- The second is to validate or compare its compliance with accessibility guidelines after being adapted by developers and posterior to its launch.
- In the third situation, developers knowing nothing about the website have to evaluate, while developers are obliged to start from the initial steps to check the real state of the website, which means using the analytical evaluation methods.

The first situation can be resolved by applying common analytical automatic evaluation methods. In the second case, it is more complicated and the best practice is to develop an empirical evaluation test. While in the third situation, many evaluators lose time to determine how to assess it.

Therefore, taking into consideration the three possible cases, we propose a combined evaluation method that fits all three objectives commonly or separately. If the duty is to assess a website in the formative stage or a website that does not meet accessibility guidelines or they do not have idea about its situation, users can always use the first step of the method. Otherwise, they can use the whole method after checking its minimum compliance with W3C recommendations. In addition, in our approach, we include a new approach for the qualification method that permits objective conclusions for evaluations.

In summary, we propose an accessibility evaluation method based on two steps:

- The analytical evaluation step.
- The empirical evaluation step.

The Analytical Evaluation Step

This step consists of two stages:

- **The Pre-Analysis Stage:** this stage helps to illustrate a rapid and clear idea about the website that is going to be verified.
- **The Automatic Stage:** this stage helps to check the compliance to W3C recommendations and the WCAG guidelines, while also permitting validation of mark-up languages (HyperText Markup Language) and style languages (Cascading Style Sheet) used in the design. After finishing this stage, the evaluator will have a complete idea about the website being evaluated and the evaluator will be able to determine whether the website is qualified to directly continue to the next step or not.

The Pre-Analysis Stage

This stage is composed of the following main activities:

- Localize and verify the identity of the website under evaluation to make sure is the correct one.
- Compare the current version of the website with the oldest version (Internet Archive,

2001), thus avoiding unexpected mistakes and having an idea about the websites evolution.

- Navigate the website in order to have a more complete idea about the design and content of the website.
- Navigate the website with different browsing technologies: remember that not everyone is using the latest version of the most popular browsers with all the plug-ins and programs that may be required.
- Check the contrast between the text and the background is well defined.
- Disable images, videos, and animations to check if the alternative text is sufficient and understandable.
- Disable style sheets and check if the content of the Web page is correctly arranged.
- Disable Java, JavaScript, and check if the Web page still works properly.
- Check if the language is clear and understandable.
- Apply software tools like Web Developer (Pederick, 2003) to revise HTML cleanlincss and detect image volume used in the Web pages.
- Apply text and sound browsers like WebbIE (King, 2011) to verify if information is accessible and equivalent to that provided by a graphical user interface browser.

Here are some of the most common errors evaluators find in this stage:

- Non-recommended presentation layouts. For example, a fixed design.
- Non-coherent presentation design. For example, a home page design different from the design of the linked Web pages.
- Non-coherent navigational mechanisms.
- Use of non-recommended HTML technical features like frames, image links, and animated links.
- Incorrect use of HTML code.
- Use of deprecated or disapproved HTML features.
- Incorrect use of map and image elements.
- Incorrect use of tables and undefined tables attributes.
- Use of strongly discouraged HTML tags, etc.

For this stage, we also propose the use of a simplified version of the Accessibility Test Suite (Thatcher, et al., 2006) due to the excellent results provided:

1. **Alt-Text:** Every image element must have an alt (alternative text) attribute.
2. **ASCII Art for Alt-Text:** ASCII art must not be used in an alt attribute.
3. **Object Requires Default Content:** An object should have content that is available if the object is not supported.
4. **Image Button:** Every image button must have an alternative text that specifies the purpose of the button.
5. **Long Alt-Text:** Alternative text should be short, succinct, and to the point.
6. **Image Map Areas:** Every area element of a map for a client-side image map needs to have an alternative text.
7. **Server-Side Image Maps:** They should never be used, it is better to use client-side image maps.
8. **Frames Title:** Each frame in a frameset needs a title attribute indicating the purpose of the frame.
9. **Quality of Frame Titles:** The title attributes on the frame elements must be meaningful.
10. **Input Element needs Label:** Every form control must be described with a label element.
11. **Use of Title Attribute for form Control:** Another way to explicitly specify prompting text for a form control.

12. **Text Intervenes between Label and Control:** The text that is enclosed by the label element does not need to be right next to the control.
13. **Prompting Text from Two Places:** It is perfectly legitimate to have two label elements with the same for attribute.
14. **An Invisible GIF holds the Prompting Text:** Although this technique is acceptable, the title attribute is much better.
15. **Label that Matches no Control:** Every label must match a control.
16. **Two Input Elements with same ID:** It is an accessibility and HTML error.
17. **Inadequate Link Text:** Link text must be significant and meaningful.
18. **Same Link Text, Different Targets:** If two links have the same link text but different targets, then the links cannot be distinguished in a list of links.
19. **Page Title:** Every page should have a non-empty title element.
20. **Structure of Headings Markup:** The use of headings is crucial for page navigation.

An example of this evaluation stage with more details can be found in (Luján-Mora, 2011).

The Automatic Evaluation Stage

This stage is composed of the following main activities:

- Validation of mark-up language using the HTML validation service of W3C (1995a).
- Validation of CSS style sheet using the CSS validation service of W3C (1995b).
- Automated accessibility evaluation using Cynthia Says (HiSoftware, 2003), TAW (Fundación CTIC, 2011), or other available automatic evaluation software.

Some of the common detected errors in this stage are:

- Non-defined <!DOCTYPE>, <meta content="charset">, and language attribute.
- Absence of <title> element.
- Too short or too long alternative text.

An example of this evaluation stage with more details can also be found in Luján-Mora (2011).

Although it can seem that the analytical evaluation step is sufficient to achieve a very accessible website, some researchers (Dey, 2004; Disability Rights Commission, 2006) claim that applying the WCAG, manual and automatic evaluations are not sufficient. Therefore, other techniques should be applied. For this, some researchers claim that user tests are also needed to guarantee Web accessibility. In the next subsection, the use of empirical user tests for evaluating Web accessibility is explained.

The Empirical Evaluation Step

Current international recommendations and laws (European Union, 2002; Jefatura del Estado de España, 2002; Ministro per l'Innovazione e le Tecnologie de la Repubblica Italiana, 2005; US Government, 1998) establish priority/level AA of WCAG as mandatory for all public and electronical products; therefore, in our method we consider two levels of test:

- Test level 1.
- Test level 2.

Test level 1 has to do with the degree of conformance with A and AA, where the biggest user groups will access information without troubles. On the other hand, Test level 2 is related to the degree of conformance with AAA that is a higher level of fulfillment in which all user groups will access without any kind of barriers.

In the following subsections the test process is presented in detail.

Users and Staff

The following users and staff are involved in an empirical evaluation:

- Users with different disabilities (vision, hearing, motor, and cognitive impairments): five are enough according to Nielsen (2000), one or two per each disability. Users must have some experience with Internet and assistive tools.
- Facilitator, one person is enough to explain to users the sequence of steps.
- Assistant or observer. An observer per disability is enough to guarantee the understanding and observation of the process.

Room and Equipment

It can be carried out in a small office or laboratory supplied with two computers for each selected disability. The computer must have the specific devices and software for the disability in question, for example: a voice browser such as Fire Vox, a screen reader such as JAWS, a screen magnifier such as ZoomText, a Braille display, etc. There are some reports (Pernice & Nielsen, 2001) that provide information about the necessary devices and software for testing disabled users.

Test Planning

In our method, we suggest doing the test in just one day and with all the selected disabilities in order to reduce costs, increase the test control, and increase the test systematization.

The required accessibility guidelines for the different guides are selected and revised according to the test level for the website being evaluated. In case of Test level 1, priorities A and AA or the guidelines representing them to be taken into consideration for the evaluation. In case of Test level 3, priority AAA and their corresponding guidelines all must be considered. Note that Test level 2 will be only applied in case evaluators are sure about website compliance of A and AA priorities.

The website must be studied to identify functionalities and detect potential barriers. For best results, it is advisable to take into account all predominant disabilities and to assign the tasks focused on the fulfillment of the most essential guidelines for easy access. While combining and taking into account tasks related to guidelines for different national and international accessibility guides or any suggestions reducing the barriers not conceived in the international and national guides should be the ones presented (Romen & Svanaes, 2008).

Material Preparation

The preparation of all tasks to be developed by the disabled users is undertaken based on the test planning with its most interesting correspondent guides, always taking into account the user's disability.

The obtained material must be attractive and not exhausting for the user, with questions that put the website accessibility at stake. To do a test with the different disabilities, it is indispensable to prepare tasks with an implicit orientation towards each disability. Users are assigned tasks that put diverse characteristics and essential accessibility actions at stake.

For example, the following is a possible list of tasks:

1. Logging on the website being evaluated.
2. Browsing mainly through the main menus, the most attractive links (suggesting some of them and covering most of them).

3. Browsing through the main menu locating a specific subject.
4. Searching and following a link predetermined by the evaluator.
5. Developing a search, this can be interesting for a user in the context of the website.
6. In case there is multimedia content, asking for a summary of the multimedia subject matter.
7. Locating some essential functionality and ordering user to access these functions through the keyboard.
8. Filling up a form or sending a message through the contact section.
9. In case the website sells products, asking the user to select one product and proceed to the purchase step.
10. In case the website allows managing legal documents, asking the user to proceed to fill in the application form and other similar procedures.

Required Tests

We propose to carry out two types of tests:

- **Pilot Test:** Performed to make sure everything will turn out just as been planned and the media works properly. The test subject user does not necessarily have to be a disabled person, but it is much better if he/she has a disability, no matter if it is visual, hearing, or motor disability.
- **Final Test:** The test is performed by the selected users in a familiar environment, free of tension. Everyone will be guided by the facilitator as to what to do. Every test subject user with an actual disabled person is accompanied by an assistant or observer to guide them during the test, at the same time they will observe their actions in detail. Test subject users are asked to use the protocol "think aloud."

Qualification Method

The qualification of the Web accessibility is accomplished with the analysis of all the collected evidence, and with a quantitative metric that establishes a value summarizing the results. The observer plays quite an important role in this qualification that we propose:

- **User Interviews:** After the user tests, the users are interviewed about their experiences and level of satisfaction. Their comments and opinions about the fulfillment of the objectives are collected. The interview results are compared with the corresponding observer's notes.
- **Recording Analysis:** Recordings of the tests are analyzed in order to find errors omitted by the observers.
- **Data Analysis:** Errors found are analyzed and classified, taking into account the priorities and corresponding checklists.
- **Conclusion:** A final report on the data analysis results is drawn up.
- **Evaluations Metric:** The evaluation metric criteria is the test measure allowing providing concrete results. Moreover, the evaluation metric is also important because it can be used anytime one wants to compare two or more websites. There are several metrics that study Web accessibility from a quantitative point of view (Vigo, et al., 2007; Vigo & Brajnik, 2011).

Evaluation Metric

In the case of the user tests, we are interested in evaluating the fulfillment of the checklist of the selected guidelines considering all user groups and all levels of accessibility. Therefore, metrics containing these two characteristics are needed.

Web accessibility metrics synthesize a value that is assumed to represent the accessibility level of a website. In the last few years, several accessibility metrics have been defined: some are totally automatic, such as the Web Accessibility Quantitative Metric (WAQM) (Vigo, Arrue, Brajnik, Lomuscio, & Abascal, 2007), Page Measure (PM) (Bailey & Burd, 2007), Unified Web Evaluation Methodology (UWEM) (Velleman, et al., 2007), A3 (Bühler, et al., 2006), etc. Others are totally manual, such as Accessibility Internet Rally (AIR) metric (Knowbility, 2011). And finally, some are hybrid (i.e. based on data produced by tools, somehow later interpreted and graded by humans, and then synthesized into a value), such as SAMBA (Brajnik & Lomuscio, 2007).

From a large set of Web accessibility metrics (Vigo & Brajnik, 2011), we selected a weighted metric-based formula called the Web Accessibility Barrier (WAB) score presented by Parmanto and Zeng (2005). WAB fulfills the necessary assessment requirements and includes all user groups. WAB allows the definition of a weight that represents the severity of an error. Besides, WAB is described as the sum of all real errors in a priority of all the evaluated Web pages divided by the sum of all potential errors of each priority multiplied by the severity of each error, where the severity is an inverse number to the priority defined as 1, 2, or 3. The formula of this metric is:

$$WAB = \frac{\sum_{j=1}^{T} \sum_{i=1}^{n} \left(\frac{b_{ij}}{B_{IJ}} \right) (W_i)}{T}$$

where,

b_{ij} : Represents the real errors and barriers in a priority.

B_{ij} : Represents the potential barriers of each priority.

W_i : Represents the gravity of the real error inverse to each priority.

T : Represents the number of Web pages of the whole website.

This metric is applied at the end of the evaluation process, and is considered to provide the best result when the results are close or equal to zero.

Final Evaluation Report

Currently, two formats of report are mainly used to present the result of Web accessibility evaluations. The first format is defined by the W3C (2002b) and the second format by the Web Accessibility Benchmarking Cluster (2007). The two formats are quite similar, but the second one introduces small changes, adds an introduction and a first section defining the context of the evaluation. Based on these two formats, we propose our own format adding a new section about the methods used during the evaluation. The template of our final evaluation report is structured in ten sections as follows:

1. Introduction.
2. Executive summary.
3. Background about evaluation.
4. Method applied.
5. Website reviewed.
6. Evaluators.
7. Evaluation process.
8. Results and recommended actions.
9. References.
10. Appendices.

CONCLUSION

Web accessibility aims to provide access to Web content for people with different abilities using different devices. Accessible websites can significantly improve the quality of life of people with disabilities by increasing their independence.

Different guidelines and laws encouraging the development of accessible websites have been proposed during the last ten years. In order to assure Web accessibility, several studies have suggested numerous evaluation methods as a mean to verify, measure, and certify the fulfillment of accessibility guidelines and therefore to provide full accessibility to disabled people. Many of these evaluation methods were implemented, thus creating a number of automatic tools to simplify the evaluation process and, by that way, providing a technical infrastructure for all software developers to guarantee minimum levels of accessibility. Despite these technical resources and the fulfillment of laws and moral obligations towards disabled persons, unfortunately many public administrations and people representations still do not apply the minimal accessibility condition for their websites.

In this chapter, we reviewed a number of methods proposed in the last years to help evaluate Web accessibility levels. Automatic evaluation tools and accessibility guidelines by themselves fail to result in fully accessible websites, because they demand accessibility expertise on the part of the developer beyond what the majority of developers currently have, and are unable to identify all problems.

It is clear that measuring the level of Web accessibility is essential for rating accessibility implementation and improvements over time. However, it is also clear that finding appropriate measurement tools is non-trivial. Any accessibility evaluation method must be considered as a process that combines automated evaluation tools with human judgment. Currently, there is no automated tool that can assert that a website is accessible and complies with the Section 508 provisions of the U.S. Government or the Web Content Accessibility Guidelines (WCAG) of the World Wide Web Consortium.

The combined evaluation method we propose in this chapter is a clear guide for Web accessibility evaluations. Our method also includes a clear and objective quantification based in WAB score weighted metric that summarizes the evaluation process. Our method is characterized by its systematic and objective results, with a user centered evaluation focus. It is systematic because all of its steps are developed formally under observation, and it is objective because it summarizes the results by means of an evaluation metric that is not pre-determined by subjective human opinion.

REFERENCES

ATutor. (2011). *AChecker: Web accessibility checker*. Retrieved February 15, 2011, from http://achecker.ca/checker/index.php

Bailey, J., & Burd, E. (2005). Tree-map visualisation for web accessibility. In *Proceedings of the Computer Software and Applications Conference,* (pp. 275-280). IEEE Press.

Brajnik, G. (2006). *Web accessibility testing with barriers walkthrough.* Retrieved February 10, 2011, from http://www.dimi.uniud.it/giorgio/projects/bw

Brajnik, G. (2008a). Beyond conformance: The role of accessibility evaluation methods. In S. Hartmann, et al. (Eds.), *2nd International Workshop on Web Usability and Accessibility,* (pp. 63-80). Auckland, New Zealand: Springer.

Brajnik, G. (2008b). Measuring web accessibility by estimating severity of barriers. In S. Hartmann, X. Zhou, & M. Kirchberg (Eds.), *2nd International Workshop on Web Usability and Accessibility,* (pp. 112-121). Auckland, New Zealand: Springer.

Brajnik, G., & Lomuscio, R. (2007). SAMBA: A semi automatic method for measuring barriers of accessibility. In *Proceedings of the ACM SIGACCESS Conference on Computers and Accessibility,* (pp. 43-49). ACM Press.

Bühler, C., Heck, H., Perlick, O., Nietzio, A., & Ulltveit-Moe, N. (2006). Interpreting results from large scale automatic evaluation of web accessibility. In Miesenberger, K., Klaus, J., Zagler, W., & Karshmer, A. (Eds.), *Computers Helping People with Special Needs* (pp. 184–191). Berlin, Germany: Springer. doi:10.1007/11788713_28

Daskalantonakis, M. K. (1992). A practical view of software measurement and implementation experiences within Motorola. *IEEE Transactions on Software Engineering*, *18*(11), 998–1010. doi:10.1109/32.177369

Dey, A. (2004). *Accessibility evaluation practices–survey, results*. Retrieved February 15, 2011, from http://deyalexander.com/publications/accessibility-evaluation-practices.html

Díaz, B., & Cachero, C. (2009). Fiabilidad de herramientas automáticas para el aseguramiento de la accesibilidad en Web. In *Proceedings of IADIS Ibero-Americana WWW/Internet 2009*. Madrid, Spain.: IEEE.

Disability Rights Commission. (2006). *Guide to good practice in commissioning accessible websites*. London, UK: British Standards Institution.

European Union. (2002). *European parliament resolution on the commission communication eEurope 2002: Accessibility of public web sites and their content (COM(2001) 529 - C5-0074/2002 - 2002/2032(COS))*. Retrieved February 5, 2011, from http://www.europarl.europa.eu/omk/omnsapir.so/pv2?PRG=DOCPV&APP=PV2&LANGUE=EN&SDOCTA=18&TXTLST=1&POS=1&Type_Doc=RESOL&TPV=DEF&DATE=130602&PrgPrev=PRG@TITRE%7cAPP@PV2%7cTYPEF@TITRE%7cYEAR@02%7cFind@%2577%2565%2562%2520%7cFILE@BIBLIO02%7cPLAGE@1&TYPEF=TITRE&NUMB=1&DATEF=020613

European Union. (2010). *Digital agenda 2010*. Retrieved February 14, 2011, from http://ec.europa.eu/information_society/digital-agenda/index_en.htm

eXaminator. (2005). *eXaminator: Evaluación automática de la accesibilidad*. Retrieved February 12, 2011, from http://examinator.ws/

Fenton, N. E., & Pfleeger, S. L. (1998). *Software metrics: A rigorous and practical approach*. Boston, MA: PWS Publishing Co.

Freire, A. P., Fortes, R., Turine, M., & Paiva, D. (2008). An evaluation of web accessibility metrics based on their attributes. In *Proceedings of the 26th Annual ACM International Conference on Design of Communication,* (pp. 73-80). Lisbon, Portugal: ACM Press.

Fundación, C. T. I. C. (2011). *TAW*. Retrieved February 10, 2011, from http://www.tawdis.net/

HiSoftware. (2003). *Cynthia says*. Retrieved February 15, 2011, from http://www.cynthiasays.com/

Internet Archive. (2001). *The wayback machine*. Retrieved February 5, 2011, from http://www.archive.org/web/web.php

Jefatura del Estado de España. (2002). *LEY 34/2002, de 11 de julio, de servicios de la sociedad de la información y de comercio electrónico (BOE nº 166)*. Retrieved February 5, 2011, from http://www.boe.es/boe/dias/2002/07/12/pdfs/A25388-25403.pdf

King, A. (2011). *WebbIE*. Retrieved February 5, 2011, from http://www.webbie.org.uk/

Knowbility. (2011). *Accessibility internet rally (AIR)*. Retrieved January 16, 2012, from http://www.knowbility.org/v/air/

López, L. (2010). *AWA: Marco metodológico específico en el dominio de la accesibilidad web para el desarrollo de aplicaciones web*. (Unpublished Doctoral Dissertation). Universidad Carlos III. Madrid, Spain.

Luján-Mora, S. (2011). *Análisis de la accesibilidad del sitio web del Senado de España*. Retrieved February 12, 2011, from http://accesibilidadweb.dlsi.ua.es/?menu=ej-analisis-senado-parte-1

Mankoff, J., Fait, H., & Tran, T. (2005). Is your web page accessible? A comparative study of methods for assessing web page accessibility for the blind. In *Proceedings of the SIGCHI Conference on Human Factors in Computing Systems,* (pp. 41-50). Portland, OR: SIGCHI.

Masri, F., & Luján-Mora, S. (2010a). *Análisis de los métodos de evaluación de la accesibilidad web*. Paper presented at Universidad 2010. Havana, Cuba.

Masri, F., & Luján-Mora, S. (2010b). *Test de usuario: Un método empírico imprescindible para la evaluación de la accesibilidad web*. Paper presented at Actas de la 5ª Conferencia Ibérica de Sistemas y Tecnologías de Información. Santiago de Compostela, Spain.

Ministro per l'Innovazione e le Tecnologie de la Repubblica Italiana. (2005). *Requisiti tecnici e i diversi livelli per l'accessibilit`a agli strumenti informatici, (G.U.n.1838/8/2005)*. Retrieved February 10, 2011, from http://www.pubbliaccesso.it/normative/DM080705.htm

Ministry of Justice. (1995). *Disability discrimination act*. Retrieved February 20, 2011, from http://www.legislation.gov.uk/ukpga/1995/50/contents

Nielsen, J. (2000). *Why you only need to test with 5 users.* Retrieved March 14, 2011, from http://www.useit.com/alertbox/20000319.html

Parmanto, B., & Zeng, X. M. (2005). Metric for web accessibility evaluation. *Journal of the American Society for Information Science and Technology, 56*(13), 1394–1404. doi:10.1002/asi.20233

Pederick, C. (2003). *Web developer for Firefox toolbar*. Retrieved February 12, 2011, from http://chrispederick.com/work/web-developer/

Pernice, K., & Nielsen, J. (2001). *Beyond alt text: Making the web easy to use for users with disabilities*. Retrieved Feb 5, 2011, from http://www.nngroup.com/reports/accessibility/testing/

Petrie, H., & Kheir, O. (2007). The relationship between accessibility and usability of websites. In *Proceedings of the SIGCHI Conference on Human Factors in Computing Systems,* (pp. 397–406). San Jose, CA: SIGCHI.

Romen, D., & Svanaes, D. (2008). Evaluating web site accessibility: Validating the WAI guidelines through usability testing with disabled users. In *Proceedings of the 5th Nordic Conference on Human-Computer Interaction: Building Bridges,* (pp. 535-538). Lund, Sweden: IEEE.

Sirithumgul, P., Suchato, A., & Punyabukkana, P. (2009). Quantitative evaluation for web accessibility with respect to disabled groups. In *Proceedings of the 2009 International Cross-Disciplinary Conference on Web Accessibililty,* (pp. 136-141). Madrid, Spain: IEEE.

Slatin, J., & Rush, S. (2003). *Maximum accessibility: Making your web site more usable for everyone*. Boston, MA: Addison-Wesley.

Thatcher, J., Burks, M., Heilmann, C., Henry, S., Kirkpatrick, A., & Lauke, P. … Waddell, C. (2006). *Web accessibility: Web standards and regulatory compliance*. Berkeley, CA: Apress.

Universidad del País Vasco. (2011). *EvalAccess 2.0: Web service tool for evaluating web accessibility*. Retrieved February 10, 2011, from http://sipt07.si.ehu.es/evalaccess2/index.html

University of Toronto. (2011). *A-prompt: Web accessibility verifier*. Retrieved February 10, 2011, from http://aprompt.snow.utoronto.ca/download.html

US Government. (1998). *Section 508 standards guide: 1194.22 web-based intranet and internet information and applications*. Retrieved February 10, 2011, from http://www.section508.gov/index.cfm?fuseAction=stdsdoc#Web

Velleman, E., Meerbeld, C., Strobbe, C., Koch, J., Velasco, C. A., Snaprud, M., & Nietzio, A. (2007). *D-WAB4, unified web evaluation methodology (UWEM 1.2 Core)*. Retrieved January 14, 2012, from http://www.wabcluster.org/uwem1_2/

Vigo, M., Arrue, M., Brajnik, G., Lomuscio, R., & Abascal, J. (2007). Quantitative metrics for measuring web accessibility. In *Proceedings of the 2007 International Cross-Disciplinary Conference on Web Accessibility*, (pp. 99-107). Banff, Canada:IEEE.

Vigo, M., & Brajnik, G. (2011). Automatic web accessibility metrics: Where we are and where we can go. *Interacting with Computers*, *23*(2), 137-155. doi:10.1016/j.intcom.2011.01.001

Villegas, E., Pifarré, M., & Fonseca, D. (2010). *Diseño metodológico de experiencia de usuario aplicado al campo de la accesibilidad*. Paper presented at Actas de la 5ª Conferencia Ibérica de Sistemas y Tecnologías de Información. Santiago de Compostela, Spain.

W3C. (1994). *Evaluation, repair, and transformation tools for web content accessibility*. Retrieved February 10, 2011, from http://www.w3.org/WAI/ER/existingtools.html

W3C. (1995a). *The W3C markup validation service*. Retrieved February 15, 2011, from http://validator.w3.org/

W3C. (1995b). *The W3C CSS validation service*. Retrieved February 15, 2011, from http://jigsaw.w3.org/css-validator/

W3C. (1999a). *Web content accessibility guidelines 1.0*. Retrieved February 10, 2011, from http://www.w3.org/TR/WCAG10/

W3C. (1999b). *Checklist of checkpoints for web content accessibility guidelines 1.0*. Retrieved February 10, 2011, from http://www.w3.org/TR/WCAG10/full-checklist.html

W3C. (2000). *Authoring tool accessibility guidelines 1.0*. Retrieved February 10, 2011, from http://www.w3.org/TR/ATAG10/

W3C. (2002a). *User agent accessibility guidelines 1.0*. Retrieved February 10, 2011, from http://www.w3.org/TR/UAAG10/

W3C. (2002b). *Template for accessibility evaluation reports*. Retrieved February 10, 2011, from http://www.w3.org/WAI/eval/template.html

W3C. (2006). *Web accessibility evaluation tools: Overview*. Retrieved February 20, 2011, from http://www.w3.org/WAI/ER/tools/

W3C. (2008). *Web content accessibility guidelines 2.0*. Retrieved February 10, 2011, from http://www.w3.org/TR/WCAG20/

W3C. (2011). Web *accessibility initiative (WAI)*. Retrieved February 15, 2011, from http://www.w3.org/WAI/

Web Accessibility Benchmarking Cluster. (2007). *Unified web evaluation methodology 1.2*. Retrieved February 5, 2011, from http://www.wabcluster.org/uwem1_2/

Web Accessibility in Mind. (2011a). *WebAIM: World laws*. Retrieved January 14, 2012, from http://www.webaim.org/articles/laws/world/

Web Accessibility in Mind. (2011b). *Web accessibility evaluation tool (WAVE) 4.0*. Retrieved January 14, 2012, from http://wave.webaim.org/

Section 5
Evaluation Research Approaches

Chapter 17
Information Systems Evaluation:
Methodologies and Practical Case Studies

Si Chen
University of Sheffield, UK

Nor Mardziah Osman
University of Sheffield, UK

Guo Chao Peng
University of Sheffield, UK

ABSTRACT

Due to the prevalent use of Information Systems (IS) in modern organisations, evaluation research in this field is becoming more and more important. In light of this, a set of rigorous methodologies were developed and used by IS researchers and practitioners to evaluate the increasingly complex IS implementation used. Moreover, different types of IS and different focusing perspectives of the evaluation require the selection and use of different evaluation approaches and methodologies. This chapter aims to identify, explore, investigate, and discuss the various key methodologies that can be used in IS evaluation from different perspectives, namely in nature (e.g. summative vs. formative evaluation) and in strategy (e.g. goal-based, goal-free, and criteria-based evaluation). Six case studies are also presented and discussed in this chapter to illustrate how the different IS evaluation methodologies can be applied in practices. The chapter concludes that evaluation methodologies should be selected depending on the nature of the IS and the specific goals and objectives of the evaluation. Nonetheless, it is also proposed that formative criteria-based evaluation and summative criteria-based evaluation are currently among the more widely used in IS research. The authors suggest that the combined used of one or more of these approaches can be applied at different stages of the IS life cycle in order to generate more rigorous and reliable evaluation outcomes. Moreover, results and outcomes of IS evaluation research will not just be useful in practically guiding actions to improve the current system, but can also be used to generate new knowledge and theory to be adopted by future IS research.

DOI: 10.4018/978-1-4666-2491-7.ch017

Copyright © 2013, IGI Global. Copying or distributing in print or electronic forms without written permission of IGI Global is prohibited.

INTRODUCTION

Evaluation research can be defined as a form of "disciplined inquiry" (Guba & Lincoln, 1981, p. 550), which "applies scientific procedures to the collection and analysis of information about the content, structure and outcomes of programmes, projects and planned interventions" (Clarke, 1999, p. 1). Both quantitative and qualitative methods, or even a mixed-methods approach, can be adopted in evaluation research. Clarke (1999, p. 2) highlights that the key to distinguish evaluation research from other forms of research is not the data collection methods being employed but the purpose for which these methods are used. In particular, it is important to note that the primary purpose or objective of evaluation research is not to explore new knowledge as other forms of research do (Clarke, 1999, p. 2). Rather, it aims at using current knowledge to assess and study the effects, effectiveness and outcomes of "some innovation, intervention, policy, practice or service" (Robson, 2002, p. 202), and then to inform decision making to guide practical actions (Clarke, 1999, p. 2; Lagsten & Goldkuhl, 2008).

This type of research started receiving substantial attention from academics since the 1960s (Robson, 2002, p. 203). Specifically, in the 1960s the US government invested a large amount of money in developing various new social programmes in education, income maintenance, housing, and health (Dart, et al., 1998). These vast investments raised the issue and need of evaluating the outcomes and impact of the developed social programmes, which subsequently turned into an interest in evaluation in Social Sciences research (Robson, 2002, p. 203; Dart, et al., 1998). In other words, evaluation research has its root in the field of Social Sciences.

In terms of Information Systems (IS) research, evaluation is particularly important. In fact, and according to the International Data Corporation 2007 report (IDC, 2008), the global software market reached US$229,946 million in 2007. This figure clearly indicates the prevalence and heavy investments of IS in modern organizations. However, and despite this apparent success in the IS market, failure rates of IS implementation and exploitation have been continuously high (Chen, et al., 2011; Peng & Nunes, 2009; Lycett & Giaglis 2000). For example, and according to a recent Standish Group Chaos Report (Standish Group, 2009), 44% of IS projects were considered as challenged and 24% were identified as a complete failure in 2008. Giving the large investment and high failure rate of IS implementation, evaluation is now recognized as an increasingly important task that can directly contribute to IS success (Ammenwerth, et al., 2003; Lycett & Giaglis, 2000).

In particular, Lycett and Giaglis (2000) argue that evaluation is very useful in predicting and assessing potential costs, benefits and risks associated with the development, implementation and use of IS, as well as assisting decision makers to take proper actions to mitigate the identified risks. Moreover, other IS researchers reinforce that in order to inform decision making and increase the possibility of IS success, evaluation should be carried out at different phases throughout the entire system's lifecycle, from feasibility study, to system development, implementation, post-implementation and even system replacement (Willcocks & Lester, 1996; Smithson & Hirschheim, 1998; Seddon, et al., 2002).

However, and despite its importance in guaranteeing IS success, evaluation is never an easy and straightforward task (Cronholm & Goldkuhl, 2003a). In particular, there is a range of IS evaluation methodologies, each one having its own strengths and limitations. Moreover, different stages of the IS lifecycle are associated with different goals, changes and outcomes. As a result, the aims and focuses of evaluation at different stages will also vary. Faced with this diversity and complexity, practitioners and evaluators may often find it difficult to select which methodology is the most suitable one for evaluating a particular IS project or a particular stage of the project.

This chapter provides a comprehensive summary and an in-depth discussion on the various key methodologies that can be used in IS evaluation, namely in terms of the nature of the evaluation (summative vs. formative) and the strategy to be adopted in the evaluation (goal-based, goal-free, and criteria-based). A set of four practical case studies are also presented and discussed to illustrate our discussion. The chapter aims to provide rich insights and practical guidelines in helping practitioners and evaluators to choose and apply a suitable evaluation methodology in their IS development projects.

FORMATIVE AND SUMMATIVE EVALUATION IN IS RESEARCH

One of the most prevalent and fundamental classifications between types of evaluation was introduced by Scriven in 1967 as acknowledged by Clarke (1999, p. 7). In particular, Scriven (1967) used the terms 'formative' and 'summative' to describe the two distinct approaches being applied in the evaluation of educational curricula. Formative evaluation (also known as process or progress evaluation) refers to a particular type of evaluation activity that aims to acquire feedback during the process of development and implementation of the IS, in order to suggest ways of improvement and help in the development of the change, innovation or intervention (Clarke, 1999, p. 7; Robson, 2002, p. 208; Bennett, 2003, p. 10). On the other hand, summative evaluation (also known as outcome or impact evaluation) refers to a different type of evaluation that is carried out after the process of development and implementation is finished, and aims to gather information and feedback to assess the effects, effectiveness, impacts and outcomes of the developed IS (Clarke, 1999, p. 7; Bennett, 2003, p. 10). A further comparison on the key features and differences between formative and summative evaluation is provided in Table 1.

Since their emergence, the approaches of formative and summative evaluation have been continuously and widely used in many other fields, especially in IS evaluation (Hamilton & Chervany, 1981; Kumar, 1990; Kushniruk, et al., 1997; Cronholm & Goldkuhl, 2003a; Karoulis, et al., 2006). Specifically, and in light of the discussion above, formative evaluation is typically used throughout the IS design, development and implementation process, with the aim to provide systematic feedback and suggestions to system designers and implementers during the project (Hamilton & Chervany, 1981; Cronholm & Goldkuhl, 2003a). In contrast, summative evaluation is normally carried out at the end of the IS project

Table 1. A comparison of key differences between summative and formative evaluation

Dimensions	Formative	Summative
Target audience	Programme managers, practitioners	Decision-makers, funders or the public
Focus of data collection	Qualitative evidence to clarify aims, content and structure of the programme	Quantitative outcome measures
Role of evaluator	Two way interaction	Independent and one-way communication
Methodology	Heavy use of qualitative design	Experimental and quantitative design
Frequency of data collection	Continuous monitoring	Limited or one round of data collection
Reporting procedures	Informal via group discussion and meetings	Formal reports
Frequency of reporting	During the overall process of evaluation	After completion of evaluation

Source: adapted from Herman et al. (1987, p. 26) and Clarke (1999, pp. 8-10)

or at the post-implementation stage, in order to inform CEOs or managers about the quality, adequacy and impact of the implemented IS and the overall effectiveness and outcomes of the project (Hamilton & Chervany, 1981; Kumar, 1990; Cronholm & Goldkuhl, 2003a).

It clearly emerges from the above discussion that in order to improve the quality of the system and enhance the possibility of success, both formative and summative evaluation should be carried out in IS projects. Nonetheless, it should be highlighted that the selection of the use of either approach is related to actual stages of the IS lifecycle. That is, the use of formative or summative evaluation is closely related to when evaluation is conducted in the IS project.

GOAL-BASED, GOAL-FREE, AND CRITERIA-BASED EVALUATION IN IS RESEARCH

Although formative and summative approaches provide clear indication about when assessment should be carried out, these two methodologies do not contain sufficient guidelines on how evaluation can be done (e.g. what strategy to adopt in the evaluation? what methods to use? should any measurement criteria be set up prior to evaluation? If so, how can these criteria be set up, and more importantly, how can they be applied in the evaluation process?). In response to these limitations, this chapter proposes to use an alternative set of evaluation methodologies, as proposed by Cronholm and Goldkuhl (2003a), in conjunction with formative and summative approaches, namely goal-based evaluation, goal-free evaluation, and criteria-based evaluation.

Goal-Based Evaluation

Evaluation researchers traditionally believe that "a social welfare programme [or for that sense any programme…] cannot be evaluated without specifying some measureable goals" (Rossi & Williams, 1972, p. 18). Weiss (1972, p. 24) reinforces that "the goal must be clear so that the evaluator knows what to look for." The goal-based approach evaluation was first developed by Tyler (1942) as a deductive methodology, in which a set of clear, specific and measurable goals are derived from an organizational context prior to evaluation (Cronholm & Goldkuhl, 2003a; Patton, 2005). The evaluators will then need to measure to which extent these predefined goals are achieved in the program or intervention (Cronholm & Goldkuhl, 2003a).

Quantitative data collection methods are traditionally adopted in goal-based evaluation (Patton, 1990, p. 117; Cronholm & Goldkuhl, 2003a). Nonetheless, it has been extensively criticized (Hirschheim & Smithson, 1988; Cronholm & Goldkuhl, 2003a) since if only quantitative methods are used, goal-based evaluation often mainly focuses on technical and economical aspects, rather than on human and social dimensions. As a consequence, the result of the evaluation may over-emphasize on the quantitative value of the innovation (e.g. a newly implemented IS, but neglect important social, organisational, and human effects (Hirschheim & Smithson, 1988). Therefore, Cronholm and Goldkuhl (2003a) suggest that when quantitative methods can be used to assess hard measurable goals, qualitative methods should actually also be adopted in goal-based evaluation in order to examine goals of more social or human nature.

Goal-Free Evaluation

In contrast with the traditional goal-based approach, some researchers argue that evaluators may come up with more interesting and unbiased results by "undertaking fieldwork in a programme without knowing the goals of the programme or at least without designing the [evaluation] study with goal attainment as the primary focus" (Patton, 1990, pp. 115-116). A very similar line of thought has led Scriven (1972) to propose much earlier on an alternative evaluation methodology, namely goal-free evaluation.

Goal-free evaluation is an inductive methodology, which aims at gathering data on a large amount of actual effects and then assessing the importance of these effects in meeting demonstrated needs of the socio-technical environment in which the IS is to produce change or innovation (Scriven, 1972; Patton, 1990, p. 116; Cronholm & Goldkuhl, 2003a). Both quantitative and qualitative methods can be used in this evaluation approach (Patton, 1987, p. 36). Scriven (1972) and Patton (1990, p. 116) highlight a number of reasons and advantages for doing goal-free evaluation, such as avoiding the risk of narrowly studying the pre-specified goals and thus missing unanticipated aspects, eliminating evaluation biases introduced potentially by knowledge of goals, and maintaining evaluator objectivity and independence through goal-free conditions.

Criteria-Based Evaluation

Criteria-based evaluation means the evaluation is conducted according to predefined checklists, heuristics, or principles. These criteria mainly stem from some specific theories as well as sets of guidelines, standards, or even legal requirements. The selected criteria for evaluation indicate that evaluators emphasize and focus on certain characteristics more than others. Therefore, the criteria used for evaluation determine the types of outcomes that can be acquired (Cronholm & Goldkuhl, 2003a).

From reviewing previous studies, criteria-based evaluation emerges as one of the most frequently used evaluation approach in the field of IS, namely in usability, accessibility and standard verification studies. Usability usually refers to the assessment of how users react to and interact with the IS (Bertot, et al., 2006). On the other hand, accessibility evaluates how well systems allow users with disabilities to have equal or equivalent use of information and services (Jaeger, 2006). Standards emerge from national, international, and professional accrediting boards and are usually available in the form of very purposefully structured documents. Evaluation based on one or more of these three types of criteria are increasingly common in the IS field.

One acknowledged major disadvantage of this type of evaluation is that, since the focus is on criteria that aim at evaluating a specific perspective, it is conceivable that some important factors about the IS and its exploitation may be ignored (Cronholm & Goldkuhl, 2003a). Another potential controversy that often surrounds criteria-based evaluation is related to the fact that evaluators with different backgrounds, specializations or even knowledge may differ in opinion on the criteria. This makes the acceptance of the results of criteria-based evaluation more difficult (Jiang, 1996).

Summary of IS Evaluation Approaches

In summary, IS evaluation research processes may vary in the nature of the process, that is, evaluation may be formative or summative. This distinction results from a difference in the implementation of the evaluation in terms of the point in time in relation to the design and development cycle of the IS: formative during the process of design and development; summative at the end of this process. Nonetheless, each of these types of evaluation can in turn use different strategies, namely goal-free evaluation, goal-based evaluation and criteria-based evaluation depending on the motivation for evaluation. Therefore, this results in six basic types of evaluation methodologies: goal-free summative methodology, goal-free formative methodology, goal-based summative methodology, goal-based formative methodology, criteria-based summative methodology and criteria-based formative methodology.

DISCUSSION OF THE BASIC TYPES OF IS EVALUATION

As has mentioned before, the field of IS has recently experienced and unprecedented rapid development. Therefore, the study of IS socio-technical environments requires the consideration of increasingly complicated factors that need to be embedded in the evaluation processes. This section of the chapter aims to discuss the six basic types of evaluation methodologies mentioned above and presented in Table 2

We will discuss each of these methodologies in detail in the following sections by describing its basic structure and criticizing its use and applicability.

Goal-Free Formative Evaluation

This type of evaluation methodology emerges from combining goal-free evaluation and formative evaluation. It means the evaluation is undertaken without clear goals during the development of information systems. Theoretically, IS can be evaluated using goal-free formative evaluation methods, but this is actually seldom used in practice. When it is used, it is with an exploratory attitude in mind, that is, to detect, identify and explore the possibility of the occurrence of unpredicted events that may have an undesirable impact in the IS under development.

Usually, external evaluators are asked to become involved in goal-free formative evaluation in order to avoid internal evaluators biases, preconceived ideas and even acquired prejudices about the IS under development (Scriven, 1991). This type of evaluation can be performed using joint application design workshops, cognitive walkthroughs, prototyping, or even interpretive observation.

Joint Application Design (JAD) workshops involve users and technical developers that work together on a variety of activities related with system design, such as requirements definition, test specification design, and user interface design (Davidson, 1999). The traditional methods of interviewing individual users and writing text

Table 2. A comparison of key differences between summative and formative evaluation

Strategy / Nature	Goal-Based Evaluation	Goal-Free Evaluation	Criteria-Based Evaluation
Formative	Methods: • Joint Application Design Workshops • Cognitive Walkthroughs • Prototyping • Observation • Mixed method approaches • Etc.	Methods: • Joint Application Design Workshops • Cognitive Walkthroughs • Prototyping • Observation • Etc.	Methods: • Feature inspection • Consistency inspection • Standard inspection • Guideline checklist inspection • Cognitive walkthroughs • Heuristic evaluation • Eye tracking • Etc.
Summative	Methods: • Cognitive Walkthroughs • Formal Specification Testing • Observation • Mixed method approaches • Cost benefit analysis • Etc.	Methods: • Cognitive Walkthroughs • Observation • Semi-structured interviews • Focus Groups • Etc.	Methods: • Cognitive walkthroughs • Heuristic evaluation • Etc.

specification have been identified as less efficient in understanding user requirements in a complicated socio-technical environment (Martin, 1991, p. 156). Therefore, JAD workshops are generally carried out in the form of focus group discussion, which is facilitated by a well-trained session leader/moderator (Davidson, 1999). This approach aims to encourage user participation, expedite system development, and lead to better quality of specifications (Davidson, 1999).

Cognitive walkthroughs are traditionally used as usability evaluation methods, with special attention to how well the interface supports "exploratory learning," i.e., first-time use without formal training (Rieman, et al., 1995). It focuses on evaluating user interfaces of a system as to attribute ease of learning, particularly by exploration, i.e. guessing what to do using the signals provided by the system. Based on early propositions by Wharton et al. (1994), cognitive walkthrough simulates users performing navigation tasks on a website by assuming that users perform goal-driven exploration (Blackmon, 2002).

Prototyping is an iterative process of design and development that aims to evaluate the design of the IS through asking users' actually trying rather than evaluating according to description (Lycett & Giaglis, 2000). A prototype can serve as a communication vehicle that allows users to get a feeling about what the new IS would be like, as well as to review how users can interact with the system (Martin, 1991, p. 172). It is particular useful in exploring the functions and design of IS, especially when the detailed design of the system is not fully understood and developed (Martin, 1991, p. 172). However, applying this type of empirical testing is expensive in formative evaluation because it requires a number of users involved in the evaluation at different important points in development (Nielsen, 1994).

Observation is a naturalistic approach in which activities and interactions with systems are monitored or recorded, using audio or video (Kushniruk, 2002). In the case of goal free evaluation the observation process is designed to be as unobtrusive as possible, with little or no experimental control. Usually this type of evaluation is either performed by free roaming and use of the system or through the use of simulation, that is, the evaluation is conducted through observing the simulative operation by potential users (Lycett & Giaglis, 2000).

Goal-Free Summative Evaluation

This type of evaluation methodology is a result of the combination of goal-free evaluation and summative evaluation. By adopting this methodology, evaluation is undertaken without clear goals after the IS development process is finished. Summative evaluations are increasingly becoming goal-free in many social sciences project evaluations (Scriven, 1991). However, and similar to goal-free formative evaluation, a literature review of the IS field revealed that this type evaluation is scarcely applied.

Methods used in this type of evaluation may be very similar to the ones used in goal-free formative evaluation, that is, cognitive walkthroughs and observation. However, due to the summative nature of this approach methods such as interviews and focus groups are also often used.

Interviews can be defined in very general terms as "conversations with a purpose" (Dexter, 1970, p. 136). They are a very powerful tool for information gathering within organizational human activity environments (Warner, 1996, p. 183). Interviews used in goal free summative evaluation are usually semi-structured in nature, that is, the questions emerge from lists of themes to be covered in an open conversation, rather using the very closed structured interview script.

A focus group is, in fact, a group semi-structured interview in which a moderator keeps the direction of discussions under control by utilizing a predefined set of questions or script (McPherson

& Nunes, 2008). Focus groups are particularly useful in goal-free summative evaluation, since this method offers a unique and comprehensive form of discussion in which IS stakeholders could use the full range of their sensibilities, knowledge and experiences to discuss and negotiate the different understanding and aspects of the implemented IS (McPherson & Nunes, 2008).

Goal-Based Formative Evaluation

Goal-based evaluation and formative evaluation are combined in this type. The aim of goal-based evaluation is to investigate whether the project has achieved its goals, this means the evaluation is carried out to assess if specific and pre-established business goals are achieved during the development of the IS. When carrying out the information systems design, the functionality of the information systems results from these business goals. These goals are expressed in terms of organizational goal descriptions, requirement specifications and IT specifications. This process of goal-based business modeling is a very important and complex one (Kueng, 1996) and is necessary before engaging with goal-based evaluation. Therefore, this goal-based evaluation is a deductive research approach as discussed above.

In practice, during the development process, a number of information systems development techniques are used to ensure the match of software functionality and business goals. Thus, goal-based formative evaluation is mainly used during the design and development of organizational IS and it provides a crucial contribution to ensure quality, usefulness, and acceptance of the IS. This type of evaluation is often connotated with IT- and SW-centred evaluation processes.

The most useful goal-based formative evaluation methods are prototyping and simulation (Lycett & Giaglis, 2000). However, the above-mentioned joint application design workshops, observation and interviews are also often used. Cronholm and Goldkuhl (2003b) propose that some of these methods can actually be combined in order to perform the evaluation. By way of an example, observation and interviews could be used, where the evaluator observes users actions and contrasts these with their perceptions in order to evaluate if goals and business actions defined in requirement specifications have been attained. Monk et al. (1993) propose that the "think aloud" observation method would be particularly adequate in this case.

Goal-Based Summative Evaluation

This type of evaluation results from the combination of goal-based evaluation and summative evaluation. This means the main aim of the evaluation is to assess if the implemented IS fulfils the business goals. Apart from evaluating the attainment of business goals and systems requirements, this type of evaluation is often also used to assess the costs and benefits of implementing the IS in order to assist decision making. Irani (2008) summarizes, that in this case, the costs and benefits should be considered including financial and non-financial measures as well as tangible and intangible factors:

- **Financial Measures:** Evaluations with financial measures are carried out in terms of cost-benefit assessment based on the traditional capital investment measure analysis.
- **Non-Financial Measures:** Information systems investments contribution can also evaluated in non-financial aspects. It is indicated that decision-makers should consider non-financial costs and benefits of information system implementation along with the rapid development of information systems. Not only the information technology, but also the interaction between users and information systems should be considered in evaluation such as the opinions from the users.

- **Tangibles:** The tangible performance measures are usually from operation or tactical levels of information systems such as sales in a period, cycle producing time and so on.
- **Intangibles:** When evaluating organizational information systems, the intangible measures such as the reputation of the company, the technological factors are also need to be considered.

Lycett and Giaglis (2000) mention various methods for evaluating different types of costs and benefits. The classical financial methods include Net Present Values (NPV), Return On Investment (ROI), and Internal Rate of Return (IRR). These methods are widely used for investment evaluation through comparing the estimations of cash flow costs and benefits, however, very difficult to perform in the IS field. However, some cost benefit analysis methods have been developed bearing IS specifically in mind, such the index "return on management."

In any case, goal-based summative evaluation methodology mainly depends on the characteristics of the stated business goals (Cronholm & Goldkuhl, 2003a). Irani (2008) provides a characterization of the main types of business goals for the implementation of IS as follows:

- **Strategic Significance:** In this aspect, research mainly concerns the evaluation of the contribution of information systems to business strategy in organizations. For example, Wagner (2002) assesses the support of strategic information systems to organization performance and also indicates the new requirements for strategic information systems.
- **Tactical Impact:** IS are evaluated at a tactical level in organizations to help their selection or implementation at tactical and decision-making levels. This perspective of evaluations has a positive impact on innovative changes.
- **Operational Consideration:** Evaluations may be carried out on the IS itself from a technological point of view or on user behavior and exploitation of the IS from an operational perspective. The assessments emphasize on how the information systems are implemented or the contribution of the operational information systems in functional areas, such as sales or human resource areas.

Criteria-Based Formative Evaluation

This type of evaluation is the combination of criteria-based evaluation and a formative approach. After reviewing the previous studies, the main criteria-based formative evaluation approaches are usability, accessibility, and standard verification studies. The criteria standards for evaluation stem from the theories as well as precise guidelines or standards.

Usually, this type of evaluation is better performed by expert evaluators, who in are much more efficient than users with less experience. Experts are much more adept at assessing possibilities, judging problems and proposing solutions (Karoulis, 2006). Moreover, experts in usability, accessibility and specific standards are bound to improve acceptance and quality assurance of the development process. Therefore, rapid and efficient interventions by experts in a formative stage are ideal.

Another advantage of expert-based evaluation is that it can be applied in very early stages of the systems design life cycle. Experts are able to evaluate the systems that are being constructed even if only very basic prototypes are available. Methods used in criteria based formative evaluation include feature inspection, consistency inspection, standard inspection, and guideline checklist inspection. All of these are usually performed against very detailed and precisely stated documented criteria using methods such cognitive walkthroughs, heuristic evaluation, or eye tracking.

Criteria-Based Summative Evaluation

This type of evaluation research combines criteria-based principles with a summative approach. Similarly with the previous summative approaches described it is usually carried out after the development of the IS is completed. Similarly to its formative counter-part, this type of criteria-based evaluation also focuses on usability, accessibility and standard verification studies. The summative nature however, gives it a very different character. This type of evaluation usually aims at certification with accrediting bodies, acceptance testing and quality assurance. It is an exercise also mostly undertaken by experts, but with a much less constructive purpose than in the formative stages of the IS design and development. Methods used here are usually methods such cognitive walkthroughs and heuristic evaluation.

CASE STUDIES OF APPLYING IS EVALUATION METHODOLOGIES

This section aims to present and discuss a series of case studies as examples to illustrate how the various IS evaluation methodologies can be applied in practices. However, an extensive literature review indicates that while goal-based and criteria-based approaches have been widely used in practices, examples of goal-free approach are rarely found in the literature. In fact, ever since its emergence, the goal-free approach has received many criticisms from social sciences researchers. When this approach sounds interesting in theoretical terms, researchers and practitioners generally criticize it to be difficult to work in practice (Scriven, 1991). Therefore, although Scriven (1991) attempted to defend his approach, goal-free evaluation is still seldom used in practice. On the other hand, cases studies for using goal-based and criteria-based approaches have been frequently reported in the literature. These two types of evaluation approaches prove to be particularly prevalent in current IS evaluation practice. Therefore, a set of four case studies adopted these evaluation approaches were identified and selected to be discussed in detail in this section:

- Case study 1 adopted goal-based formative evaluation.
- Case study 2 used goal-based summative evaluation.
- Case study 3 used criteria-based formative evaluation.
- Case study 4 adopted criteria-based summative evaluation.

By discussing these case studies, we aim to provide more specific and useful practical guidelines to help IS researchers and practitioners to carry out IS evaluation more effectively.

Case Study 1: Goal-Based Formative Evaluation

Context of the Case Study

This first case study was summarized from the paper published by Kushniruk and Patel (2004). By adopting a goal-based formative approach, Kushniruk and Patel evaluated a clinical information system which was being developed for the Columbia Presbyterian Hospital and was called Doctor's Outpatient Practice (DOP) System. The DOP system was developed to support the recording of patient information, such as problems, allergies, adverse reactions, and current medication of patients. The usability of a specially designed interface for users to enter patient's problems was tested in the case study.

Aims of the Evaluation

In light of the above discussion, the aim of the evaluation performed in this case study was to assess the usability of the user interface design of the DOP system. As a goal-based evaluation, the evaluators mainly aimed to assess whether users could achieve the goal of entering patent problems into the DOP system effectively and easily. The evaluators also attempted to identify any possible errors that might exist in the interface design of the system and could have negative influences on the use and operation of the system. In order to achieve these objectives, cognitive walkthrough was applied in this case study as the technique to evaluate the system during its development life cycle, and thus give formative feedback to the IS development team in order to improve the system further.

Phases of the Evaluation

It should be noted that the evaluation of the DOP system by using the cognitive walkthrough technique did not involve end users (that is, the entire walkthrough process was performed by the evaluators/analysts). It was carried out in five phases.

Before conducting the actual walkthrough evaluation, the evaluators first of all attempted to define and identify the potential end users of the system. During this process, the evaluators also developed a good understanding on the users' background and prior experience that might have impact on their interactions with the systems. This can help the evaluators to better predict and understand what potential problems end users may have when using the system.

The second phase was to define the tasks around which the walkthrough evaluation would be conducted. That is, evaluators need to identify the functions or the components of the systems that are going to be assessed with walkthrough. Subsequently, evaluators need to design a set of tasks associated with these specific functions/components of the system. In this case study, the evaluators designed a task for entering patient's information (e.g. problems, allergies, or medications) in the tested DOP system. As a goal-based evaluation, this designed task is directly related to the goal that needs to be tested (i.e. whether the DOP system would allow users to enter patent problems effectively and easily).

Subsequently, and in the third phase, a group of analysts wearing "the users' shoes" performed the designed task by using the DOP system. During this process, video recording equipments were also used to record the entire test. By reviewing these video records, evaluators were able to make a detailed analysis and examination of each action and step undertaken to complete the task by using the system under evaluation. By following guidelines given by Polson et al. (1992), the evaluators particularly attempted to identify the following aspects during the data analysis:

- The goal and subgoals that are involved in the tasks (e.g. the goal in this case study was to enter a patent's problem into the DOP system).
- The users' actions that need to be taken in order to achieve the goals (e.g. one of the actions in this case study was to click on the system button labeled "Add New Problem").
- The behavior of the system in response to the users' actions (e.g. after clicking on the "Add New Problem," a "Keyword Search Window" would appear for the user to enter the problem).
- Potential problems that users may experience, given considerations about their background and prior experience as identified from Phase 1 (e.g. when the "Keyword Search Window" shows up, users may not realize that they now need to enter a term in the search window).

In phase four, the evaluators summarized the walkthrough results. The data and results derived from the walkthrough evaluation can generally provide 1) a measure of the number of actions that need to be undertaken by actual users to complete a specific task, and 2) a list of potential problems that end users may experience when they use the system to perform the same task. This information can be very useful for improving the design of the system, especially when the IS team needs to compare the complexity of two alternative system designs for carrying out the same task (e.g. entering patent data). Therefore, and in phase five, a set of recommendations about changing the system design or selecting one of the few alternative system designs could be proposed based on the results of the walkthrough evaluation.

Discussion

The above formative walkthrough evaluation can provide useful insights into problems that may be encountered by actual users and thus propose possible recommendations about changes. However, Kushniruk and Patel (2004) stress that it cannot be used to replace an end user testing of the system. We agree with this comment, and think that a further system usability test that involves actual users in the system development stage will allow the IS team to further improve the system to satisfy user needs. Nevertheless, a well-conducted formative walkthrough evaluation can shed light on a later usability testing (Kushniruk & Patel, 2004), and can even provide useful insights to carry out summative IS evaluation at the end of the project. Furthermore, it should be pointed out that the technique of cognitive walkthrough can be used in not just a goal-based evaluation strategy. In fact, cognitive walkthrough can also be applied in other evaluation strategies, e.g. in criteria-based evaluation, as discussed further in section 5.4.

Case Study 2: Goal-Based Summative Evaluation

Context of the Case Study

The case study is summarized from the paper published by Lagsten and Goldkuhl (2008). Based on their previous research and practical experience, Lagsten and Goldkuhl developed an interpretive IS evaluation methodology, namely VISU (this is a Swedish acronym for 'IS evaluation for work-practice development'). In this case study, the authors adopted their VISU approach to carry out goal-based evaluation on an information system that supports social welfare services in Sweden, namely Procapita.

Aims of the Evaluation

At the time of the study, the Procapita system had been used for almost 10 years by more than 150 municipalities in Sweden. In the local municipality where this evaluation was carried out, the system was used by more than 350 social workers to conduct their daily work. This evaluation aimed to investigate if this Procapita system still satisfied the requirements of the users and the organization as a whole, and if not, what post-implementation improvements should be needed, or whether Procapita should be replaced by a more advanced system.

Phases of the Evaluation

By following the VISU approach, the interpretive goal-based evaluation conducted in this study consisted of four main phases.

At the Initiation phase, the evaluators worked together with the IS manager and IS operations manager of the municipality to decide the preconditions of the evaluation. As an important outcome of this phase, a precondition document was developed to highlight the evaluation objectives and aims, the problems that should be resolved through the evaluation, and the evaluation methods to be used. A general goal (i.e. whether or not the system still satisfied the organizational needs) to be measured in the evaluation was also identified at this stage.

At the Arrangement phase, the evaluators tried to identify and select suitable stakeholders to be involved in the evaluation. In order to do so, the evaluators first of all needed to develop a sound knowledge of the services supported by the system, by participating in regular internal meetings and reading central documents. A theory-based model for the case handling process was also created in order for the evaluators to better understand current work practices in the municipality. After acquiring a better understanding of the organization and the people inside, the evaluators were able to select a group of stakeholders, who had crucial interests and knowledge on the system, to participate in the evaluation. In order to generate more meaningful outcomes, as well as to facilitate personnel arrangement during the evaluation, an evaluation board that composed by the IS manager, the IS operations manager and maintenance people, was also set up in this phase.

In the Evaluation phase, a set of dialogue-seminars was conducted. As defined by Lagsten and Goldkuhl (2008), a dialogue-seminar is similar to a focus group, in which a specific and predefined set of questions are addressed, discussed, and examined by a group of stakeholders. In this case study, 8 groups of stakeholders were selected (including 5 groups of users, 1 group of units managers, 1 group of maintenance staff, and 1 group of IS managers). Each group consisted of 3-7 individuals, and was involved in two 2-hour seminars. Consequently, a total of 16 dialogue-seminars were carried out. By bearing in mind the general evaluation goal (i.e. if the system still satisfied the organizational needs) as set up above, the evaluators particularly focused on the following four questions in the seminars:

- "What do you do while using Procapita?"
- "What problems do you perceive?"
- "What good does the system do for you?"
- "What are the goals you try to achieve?"

Data collected from each seminar was transcribed into a working report that captured stakeholders' answers and concerns into four categories, namely "activities, problems, strengths, and goals." Moreover, the working report was also sent to each participant for further refinements and validation. In paralleled with the seminars, the reports were also sent to the evaluation board, who made initial analysis and interpretations on the results. These seminars resulted in the collection of a large amount of valuable statements on, e.g. problems (400), strengths (50) and goals (70). Finally, and after all the seminars, the collected data were analyzed by the evaluators by using a Grounded Theory approach. An account was written for each stakeholder group. By joining the different accounts, a final evaluation report was written and was then used by the IS manager as the basis to draw a detailed action plan to improve the Procapita system.

Discussion

As mentioned above, the goal-based evaluation discussed in this case study focused on one main goal (i.e. whether the Procapita system that was implemented 10 years ago still satisfied current user needs). Although only one goal was measured, the interpretive evaluation approach resulted in a very in-depth study that covered the perspectives of multi-stakeholder groups. Such rich set of evidence is crucial for user organization to make strategically important IT decisions (e.g. replace the legacy system). Moreover, there is a prerequisite for using such an interpretive evaluation approach. That is, the stakeholders/users involved in the study need to have in-depth insights and prior knowledge on the specific aspects/issues being evaluated. In order for stakeholder/users to give rich insights to the evaluation, evaluators need to keep this principle in mind when designing either a formative or summative study by using this interpretive evaluation approach.

Case Study 3: Criteria-Based Formative Evaluation

Context of the Case Study

This case study was also summarised from the paper published by Kushniruk and Patel (2004). As such, it has a similar context to the first case study presented in section 5.1. In this case study, Kushniruk and Patel aimed to evaluate the same clinical information system, i.e. DOP, developed for the Columbia Presbyterian Hospital. However, rather than using goal-based evaluation, the criteria-based formative evaluation strategy based on a heuristic method was employed here.

Aims of the Evaluation

The heuristic method was originally developed by Nielsen (1993). By using the heuristic method, this evaluation aimed to inspect the usability of the DOP system based on a set of well-tested system design criteria (such as, ease of use, system flexibility, visibility of system status, efficient user control, and content consistency, etc.). The results of this formative evaluation could then be used to inform further improvements on the design of the system and/or the user interface during the development life cycle.

Phases of the Heuristic Evaluation

By following guidelines given by Nielsen (1993), this heuristic evaluation was carried out through a few stages. Firstly, the evaluators were given a list of heuristics/criteria for the system or the user interfaces that were to be tested. It should be note that the list of heuristics applied in this case study was initially proposed by Nielsen (1993), and has then been widely used, tested and refined in the IS field (Kushniruk & Patel, 2004):

- **Heuristic 1:** "Visibility of system status." This heuristic refers to whether or not the user is clearly informed about the status of the system at any given point of time (e.g. a sign should be shown to users when the system is processing a user request for data retrieval, or when a system operation was successfully completed).
- **Heuristic 2:** "Match the system to the real world." This heuristic involves assessing the system in two aspects. First, the system should be designed using the "natural" language of the users (e.g. no jargon or technical terms). Second, the system should be designed by using as many real-world conventions as possible (e.g. having a "rewind" button in the system to indicate backwards navigation can map to the physical rewind button on a common VCR recorder).
- **Heuristic 3:** "User control and freedom." This heuristic indicates the requirement about whether or not users feel they are in control when operating the system (e.g. the system should provide a way for users to back out of current tasks, as well as to carry out undo and redo actions).
- **Heuristic 4:** "Consistency and standards." This principle means that the user interface and the system operations should be designed consistently with the same standards (e.g. basic operational buttons, such as "menus," "open," and "close," should be displayed in the same positions on the screen throughout the system).
- **Heuristic 5:** "Error prevention." This criterion requires that the interface should be designed to prohibit errors from happen-

ing (e.g. the screen should be simplified to avoid any complicated modes that will confuse the users).

- **Heuristic 6:** "Minimize memory load-support recognition rather than recall." This heuristic indicates that the design of user interface should support "recognition rather than recall" (e.g. operational buttons of the system can be easily recognized, and so users do not need to memorize a lot of information in order to carry out work on the system).
- **Heuristic 7:** "Flexibility and efficiency of use." The designers should make the user interfaces customizable and flexible to the users (e.g. users can set up and customize the display of screens based on their preferences).
- **Heuristic 8:** "Aesthetic and minimalist design." This heuristic indicates that in order to make the system easier for user to apply, the options should be designed in the simplest and minimal way.
- **Heuristic 9:** "Help users recognize, diagnose, and recover from errors." When users make a mistake, the systems should be able to alert users about their errors and provide easy-understanding information for users to recover from the error.
- **Heuristic 10:** "Help and documentation." This principle involves that a "help" function or document should be available on the system at any time when users need it.

By adopting this comprehensive list of heuristics, a number of usability experts/analysts inspected the DOP system and its user interface thoroughly. Each analyst evaluated the system independently and noted any heuristic violations that were found in the system. These individual lists of heuristic violations were then compiled into a single list. Finally, the results of the evaluation were summarized and categorized (e.g. according to the number of types of violations), and then presented to the IS team together with formative recommendations for further system improvement.

Discussion

By comparing this case study with the above two examples of goal-based evaluation (especially with case study 1), it becomes clear that when goal-based evaluations typically focus on a limited set of goals that are of strategic importance to the user organization, criteria-based evaluations seem to cover a wider range of aspects that are closely related to the design of the system. This criteria-based approach thus seems to be able to provide a more comprehensive and helicopter view on the design of the user interface of the system. Kushniruk and Patel (2004) thus concluded that results of this type of formative evaluation can often allow the IS team to dramatically improve the usability of the system under development.

Case Study 4: Criteria-Based Summative Evaluation

Context of Case Study

This fourth case study was summarised from the paper published by Aleixo et al. (2012). In Portugal, an e-government strategy had been developed according to a more general Information Society policy established in the country in Sep 2009. As a result of the implementation of this national strategy, e-government websites have been established and used by most (if not all) local municipalities across the country. However, and owing to a lack of digital competence and information literacy skills, a considerable amount of Portuguese citizens are less capable of using the Internet and websites. This phenomenon is referred to as 'digital exclusion.' As a result of digital exclusion, these citizens are currently neither able to participate in nor enjoy the benefits provided by e-government services. It is clear that improving the usability of e-government websites is one of the efficient ways to address such digital exclusion. Therefore, in order to recommend necessary usability changes, this case study carried out a cognitive walkthrough evaluation on the e-government websites that have been designed and used by local municipalities in Portugal.

Aims of the Cognitive Walkthrough

In light of the above discussion, the aim of the evaluation performed in this case study was to explore the relationship between usability and digital literacy with the main objective of proposed ways to improve the usability of e-government websites and thus bridge the digital divide. Moreover, the cognitive walkthrough was performed in 28 selected local government websites, which represent a significant sample of the Portuguese national scene. However, it should be pointed out that this cognitive walkthrough was performed by following a criteria-based approach, which is very different from the goal-based cognitive walkthrough presented in section 5.1. The detailed phases involved in this criteria-based evaluation are discussed below.

Phases of Evaluation

The evaluation of the 28 Portuguese local government websites by using the cognitive walkthrough technique was carried out in three phases.

At the Preparatory phase, the evaluators focused on three aspects. First of all, they identified a full profile of the users. In this case, the users were Portuguese citizens who were expected to have lower levels of digital literacy skills but need to seek information or use services on the website of Portuguese municipalities. Secondly, the evaluators defined and designed a set of seven tasks (such as, register on the website, and make an online complaint, etc.) to be performed in the walkthrough. In order for the cognitive walkthrough to be carried out effectively, Aleixo et al. (2012) stressed that the number of designed tasks needs to be limited but these tasks should be representative. Thirdly, the evaluators defined and devised a detailed list of actions (e.g. find access to the registration page, fill out the registration form, and register) that are needed to complete each of the above high-level tasks (e.g. register on the website).

At the Analysis phase, the evaluators/analysts executed the tasks designed in the previous phase. It should be noted that before conducting the walkthrough, a comprehensive list of 53 usability guidelines was established and served as the criteria to evaluate the usability of the selected e-government websites. Moreover, this list of usability guidelines was also linked with a list of digital competences. The execution of the designed tasks would start from the homepage of the municipality website. During the walkthrough, the evaluators verified the compliance of the website interfaces with the identified usability guidelines/criteria. Moreover, they also attempted to explore whether the usability design of the website minimized the need for a certain digital competence (e.g. competence in locating and accessing information) of the users. Subsequently, the results of the analysis were expressed in terms of a five-point Likert scale, ranging from 1 = “very poor” to 5 = “very good.”

The Final phase involved the interpretation of results. In particular, the collected data of each municipality were analysed and interpreted at individual level, and were then combined together in order to gain a holistic profile. The evaluation results showed that there was a poor consideration of the digital inclusion issues and guidelines in designing local e-government sites in Portugal. Based on these findings, the evaluators proposed a set of recommendations for improving the usability of these websites.

Discussion

Although the same technique (i.e. cognitive walkthrough) was employed in this case study as well as in case study 1, the design of the evaluation varies essentially between a criteria-based study and a goal-based one. In particular, and as discussed above, goal-based evaluation typically focuses on a limited set of goals, which are aligned with specific business objectives and are thus important for the system to satisfy. As shown in the above example, when adopting an interpretive goal-based approach, evaluators can often assess and explore the system's capacity in achieving each goal in a very in-depth level. However, the scope of the evaluation will also be restricted by the limited set of goals. On the other hand, the design of a criteria-based evaluation generally involves the use of a more extensive list of heuristics and principles that aim at covering all key aspects of system usability. Evaluators can then generate a very holistic and helicopter picture on the usability and/or interface design of the system. However, when focusing on such a wide range of criteria, the evaluator may often not be able to study each heuristic in a very in-depth level. In fact, and in order to generate both comprehensive and in-depth results, a hybrid approach that mixes the use of various evaluation methods will always be applied by evaluators in practices, as further discussed below.

FURTHER PRACTICAL GUIDANCE FOR CONDUCTING IS EVALUATION

A set of IS evaluation approaches and methodologies have been discussed in this chapter, with support of evidence from four practical case studies. By drawn on our discussion above, this section provides a further set of practical guidelines in order to help IS researchers and practitioners to design and implement IS evaluation studies more effectively.

Specifically, given the existence of a wide range of IS evaluation strategies, one of the first important tasks when initiating an IS evaluation study, is to identify and select which particular technique and methodology should be adopted in the evaluation. As discussed above, a formative evaluation is particularly useful to generate systematic feedback and recommendation to improve the design of the user interface and the system during the development and implementation stage. In contrast, a summative evaluation can be used at the end of the project or at the post-implementation to inspect the overall quality, efficiency, and adequacy of the implemented IS. Moreover, and as illustrated in the above case studies, a goal-based evaluation can generally produce a rich set of insights and evidence to measure a limited set of important goals that the system should achieve. In contrast, a criteria-based evaluation, which often involves an extensive list of heuristics and principles, is more efficient in providing a helicopter view of the usability of the system. Overall, it becomes apparent that each type of IS evaluation methodologies discussed in this chapter indeed has its own value and aims to achieve different objectives and purposes. Therefore, IS researchers and practitioners should not simply conclude that one methodology is more advanced than the other. Instead, any decisions on selecting IS evaluation methodologies should be made according to the actual context and needs of the project, the nature of the IS under investigation, and the aims and objectives of the evaluation.

Moreover, it clearly emerges from the above case studies that a particular IS evaluation technique is not necessarily bound with a specific strategy/methodology. For instance, when cognitive walkthrough was used in the goal-based evaluation in case study 1, the same technique was also employed in the criteria-based evaluation in case study 4. These examples thus demonstrate

Figure 1. The use of evaluation results (modified from Lagsten and Goldkuhl, 2008)

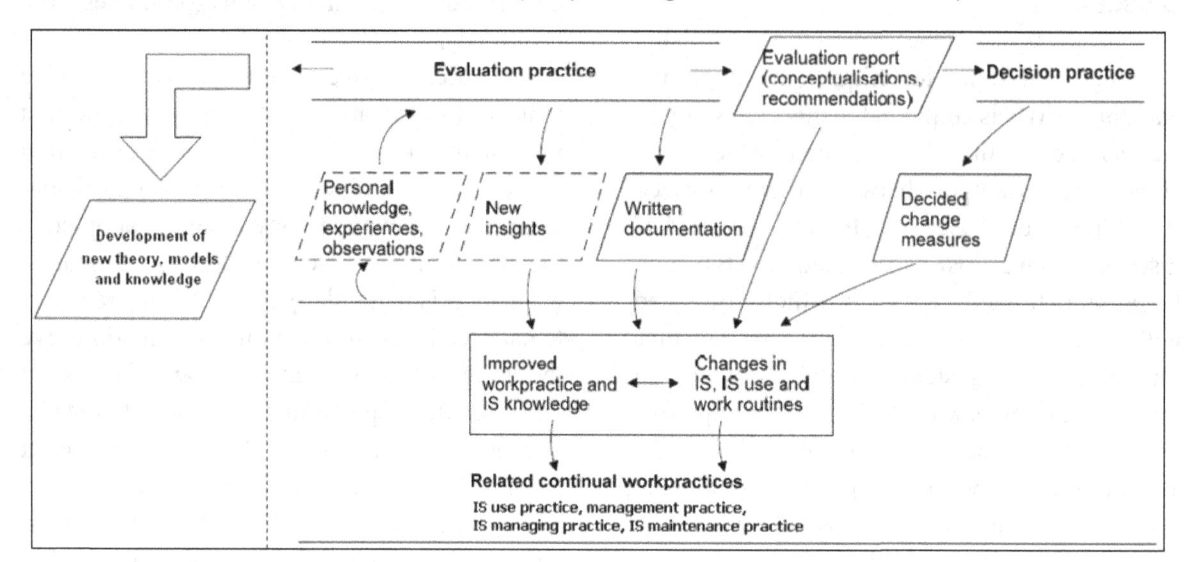

that, when the general rules embedded in the evaluation techniques are earnestly followed, IS researchers and practitioners can potentially apply the same techniques in different evaluation strategies as relevant.

Furthermore, for any given IS projects, multiple IS evaluation techniques and methodologies should be considered to employ. For instance, in the evaluation of the DOP system (as discussed in case study 1 and 3), the evaluators applied a cognitive walkthrough and a heuristic approach to generate formative feedback for system designers. Kushniruk and Patel (2004) mentioned that further summative evaluation would also be carried out on the completed system. We consider that such a hybrid evaluation approach will be very useful and beneficial for helping the IS team to iteratively improve the system across different stages of its life cycle, and should therefore be used more frequently in practice.

Finally, researchers and practitioners should consider further how the IS evaluation results can be used. It is obvious that results of IS evaluation can always be used to produce reports and recommendations to change and improve the system. Moreover, through a very elaborately developed model (Figure 1), Lagsten and Goldkuhl (2008) explained that the IS evaluation process, which is a temporary practice, is related to the continual practices (e.g. IS use practice, management practice, IS management practice, and IS maintenance practice) that already exist in the organization.

As shown in the diagram, when stakeholders from different ordinary work practices bring in their personal knowledge and experiences to the IS evaluation, they may also generate new insights about the system and their workplace during the evaluation process. These new insights may then be turned into changed behavior and changed routines in their ordinary work practices. Consequently, the results of IS evaluation will not just be used to change the system, but may also lead to improvements in individual attitudes and behavior as well as in organizational processes and management mechanism. Furthermore, one key aspect that was missing in Lagsten and Goldkuhl's original model was that, the results of IS evaluation will not just have practical values, but can also contribute to research by generating new theories, models and knowledge (as shown in the modified diagram above). For instance, in the above case study 4, Aleixo et al. (2012) proposed an extended model to investigate information literacy through their evaluation study on e-government websites. IS evaluation studies can also lead to the development of new approaches and strategies that can be applied in the evaluation process (e.g. Lagsten & Goldkuhl, 2008).

CONCLUSION

This chapter provides an overview of the main IS evaluation research methodologies. As the result of an extensive literature review and a survey of studies, the chapter proposes a classification of these methods divided by their nature—formative evaluation and summative—and the strategy followed—goal-free evaluation, goal-based evaluation, and criteria-based evaluation. Consequently, this resulted in the six-type classification discussed above. The chapter also identified potential methods to be used with each of the types of evaluation. A set of four case studies were also presented and discussed in the chapter, aimed at providing the reader with the knowledge and guidelines on how to apply the different types of evaluation and methods in practice.

Finally, the chapter draws several important conclusions. Firstly, both goal-free formative and goal-free summative evaluations are seldom used in IS. Although the researchers who support goal-free evaluation emphasize the capacity to identify unexpected opportunities, impacts and negative effects in the IS under evaluation, goal-free evaluations are still very difficult to implement in practice. Secondly, both goal-based and criteria-based formative evaluation focus on providing constructive feedback that helps in the design and development of the IS. This type of formative evaluation results in IS that are more usable, efficient and compatible with the socio-technical environment where they are to be implemented. Thirdly, a large component of current IS research evaluation focus on criteria-based evaluation. This type of methodology aims at enforcing standards and quality assurance and seems to have grown in importance with current trends and needs for standard and guideline compliance by organizational IS. Nevertheless, and despite these facts, we would suggest that IS researchers and practitioners should not be biased towards certain types of evaluation approaches and methodologies. In particular, IS evaluation techniques and methodologies should be selected and used flexibly depending on the nature of the IS and the specific goals and objectives of the evaluation. We would also suggest that the combines used of one or more of these approaches can be applied at different stages of the IS life cycle in order to generate more rigorous and reliable evaluation outcomes.

REFERENCES

Aleixo, C., Nunes, M., & Isaias, P. (2012). Usability and digital inclusion: Standard and guidelines. *International Journal of Public Administration*, *35*, 221–239. doi:10.1080/01900692.2011.646568

Ammenwerth, E., Graber, S., Herrman, G., Burke, T., & Konig, J. (2003). Evaluation of health information systems-problems and challenges. *International Journal of Medical Informatics*, *71*, 125–135. doi:10.1016/S1386-5056(03)00131-X

Bennett, J. (2003). *Evaluation method in research*. New York, NY: Continuum.

Bertot, J. C., Snead, J. T., Jaeger, P. T., & McClure, C. R. (2006). Functionality, usability, and accessibility: Iterative user-centered evaluation strategies for digital libraries. *Performance Measurement and Metrics*, *7*(1), 17–28. doi:10.1108/14678040610654828

Blackmon, M. H., Polson, P. G., Kitajima, M., & Lewis, C. (2002). Cognitive walkthrough for the web. In *Proceedings of the CHI 2002 Conference on Human Factors in Computing Systems,* (pp. 463-470). ACM Press.

Chen, H., Nunes, M. B., Zhou, L., & Peng, G. (2011). Expanding the concept of requirements traceability: The role of electronic records management in gathering evidence of crucial communications and negotiations. *Aslib Proceedings, 63*(2/3), 168–187. doi:10.1108/00012531111135646

Clarke, A. (1999). *Evaluation research: An introduction to principles, methods and practice*. Thousand Oaks, CA: Sage Publications.

Cronholm, S., & Goldkuhl, G. (2003a). Strategies for information systems evaluation: Six generic types. *Electronic Journal of Information Systems Evaluation, 6*(2), 65–74.

Cronholm, S., & Goldkuhl, G. (2003b). Strategies for information systems evaluation: Six generic types. In *Proceedings of the Tenth European Conference on Information Technology Evaluation,* (pp. 65-74). Madrid, Spain: IEEE.

Dart, J., Petheram, R. J., & Straw, W. (1998). *Review of evaluation in agricultural extension*. Retrieved from https://rirdc.infoservices.com.au/downloads/98-136.pdf

Davidson, E. J. (1999). Joint application design (JAD) in practice. *Journal of Systems and Software, 45*, 215–223. doi:10.1016/S0164-1212(98)10080-8

Dexter, L. A. (1970). *Elite and specialized interviewing*. London, UK: Colchester.

Guba, E. G., & Lincoln, Y. S. (1981). *Effective evaluation*. San Francisco, CA: Jossey-Bass.

Hamilton, S., & Chervany, N. L. (1981). Evaluating information system effectiveness-Part 1: Comparing evaluation approaches. *Management Information Systems Quarterly, 5*(3), 55–69. doi:10.2307/249291

Herman, J. L., Morris, L. L., & Fitz-Gibbon, C. T. (1987). *Evaluator's handbook*. Newbury Park, CA: Sage.

Hirschheim, R., & Smithson, S. (1988). *A critical analysis of information systems evaluation*. Amsterdam, The Netherlands: North-Holland.

Irani, Z., & Love, P. (Eds.). (2008). *Evaluating information systems: Public and private sector*. Oxford, UK: Butterworth-Heinemann.

Jaeger, P. T. (2006). Assessing Section 508 compliance on federal e-government web sites: A multi-method, user-centered evaluation of accessibility for persons with disabilities. *Government Information Quarterly, 23*(2), 169–190. doi:10.1016/j.giq.2006.03.002

Jiang, J. J., & Klein, G. (1996). User perceptions of evaluation criteria for three system types. *The Data Base for Advances in Information Systems, 27*(3), 63–69. doi:10.1145/264417.264435

Karoulis, A., Demetriadis, S., & Pombortsis, A. (2006). Comparison of expert-based and empirical evaluation methodologies in the case of a CBL environment: The "Orestis" experience. *Computers & Education, 47*(2), 172–185. doi:10.1016/j.compedu.2004.09.002

Kueng, P., & Kawalek, P. (1996). Goal-based business process models: Creation and evaluation. *Business Process Management Journal, 3*(1), 17–28. doi:10.1108/14637159710161567

Kumar, K. (1990). Post implementation evaluation of computer-based information systems: Current practices, computing practices. *Communications of the ACM, 33*(2), 203–212. doi:10.1145/75577.75585

Kushiniruk, A. (2002). Evaluation in the design of health information systems: application of approaches emerging from usability engineering. *Computers in Biology and Medicine*, *32*, 141–149. doi:10.1016/S0010-4825(02)00011-2

Kushniruk, A. W., & Patel, V. L. (2004). Cognitive and usability engineering methods for evaluation of clinical information systems. *Journal of Biomedical Informatics*, *37*, 56–76. doi:10.1016/j.jbi.2004.01.003

Kushniruk, A. W., Patel, V. L., & Cimiho, J. J. (1997). Usability testing in medical informatics: Cognitive approaches to evaluation of information systems and user interfaces. In *Proceedings of AMIA Annual Fall Symposium*. Montreal, Canada: AMIA.

Lagsten, J., & Goldkuhl, G. (2008). Interpretative IS evaluation: Results and uses. *The Electronic Journal Information Systems Evaluation*, *11*(2), 97–108.

Lycett, M., & Gialis, G. M. (2000). Component-based information systems: Toward a framework for evaluation. In *Proceedings of the 33rd Hawaii International Conference on System Sciences*. Hawaii, HI: IEEE.

Martin, J. (1991). *Rapid application development*. New York, NY: Macmillan Publishing Company.

McPherson, M. A., & Nunes, J. M. (2008). Critical issues for e-learning delivery: What may seem obvious is not always put into practice. *Journal of Computer Assisted Learning*, *24*(5), 433–445. doi:10.1111/j.1365-2729.2008.00281.x

Monk, A., Wright, P., Haber, J., & Davenport, L. (1993). *Improving your human-computer interface*. Englewood Cliffs, NJ: Prentice Hall Publishing.

Nielsen, J. (1993). *Usability engineering*. New York, NY: Academic Press.

Nielsen, J. (1994). *Usability inspection methods*. Paper presented at CHI 1994 Conference Companion on Human Factors in Computing Systems. Boston, MA.

Patton, M. Q. (1987). *Utilization-focused evaluation*. Newbury Park, CA: Sage Publisher.

Patton, M. Q. (1990). *Qualitative evaluation and research methods*. Newbury Park, CA: Sage Publisher.

Patton, M. Q. (2005). Goal-based vs. goal-free evaluation. *Encyclopedia of Social Measurement*, *2*, 141-144.

Peng, G. C., & Nunes, M. B. (2009). Identification and assessment of risks associated with ERP post-implementation in China. *Journal of Enterprise Information Management*, *22*(5), 587–614. doi:10.1108/17410390910993554

Polson, P., Lewis, C., Rieman, J., & Wharton, C. (1992). Cognitive walkthroughs: a method for theory-based evaluation of user interfaces. *International Journal of Man-Machine Studies*, *6*, 741–773. doi:10.1016/0020-7373(92)90039-N

Rieman, J., Franzke, M., & Redmiles, D. (1995). *Usability evaluation with the cognitive walkthrough*. Paper presented at CHI 1995 Mosaic of Creativity. Denver, CO.

Robson, C. (2002). *Real world research: A resource for social scientists and practitioner-researchers*. Oxford, UK: Blackwell Publishing.

Rossi, P. H., & Williams, W. (1972). *Evaluating social programs: Theory, practice, and politics*. New York, NY: Seminar Press.

Scriven, M. (1967). The methodology of evaluation. In Tyler, R. W., Gagne, R. M., & Scriven, M. (Eds.), *Perspectives of Curriculum Evaluation* (pp. 39–83). Chicago, IL: Rand MacNally.

Scriven, M. (1972). Objectivity and subjectivity in educational research. In Thomas, L. G. (Ed.), *Philosophical Redirection of Educational Research: The Seventy-First Year Book of the National Security for the Study of Education*. Chicago, IL: University of Chicago Press.

Scriven, M. (1991). Prose and cons about goal-free evaluation. *The American Journal of Evaluation, 12*(1), 55–62. doi:10.1177/109821409101200108

Seddon, P. B. (2002). Measuring organizational IS effectiveness: An overview and update of senior management perpectives. *Advances in Information Systems, 33*(2), 11–28. doi:10.1145/513264.513270

Smithson, S., & Hirschheim, R. (1998). Analysing information systems evaluation: Another look at an old problem. *European Journal of Information Systems, 7*(3), 158–174. doi:10.1057/palgrave.ejis.3000304

Standish Group. (2009). *Chaos report*. Retrieved from http://www1.standishgroup.com/newsroom/chaos_2009.php

Tyler, R. W. (1942). General statement on evaluation. *The Journal of Educational Research, 35*, 492–501.

Wagner, C. (2004). Enterprise strategy management systems: Current and next generation. *The Journal of Strategic Information Systems, 13*, 105–128. doi:10.1016/j.jsis.2004.02.005

Warner, T. (1996). *Communication skills for information systems*. London, UK: Pitman Publishing.

Weiss, C. H. (1972). *Evaluation research: Methods for assessing program effectiveness*. Englewood Cliffs, NJ: Prentice Hall. doi:10.1016/S0886-1633(96)90023-9

Wharton, C., Rieman, J., Lewis, C., & Poison, P. (1994). The cognitive walkthrough method: A practitioner's guide. In Nielsen, J., & Mack, R. (Eds.), *Usability Inspection Method* (pp. 105–141). New York, NY: John Wiley& Sons.

Willcocks, L. P., & Lester, S. (1996). *The evaluation and management of information systems investments: Feasibility to routine operations, investing in information systems*. London, UK: Chapman and Hall.

Chapter 18
Supporting Unskilled People in Manual Tasks through Haptic-Based Guidance

Mario Covarrubias
Politecnico di Milano, Italy

Monica Bordegoni
Politecnico di Milano, Italy

Umberto Cugini
Politecnico di Milano, Italy

Elia Gatti
Politecnico di Milano, Italy

ABSTRACT

This chapter presents a methodology that the authors developed for the evaluation of a novel device based on haptic guidance to support people with disabilities in sketching, hatching, and cutting shapes. The user's hand movement is assisted by a sort of magnet or spring effect attracting the hand towards an ideal shape. The haptic guidance device has been used as an input system for tracking the sketching movements made by the user according to the visual feedback received from a physical template without haptic assistance. Then the device has been used as an output system that provides force feedback capabilities. The drawn shape can also be physically produced as a piece of polystyrene foam. The evaluation methodology is based on a sequence of tests, aimed at assessing the usability of the device and at meeting the real needs of the unskilled people. In fact, the system has been evaluated by a group of healthy and unskilled people, by comparing the analysis of the tracking results. The authors have used the results of the tests to define guidelines about the device and its applications, switching from the concept of "test the device on unskilled people" to the concept of "testing the device with unskilled people."

INTRODUCTION

The aim of the present chapter is to describe the test methodology in order to control and measure the efficiency of the haptic guidance device.

As the design of the haptic guidance device is a widely inter-disciplinary project involving experts from various disciplines, including such diverse areas as pedagogy, psychology, computer science, mechanical, mechatronic, and product design, this

DOI: 10.4018/978-1-4666-2491-7.ch018

Copyright © 2013, IGI Global. Copying or distributing in print or electronic forms without written permission of IGI Global is prohibited.

chapter is written by the intent to make not only the specific methods and procedures of accuracy testing intelligible for readers from various disciplines, but also its motivations. Therefore, not only the methodology itself and the tools that form the basis of this methodology will be described in a more detailed way.

This chapter should also serve as a set of practical guidelines for the testing procedures (but not yet as a detailed manual). It is written with the intent that experts involved in the accuracy testing procedures in the design of the haptic guidance system be able to use it as a detail reference work for actual practical activities within efficiency and accurate testing: explaining the tools to be used, the measurements to be taken, the way of data collecting.

This chapter is also meant to be intelligible, explanatory and enough detailed for those who approach the haptic guidance device and its testing activities professionally somewhat from the outside, but are nevertheless deeply involved via other ways.

By implementing the haptic guidance device we need to prove the followings: 1) the haptic support work; 2) it does suit the scientific facts and the experiences about the nature of the given problem; and 3) it is safe. In other words, we have to offer evidence-based.

A point that is crucial in the logic of evaluation of the accuracy of the haptic guidance device is multiple comparisons. We assess the unskilled people initial state and then its changes due to the use of the haptic guidance device, and we compare before and after measurement results. Moreover, we create an appropriate control group.

At first sight, it seems relatively simple a procedure, but there are several further considerations in order to reach High Methodological quality and to get strong evidence. The following brief review is based on Geyman and colleagues research (Geyman, Deyo, & Ramsey, 2000).

1. A prospective "before-after" evaluation is needed. The retrospective efficiency testing is not as accurate, and could be influenced by interpretations, latter experiences.
2. We have to select the subjects carefully and advisedly. Obviously, the main principles of the selection depend on the goal of the haptic guidance system, but there is no doubt, that solid baseline assessments (e.g. formalized, standard evaluation, etc.) are needed before we started the accuracy testing of the device.
3. After the baseline assessments, we assign the subjects randomly to the experimental group.
4. The sample (the number of experimental subjects) has to be large enough for an appropriate statistical analysis. In this way, by applying and adequate level of statistical significance as a criterion, we will assure the accuracy of the haptic guidance device providing confirmatory results.
5. Testing and evaluation of the results may easily lead with possible errors.

To accomplish our aim it is necessary to assess how the sketching control movements under haptic feedback are affected in people with motor and visuo-spatial disorders, principally by Down syndrome. Sketching is one of the most complex human activities in which the hand movements are controlled by the central nervous system, which regulates the activity of the hand and arm muscles to act in synergy. The central nervous system receives dynamic feedback information from visual sensors and from other body sensors located on the skin, muscles, and joints, while regulating the motor output.

Haptic technology can among other more traditional technical applications, be a great help for a very special group of users with specific disorders, as for example, people with Down syndrome, forms of mental retardation and other development defects (Huanran & Liu, 2011) and for the treatment of motor dexterity disabilities (Prisco, Avizzano, Calcara, Ciancio, Pinna, & Bergamasco, 1998). Down syndrome people differ from normal people in their age in various dimensions. For instance in terms of cognitive capabilities, language input and usage, physical and motor abilities, as well as social and individual characteristics. In case of developmental disorders, such as Down syndrome, motor control may be greatly affected (Blank, Heizer, & von Voss, 1999). People with Down syndrome may demonstrate reduced sensory acuity, lengthened reaction time and altered postural responses to perturbation. One of the hypotheses suggests that the source of motor difficulties originates in deficit of the central representation of actions.

It is well known that everyday individuals with Down syndrome strive to accomplish the same goals as everyone else. They want self-fulfillment and inclusion into the communities' accomplishments and activities. Many children with Down syndrome who have received family support, enrichment therapies, and tutoring have been known to graduate from high school and college, and enjoy employment in the work force. As suggested by the literature (Blank, Heizer, & von Voss, 1999; Kurillo, Gregorič, Goljar, & Bajd, 2005) practice can have positive influence on the motor skill, and effective practice is greatly influenced by motivation.

The overall main objective of the research presented in this chapter is to develop a system to assist unskilled people to develop manual and visuo-spatial skills ability. Manual skill is developed by means of sketching, and unskilled people motivation is elicited by the possibility of cutting physical shapes and building simple objects with them. The system is based on the modern haptic technology, which can play an important role by supporting the manual tasks for a group of users with specific needs. Haptic guidance has demonstrated to have a significant value in many applications, such as medical training (Liu, Tendick, Cleary, & Kaufmann, 2003), hand writing learning (Teo, Burdet, & Lim, 2002), and in applications requiring precise manipulations (Ahlström, 2005).

This chapter presents the results of a research work aiming at developing a system based on haptic technology that provides assistance to unskilled users with movement control problems, as people affected by Down syndrome, during hand sketching activities. The chapter is organized as follows. First, works related to the research presented in the chapter are listed and discussed. Then, the concept of the haptic-based guidance system, including hardware and software components, is presented. Then, the overall framework for the evaluation and testing of the haptic-based guidance device is presented, considering both, qualitative and qualitative approaches. It includes procedure for recruitment and subject selection, data collection, procedures to ensure anonymity and appropriate consent, as well as the specification of proposed tests and evaluation in the three different phases of the research. Preliminary tests performed with healthy and unskilled people and structured experiment for validating our haptic-based guidance system are presented. Finally, general discussion on the results is presented including, more device applications as hatching and cutting tasks and the rehabilitation paradigm.

METHODOLOGY FOR THE DEVICE EVALUATION

This section presents the methodology for the evaluation of the haptic guidance device whose aim is to reach rigorous, valid, reliable, systematic, useful, and practical conclusions about the designed device.

In our research we have been interested in two main aspects. The first was to evaluate the haptic guidance device with selected target users; the second was to re-design the device and provide guidelines for the design of future devices by using the evaluation results. For this reason, the methodology is based on an iterative and continuous process of data collection, analysis, and interpretation. In order to obtain more information about the various aspects of the device functioning, the methodology plans to test the device with three different kinds of population. The contribution of each population to the development of the device will be extensively exposed in the three phases of the methodology.

It is important to notice that this iterative process requires a continuous relationship with the unskilled people and their teachers. This consideration demands a conceptual and methodological shift from "doing research *on* how people use the haptic guidance system" to "doing research *with* the people who use the haptic guidance system." In other words, it means that the users of the haptic guidance system are not the objects of the research, but are active subjects who can influence the design of the haptic guidance system and can contribute in making decisions about its usability, effectiveness, and practical use.

The methodology consists of three phases. The aim of each phase is providing qualitative and quantitative results, depending on the task performed and on the sample of the participants. Validating the accuracy of the device for both users and teachers also requires an appropriate collection of data and some interviews or questionnaires appropriately designed to gauge the teacher's views.

Methodology Phases

The methodology should allow capturing appropriate data that can provide answers to the research questions, such data being both valid and reliable as far as possible. In order to accomplish this objective we structured the testing session in three phases. First, we consider healthy people in order to validate the accuracy of the system (Phase 1). Then, we have considered unskilled people and their teachers as subject of the research; this means thinking of them as active subject not only in using the haptic guidance device as instructed, but also active in reformulating its use and adapting it to their needs (Phases 2 and 3).

Phase 1: Validation of Accuracy with Healthy People

This phase aims to validate the accuracy of the system in quantitative way. Several tests have been performed with healthy people in order to detect the precision of the haptic guidance device while performing drawing tasks as for example sketching and hatching, but also, while performing the foam cutting task.

Phase 2: Preliminary Test with Unskilled People

This phase aims to monitor and collect data on the evolution of users' changes and of the device development. It seeks to capture how unskilled people and their teachers cope with the new technology, what changes the new technology demands, as well as most importantly giving indications of the impact of the technology and requirements for its redesign and improvement. At the end of Phase 2 we have drawn some preliminary conclusions taking into account the results of comparing the sketching data obtained by using the device with and without the haptic support. On the basis of the results obtained in this phase, we have planned to revise the tests to perform with unskilled people.

Phase 3: Structured Test to Validate the Haptic Guidance Device (by Unskilled People)

On the basis of the lessons learned in Phase 2, the evaluation tests have been revised so as to adopt only those that have proven to be effective and successful. At the end of Phase 3, the data collected have been compared in order to highlight the advantages of using the haptic guidance system.

Figure 1 shows the schematic view of the evaluation process that has been implemented.

The design above aims to show that the process of evaluation requires an iterative and continuous process of data collection, analysis, and interpretation. This is because in the case of the haptic guidance system we are not simply testing the device, but actually developing it while we test it. This feature already determines the fact that we are looking for both, qualitative and quantitative evaluation, both in terms of the phases of research and in terms of the content of the evaluation itself.

RELATED WORKS

Within the field of virtual reality environments and simulation tools, the sense of touch is provided by haptic interfaces (Burdea & Coiffet, 2003). Haptic interfaces are based on devices that present tactile and force feedback to a human user who is interacting with a real or simulated object via a computer (Jones & Lederman, 2006) in order to feel the virtual object properties (i.e., texture, compliance or shape). The research concerning haptic technology has increased rapidly in the last few years, and results have shown the significant role that haptic feedback plays in several fields, including rehabilitation (Kayyali, Shirmohammadi, & El Saddik, 2008; Oblak, Cikajlo, & Matjačić, 2010; Lugo-Villeda, Frisoli, Sandoval-Gonzalez, Padilla, Parra-Vega, Avizzano, Ruffaldi, & Bergamasco, 2009). Specifically, the touch modality has been shown to make it possible for visually impaired people to explore and navigate virtual environments (Sjostrom, 2001). The interaction is enriched by the use of the sense of touch, since visually impaired users can identify objects and perceive their shape and texture. There are, however, very few assisted applications for unskilled people in order to promote them in a specific employment role. Force feedback guidance using the haptic point-based approach is deemed to be a good method for facilitating 2D cutting operation in unskilled people in general and Down and blind people in particular. Haptic interface technology can enable individuals who are blind to expand their knowledge by using an artificially made reality built on haptic and audio feedback. Research on the implementation of haptic technologies within Virtual Environments has reported the potential for supporting the development of cognitive models of navigation and spatial knowledge with sighted

Figure 1. Methodology for the evaluation of the haptic guidance device

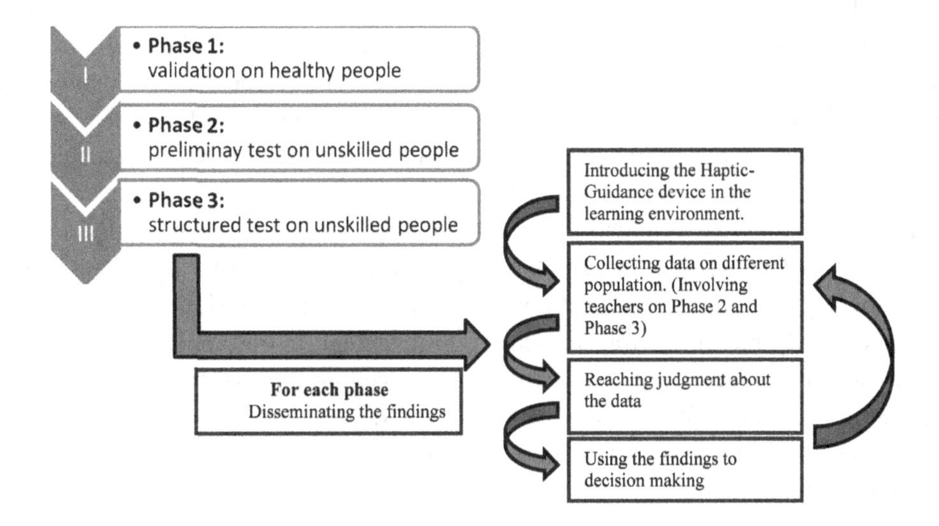

people (Witmer, Bailey, Knerr, & Parsons, 1996; Giess, Evers, & Meinzer, 1998; Gorman, Lieser, Murray, Haluck, & Krummel, 1998; Darken & Peterson, 2002) and with people who are blind as well (Colwell, Petrie, & Kornbrot, 1998; Jacobson, Kitchin, Garling, Golledge, & Blades, 1998).

THE HAPTIC-BASED GUIDANCE SYSTEM CONCEPT

The haptic-based guidance system consists of a device that allows users, assisted by a haptic device, to sketch, hatch and cut 2D shapes. The user manipulates the haptic device looking as a pen, which is a force-feedback, point-based Phantom device (PHANToM, 2011). The Phantom device generates forces that guide the user during the sketching activities. The effect is that of a magnet or a spring attracting the device in order to follow the path corresponding to the drawing. As an alternative or an add-on to the drawing function, the device can also cut a piece of foam material that is positioned under the drawing surface. The cutting operation is performed by means of a hot wire tool, which is linked through a pantograph mechanism to the Phantom device, as described in the following.

Before presenting the concept developed for the haptic-based guidance system, we describe how a haptic device of the type used for the development of our system works. In its simplest form, a point-based haptic device, like the Phantom, is a small reversible robot driven by DC brushed motors, which are equipped with encoders, and a pen held by the user. The x, y and z coordinates of the tip of the pen are tracked by the encoders, and the motors control the x, y and z forces exerted upon the user. Torques from the motors are transmitted through pre-tensioned cable reductions to a stiff, lightweight aluminum linkage. At the end of this linkage, there is a passive, three degrees of freedom gimbal attached to a thimble. Because the three passive rotational axes of the gimbal coincide at a point, there can be no torque about that point, but only a pure force. This allows the tip of the pen to assume any comfortable orientation. A haptic device allows users to feel a virtual object through the sense of touch. A fundamental problem in haptics is to detect the contact between the virtual objects and the haptic device. Once this contact is reliably detected, a force corresponding to the interaction physics is generated and rendered using the device.

The System Architecture

Figure 2 shows the system architecture of the haptic-based guidance device. A virtual shape (for example a circle) is created by used a Computer Aided Design (CAD) tool. This is used by the system to guide the young people to draw and hatch this shape by using the pen of the haptic device. Besides, at the same time a physical piece of foam material having the same shape (for example, the circle) is cut out.

The concept of the haptic-based guidance system has been developed through a series of virtual and physical prototypes in order to enable the evaluation of its potential for improving 2D drawing and cutting operations. Figure 3 represents the concept of the system, which is based on the pantograph (2) mechanism. A pantograph is a mechanical linkage, which is connected, based on parallelograms, to the Phantom device. In this way the movement of the Phantom device pen, while sketching, produces identical movements in the hot-wire tool positioned below. Therefore, if a sketch is traced by the pen, an identical copy will be cut by the hot-wire tool.

The system works as follows. The pen (5) of the Phantom device is driven under the operator's movement. In our application, the three degrees of freedom passive linkage is partially blocked in order to allow only a 2D displacement on the piece of paper where to draw. The Magnetic Geometry Effect (MGE) assists the movement. When this option is activated, a spring force tries to pull

Figure 2. System architecture

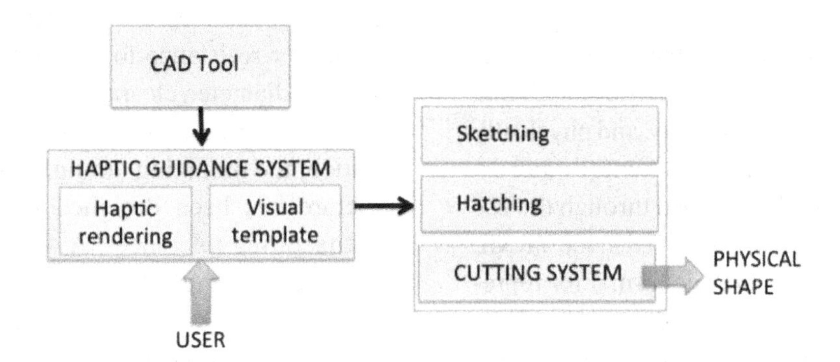

the sphere of the stylus (5) of the haptic device towards the path defined as the drawing to be reproduced (4). This effect is used in order to assist the user during the task performance. In the cutting mode, while the user follows the 2D silhouette helped by the haptic-based guidance, the wire tool (6) cuts the polystyrene foam (3).

In this way, the user can easily follow the contour of the drawing (theoretically with no effect) and feel an attractive force when deviating from the correct trajectory. This attracting force is linearly increasing with the distance from the exact trajectory by mimicking the effect of a spring attracted to the pen tip or the effect of a magnet attracting the pen along the trajectory to be followed.

The lengths of the links in the Phantom device (1) determine the mechanisms kinematic properties, such as workspace and manipulability. Therefore, the dimension of the silhouette to draw and the hand movement freedom are dependent on these properties.

Figure 3. Concept of the haptic-based guidance system

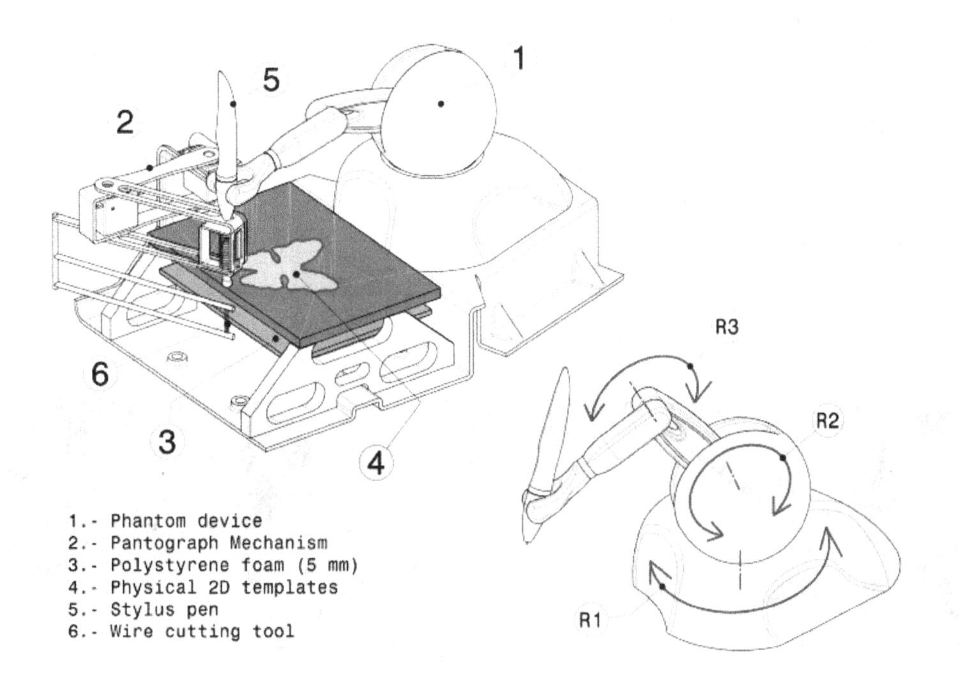

The Haptic Guiding Algorithm

The haptic-based guidance system is based on the initial definition of a set of geometric shapes that the users design in an assisted way, and physically produce thanks to the cutting system (Figure 2). The shapes are initially designed through the use of a CAD tool. The shapes are saved in the VRML format, which is a standard file format for representing 3-dimensional interactive vector graphics. This file includes the *IndexedFace set list,* which represents the 3D shape formed by constructing faces (polygons), and the *Coordinate point list,* which contains the coordinate of each single node that defines the 3D vertices of the shape. Finally, these data are imported in the H3DAPI software that is used for rendering the haptic guiding path, on the basis of the geometry of the shape. This software is an open source platform that allows us to handle both graphics and haptic data. The software also allows us to manage the Magnetic Surface constraint, which provides a force on the haptic device based on a certain distance from a virtual surface. In this way, a snap constraint is applied by considering the stiffness and damping constraints. The snap distance is a parameter that dictates the outward distance for the application of the effect.

Physical Prototype of the System

A first physical prototype has been developed for testing the concept with a group of users. In the manufacturing process, some considerations have been made: the selection of revolute joints and its stiffness or resistance to all undesired motions, the shafts diameter, clearances and tolerances and mounting configurations of the components. In addition, the mounting arrangement of the main structure has been designed to accommodate manufacturing tolerances. A 4 volts DC battery is connected to the cutting wire tool through the red wire. The 4 volts DC battery is required in order to safely produce the necessary heat to cut the polystyrene. The safety is really an important constraint that has been addressed in order to prevent any unsecured condition for the users. For this reason, we decided to use a DC low voltage instead of an AC high voltage for producing the heat. In fact, 4 volts are enough for heating the wire used as cutting tool.

The final haptic-based guidance device is shown in Figures 4a and 4b. Figures 4c and 4d show the way in which unskilled people holds the stylus while performing the 2D tasks.

The main structure used for the cutting system has been designed taking into account some important considerations related to the use of sheet metal and aluminum components, which implies: low inertia, light weight parts, and low friction. Figure 5 shows the prototype of the cutting system.

Figure 4. Haptic-based guidance system

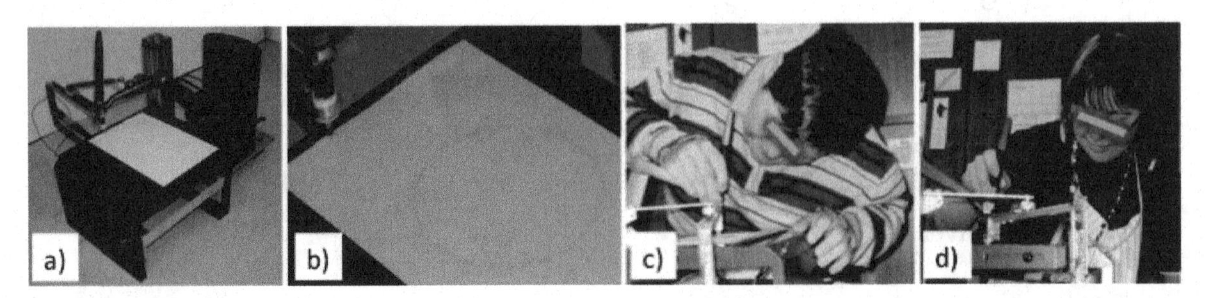

Figure 5. Physical prototype of the cutting system

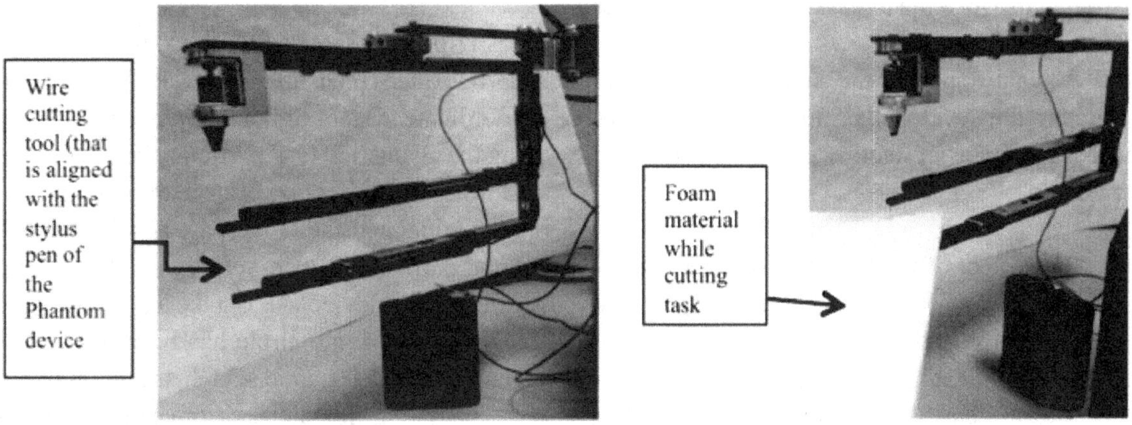

USERS TESTS

The evaluation activity regarding the developed prototype is a kind of conceptual test, which focuses on the practical implementation of the initially conceptual idea of the haptic-based guidance device in concrete form. In fact, it gives initial indications as to effectiveness, as well as to the extent to which the guidance device can or cannot make a difference to manual skills development and the extent to which it can or cannot be successfully integrated into the existing teachers' practices. These results offer initial recommendations about the specific contexts in which the haptic-based guidance device can be effective in developing manual skills with unskilled people, which can be subject to further structured investigation in following studies.

Several educational researches have reported on the positive benefits of emerging technologies on teaching and learning (Roblyer, 2003). While this body of research is still mainly focused on traditional technologies (desktops, laptops, and various educational software tools), there is also an emerging field interested in the application of haptic technology. For example a haptic guidance system has been created by Mohamad and colleagues (Eid, Mansour, El Saddik, & Iglesias, 2007) to help users to learn handwriting in a foreign language.

The evaluation tasks performed by using the guidance device represents a mixed-method testing methodology, which uses a variety of methods, quantitative and qualitative in a parallel and integrated way. It has the potential of advanced knowledge on how technology can improve the manual tasks in unskilled people, but it also has the opportunity to advance our understanding of how haptic technology can help teaching and learning in general.

Specifically, in order to test the functioning and the effectiveness of our system we have organized three different testing phases. The first phase concerns the validation of the system with healthy people, i.e. people able to well perceive the force feedback and to easily understand how to use the system to perform the task. In this phase, we focused our study on the functioning of the device itself. The second phase has been intended to be preliminary tests performed with unskilled people. The aim of these tests was to assess the system functioning also for people that have not a good motor and cognitive ability. The third phase has been structured in a more complex way with the aim to measure the efficiency of our device in a standardized way. All the three test phases gave us qualitative and quantitative results, depending on the task performed and on the sample of the participants (unskilled vs. skilled, youth worker vs. engineer).

For what concerns the evaluation activities related to the developed prototype of the haptic-based guidance system, we have decided to evaluate:

1. The effect of the introduction of a haptic-based guidance system in developing the manual skill of unskilled people.
2. The applicability of the haptic-based guidance system to the learning environment. For example, how it fits in with existing practices of teaching and learning manual activities as, sketching, hatching and cutting tasks, and what impact it has on such activities.
3. Gaining feedback on the development, improvement, and overall technical assessment of the haptic-based guidance device, and suggesting recommendations for functional changes for subsequent prototypes.

PHASE 1: VALIDATION OF ACCURACY WITH HEALTHY PEOPLE

As mentioned above we have carried out a preliminary test in order to validate the accuracy in 2D sketching that can be obtained by using the haptic-based guidance device. Ten healthy participants from the Politecnico di Milano participated in this study. The group consisted of 7 male and 3 female participants between the ages of 20 and 33. All the participants reported normal sense of touch and vision, and all of them were right-handed. The operations have been performed by tracking the pen of the Phantom device through the DeviceLog command provided by the H3D API platform that has been used for the software implementation. The tracking sample rate has been used at 25 Hz.

The 2D Sketching Task and Analysis

A 2D printed sketch has been provided to the participants, consisting of a circle. It has the same coordinate system of the 2D haptic sketch. During the evaluation test, participants practiced in a familiarization session prior to the experiment to get familiar with the haptic-based guidance device. First, participants were required to draw a circle by using the device without the haptic assistance, and then the same circle has been designed by enabling the 2D haptic sketch as a guiding path.

Sketching trajectories from the non-assisted task and the assisted one have been collected for subsequent analysis. Differences between a user trajectory and the corresponding reference trajectory have been measured in order to describe the accuracy of the device while sketching.

The difference between the two circles are described by using the equation of a circle in terms of radius and with its center in the coordinates (a, b).

$$r = \sqrt{(x-a)^2 + (y-b)^2} \quad (1)$$

The first and third columns of Figure 6 show the tracked path related to the user's sketching operation in which the haptic feedback is off.

The second and fourth columns show the same sketch operation with the haptic feedback enabled. We can observe a strong difference in the accuracy of the operation.

Under the representation of each single tracking path, we have represented the error related to the accuracy in terms of millimeters. We compare the "developable" curves of the non-assisted circle and the assisted-circle with the ideal 100 mm radius circle. In order to illustrate the validation test outcomes, the results for the ten participants are displayed in Figure 7.

The most evident advantage provided by the 2D haptic sketch is the accuracy. In fact, by using the haptic guidance the error average is 2,63 (+-0.2) mm., without the haptic guidance the error

Figure 6. Accuracy validation while sketching

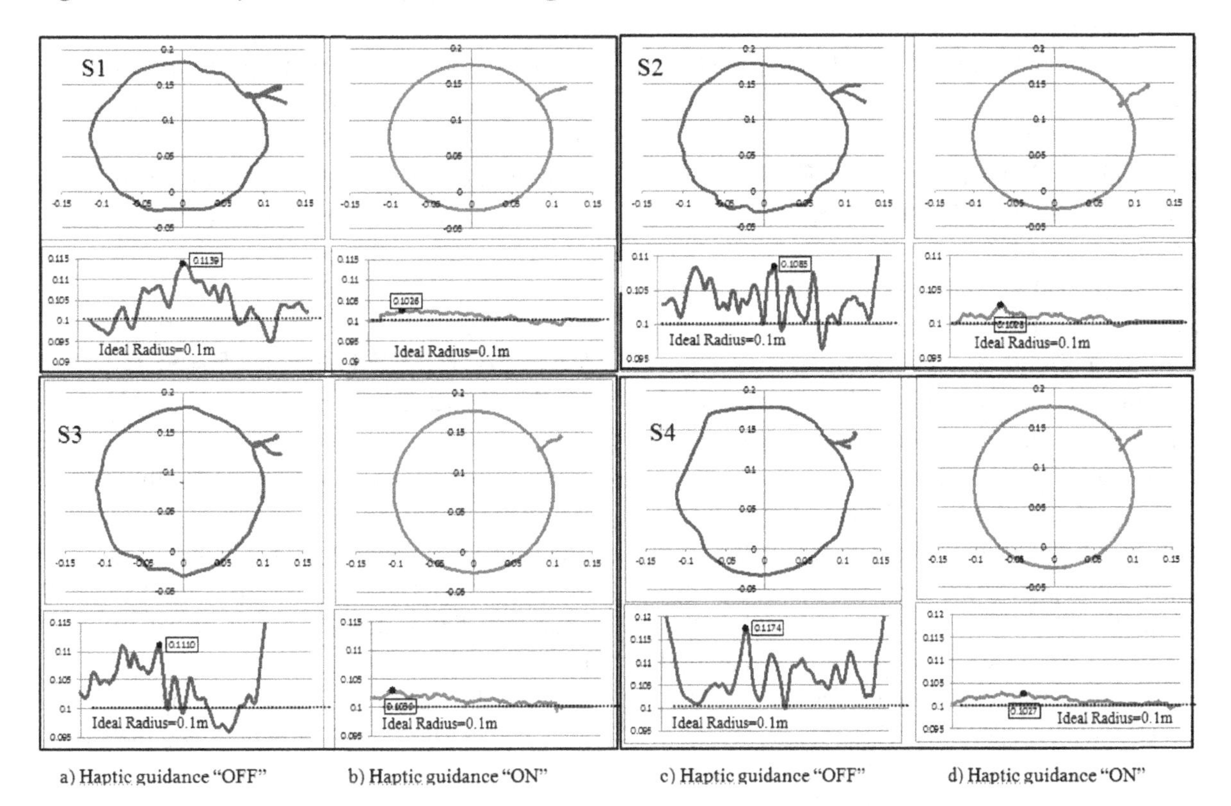

Figure 7. Error value while sketching

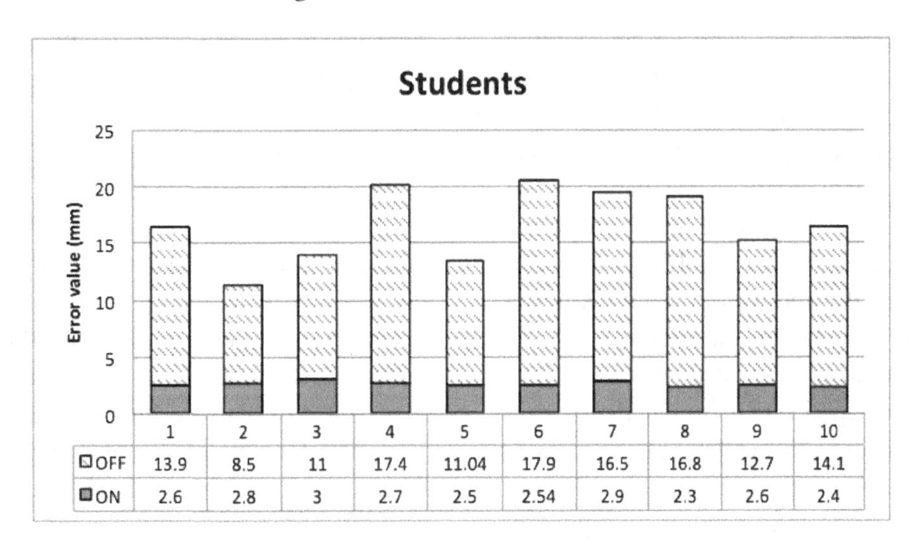

average is 13.98(+- 3) mm. Wilcoxon rank sum test showed a significant difference between the error with and without haptic guidance ($p < 0.05$).

PHASE 2: PRELIMINARY TEST WITH UNSKILLED PEOPLE

For the second phase of our validation process, we experimented our system on unskilled people.

Even if this pre-test is not compulsory and can sometimes be avoided, it is useful to have a prior experiment with a flexible experimental paradigm on the population of interest.

This provides some guidelines to the experimenter and makes him able to better suit the core experiment (phase 3) to the final user ability. For this phase we took data from 6 subjects (4 males) aged 18 to 40 years. Down syndrome and mental retardation affected all participants.

Also in this experiment a brief familiarization has been offered to the participants and to all participants it has been asked to perform a task that involved a combination of visual and haptic functions in order to design a circle with 100 mm of radius and a triangle with a 80 mm long side.

The time taken to complete the task varied considerably among participants. Figure 8 shows the tracked motion with (right) and without (left) haptic guidance while sketching the circle. In order to systematically assess the contribution of the haptic guidance we computed the error between the radius of the circles as previously described. We did not systematically analyze the triangle sketched by subjects. This shape was too difficult for them even with the haptic guidance; in particular subjects were not able to feel the end of the triangle sides and went straight by inertia without feeling the haptic constraint.

Wilcoxon rank sum test with continuity correction has been applied to the subject's errors in drawing the circle, with and without haptic guidance. Results showed that the error significantly decreases ($p < 0.05$) when subjects were guided. The overall response to the experience was positive.

Also in this experiment the usefulness of the haptic guidance is evident. In fact, by using the haptic guidance the error average is 2.66 mm, and without the haptic guidance the error average is quite high, i.e. 60.47 mm (see Figure 9).

PHASE 3: STRUCTURED TEST TO VALIDATE THE HAPTIC-BASED GUIDANCE SYSTEM

We have performed a structured test with 12 users. Participants suffer from low mental retardation and Down syndrome diagnosis as shown in Table 1. They are all right-handed, aged from 21 to 50 years. Verbal instructions and familiarization period were addressed.

As in the previous test to all participants it was asked to perform a task that involved the use of a combination of visual and haptic senses in order to design a circle with 100 mm radius. Moreover, basing on the results of the preliminary tests (phase 2) we rejected the hypothesis to test the participants' ability also on figures with shared angles.

In this testing phase, we adopted the following paradigm to test the participants only on the selected shape (circle). Subjects stand in front of the device, their position varied in order to make them able to perform the task with the maximum of the comfort. The grasping on the device has been controlled and encouraged to be similar to a pen grasping. Participants were asked to trace a circle following a draw previously printed.

The first tracing were performed without any force cue, just using the device in order to record the trajectory in the space. After this phase, a second tracing were performed, with the haptic-based guidance activated. The majority of participants were able to complete the specified tasks; the time taken to complete the task varied considerably between participants. Each participant in the evaluation has been codified in order to maintain

Figure 8. Preliminary test

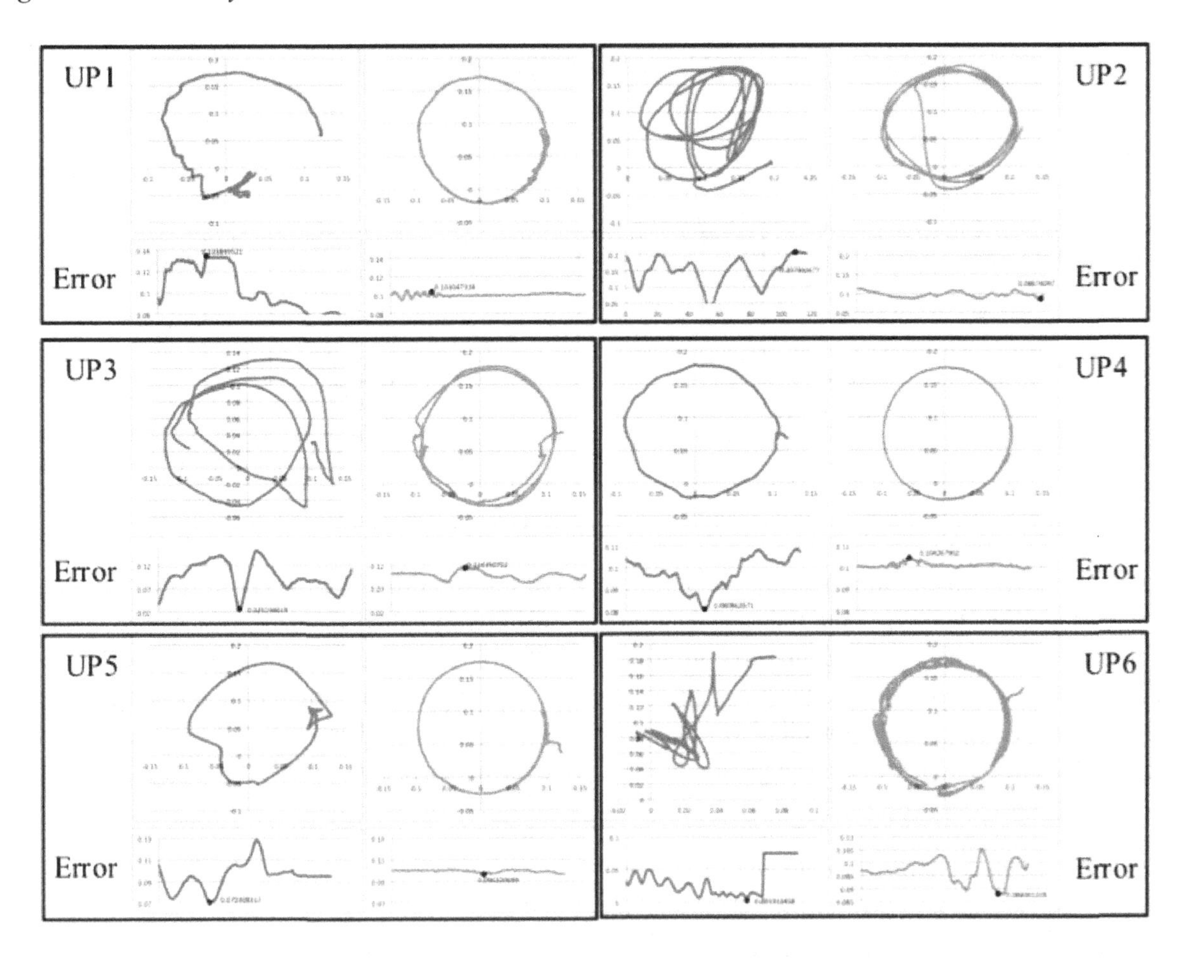

the privacy statements as agreed with the institutes that hosted the tests.

Data Analysis

On the left hand side of Figure 10 is displayed the tracked motion without haptic guidance while sketching the circle. On the right side is displayed the tracked motion with the haptic guidance. When comparing the final results on each of the twelve participants, we can conclude that haptic guidance can benefit the performance in sketching tasks. The statistically significant difference has been demonstrated by applying the Wilcoxon rank sum test to the error with and without haptic guidance ($p < 0.01$).

In summary, participants were generally pleased with the ease with which they were able to draw an almost "perfect" circle. The overall response to the experience was positive (see Figure 11).

GENERAL DISCUSSION OF THE RESULTS OF THE TESTING PHASE

The three-phase evaluation of the first prototype serves a dual purpose. In the first place, the evaluation is designed to investigate quantitative results. In the second place, the evaluation is designed to systematically collect qualitative experiences in order to expand the functionality of the haptic

Figure 9. Error value while sketching in the preliminary test

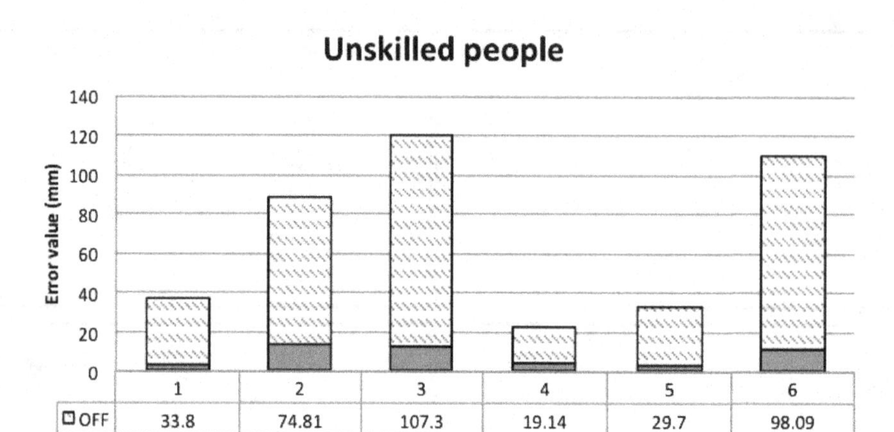

guidance device taking into account ergonomic analysis and reachability. Both these evaluation approaches are aimed at extracting design principles that may extend beyond the research.

Qualitative Approach: Qualitative analysis allows the experimenter to reach a great number of information from the environment, basing on different method (Denzin, Norman, & Lincoln, 2005). Basically, qualitative analysis offers to the enquirer general information about how something works, and overall impression by users, subject and even patients, without strictly relying on measures

Table 1. Diagnosis of the participants

Code	School	Tutor	Gender	Year of Birth	Diagnosis
30.111	Centro Diurno "Casa dei Ragazzi"	B1, B2, S1	Female	1968	Down Syndrome
30.112			Male	1968	Down Syndrome, mental insufficiency
30.113			Male	1990	Congenital cerebropathy with mental retardation, Disturb motor coordination
30.114			Female	1975	Down Syndrome
30.115			Female	1979	Mental insufficiency, medium level with psychotic personality disorder
30.116			Male	1985	Malformative cerebropathy with West Syndrome, mental insufficiency level medium-high
30.117			Female	1977	High level of mental retardation
30.118			Male	1961	Residual schizophrenia, psychotic disturbs with low level of mental retardation
30.119			Female	1988	Mental insufficiency, medium-high level with dysgenetic syndrome
30.120			Female	1978	Neonatal encephalopathy
30.121			Female	1990	Mental retardation, medium level, Down Syndrome
30.122			Male	1979	High level of Mental retardation
30.123			Female	1983	Perinatal asphyxia, mental retardation

Figure 10. Sketching task in the structured test

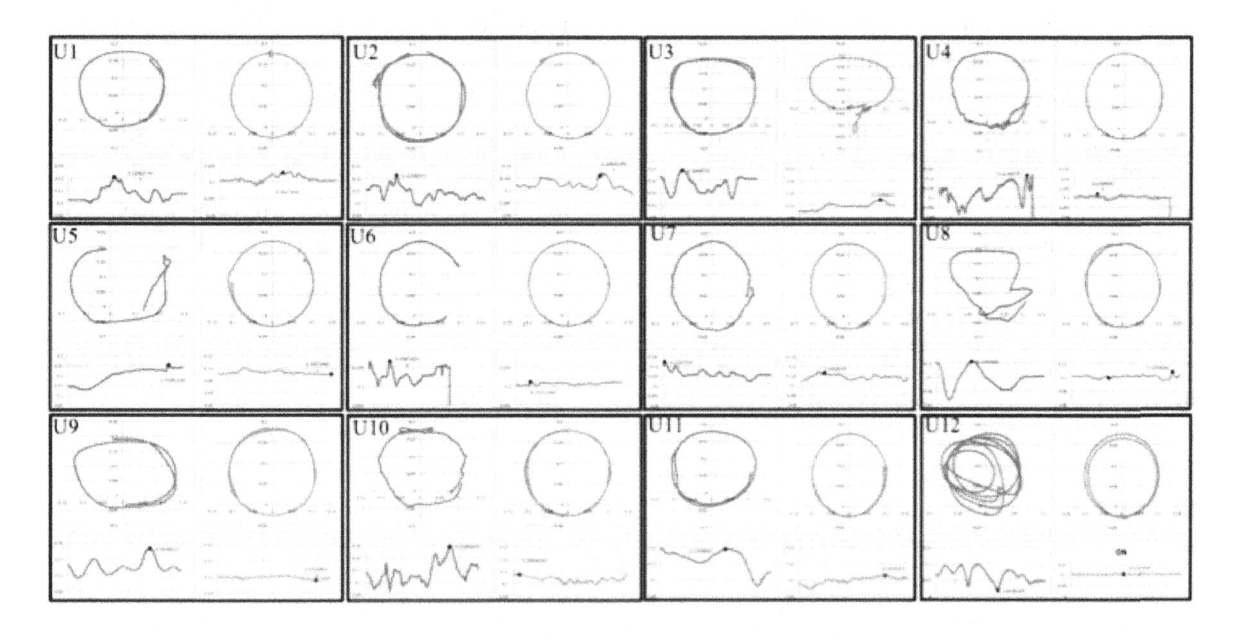

Figure 11. Error value while sketching in the structured test

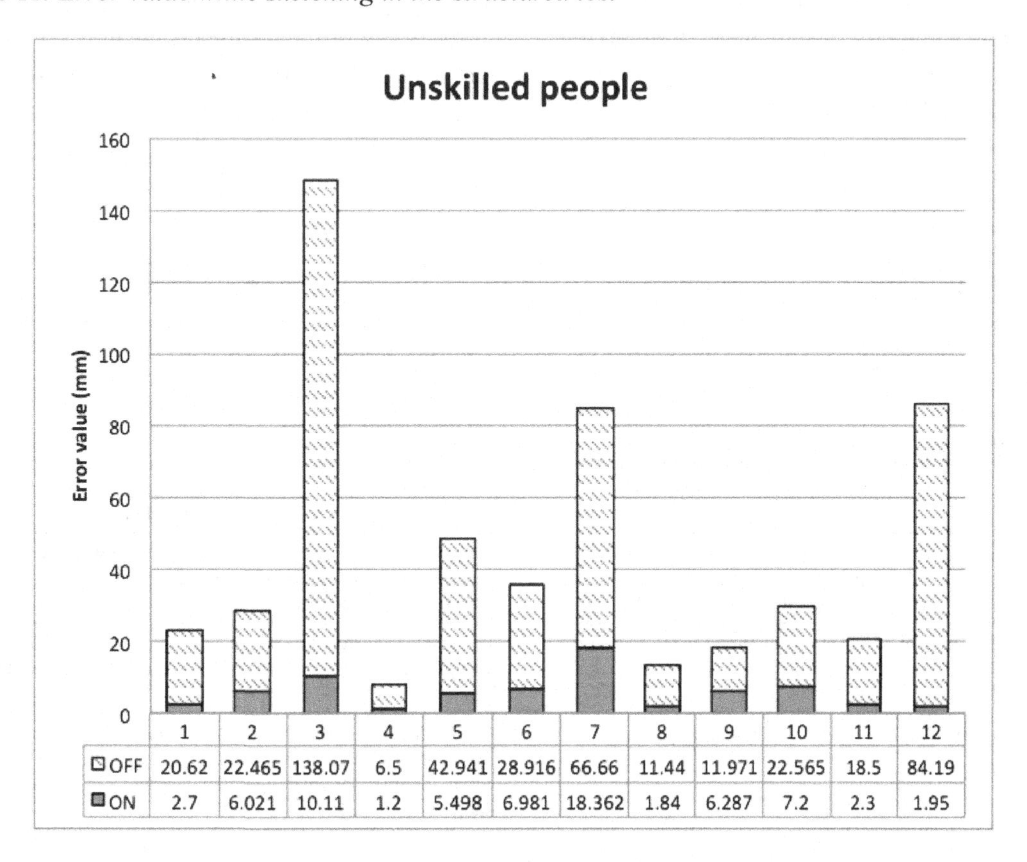

and indexes. For these kinds of analyses a large set of participants is not required, but it is important to collect as much information as possible, structuring an explanation of the overall recorded impression by organizing all the information in a simple and coherent way. During the years a large number of structured approaches have been developed, both to collect and analyze qualitative data. For a complete review it is advised the book from Kathy Charmaz (2006a). In our study, we have used the qualitative approach to evaluate the ergonomics of the system, and to reach the overall impression from users and their assistant. We decided to interpret data as complementary of the obtained quantitative data. One of the major problems of the qualitative analysis is the challenging validation of the theories that rise from the recorded impressions. Nevertheless, some approach able to switch data from a qualitative form in a quantitative and valuable one has been proposed (i.e., coding by Charmaz, 2006b). In our qualitative observation of the users, we focus our attention on some points, which are summarized in the following Table 2. The overall impression is positive. The qualitative data have been obtained by two ways. For points 1,4 and 5 in Table 2 we directly asked to users or to the educators opinions and suggestions. For the points 2 and 3 we obtained data observing the users' behavior. All the participants have found the system comfortable during the test evaluation. Despite the participants had different impairments, all of them have been able to find a position that allow them to accomplish the task in a comfortable way. The system "helping effect" has been found very interesting and rewarding by the users who have been able to focus their attention on the task.

Unfortunately in two cases out of thirteen the interest on the system itself overcame the interest on the reward obtained accomplishing perfectly the drawing task. Considering the users' behavior while interacting with the device we obtained some additional information about the usability of the system.

Indeed, we noticed that, after the task explanation, the majority of subjects grasped the haptic device's pen without any particular instruction. For three out of thirteen subjects the grasp on the pen has been corrected by the educators. Those subject had the most severe motor impairments among

Table 2. Qualitative results

Challenge:	Main questions:	Recorded results:
1. System comfort	Is the system comfortable for the users? They can use it in a comfortable position?	All the participants involved into the three phases of evaluation has been able to use the system in a comfortable way; the portability and the reduced dimensions allow all the subjects to set their position in the most comfortable way while they performed the task for both the hands.
2. Interest in the system	Is the system able to interest and to focus the user attention on the task?	For all the participants the system has been great source of interest. in the third evaluation phase 11 out of 13 of them has been able to complete the task without distraction.
3. Usability of the system	Is the system intuitive or need a deep explanation of its functioning to be utilized by subject?	The majority of the subjects started to interact whit the system in a correct fashion grasping the Phantom Pen and moving It without any instruction.
4. Utility of the system	Is the system useful in the educators' opinion?	For all the educators the system constitutes a valid help in their work, interesting and helping subjects in the drawing and coping task.
5. System completeness	Provide the system all the cues needed for a good performance of the task?	In the educators opinion more important forces are needed, depending on the user deficit. Moreover, an interactive and funny interface has been suggested in order to involve more the subjects in the task.

Table 3. Test and significance for the three experiment phases

Testing phase	Sample size	Applied test	Results	Conclusion
Phase 1	10	Wilcoxon rank sum test	P < 0.05	There is a significant difference between the mean of the error computed with and without HG.
Phase 2	6	Wilcoxon rank sum test	P < 0.05	There is a significant difference between the mean of the error computed with and without HG
Phase 3	12	Wilcoxon rank sum test	P < 0.001	There is a significant difference between the mean of the error computed with and without HG

the observed users. The opinions of educators concerning the utility of the system and further improvements have been considered. For all the educators the system constitutes a valid help in their work, interesting and helping subjects in the sketching and cutting tasks. However, some indication to improvement has been proposed, strongly supported by the previous cited qualitative observations. In fact, more important force feedback should be implemented to help people with more severe motor deficits. Moreover, an interactive and funny interface has been suggested to focus the attention of the subjects on the task, rather than on the device itself.

Quantitative Approach: Quantitative analyses are typically used to evaluate the functioning of a device, or also to measure the behavior of a subject under pre-determinate restricted conditions. These kinds of analyses usually are based on the choice of a measureable index, closely related with the aspect that the experimenter wants to test. The power of the quantitative approach is its capability in building prediction and testing hypothesis based on powerful statistical tests. Several tests exist to verify a large range of hypotheses, and their power varies depending on the number of the sample and on the complexity of the experimental setup. The most used statistical tests are the t-student test and the analysis of the variance (ANOVA), despite when the sample is composed by a restricted number of participants the use of non-parametric test is required. More complex data analysis involves the use of mixed models and other methodologies to analyze data ant to extract information from datasets. For a more accurate review we advise the book "Mathematical Statistics and Data Analysis" (Rice, 2007).

Because of the nature of our test, we decided to choose as quantitative index able to describe the functioning of the cutting system the length of the copied radius, which allowed us to compute the error in the copying task. Other index could be chosen, such as the time to complete the task, or the results from a procrustes analysis of the drawn shapes (Gower, 1975). Nevertheless, we chosen the simplest and intuitive index, more directly related to the aim of our study. In fact, we were not interested in the performance speed, but in the drawing accuracy. The resumed results for each test phase are shown in the following Table 3.

ADDITIONAL APPLICATIONS OF THE SYSTEM

This section describes some other applications of the haptic-based guidance systems that have been developed and also prelimiary tested with users. They consist of cutting out shapes for assemblying objects and hatching a shape.

Cutting and 3D Assembly Task

The original aim of our system is to allow people to perform a cutting task. Therefore, we have performed a preliminary test to asses the functionality of the system in a cutting condition. In this test, the task consisted of manufacturing some selected

Figure 12. Cutting and assembly operations

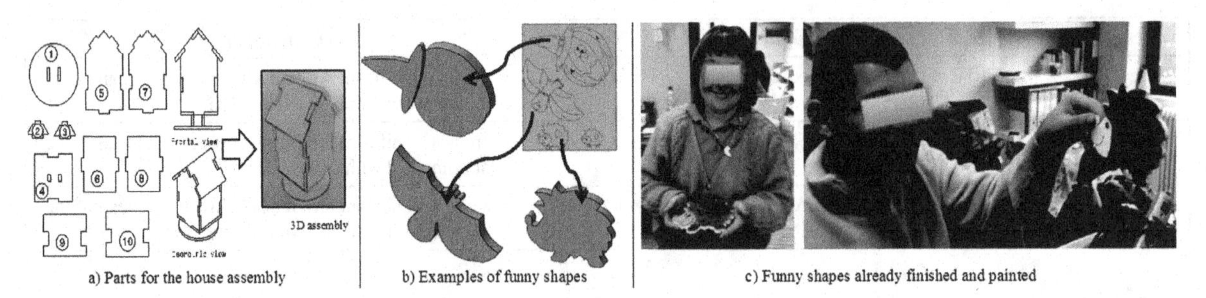

a) Parts for the house assembly b) Examples of funny shapes c) Funny shapes already finished and painted

objects by using the haptic cutting modality to cut out each single component. The target object that has been selected is a house, composed of several parts in order to allow users to build up an assembly by using the parts.

Figure 12a shows the ten polystyrene foam components that are required in the house assembly task. The house shape has been designed enabling both easy assembly and disassembly operations. In fact, all the components are symmetric. Besides, we have integrated several slots, which are used for the correct component positioning in the final assembly. Figure 12b shows several funny shapes in which the users were able to cut the sketches, by using the cutting modality of the system.

Hatching Task

Another application of the system is allowing the users to perform hatching task. A preliminary test has been also performed to validate the effectiveness of the system with this kind of help. In the hatching task we request a user to fill the internal surface of the butterfly sketch as can be seen in Figure 13a. Figure 13b shows the hatching operation performed by the user without the haptic feedback, in fact, the 2D haptic sketch has been disabled. Figure 13c shows the hatching operation with the haptic feedback enabled. Figure 13 also shows the circle sketch used in the hatching accuracy task. In fact, the blue-hatched circles were performed without the haptic assistance; conversely, the red circles have been hatched with the assistance of the haptic device.

Also in this operation, it is evident the advantage of using the haptic feedback as a virtual guide for assisting the user's hand. In this case, the sketch has been used as an external wall, without the Magnetic Geometry effect.

Guidelines for the Design of the Haptic-Based Guidance System

The following guidelines have been considered in the design of the haptic-based guidance system:

1. **Free space must be perceived as free:** The perception of free space is related to the fact that the device must not encumber the users. That is, the device should exert no external forces on a user moving through both, the virtual guide and the real space. Translated into engineering requirements, this means that there should be little back-drive friction, low inertia at the human-machine interface, and no unbalanced weight.
2. **The virtual guidance path must feel stiff:** One parameter to evaluate for a force-reflecting interface is the maximum stiffness of the virtual guide that it is capable of representing. Because no structure or control loop is perfectly stiff, each virtual path is not limited by the stiffness of the structure, but rather by the stiffness of stable control that can be achieved.

Figure 13. Hatching task

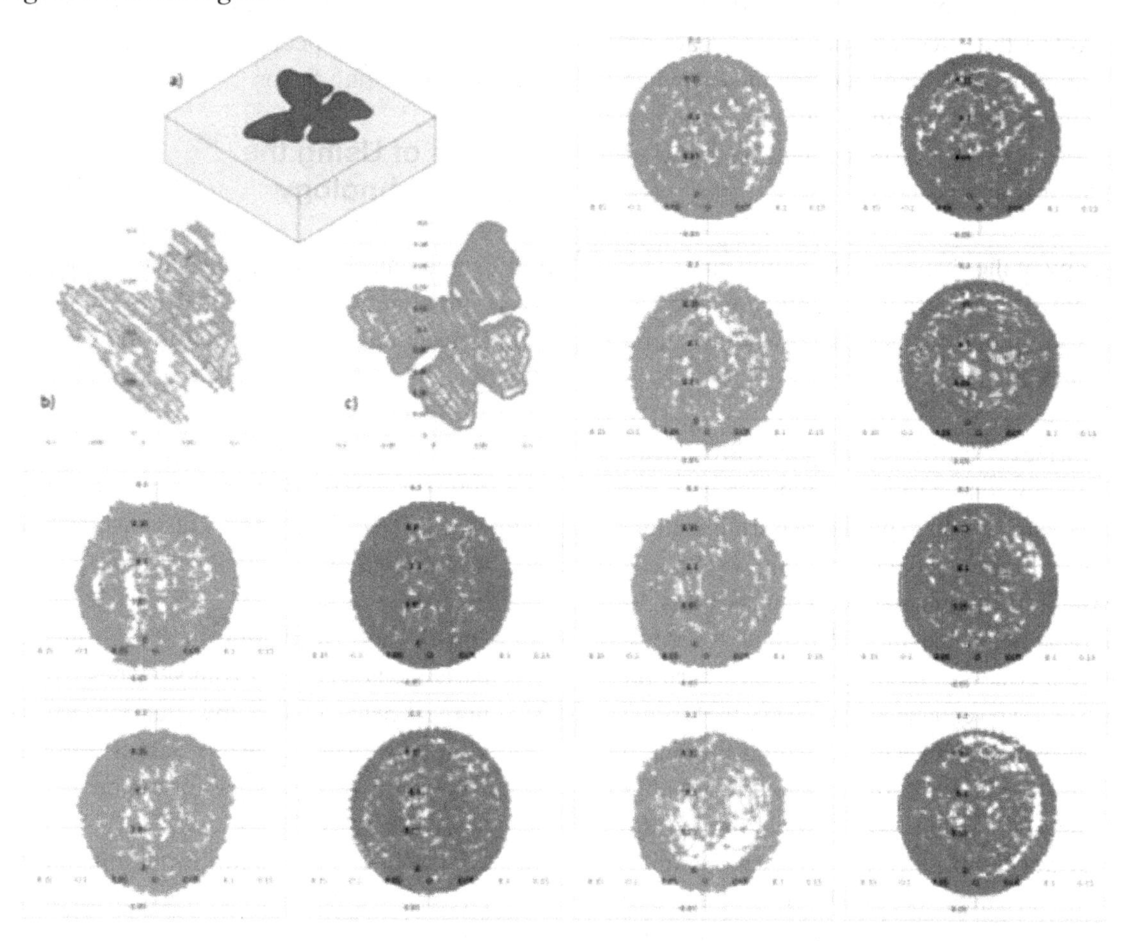

3. **Virtual constraints must not be easily saturated:** The maximum exertable force for the human finger is on the order of 40 N [10], but during precise manipulation, people rarely exert more than 10 N of force, which is exactly the peak maximum for the Phantom device. In fact, the time average force exerted during normal operations is on the order of 1 N, while the maximum continuous force capability for the Phantom device is about 1.5 N.
4. **The magnetic geometry effect must be stiff:** The forces required to maintain the reference point of the Phantom on the virtual surfaces should be the same than the ones provided to feel stiff a virtual object.

Rehabilitation Paradigm

Several studies demonstrated that the haptic guidance systems can help people in recovering cognitive function at different levels of complexity and impairment (Liu, Cramer, & Reinkensmeyer, 2006). We speculate that the applications supported by our device could also have an important role in supporting physical therapist and cognitive psychologist in helping patients to recover motor and visuo-spatial abilities. To assess this speculation we propose a training paradigm involving the copy of the circles and more complex shapes, for example spirals, (shapes that have already been used in tracing/ copying task (Liu, Carroll, Wang, Zajicek, & Bain, 2005), with and without haptic guidance. We are planning to perform experiments to test the effect of priming and learning in drawing shapes. The experimental set-up has been thought of comprising 7 different sessions. During each session, a subject will be involved in the following tasks:

- Copy of the shape without haptic guidance.
- Copy of the shape with haptic guidance.
- Further copy of the shape without haptic guidance.

To test the effect of the priming we will compare the first copy without haptic guidance with the last copy without haptic guidance during a single session. If a priming effect is present, we expect a decrease in the copying error in the last copy. In fact, it is plausible imagining that the motor faculties and the visuo-spatial ability needed to perform the task could be enhanced by the trial performed with the haptic guidance, that could provide a simplified conditions allowing subjects to better understand the shape to trace and the motor scheme required to accomplish the required task. A similar effect has been demonstrated by Bluteau and colleagues for the tracking of trajectories (Bluteau, Coquillart, Payan, & Gentaz, 2008). Moreover, a learning effect is expected between the sessions. In particular, we expect to see a gradual decrease of the error in the copy task from the last to the seventh session. This prevision is supported by literature: indeed haptic guidance has been also used with noticeable results in a training to enhance perceptual motor skills (Feygin, Keehner, & Tendick, 2002). Further research and test of the subject could confirm our hypothesis.

DISCUSSION

The results of our study show that the haptic-based guidance device helps people during sketching, hatching, and cutting operations. The opportunity to create a haptic system that would make real difference in Down people's life appeared to be a highly motivating factor.

Benefits of Using the Haptic Technology

A few researches have concluded that technology is very important to students with Down syndrome. In fact, technology can be used to improve the effectiveness and the speed of the learning processes (Galanouli, Murphy, & Gardner, 2004; Andrews, Freeman, Hou, McGuinn, Robinson, & Zhu, 2007). The following advantages have been suggested by (Black & Wood, 2003), which have also been proved by our work.

- **Improving Motivation:** The learning experience is enhanced with sketches, which may increase interest and attention. People with Down syndrome are "visual learners" who learn best when information is presented visually (Fidler, 2002).
- **Immediate Feedback:** From qualitative analysis we noticed that users with Down syndrome are rewarded for their successes immediately, e.g. by creating a sketch or by cutting some funny objects. The haptic system never gets impatient or frustrated by repeated errors, and feedback is non-threatening and non-judgmental.
- **Opportunities for Practice:** Users with Down syndrome need much more practice to acquire new skills and the haptic device can provide as many opportunities as necessary to repeat the same objective in exactly the same way allowing to proceed as fast or as slow as he or she wishes (Pueschel, 1987); the haptic system will "wait" for the child to respond without prompting them before they have time to fully process the information and construct their response.

Involvement of Teachers and Unskilled People in the Tests

The application of the capability approach to the evaluation of the effectiveness of the haptic guidance device while sketching, hatching and cutting tasks has the potential of introducing and innovative and growing interest in alternative approaches to understanding the relationship between the use of haptic technology and the assessment of manual tasks. It also has the potential of bringing new and powerful insights into the nature of the unskilled people. A focus on how individuals speak about their aspirations and progress and evaluate the usefulness of the haptic guidance device will cast them as active subject, rather than passive recipients of haptic technology. While the unskilled people are the end-users of the technology, aspects of the evaluation have aspects as "research objects," it is possible to consider a more "person-centered" approach to both the evaluation of the haptic guidance and to the research.

In order to gain a rich and accurate description of the impact of the haptic guidance device, it is also important in the evaluation to gain information from the teachers and care assistants.

CONCLUSION

This chapter proposes a methodology that has been designed to delineate an overall strategy for implementing the evaluation of the use of the haptic guidance system, its effectiveness and usability. In doing so, a number of aspects have been taken into consideration. While primarily focused on supporting the design of a rigorous evaluative design, we have noted that the complexity of the task, the multi-perspectival and multi-disciplinary approach requires both a clear and practical methodology.

Therefore, we suggested that a mixed-method research design is the most suited for the evaluation. This implies the use of a method suitable for collecting valid and reliable data useful for the integration of the research goals and for the respective evaluation.

In applying the methodology, we have used the haptic guidance device based on the point-based approach. The main application of the haptic device is related to assisting unskilled people in the assessment and training of hand movements while performing manual tasks. The device has been tested with unskilled people, and the preliminary results are very encouraging. Results show that the effect of using the haptic guidance system increases the accuracy in the tasks operations. We can resume that the system leads to the satisfaction of the following objectives:

- The force feedback is an additional perceptual channel for enhancing the interaction between user and computer;
- The use of haptic cutting device for enhancing human skills and/or as a rehabilitation tool for disable people.

We are currently performing an evaluation with unskilled people in order to measure their learning improvements in 2D operation skills.

From the technical point of view, the system has assured the coherence and collocation between the haptic and the physical 2D template. Anyway, further research, however, is still needed to improve the performance of the cutting haptic device by increasing the working area and by integrating sound as an additional perceptual channel in order to provides some inputs related to: start and/or finish the manual task; by hearing the sound, the user will also control the velocity of his hand movements while interacts with the haptic guidance device.

Despite the limitation of the haptic guidance device, mainly related to the fact that the working area is quite small, it is indeed a step forward in the development of haptic guidance devices for the assessment and improvements of manual tasks in unskilled people.

ACKNOWLEDGMENT

The authors would like to thank all the instructors at Laboratorio Artimedia (Calolziocorte, Italy), Casa Dei Ragazzi "Treves De Sanctis" O.N.L.U.S., and Centro Diurno Disabili Di Barzanò for their support in the preliminary evaluation of the haptic guidance device and for giving us the opportunity to test the system.

REFERENCES

Ahlström, D. (2005). Modeling and improving selection in cascading pull-down menus using fitts' law, the steering law and force fields. In *Proceedings of the SIGCHI Conference on Human Factors in Computing Systems, CHI 2005*, (pp. 61-70). ACM.

Andrews, R., Freeman, A., Hou, D., McGuinn, N., Robinson, A., & Zhu, J. (2007). The effectiveness of information and communication technology on the learning of written English for 5 To 16-year-olds. *British Journal of Educational Technology, 38*(2), 325–336. doi:10.1111/j.1467-8535.2006.00628.x

Black, R., & Wood, A. (2003). Utilising information communication technology to assist the education of individuals with Down syndrome. In *Down Syndrome Issues and Information*. Portsmouth, UK: The Sarah Duffen Center.

Blank, R., Heizer, W., & Von Voss, H. (1999). Externally guided control of static grip forces by visual feedback age and task effects in 3-6 year old children and in adults. *Neuroscience Letters, 271*, 41–44. doi:10.1016/S0304-3940(99)00517-0

Bluteau, J., Coquillart, S., Payan, Y., & Gentaz, E. (2008). Haptic guidance improves the visuo-manual tracking of trajectories. *PLoS ONE, 3*(3), e1775.

Burdea, G. C., & Coiffet, P. (2003). *Virtual reality technology* (2nd ed.). Mahwah, NJ: John Wiley & Sons Inc.

Chamaz, K. (2006a). *Constructing grounded theory: A practical guide through qualitative analysis*. Thousand Oaks, CA: SAGE Publication Ltd.

Charmaz, K. (2006b). Coding in grounded theory practice. In Charmaz, K. (Ed.), *Constructing Grounded Theory: A Practical Guide through Qualitative Analysis* (pp. 42–71). Thousand Oaks, CA: SAGE Publication Ltd.

Colwell, C., Petrie, H., & Kornbrot, D. (1998). *Haptic virtual reality for blind computer users*. Paper presented at the Assets 1998 Conference. Los Angeles, CA.

Darken, R. P., & Peterson, B. (2002). Spatial orientation, wayfinding, and representation. In Stanney, K. M. (Ed.), *Handbook of Virtual Environments Design, Implementation and Applications* (pp. 493–518). Mahwah, NJ: Lawrence Erlbaum Associates.

Denzin, J., Norman, K., & Lincoln, Y. S. (2005). *The Sage handbook of qualitative research* (3rd ed.). Thousand Oaks, CA: Sage.

Feygin, D., Keehner, M., & Tendick, F. (2002). Haptic guidance: Experimental evaluation of a haptic training method for a perceptual motor skill. In *Proceedings of the Sympathetic Haptic Interfaces for Virtual Environment and Teleoperator Systems*, (pp. 40-47). IEEE.

Fidler, D. J., Hodapp, R. M., & Dykens, M. E. (2002). Behavioral phenotypes and special education: Parent report of educational issues for children with Down syndrome, Prader-Willi syndrome, and Williams syndrome. *The Journal of Special Education, 36*, 80–88. doi:10.1177/00224669020360020301

Galanouli, D., Murphy, C., & Gardner, J. (2004). Teachers' perceptions of the effectiveness of ICT-competence training. *Computers & Education*, *43*(1-2), 63–79. doi:10.1016/j.compedu.2003.12.005

Geyman, J. P., Deyo, R. A., & Ramsey, S. D. (Eds.). (2000). *Evidence-based clinical practice: Concepts and approaches*. Boston, MA: Butterworth-Heinemann.

Giess, C., Evers, H., & Meinzer, H. P. (1998). *Haptic volume rendering in different scenarios of surgical planning*. Paper presented at the Third PHANTOM Users Group Workshop. Cambridge, MA.

Gorman, P. J., Lieser, J. D., Murray, W. B., Haluck, R. S., & Krummel, T. M. (1998). *Assessment and validation of force feedback virtual reality based surgical simulator*. Paper presented at the Third PHANToM Users Group Workshop. Cambridge, MA.

Gower, J. (1975). Generalized procrustes analysis. *Psychometrika*, *40*, 33–51. doi:10.1007/BF02291478

Jacobson, R. D., Kitchin, R., Garling, T., Golledge, R., & Blades, M. (1998). *Learning a complex urban route without sight: Comparing naturalistic versus laboratory measures*. Paper presented at the International Conference of the Cognitive Science Society of Ireland. Dublin, Ireland.

Jones, L. A., & Lederman, S. J. (2006). *Human hand function*. Oxford, UK: Oxford University Press. doi:10.1093/acprof:oso/9780195173154.001.0001

Kayyali, R., Shirmohammadi, S., & El Saddik, A. (2008). Measurement of progress for haptic motor rehabilitation patients, In *Proceedings of the IEEE International Workshop on Medical Measurements and Applications*, (pp. 108-113). IEEE.

Kim, S.-C., Kim, C.-H., Yang, G.-H., Yang, T.-H., Han, B.-K., Kang, S.-C., & Kwon, D.-S. (2009). Small and lightweight tactile display (salt) and its application. In *Proceedings of the World Haptics Conference*, (pp. 69–74). World Haptics.

Kurillo, G., Gregorič, M., Goljar, N., & Bajd, T. (2005). Grip force tracking system for assessment and rehabilitation of hand function. *Technology and Health Care*, *13*, 137–149.

Liu, A., Tendick, F., Cleary, K., & Kaufmann, C. (2003). A survey of surgical simulation: Applications, technology, and education. *Presence: Teleoperating a Virtual Environment*, *12*, 599–614. doi:10.1162/105474603322955905

Liu, J., Cramer, S. C., & Reinkensmeyer, D. J. (2006). Learning to perform a new movement with robotic assistance: Comparison of haptic guidance and visual demonstration. *Journal of Neuroengineering and Rehabilitation*, *3*(1), 20. doi:10.1186/1743-0003-3-20

Liu, X., Carroll, C. B., Wang, S., Zajicek, J., & Bain, P. G. (2005). Quantifying drug-induced dyskinesias in the arms using digitised spiral-drawing tasks. *Journal of Neuroscience Methods*, *144*, 47–52. doi:10.1016/j.jneumeth.2004.10.005

Lugo-Villeda, L. I., Frisoli, A., Sandoval-Gonzalez, O., Padilla, M. A., Parra-Vega, V., & Avizzano, C. A. … Bergamasco, M. (2009). Haptic guidance of light-exoskeleton for arm-rehabilitation tasks, In *Proceedings of the 18th IEEE International Symposium on Robot and Human Interactive Communication*, (pp. 903-908). IEEE.

Mohamad, A., Eid, M. M., & Abdulmotaleb, H. El Saddik, & Iglesias, R. (2007). A haptic multimedia handwriting learning system. In *Proceedings of the International Workshop on Educational Multimedia and Multimedia Education (Emme 2007)*, (pp. 103-108). New York, NY: ACM.

Oblak, J., Cikajlo, I., & Matjačić, Z. (2010). Universal haptic drive: A robot for arm and wrist rehabilitation. *IEEE Transactions on Neural Systems and Rehabilitation Engineering*, *18*(3), 293–302. doi:10.1109/TNSRE.2009.2034162

PHANToM device. (2011). *SenSable technologies inc*. Retrieved from http://www.sensable.com

Prisco, G. M., Avizzano, C. A., Calcara, M., Ciancio, S., Pinna, S., & Bergamasco, M. (1998). A virtual environment with haptic feedback for the treatment of motor dexterity disabilities. In *Proceedings of the International Conference on Robotics and Automation, 1998,* (vol. 4, pp. 3721-3726). IEEE Press.

Pueschel, S. M., Gallagher, P. L., Zartler, A. S., & Pezzullo, J. C. (1987). Cognitive and learning processes in children with Down syndrome. *Research in Developmental Disabilities*, *8*(1), 21–37. doi:10.1016/0891-4222(87)90038-2

Rice, A. J. (2007). *Mathematical statistics and data analysis* (3rd ed.). New York, NY: Cengage Learning, Inc.

Roblyer, M. D. (2003). *Integrating technology: Educational technology into teaching*. Upper Saddle River, NJ: Merrill Prentice-Hall.

Sjostrom, C. (2001). Designing haptic computer interfaces for blind people. *Proceedings of IS-SPA*, *1*, 68–71.

Teo, C., Burdet, E., & Lim, H. (2002). A robotic teacher of Chinese handwriting. In *Proceedings of the 10th Symposium on Haptic Interfaces for Virtual Environment and Teleoperator Systems, HAPTICS 2002*. IEEE Computer Society.

Wang, H., & Liu, X. P. (2011). Haptic interaction for mobile assistive robots. *IEEE Transactions on Instrumentation and Measurement*, *60*(11), 3501–3509. doi:10.1109/TIM.2011.2161141

Witmer, B. G., Bailey, J. H., Knerr, B. W., & Parsons, K. C. (1996). Virtual spaces and real world places: Transfer of route knowledge. *International Journal of Human-Computer Studies*, *45*, 413–428. doi:10.1006/ijhc.1996.0060

Chapter 19
Integrated Methods for a User Adapted Usability Evaluation

Junko Shirogane
Tokyo Woman's Christian University, Japan

Hajime Iwata
Kanagawa Institute of Technology, Japan

Yuichiro Yashita
Waseda University, Japan

Yoshiaki Fukazawa
Waseda University, Japan

ABSTRACT

For software development, methods must be able to effectively perform evaluations with respect to financial and time considerations. Usability evaluations are commonly performed to ensure software is usable. Most evaluations are individually performed, leading to some significant disadvantages. Although individual evaluations identify many usability problems, efficient modifications in terms of cost and development time are difficult. Additionally, usability problems in only specific perspectives are identified in individual usability evaluations. It is important to identify comprehensively usability problems in various perspectives. To improve these situations, the authors have proposed a method to automatically integrate various types of usability evaluations.

Their method adds functions to record the operation histories of the target software. This information is then used to perform individual usability evaluations with an emphasis on usability categories, such as efficiency, errors, and learnability. Then the method integrates these individual evaluations to identify usability problems and subsequently prioritize these problems according to usability categories determined by the software developers and end users.

Specifically, the authors' research focuses on employing automatic usability evaluations to identify problems. For example, they analyze the operation histories, but do not focus on manually performed evaluations such as heuristic ones. They assume their research can aid software developers and usability engineers because their work allows them to recognize the more serious problems. Consequently, the software can be modified to resolve the usability problems and better meet the end users' requirements. In the future, the authors strive to integrate more diverse usability evaluations, including heuristic evaluations, to refine integration capabilities, to identify problems in more detail, and to improve the effectiveness of the usability evaluations.

DOI: 10.4018/978-1-4666-2491-7.ch019

Copyright © 2013, IGI Global. Copying or distributing in print or electronic forms without written permission of IGI Global is prohibited.

INTRODUCTION

To develop usable software, usability of software GUIs are very important, because end users directly interact with GUIs. For this purpose, there are various methods. For example, GUIs are developed along with usability guidelines and patterns, and developed GUIs are evaluated and improved. Especially, whether end users feel GUIs usable differs among end users. Thus, developed GUIs are often evaluated and improved iteratively in terms of usability.

Various usability evaluation methods have proposed, including experimental methods in usability testing (Barnum, 2001), analytic analytic evaluations and heuristic evaluations (Leventhal & Barnes, 2008), and each method has a different perspective. That is, when several types of usability evaluations are performed for a specific software package, numerous problems are found according to each usability evaluation. Consequently, it is difficult to resolve all problems due to software development costs and schedule. Thus, the identified issues should be prioritized or grouped together to elucidate the more serious problems.

Herein we propose a method to integrate various types of usability evaluations and to identify the more serious usability problems. Using our method, usability problems are prioritized. Our method initially generates functions to record operation histories of end users in software. This information is then used for usability evaluations. Finally, the results of each usability evaluation are integrated based on the priorities of usability categories (Nielsen, 1994), which are determined by software developers and end users.

Our chapter is divided as follows. The second section describes the features of our method, while the third discusses NEM (Novice Expert radio Method) and AHP (Analytic Hierarchy Process), which are important techniques used in our method. The fourth section provides an overview of our method with an emphasis on operation histories for usability evaluations; in addition to generating and adding functions to record operation histories, the type, acquisition, and analysis of operation histories are discussed. The fifth section describes the types of usability evaluations that our method is applicable to as well as the integration results of the evaluations. Our method adopts three types: efficiency, error, and learnability evaluations. The sixth section shows the results of applying our method to two examples. The seventh section describes related works, and finally the eighth section provides a conclusion and future research.

BACKGROUND

There are many researches focused on usability evaluations.

Babaian et al. proposed a method to use operation histories of end users to design interfaces (Babaian, Lucas, & Topi 2006). In every component, various types of information such as operation time and keystrokes are recorded. Then usability assessments are performed in terms of efficiency of UI operations and work achievements. However, the results are not integrated based on criteria that software developers and end users determine. In our method, the results of usability evaluations can be integrated based on specific criteria.

Fukuzumi et al. proposed a method to evaluate usability of a system via checklists (Fukuzumi, Ikegami, & Okada, 2009). The criteria are clear in the checklists and the evaluation can reflect the evaluator's intent. However, evaluations of a large-scale system are a heavy burden and highly skilled evaluators are necessary to appropriately assess usability. In contrast, usability evaluations in our method only require the evaluators to operate the software.

Fiora et al. proposed to evaluate usability of software using component information and operation histories of end users (Fiora, Baker, & Warren, 2008). GUIs are evaluated based on operating time, number of times of component usage and users' behaviors. Based on this method, a usability testing framework for mobile terminals is developed. However, usability evaluations in this method are performed for one screen. In our method, usability of many windows is evaluated, and more serious usability problems can be identified.

Atterer et al. proposed a method to evaluate usability of Web pages (Atterer, Wnuk, & Schmidt, 2006). In this method, operation histories, such as mouse and keyboard events in Web pages, are recorded using JavaScript. Based on the recorded information, usability evaluations are performed by analyzing whether end users finish a specific task. However, evaluations of various perspectives are not performed in this method.

Thus, there are few methods to perform various types of usability evaluations and integrate the results of them. In such situations, usability evaluations are performed specific perspectives, and serious usability problems for end users may not be found. In addition, when these proposed usability evaluations are performed individually, many usability problems are found, and all problems may not be able to be modified because of cost and development schedule. To resolve these problems, in our method, various types of usability evaluations are performed, and the results are integrated. When integrating results of individual usability evaluations, end users can determine priorities of usability categories. According to the priorities, the more serious usability problems for end users can be found.

PRELIMINARY

Our method employs two existing techniques. The first technique is NEM (Novice Expert ration Method), which is used as an individual usability evaluation. The second is AHP (Analytic Hierarchy Process), which is used to integrate the results of individual usability evaluations. By using both these techniques, the more serious usability problems can be identified from the numerous usability problems found by the individual usability evaluations. Below these two techniques are described in detail.

NEM

NEM is a method that focuses on the differences between end users and expert users (Kurosu, Urokohara, & Sato, 2002). When design models by developers and system images by end users differ, operation problems occur (Norman, 1988). In NEM, problems are identified by clarifying these differences in terms of operation time.

The ratio of the operation time of an expert user (Expert) to that of an end user (Novice) is calculated in NEM. This ratio is called the "NE ratio." For *i*-th step (i = 1, 2, ..., m) of *m* steps in a task, the operation time of an expert user is T_{ei}, and the operation time of an end user is T_{ni}. The NE ratio (R_i) for the *i*-th step is calculated by Formula 1.

$$\frac{T_{ni}}{T_{ei}} \qquad (1)$$

A high NE ratio indicates large differences in the system images between expert users and end users are present. According to various experiments, an NE ratio greater than 4.5 indicates the step of the task is problematic.

AHP

To integrate usability evaluation results, AHP (Saaty, 1980), which is a requirement analysis method, is used. When a resolution must be selected from candidates that satisfy specific requirements, "purpose" and "indicators" are defined where the former is the purpose of the specific requirements and the latter is the criteria

Figure 1. Hierarchical structure of AHP

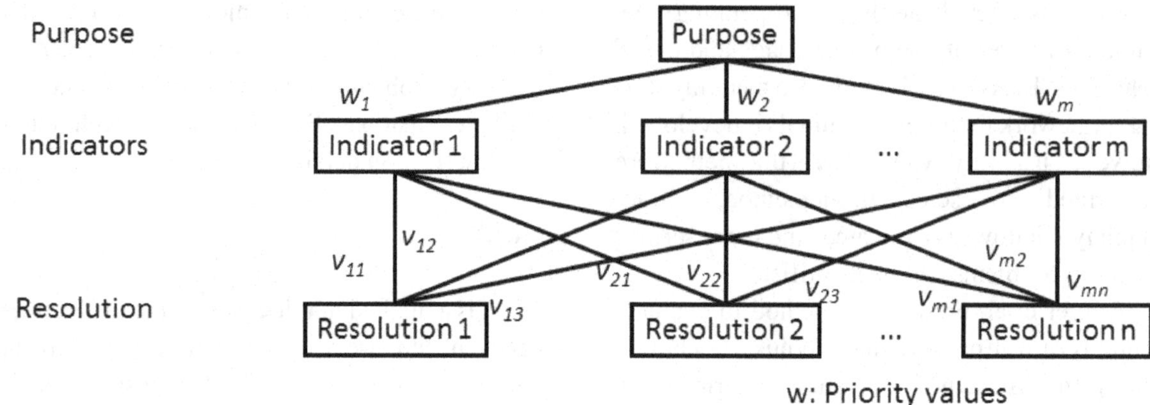

to achieve the "purpose." Then an evaluation value for each resolution is calculated, and the optimal resolution can be obtained.

For example of purchasing an airplane ticket, there are three types of tickets: ticket A, ticket B, and ticket C. In this case, the purpose is defined as buying an airplane ticket. Selection of a ticket is defined as the resolution. To evaluate these tickets, indicators include cost, flight duration, and departure and arrival dates.

AHP is depicted as a hierarchical model like Figure 1. In this figure, there are n resolutions, and m indicators are defined.

The priority values of the indicators are set after defining the purpose, indicators, and resolutions. Priority values represent the contribution degrees of the indicators necessary to achieve the purpose. Priority values are defined by comparing the importance of two indicators. When the importance is high, a large value is assigned as the priority value. In this section, a_{ij} indicates the priority value of indicator i ($i = 1, 2, ..., m$) compared to indicator j ($j = 1, 2, ..., m$). In this case, a_{ji} is defined as Formula 2.

$$a_{ji} = \frac{1}{a_{ij}} \tag{2}$$

Then, a pairwise matrix (3) is created.

$$\begin{pmatrix} 1 & a_{12} & a_{13} & \cdots & a_{1m} \\ a_{21} & 1 & a_{23} & \cdots & a_{2m} \\ a_{31} & a_{32} & 1 & \cdots & a_{3m} \\ \vdots & \vdots & \vdots & \ddots & \vdots \\ a_{m1} & a_{m2} & a_{m3} & \cdots & 1 \end{pmatrix} \tag{3}$$

Consider the example of an airplane ticket. Indictors include cost, flight time, and departure and arrival dates. Then these indicators are compared to each other.

- Cost and flight duration: Cost is very important.
- Cost and departure and arrival dates. Cost is a somewhat important.
- Flight duration and departure and arrival dates: Departure and arrival dates are important.

As shown in Table 1, the relationships among indicators are considered to create pairwise matrix (4).

Table 1. Example of defining priority values

	Cost	Flight duration	Departure and arrival date
Cost	1	7	3
Flight duration	1/7	1	1/5
Departure and arrival date	1/3	5	1

$$\begin{pmatrix} 1 & 7 & 3 \\ 1/7 & 1 & 1/5 \\ 1/3 & 5 & 1 \end{pmatrix} \quad (4)$$

Using these priority values, the "w" values are calculated where "w" means the weight of a specific indicator compared to the other indicators. According to pairwise matrix (3), the value "w" of indicator x (w_x) ($x = 1, 2, ..., m$) is calculated by Formula 5.

$$w_x = \frac{\sqrt[m]{\prod_{j=1}^{m} a_{xj}}}{\sum_{i=1}^{m} \sqrt[m]{\prod_{j=1}^{m} a_{ij}}} \quad (5)$$

Using the airplane ticket example in pairwise matrix (4), "w" is calculated as shown in Table 2.

Next, the resolutions are evaluated by every indicator. In terms of indicator k (k = 1, 2, ..., m), the resolutions are compared each other. Then the evaluated values are defined. Evaluated value "b_{kij}" (i = 1, 2, ..., n) (j = 1, 2, ..., n) indicates the evaluation of resolution i compared to resolution j in terms of indicator k. When the evaluation is high, the evaluation item has a large value. In this case, bk_{ii} is defined as Formula 6.

Table 2. Example of value "w" for each indicator

	Cost	Flight duration	Departure and arrival date
Value of "w"	0.649	0.0719	0.279

Table 3. Example of evaluated values in terms of costs

	Ticket A	Ticket B	Ticket C
Ticket A	1	5	5
Ticket B	1/5	1	1
Ticket C	1/5	1	1

$$b_{kji} = \frac{1}{b_{kij}} \quad (6)$$

Then, a pairwise matrix (7) for indicator k is created.

$$\begin{pmatrix} 1 & b_{x12} & b_{x13} & \cdots & b_{x1m} \\ b_{x21} & 1 & b_{x23} & \cdots & b_{x2m} \\ b_{x31} & b_{x32} & 1 & \cdots & b_{x3m} \\ \vdots & \vdots & \vdots & \ddots & \vdots \\ b_{xm1} & b_{xm2} & b_{xm3} & \cdots & 1 \end{pmatrix} \quad (7)$$

For the airline ticket example, the three situations are evaluated in terms of cost. If tickets A is the cheapest, and tickets B and C are almost the same, then Table 3 depicts the evaluated value of each resolution, and pairwise matrix (8) is created.

$$\begin{pmatrix} 1 & 5 & 5 \\ 1/5 & 1 & 1 \\ 1/5 & 1 & 1 \end{pmatrix} \quad (8)$$

In terms of flight duration, if ticket A is the shortest, ticket B is the middle and ticket C is

Table 4. Example of evaluated values in terms of flight duration

	Ticket A	Ticket B	Ticket C
Ticket A	1	5	7
Ticket B	1/5	1	5
Ticket C	1/7	1/5	1

Table 5. Example of evaluated values in terms of departure and arrival date

	Ticket A	Ticket B	Ticket C
Ticket A	1	1	3
Ticket B	1	1	3
Ticket C	1/3	1/3	1

the longest, then Table 4 depicts the evaluated value of each resolution, and a pairwise matrix (9) is created.

$$\begin{pmatrix} 1 & 5 & 7 \\ 1/5 & 1 & 5 \\ 1/7 & 1/5 & 1 \end{pmatrix} \quad (9)$$

In terms of departure and arrival date, if ticket A and B is almost the same and appropriate for the schedule, and ticket C is a little tight schedule, then Table 5 depicts the evaluated value of each resolution, and a pairwise matrix (10) is created.

$$\begin{pmatrix} 1 & 1 & 3 \\ 1 & 1 & 3 \\ 1/3 & 1/3 & 1 \end{pmatrix} \quad (10)$$

According to pairwise matrix (7), the value "v" of resolution x (v_{kx}) ($x = 1, 2, ..., n$) in terms of indicator k is calculated by Formula 11. In Figure 1, v is the weight of a specific resolution compared to other resolutions.

$$v_{kx} = \frac{\sqrt[n]{\prod_{j=1}^{n} b_{kxj}}}{\sum_{i=1}^{n} \sqrt[n]{\prod_{j=1}^{n} b_{kij}}} \quad (11)$$

For example, using pairwise matrixes (8), (9), and (10), the "v" is calculated as shown in Table 6.

Eventually, the final value is calculated for each resolution using Formula 12. In this formula, p_i is the final value of resolution i ($i = 1, 2, ..., n$). When p_i is high, resolution i is optimal.

Table 6. Example of evaluated values in terms of flight duration

	Cost	Flight duration	Departure and arrival date
Ticket A	0.714	0.714	0.429
Ticket B	0.143	0.218	0.429
Ticket C	0.143	0.067	0.143

$$p_k = \sum_{x=1}^{n} w_x v_{xk} \quad (12)$$

For the airplane ticket example, the final value is calculated as shown in Table 7. According to this result, the value of ticket A is the largest. Consequently, ticket A is considered to be optimal solution.

INTEGRATION OF USABILITY EVALUATIONS

Figure 2 depicts a process of our method.

Our method consists of five steps. First, source programs of the target software are analyzed, and functions to record operation histories are generated and added to the target software. Next, end users operate the target software with the functions, and their operation histories are analyzed via usability evaluations. Then the usability of the target software is evaluated using three types of individual usability evaluations. Finally, the results of the individual usability evaluations are

Table 7. Example of evaluated values in terms of flight duration

	Ticket A	Ticket B	Ticket C
Final value	0.634	0.228	0.137

Figure 2. Process of our method

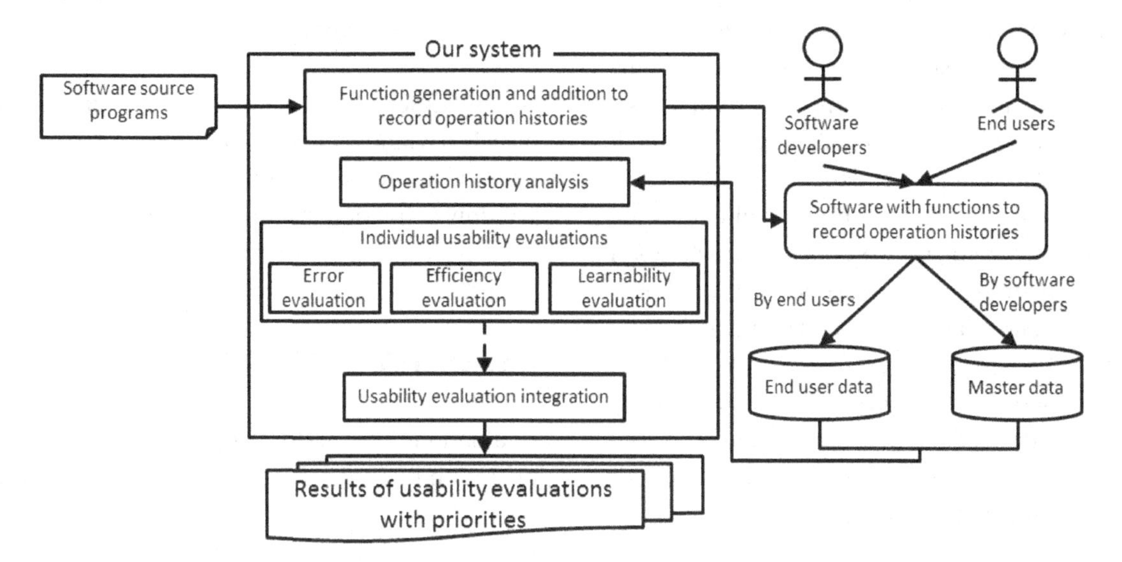

integrated, and the more serious usability problems are identified from all the usability problems found in the individual usability evaluations.

In this section, we describe the first three steps of our method: the generation and addition of functions to record operation histories, usages of the target software with the functions, and analysis of the operation histories.

Function Generation and Addition

Functions to record operation histories are generated by analyzing source programs of the target software as programs of AspectJ (Miles, 2005). AspectJ is a tool that realizes aspect oriented programming for Java programming language.

Functions, which are not present in developed software, can be easily added using AspectJ, which adds functions without modifying the source programs. In AspectJ, a specific point in the source code, such as method invocation, method execution, instance initialization, reference of value from a field, or assignment of a value to a field, is called a "join point." A process in a function added to software is called "advice." The definition of points where an advice is executed is called a "point cut." A point cut consists of one or more join points. When a join point is executed, the advice defined in the corresponding point cut to the join point is executed. In a point cut, whether the advice is executed before or after executing the join points can be determined. Using AspectJ, extra functions can be added to the target software without modifying the source programs. Consequently, AspectJ is widely used to add extra functions such as ones to record operation histories.

To generate functions to add to software using AspectJ, the following is extracted from the source programs of the target software.

- Variable names of the widgets in a window.
- Class names of the widgets in a window.
- Superclass name of each class.

The generated functions are then added to the target software. Consequently, operating the software, the operation histories are recorded.

Usage of the Target Software

The operation histories are recorded for usability evaluations as the target software with functions to record histories is used. When operating software for usability evaluations, software developers order a certain task, and end users operate the target software to achieve the task. In this case, two types of operation histories must be recorded: end user data and master data.

In our method, the usability evaluations are performed by comparing the operation histories of end users to those of expert users. Operation histories by end users are called "end user data," while those by expert users are called "master data." Expert users may include software developers. Ideally, end user data for a task will be comprised of data from many end users. In contrast, master data requires input from an expert.

Analysis of Operation Histories

Our method analyzes operation histories for usability evaluations because the evaluations are automatically performed and the results are obtained as objective values. Thus, software developers can devise plans to modify usability problems according to financial and time constraints of software development.

In our usability evaluations, the usability of the target software is evaluated in terms of efficiency according to the operation time of a task by end users, errors in a task by end users based on the operation sequences, and learnability using the operation time. To evaluate these, the following items for a widget or a window are extracted from the operation histories.

- Window in which widgets are located.
  - For error evaluation.
- Window activation.
  - For all evaluations.
- User events associated with the widgets such as mouse clicks and key strokes.
  - For all evaluations.
- Time that the user events occurs in the widget and time windows are activated.
  - For efficiency and learnability evaluations.

In addition, usability evaluations are performed based on windows, as described below. Therefore, operation histories are divided into windows using window activation information. By this division, sequences of window switching are extracted.

USABILITY EVALUATION

Evaluations are performed using the obtained operation histories. First, individual evaluations are performed, and then the results are integrated into an overall usability evaluation.

Individual Usability Evaluations

Based on the usability categories defined by Nielsen (1994), the individual usability evaluations include efficiency, errors, and learnability. "Efficiency" indicates how effectively end users can use the target software. "Errors" indicates how few errors occur. "Learnability" indicates how easily end users can become proficient in the software.

In each evaluation, a window in a sequence of extracted window switching is recognized as a step in a performed task. Every step is evaluated and assigned a score. In the evaluations, when the score is high, the step has usability problems. Below, we describe each usability evaluation considering a three-step task as an example.

Efficiency Evaluation

Considering operation time to achieve a task, end users usually require more time than expert users. In this evaluation, NEM is used.

In our method, the operation time is defined as the time each window is operated. The operation time for a window is from activation of the win-

Figure 3. Example of the NE ratio

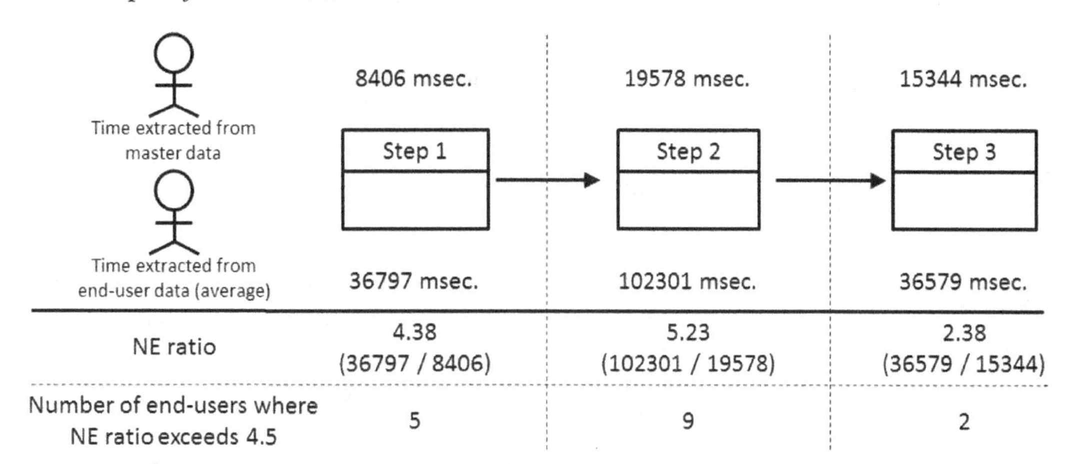

dow to activation of the next window in window switching. The end user operation time for a task is the average time of all the end users. Figure 3 shows an example of the NE ratio. In this case, the NE ratio of window B exceeds 4.5. Consequently, window B is considered to have problems. It is noteworthy that window B also has the most end users with an individual NE ratio above 4.5.

Error Evaluation

Software where end users often make mistakes is considered to have usability problems. Our method focuses on differences between operations of a task in the master and end user data by comparing the two. In this evaluation, information of user events occurring in widgets and window switching is used. If the operations recorded in the end user data differ from those recorded in the master data, then the score is added. In this evaluation, the sum of scores for each step of the task is calculated, and a high score indicates usability problems. Table 8 shows the score of each operation.

Table 8. Scores defined for error evaluation

Target operations		Score
Unintentional window switching		5
Unnecessary widget operation	With window switching	3
	Without window switching	1

Using the scores defined in Table 8, the sum of the scores for each step in a task is calculated from the end user data. Table 9 shows an example of the calculated scores for each step. In this example, the sum of the scores for the step 1 is higher than the other steps.

Learnability Evaluation

The first time end users operate the software, it takes a long time, but once they become proficient, the operation time becomes shorter until reaching a constant level. If end users have to operate the software many times to become proficient, then

Table 9. Example of calculated scores for each step

Target operations		Step 1	Step 2	Step 3
Unintentional window switching		6	2	2
Unnecessary widget operation	With window switching	3	0	2
	Without window switching	57	34	23
Sum of scores for each step		66	34	27

Table 10. Example of calculated values for learnability

	Step 1	Step 2	Step 3
Operation time in master data (msec)	8406	19578	15344
Average time of first operation by end users (msec)	30797	77301	36579
Average time of second operation by end users (msec)	15484	30953	20375
Calculated values	0.316	0.197	0.237

the software has usability problems. In our method, the operation time of each step in a task extracted from the master data recorded in Section 3.2 is considered to be the operation time that end users become proficient. Thus, when end users perform the same operation twice, the step time of the two operations are compared. In cases where the step time of the second operation is not close to that of the master data for the first operation, the step is deemed to have usability problems.

In this evaluation, Formula 9 is used to evaluate learnability. T_e indicates the operation time of a step recorded in the master data, T_{u1} is the average time of the step in the first operation by all end users, and T_{u2} denotes the average time of the step in the second operation by all end users. If this calculated value is high, then the step has usability problems.

$$\frac{T_{u2} - T_e}{T_{u1} - T_e} \tag{9}$$

Table 10 shows an example of a learnability evaluation based on master and end user data. In this example, the calculated value of step 1 is higher than the other steps.

Integration of Usability Evaluations

After performing three types of usability evaluations, the results are integrated to extract the more serious usability problems using AHP.

Our method aims to elucidate the more serious usability problems. Thus, the steps of a task are considered to be resolutions of AHP. Three usability categories, efficiency, errors, and learnability, are considered as indicators of AHP because their importance differs according to software developers, end users and software characteristics. Figure 4 shows an example of a hierarchical structure of the relationship between usability categories and steps of a task along with AHP.

Figure 4. Example of hierarchical structure

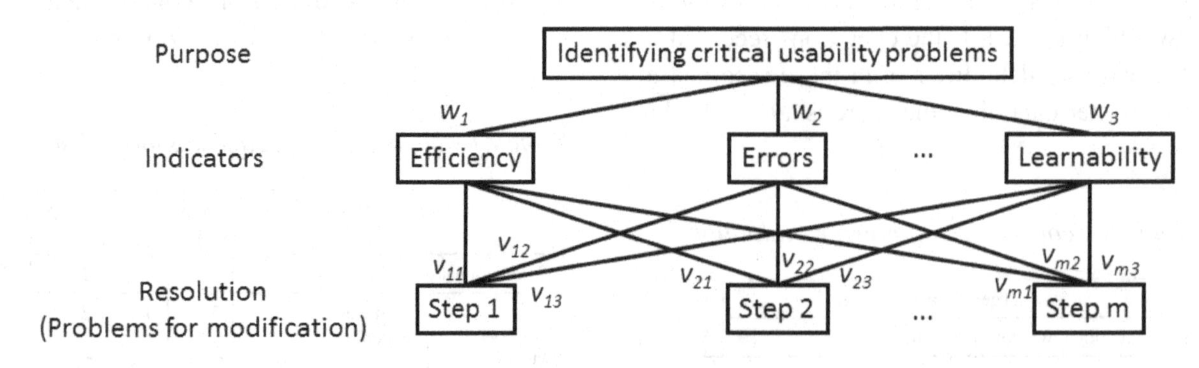

In Figure 4, "*w*" is a value calculated using a pairwise comparison matrix, which is derived using priority values for indicators determined by software developers and end users. Additionally, "*v*" is a value calculated using a pairwise comparison matrix, which is created using the scores of individual usability evaluations. Applying the Formula 3 and 5 to Figure 4, the pairwise matrix and the formula for calculating "*w*" are (13) and (14), respectively. In these, a_{ij} is the priority value of indicator i (i = 1, 2, 3) compared to indicator j (j = 1, 2, 3). Indicators 1, 2, and 3 represent "efficiency," "errors," and "learnability," respectively. End users and software developers assign the priority values.

$$\begin{pmatrix} 1 & a_{12} & a_{13} \\ a_{21} & 1 & a_{23} \\ a_{31} & a_{32} & 1 \end{pmatrix} \tag{13}$$

$$w_x = \frac{\sqrt[3]{\prod_{j=1}^{3} a_{xj}}}{\sum_{i=1}^{n} \sqrt[3]{\prod_{j=1}^{n} a_{ij}}} \tag{14}$$

In addition, applying the Formula 7 and 11, the pairwise matrix and the formula for calculating "*v*" are (15) and (16), respectively. Value v_{kx} indicates the weight of resolution k (k = 1, 2, ..., m) compared to other resolutions in terms of indicator x (x = 1, 2, 3). In these, b_{xij} is a evaluated value of resolution i (i = 1, 2, ..., m) comparing resolution j (j = 1, 2, ..., m) in terms of indicator x (x = 1, 2, 3).

$$\begin{pmatrix} 1 & b_{x12} & b_{x13} & \cdots & b_{x1m} \\ b_{x21} & 1 & b_{x23} & \cdots & b_{x2m} \\ b_{x31} & b_{x32} & 1 & \cdots & b_{x3m} \\ \vdots & \vdots & \vdots & \ddots & \vdots \\ b_{xm1} & b_{xm2} & b_{xm3} & \cdots & 1 \end{pmatrix} \tag{15}$$

$$v_{kx} = \frac{\sqrt[m]{\prod_{j=1}^{m} b_{kxj}}}{\sum_{i=1}^{n} \sqrt[3]{\prod_{j=1}^{m} b_{kij}}} \tag{16}$$

Normally, an integral number assigned to evaluated value b_{xij} (or b_{xji}) and b_{xji} (or b_{xij}) is defined as $1/b_{xij}$ (or $1/b_{xji}$) in AHP, However, our method directly reflects the scores of individual usability evaluations in this integration process by defining b_{xij} as Formula 17. In this formula, r_{xi} indicates the score of step i (resolution i) in usability evaluation x (indicator x) and r_{xj} indicates the score of step j in usability evaluation x. Due to the individual usability evaluations, the score for each step in a target task is calculated. However, the basis of the scores in each usability evaluation differs. Thus, these scores are modified as deviation scores. r_{xi} and r_{xj} are the deviation scores.

$$b_{xij} = \frac{r_{xi}}{r_{xj}} \tag{17}$$

Using the calculated valued "*w*" and "*v*" by the Formula 14 and 16, the final scores of AHP are calculated using the Formula 18. If this formula yields a high score, the step has more serious usability problems compared to other steps.

$$p_i = \sum_{x=1}^{3} w_x v_{xi} \tag{18}$$

This integration process is applied to the example of the individual usability evaluation results in 5.1. First, the priority value of each indicator is assigned as shown in Table 11, and pairwise matrix (19) is created. Table 12 show the calculated "w" values.

Table 11. Example of priority values for usability evaluations

	Efficiency	Errors	Learnability
Efficiency	1	1/7	1/5
Errors	7	1	3
Learnability	5	1/3	1

Table 12. Example of priority values of indicators and "w"

	Efficiency	Errors	Learnability
Value of "w"	0.072 (w_1)	0.649 (w_2)	0.279 (w_3)

Table 13. Modified deviation scores of individual usability evaluations

	Efficiency	Errors	Learnability
Step 1	52.618	61.382	60.899
Step 2	58.430	45.992	41.248
Step 3	38.951	42.626	47.853

Table 14. Evaluated values in terms of efficiency

	Step 1	Step 2	Step 3
Step 1	1	52.618/58.430	52.618/38.951
Step 2	58.430/52.618	1	58.430/38.951
Step 3	38.951/52.618	38.951/58.430	1

Table 15. Evaluated values in terms of errors

	Step 1	Step 2	Step 3
Step 1	1	61.382/45.992	61.382/42.626
Step 2	45.992/61.382	1	45.992/42.626
Step 3	42.626/61.382	42.626/45.992	1

Table 16. Evaluated values in terms of learnability

	Step 1	Step 2	Step 3
Step 1	1	60.899/41.248	60.899/47.853
Step 2	41.248/60.899	1	41.248/47.853
Step 3	47.853/60.899	47.853/41.248	1

Table 17. Example of values of "v"

	Step 1	Step 2	Step 3
Result of efficiency evaluation	0.351 (v_{11})	0.390 (v_{21})	0.260 (v_{31})
Result of error evaluation	0.409 (v_{12})	0.307 (v_{22})	0.284 (v_{32})
Result of learnability evaluation	0.406 (v_{13})	0.275 (v_{23})	0.319 (v_{33})

$$\begin{pmatrix} 1 & 1/7 & 1/5 \\ 7 & 1 & 3 \\ 5 & 1/3 & 1 \end{pmatrix} \tag{19}$$

Next, the scores of individual usability evaluations are modified as deviation scores. Table 13 shows the modified deviation scores of examples of 5.1.1, 5.1.2, and 5.1.3.

Based on these scores, the evaluated values are assigned, and pairwise matrixes are created. For efficiency, errors, and learnability indicators, the evaluated values are shown in Tables 14, 15, and 16, and pairwise matrixes are shown in (20), (21), and (22), respectively.

$$\begin{pmatrix} 1 & 52.618/58.430 & 52.618/38.951 \\ 58.430/52.618 & 1 & 58.430/38.951 \\ 38.951/52.618 & 38.951/58.430 & 1 \end{pmatrix} \tag{20}$$

$$\begin{pmatrix} 1 & 61.382/45.992 & 61.382/42.626 \\ 45.992/61.382 & 1 & 45.992/42.626 \\ 42.626/61.382 & 42.626/45.992 & 1 \end{pmatrix} \tag{21}$$

$$\begin{pmatrix} 1 & 60.899/41.248 & 60.899/47.853 \\ 41.248/60.899 & 1 & 41.248/47.853 \\ 47.853/60.899 & 47.853/41.248 & 1 \end{pmatrix} \tag{22}$$

Based on pairwise matrixes (20), (21), and (22), the values of "v" are calculated. Table 17 shows an example of the "v" values.

Table 18. Example of the final result

	Step 1	Step 2	Step 3
Integrated result	0.404	0.304	0.292

As shown in Table 18, the final values are calculated using Formula 18. According to this table, step 1 has the highest score focusing errors. Consequently, step 1 is deemed to have the most serious usability problems

EVALUATION

To confirm effectiveness of our method, we evaluated the usability of two software packages: jWorkSheet (Ponec, 2009) and task management system. The jWorkSheet is used to manage work and measure the time within a project, while the task management system is used to manage work and deadlines in a project.

Participants

Fifteen participants performed the evaluation. The participants were around 24 years old and all were master course students or company workers. All had high information literacy, but were not expert users of these two software packages.

Process

We prepared two software packages, which record operation histories. These were added to jWorkSheet and the task management system. All of the participants operated both jWorkSheet and the task management system. Prior to the participants' operations, we determined specific tasks for the usability evaluations. The instructions for these tasks, which were provided to all participants, were for the flow of tasks, but not for the operation methods of the target software. Then the participants operated the target software packages based on the given tasks.

The given tasks for usability evaluations were as follows:

- **jWorkSheet.**
  - **Step 1-1** Select a project and task previously registered, and begin timing measurement.
  - **Step 1-2** Register a new project and a task.
  - **Step 1-3** Select the project and task registered in Step 1-2, and output a report.
- **Task management system.**
  - **Step 2-1** Login using ID and password.
  - **Step 2-2** Add a task.
  - **Step 2-3** Register completion of a task.
  - **Step 2-4** Register deferment of a task.
  - **Step 2-5** Change the name of a task.
  - **Step 2-6** Print a list of all tasks.

In this case, end user data was the operation histories operated by the participants. For master data, one of authors became proficient in the target software packages and created the master data.

Result

Table 19 shows the results of performing the above two tasks, while Table 20 shows the sums of scores calculated in individual usability evaluations. The parentheses in Table 20 indicate the values calculated using a pairwise comparison matrix based on the scores on the left.

In Tables 19 and 20, the results of individual usability evaluations were integrated based on priority values of the indicators. The priority values for indicators and "*w*" in Section 5.2 are shown in Tables 21, 22, and 23.

Based on Table 20, our system calculated "*v*" for each step. Table 24 shows the calculated "*v*."

Table 19. Results of performing tasks

	Operation time (msec)			Unintentional window switching	Unnecessary widget operations	
	Master data	First operation by end users (average)	Second operation by end users (average)		With window switching	Without window switching
Step 1-1	8406	30797	15484	6	3	57
Step 1-2	19578	77301	30953	2	0	34
Step 1-3	15344	36579	620375	2	2	23
Step 2-1	5578	20375	15407	1	1	0
Step 2-2	11735	36766	15781	9	3	14
Step 2-3	18500	54888	24172	2	0	9
Step 2-4	11625	54406	17125	2	1	11
Step 2-5	17109	37613	31563	2	2	13
Step 2-6	3641	16953	6969	4	3	7

Table 20. Results of individual usability evaluations

	Score of efficiency	Score of errors	Score of learnability
Step 1-1	5 (0.327)	96 (0.41)	0.316 (0.406)
Step 1-2	9 (0.403)	44 (0.3)	0.197 (0.275)
Step 1-3	2 (0.27)	39 (0.29)	0.237 (0.319)
Step 2-1	6 (0.324)	8 (0.263)	0.664 (0.414)
Step 2-2	4 (0.277)	68 (0.443)	0.162 (0.28)
Step 2-3	5 (0.342)	19 (0.335)	0.156 (0.323)
Step 2-4	12 (0.342)	21 (0.335)	0.129 (0.323)
Step 2-5	3 (0.26)	29 (0.324)	0.705 (0.416)
Step 2-6	9 (0.366)	36 (0.337)	0.25 (0.297)

Table 21. Priorities of indicators (focused on efficiency)

	Efficiency	Errors	Learnability
Efficiency	1	9	5
Errors	1/9	1	1/7
Learnability	1/5	7	1
Value of "*w*"	0.722	0.051	0.227

Then, Based on Tables 21, 22, 23, and 24, the individual usability evaluation results were integrated using AHP. Figures 5 and 6 show the integrated results for jWorkSheet and the task management system, respectively.

Discussion

Evaluation results in 6.3 imply that the focused steps with usability problems depend on the priorities of the indicators. For example, Figure

Table 22. Priorities of indicators (focused on errors)

	Efficiency	Errors	Learnability
Efficiency	1	1/5	7
Errors	5	1	9
Learnability	1/7	1/9	1
Value of "*w*"	0.227	0.722	0.051

Table 23. Priorities values of indicators (focused on learnability)

	Efficiency	Errors	Learnability
Efficiency	1	1/7	1/9
Errors	7	1	1/5
Learnability	9	5	1
Value of "*w*"	0.051	0.227	0.722

5 for jWorkSheet shows that step 1-1 had usability problems for errors and learnability because participants were unfamiliar with the time and work management system. However, even if we explained JWordkSheet prior to the evaluations, many participants seemed unable to concretely recognize their ability. Figure 5 also shows that step 1-2 had usability problems. Step 1-2 required the same type of data to be inputted, but operation flow had some limitations. Thus, participants appeared to have difficulty understanding the operation flow. However, step 1-2 had a good value of learnability, which is likely because the participants spent too much time on the first operation. That is, the participants could spend much less time on the second operation than the first operation. Like this, when there are big differences of operation time between the first step and the second step, the results of evaluations for learnability become good. However, in this case, the step is considered to have usability problems according to the operation time of the first step. It is necessary to develop methods to find the usability problems in this case.

Figure 6 depicts large differences in the focused steps based on the priorities of the indicators. In the figure, the error values decrease from step 2-2 to step 2-5. In this task management system, after step 2-2, almost all the same windows were displayed, but the a few widgets differed in the windows. Hence, the values decreased because the participants became familiar with the system from step 2-2 to step 2-5.

In our method, a step is operations in a window. Consequently, it may be difficult to identify detailed usability problems in a window. That is, in this task management system, even if errors decreased from step 2-2 to 2-5, detailed usability

Table 24. Values "v" for the steps

	Efficiency	Errors	Learnability
Step 1-1	0.327	0.369	0.366
Step 1-2	0.403	0.270	0.248
Step 1-3	0.270	0.261	0.287
Step 2-1	0.162	0.072	0.119
Step 2-2	0.142	0.125	0.083
Step 2-3	0.152	0.082	0.082
Step 2-4	0.221	0.084	0.080
Step 2-5	0.132	0.091	0.122
Step 2-6	0.191	0.097	0.089

Figure 5. Integrated result of jWorkSheet

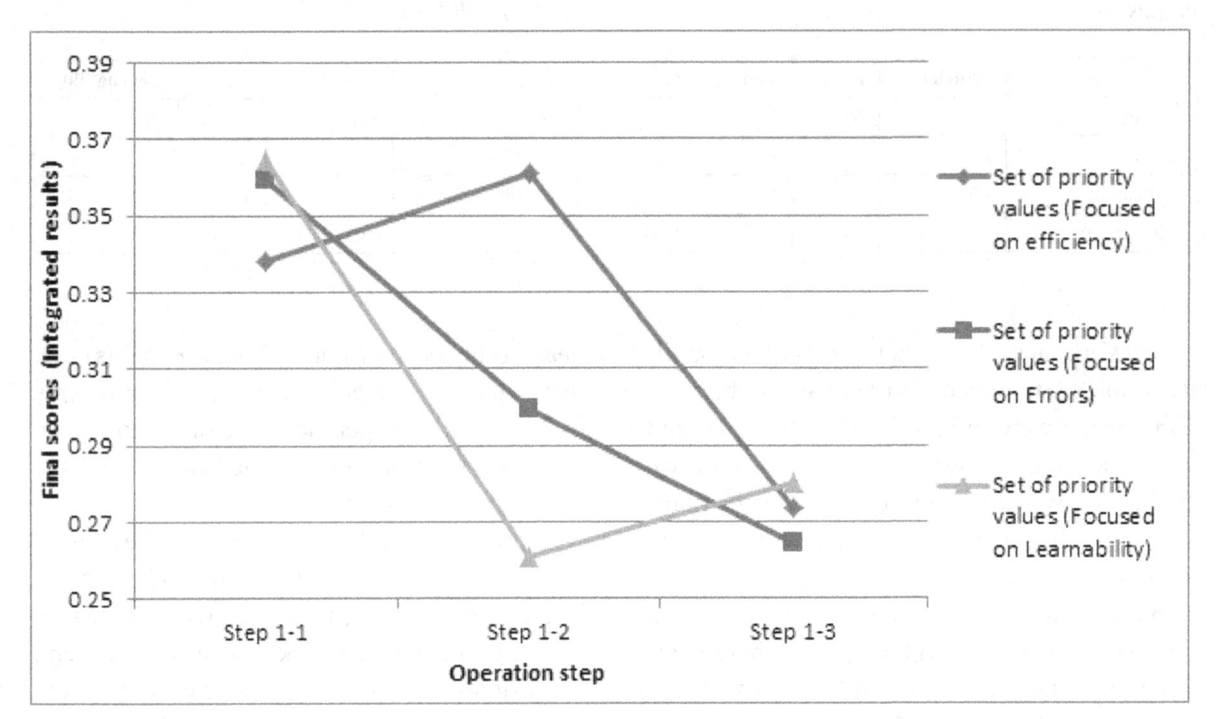

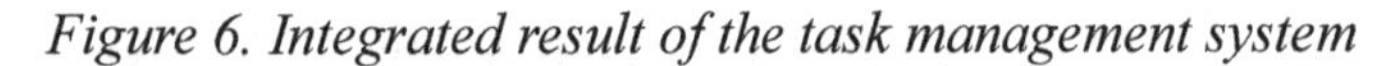
Figure 6. Integrated result of the task management system

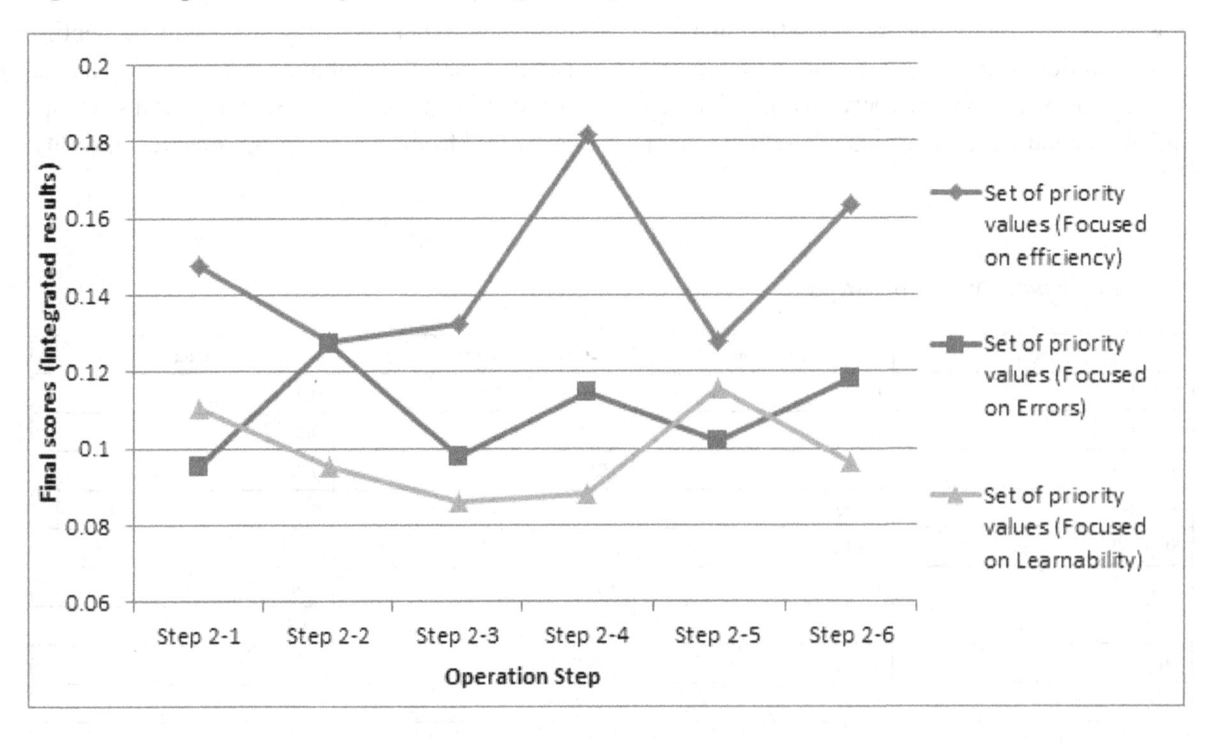

problems may be present in the windows. However, our method needs to be improved so that steps can be subdivided, and usability evaluations can be performed at a level of widgets.

According to Figures 5 and 6, the more serious usability problems depend on the priority values of the indicators, which are determined by the end users and software developers. Figure 6 demonstrates that the indicators can identify the more serious usability problems. Thus, our method allows end users and software developers to preferentially recognize which usability problems should be modified.

According to this evaluation, our method can integrate individual usability evaluations, and steps with usability problems can be extracted based on the priorities. Moreover, the integrated results are appropriate, implying that our method can effectively extract more serious usability problems based on indicators established by software developers and end users.

BENEFITS AND DISADVANTAGES

Numerous methods have been employed to identify usability problems. Because software packages should perform usability evaluations from various perspectives, typically numerous problems are identified. Due to financial and time restrictions, software developers cannot modify every identified issue. Thus, the usability problems must be prioritized, so the more serious ones can be resolved. To this end, we have developed a method to support usability evaluations and to integrate the results.

Target of Our Method

We assume our method can support software developers and usability engineers. Using our method, they can recognize the more serious usability problems. In addition, the more serious usability problems can be identified based on end users' requirements and modified. Consequently, the target software can satisfy end users' requirements.

Benefits of Our Method

Based on the evaluation in Section 6, the assumed advantages of our method previously could be confirmed, and additional advantages could be found. They are described below.

Integration of Usability Evaluations

Many methods exist to evaluate usability, but usability evaluations are performed from different perspectives. Consequently, applying these methods to a software package results in numerous issues, and due to cost and time constraints of software development, it is difficult to resolve every problem. Thus, the more serious usability problems must be identified. Our method integrates the results of usability evaluations to identify critical usability problems.

Prioritization of Usability Problems

Additionally, our method can integrate various types of usability evaluations, allowing usability problems to be prioritized based on the categories determined by software developers and end users. This will enable software developers to determine which issues to address first. Consequently, the costs and schedule of software development are easy to estimate. Moreover, the usability along with the opinions of software developers and end users can be easily realized.

Automatic Evaluation of Usability

In our method, our system automatically adds functions to record operation histories of target software using its source programs. In this way, only by operating our system, functions to obtain operation histories can be added to the software. Then our system performs usability evaluations based on these histories. Because software developers and end users only operate the target software and determine the priorities of the usability categories, the usability evaluation process is mostly automatic. Thus, regardless of skills of software developers for usability evaluations, usability evaluations can be performed easily and efficiently.

Reduction of Costs for Usability Evaluations

Currently usability evaluation methods have manual processes. However, as described above, our method almost automatically performs usability evaluations. Like other usability evaluation methods, our method requires the end user to operate the target software. However, our method automatically performs usability evaluations and integrates the results using the source programs of the target software, operation histories, and by prioritizing the indicators. Moreover, when end users evaluate the usability of the software, the source program does not have to be specially prepared for our method. That is, the end users and software developers only have to assign priority values of the indicators in our method, which is an easy task. Thus, our method enables end users and software developers to easily and effectively evaluate usability. That is, cost and burden of usability evaluations can be reduced using our method.

Disadvantages of Our Method

Due to the evaluation in Section 6, some disadvantages were found. They follow.

Difficulty to Identify Widget Level with Usability Problems

Our method can identify windows with usability problems because the usability evaluations occur at the window level and not at the widget level. However, specific problems in the windows cannot be identified. To rectify this, window structures, including arranged widgets and layouts as well as widget operations, must be analyzed.

Effect of Windows with a Few Operations

Our method can find usability problems for windows that contain many widget operations. In windows with many operations, errors occur more often as it takes longer operation times. Therefore, compared to windows with many operations, our method is not very effective for windows with a few operations.

Problems of Function Addition

Functions could not be added to some widgets when generating and adding functions to record operation histories because some widgets are created dynamically in running software, such as using a loop statement in source programs. To resolve this, we must modify the analysis of the source program or develop a system to record operation histories from running software that does not require the addition of functions.

CONCLUSION

In this chapter, we proposed a method to integrate various types of usability evaluations. Using our method, three types of usability evaluations are performed individually, and the results are integrated. By integrating the results of individual usability evaluations, more serious usability problems can be extracted based on the priorities of usability categories.

Additionally, we evaluated our method and usability problems that could not be identified. For example, if a task has similar windows, appropriately evaluating the differences of usability among the windows is difficult. One reason is the usability evaluations in our method are performed at the window level and not widget level. If usability evaluations are performed at the widget level, then usability problems can be identified and analyzed in more detail. Performing evaluations at the widget level would improve our method. Although we confirmed the effectiveness of our method, our future work aims to employ usability evaluations at the widget level.

Future projects will focus on:

- Extracting more detailed usability problems.
- Defining more appropriate scores for each individual usability evaluation.
- Proposing various types of integrating usability problems.
- Developing another usability evaluation method.

REFERENCES

Atterer, R., Wnuk, M., & Schmidt, A. (2006). Knowing the user's every move - User activity tracking for website usability evaluation and implicit interaction. In *Proceedings of the 15th International Conference on World Wide Web (WWW 2006)*, (pp. 203-212). IEEE.

Babaian, T., Lucas, W., & Topi, H. (2006). Making memories: Apply user input logs to interface design and evaluation. In *Proceedings of CHI 2006 Extended Abstracts on Human Factors in Computing Systems*, (pp. 496-501). ACM.

Barnum, C. M. (2001). *Usability testing and research*. London, UK: Longman.

Fiora, T. W. A., Baker, S. W., & Dobbie, G. (2008). Automated usability testing framework. In *Proceedings of the Ninth Conference on Australasian User Interface (AUIC 2008)*, (pp. 55-64). AUIC.

Fukuzumi, S., Ikegami, T., & Okada, H. (2009). Development of quantitative usability evaluation method. In *Proceedings of the 13th International Conference on Human-Computer Interaction (HCI International 2009)*, (pp. 252-258). ACM.

Kurosu, M., Urokohara, H., & Sato, D. (2002). A new data collection method for usability testing. In Leventhalm, L., & Barnes, J. (Eds.), *Usability Engineering: Process, Products and Examples*. Upper Saddle River, NJ: Pearson Prentice Hall.

Miles, R. (2005). *Aspectj cookbook*. San Francisco, CA: O'Reilly & Associates Inc.

Nielsen, J. (1994). *Usability engineering*. Boston, MA: Morgan Kaufmann.

Norman, D. A. (1988). *The psychology of everyday things*. New York, NY: Basic Books.

Ponec, P. (2009). *jWorkSheet*. Retrieved from http://jworksheet.ponec.net/

Saaty, T. L. (1980). *Analytic hierarchy process*. New York, NY: McGraw-Hill.

Chapter 20
A Protocol for Evaluating Mobile Applications

Clare Martin
Oxford Brookes University, UK

Derek Flood
Dundalk Institute of Technology, Ireland

Rachel Harrison
Oxford Brookes University, UK

ABSTRACT

The number of applications available for mobile phones is growing at a rate that makes it difficult for new application developers to establish the current state-of-the-art before embarking on new product development. This chapter is targeted towards such developers (who may not be familiar with traditional techniques for evaluating interaction design) and outlines a protocol for capturing a snapshot of the present state of the applications in existence for a given field in terms of both usability and functionality. The proposed methodology is versatile in the sense that it can be implemented for any domain across all mobile platforms, which is illustrated here by its application to two dissimilar domains on three platforms. The chapter concludes with a critical evaluation of the process that was undertaken and suggests a number of avenues for future research including further development of the keystroke level model for the current generation of smart phones.

INTRODUCTION

Portable devices such as mobile phones and music players are capable of running a wide variety of applications (or 'apps') which enable users to perform a multitude of tasks while they are away from traditional computing devices. This has contributed to the staggering growth of the mobile phone market in recent years. In 2011 it was estimated that there were almost 6 billion mobile subscriptions worldwide, with a global penetration of 87% as well as a high take up in the developing world (79%) (ITU, 2011). The combined revenue of applications funded either

DOI: 10.4018/978-1-4666-2491-7.ch020

Copyright © 2013, IGI Global. Copying or distributing in print or electronic forms without written permission of IGI Global is prohibited.

through payment for downloading, advertising, or value added services such as subscriptions is expected to rise from just below $10 billion in 2009 to $32 billion in 2015 (Holden, 2010), which is a significant increase in a very short space of time.

The recent growth in mobile application development is partly due to the relatively low cost and high speed of such development. One reason for this is the simplicity of the dominant mobile platforms, such as iOS from Apple and Android from Google, and also because the associated software development tools are freely available and easy to learn. Consequently, users are often faced with a broad choice of applications to help them complete a given task. A recent survey (Flood, et al., 2011) identified four factors, which influence users when choosing an application: function, price, opinions of others and usability. Developers need to consider all of these factors when designing a new application, but it can still be difficult to evaluate the current status of the market for any given domain, since each one is so densely populated.

Before any project can be started, some groundwork is needed to determine the necessary requirements and the context in which the work will take place (Finkelstein, 1993). The context is particularly important for mobile applications development since it is essential to consider the socio-technical environment in which the applications will be developed and used. The groundwork may include an analysis of the present state of the art through a systematic survey of existing applications. Accordingly, this chapter describes a flexible, cost-effective protocol that can be used to perform such a survey quickly and thus eliminate a vast number of the irrelevant applications in a domain.

The proposed evaluation methodology is performed by experts rather than by users. It comprises a series of steps that can be used to filter through a collection of applications by comparing the features that they offer, as well as efficiency and other attributes affecting usability such as personalization, ergonomics, flexibility, security, and error management. In addition, the process was designed to elicit functional requirements by generating a list of features offered by existing applications. The protocol is both platform and task independent which is demonstrated here through its application to two separate task domains on three platforms.

This chapter is organised as follows. Section 2 contains some background information about usability evaluation, with particular reference to mobile devices. The evaluation protocol is then introduced in Section 3, followed by a summary of the results of the two case studies in Section 4. A critical evaluation of the protocol is discussed in Section 5, which is followed by some suggestions for future research in Section 6. Section 7 then concludes this chapter.

BACKGROUND

Mobile devices are hand-held tools, which typically have a graphical display with input via touch, stylus, miniature keyboard, or some combination of these methods. Examples include Personal Digital Assistants (PDAs), traditional mobile phones, smart phones, music players such as the iPod Touch and tablet computers such as the iPad and Kindle Fire. The study of the usability and design issues associated with such devices is still in its infancy, since they are very different from desktop computers both in terms of interaction mechanisms and other attributes such as context, connectivity, screen size, display resolution and processing capability (Zhang, et al., 2005). The major platform providers (including Apple and Google) have produced extensive guidelines (Apple, 2012; Android, 2012) for developers of mobile applications, and there are also recent independent guidelines that focus specifically on improving the user experience (Nielsen, 2012). However, previous research suggests that current techniques for the evaluation of such technology lack structure and that there is a need for a systematic approach using a combination of methods, particularly because no single technique can give answers to all design questions (Streefkerk, et al., 2008).

Standard techniques for usability evaluation can be grouped into three categories (Rogers, et al., 2011): controlled settings involving users (such as tests in designated laboratories), natural settings involving users (such as workplace evaluations), and settings which do not involve users, but instead use consultants and researchers. The latter method is usually the least expensive and easiest to perform, but the results can be biased by the preconceptions of the evaluators and lack input from genuine users. There are a number of different ways to test user interfaces in each of the three categories, including observation, interviews, questionnaires, logging, heuristics and walkthroughs. The protocol described in this chapter falls into the third category: it is designed to be conducted by experts rather than users, and the standard procedures that it utilizes are *keystroke level modelling* and *heuristic evaluation*. The first of these techniques is useful for giving a quick and accurate quantitative measure of the efficiency of an application, and the second is useful for giving qualitative information about usability problems. Both are described in more detail below.

The Keystroke Level Model

The keystroke level model (KLM) was originally devised as a way of allowing individuals or companies to give quantitative predictions of the time expert users would take to complete certain tasks, thus reducing the need for expensive user studies (Card, et al., 1980). It is closely associated with the GOMS (Goals, Operators, Methods, and Selection) model (Card, et al., 1983), but is simpler to use. The GOMS model operates by first selecting a goal, or task, that is to be achieved, such as inserting a formula into a spreadsheet. There are often several methods for achieving such a goal, and each one is broken down into steps. The operators involved in each step, such as mouse clicks and key presses, are then identified, and selection rules are used to decide which method to choose. For example, when summing a small group of cells in a spreadsheet it might be quicker to select the group using the mouse, but for a larger group it might be quicker to type in the formula. KLM is closely related to GOMS in the sense that it can be used to give a quantitative comparison of different ways of achieving the same goal using different methods, applications or devices by predicting the time that would be taken to achieve the goal using each particular method. It is limited in scope in the sense that the time predictions assume that the tasks will be carried out by a skilled user, using a specific method, and will be error free.

KLM was first developed through empirical studies (Card, et al., 1980) to determine the expected times for the most commonly used operators. The original KLM model considered six operators, four of which represent physical actions:

- **K:** Keystroke or button press.
- **P:** Pointing to a target on the display with a mouse.
- **H:** Homing the hand(s) on the keyboard or other device.
- **D:** Drawing (manually) straight line segments.

The remaining two operators are concerned with times for mental preparation and system response. Once a goal has been broken down into steps involving operators via GOMS, the times associated with the keystroke level model can be used to give a consistent prediction of the time it will take to carry out the task using different methods or with different applications. This method is therefore primarily used to estimate task times at an early stage of interaction design.

One limitation of the KLM method is that it is device dependent, in the sense that its timings are associated with the interaction mechanisms of the device being evaluated. The operators used in the original model were all associated with desktop computing, and therefore are not necessarily transferrable to the mobile domain, but there have been a number of studies that have proposed extensions of the original model to include operators that have been introduced by mobile computing. Dunlop reported on using a keystroke level analysis to compare three different methods of typing on a traditional numeric phone keypad where each number key corresponds to three characters (Dunlop, et al., 2000),. The three possible methods were typing a whole word, using predictive text, and automatic word completion. The authors concluded that the execution time parameters of the original model were not applicable for mobile devices. A later study (Lou, et al., 2005) investigated the accuracy of KLM predictions for tasks carried out using a stylus on a Palm PDA. This work added a new *stroke* operator to the original model in order to make accurate predictions, but concluded that the system response time of the original model needed to be estimated more precisely since it is more significant on mobile devices. (Holleis, et al., 2007) took this work much further by extending the traditional KLM to include a number of basic interaction elements for mobile phones. Timing predictions were estimated using empirical methods based on several user tests.

The new generation of smart phones has introduced many more interaction mechanisms which further complicate the use of KLM predictions and render the work of (Holleis, et al., 2007) incomplete. The problems are compounded by the fact that the interaction methods vary from one platform to another, each of which has its own collection of touch gestures (Wroblewski, 2010). One study that suggests a way of mitigating against these changes is Schulz (2008), which introduced a tool that can generate keystroke level models and used it to evaluate some mobile phones, including the iPhone, which was found to be the most efficient device in the study. More recent work (Li, et al., 2010) proposes an extension of the original model to include fourteen new operators, but this has not yet been validated with empirical measurements. A more developed user study is given in Holleis et al. (2011), which proposes an update of KLM that focuses on NFC-based applications., The rapid evolution and variety of interaction mechanisms on mobile devices means that this is still an ongoing area of research.

Heuristic Evaluation

Heuristic evaluation (Nielsen, 1990) is another way of measuring the usability of an interface, in which a small group of usability experts examine different aspects of the interface in relation to standard design principles, or heuristics. The number of experts involved in such an evaluation can be as small as three since it has been shown (Nielsen, 1990) that three to five experts are sufficient to uncover most usability problems. The evaluations are performed independently, and the results are then analysed to detect the range and severity of the various usability problems. It is a cost-effective evaluation method in comparison with some other popular techniques, but it does have limitations since the participants are not real users and it does not fully capture the context of use. The set of heuristics to use when evaluating an interface can depend on the domain of use; the list originally developed (Nielsen, 1990) provides a good starting point, and can be supplemented by extra heuristics developed from design guidelines, or from knowledge of the domain. The study performed in (Bertini, et al., 2009) of the usability issues associated with mobile applications produced the set of mobile usability heuristics shown in Table 1.

These heuristics have been systematically developed and empirically validated to take into account some of the differences between mobile devices and traditional computing devices. For example, a small screen size implies that only

Table 1. Mobile usability heuristics (Bertini, et al., 2009)

Heuristic	Description
A	Visibility of system status and losability/findability of the device
B	Match between system and the real world
C	Consistency and mapping
D	Good ergonomics and minimalist design
E	Ease of input, screen readability and glancability
F	Flexibility, efficiency of use and personalization
G	Aesthetic, privacy and social conventions
H	Realistic error management

crucial information should be displayed (heuristic D). This contrasts with typical desktop design, which can be lazy in the sense that everything can be put on the screen, leaving the user to determine what features are important. For the purposes of the two case studies described here, each mobile heuristic was broken down into subdivisions to be assessed by each evaluator, some of which were specific to a domain; more details are given in Section 4.2.

THE EVALUATION PROTOCOL

The methodology used here was devised as a way of allowing experts to filter the large number of mobile applications in each domain to obtain a set that was small enough to be evaluated in greater depth through an adapted form of KLM and heuristic evaluation. The purpose of the method is not just to compare the usability of each product but also to elicit functional requirements. The protocol was influenced by the usability standard for medical devices ISO/IEC 62366, which describes a usability engineering process with nine stages, two of which involve identifying key functionality (Red Route Usability, 2012). Our protocol is divided into the following steps:

1. **Identify all potentially relevant applications:** This step consists of searching the applications related to a particular keyword associated with the domain of interest. For example, one of the domains we used to validate our protocol in Section 4 is diabetes management, and so the keyword *diabetes* was chosen. This task is facilitated by the search engines of the current online stores such as the App Store™ from Apple, the Android Market from Google and the App World from BlackBerry. The search should be performed through a Web-based interface rather than on the mobile device as some on-device store searches do not present all current mobile applications.
2. **Remove light or old versions of each application:** Most searches will return a number of trial versions of an application that offer only a subset of the functionality available in the corresponding full application or restrict access to the full application for a limited period of time. These applications should be removed for efficiency.
3. **Identify the essential functional requirements and exclude all applications that do not offer this functionality:** This step involves careful consideration of the essential functional requirements of the application domain. Mobile applications tend to have much more limited functionality than their desktop counterparts and so the requirements may be quite restricted. For example, on a mobile device spreadsheet users are more likely just to view spreadsheets and make minor updates than to enter complex formulae. Functionality can be categorised into frequently used functions and functions that are essential but infrequently used, such as language and unit settings. Only the applications that meet all the essential requirements are carried forward to subsequent steps of the protocol.

4. **Identify all desirable requirements:** This step consists of identifying any functionality offered by an application that was not identified in the previous step. This might include novel functions that are not common to many applications but are worth considering in order to make a unique product. For example, some of the diabetes management applications included in this study had chat forums that were not essential, but added potentially useful value to users. The purpose of this desirable list is to gather a comprehensive list of desirable functionality for a developer to analyse during the requirements elicitation process for any new application.
5. Construct tasks to test the essential functional requirements using method (a) below, and for the most efficient applications method (b) as well:
   a. Keystroke Level Modelling (KLM) to determine a quick and accurate quantitative measure of the efficiency of each application when executing core tasks. Since the new interaction methods provided by mobile devices have not all been incorporated into KLM, it is not possible to predict efficiency in terms of time, however it is possible to use the number of interactions as an approximate measure of the efficiency of each application.
   b. Heuristic evaluation to give a qualitative analysis of the possible usability. This process is conducted independently by experts, and the results are collated to determine design guidelines. The heuristics in Table 1 are appropriate for this analysis. In the study described here, each heuristic was split into sub-heuristics, *which* are described in Section 4, and these were used to rank each of the applications.

PROTOCOL VALIDATION

The evaluation protocol was validated through its application to two very different domains (healthcare and business) which are described in more detail in Sections 4.1 and 4.2, respectively. These particular domains both require *productivity applications* (Fling, 2009) that facilitate tasks that are based on organising and manipulating detailed information. The two domains were deliberately chosen to be very different from each other in the sense that the particular healthcare applications are used to enable a small set of tasks whereas the business applications offer a lot more flexibility. We used three different mobile device platforms (iOS, Android, and Blackberry OS). The results of the validation are described in Section 4.3.

Healthcare: Diabetes Management

We chose diabetes management applications for one of our validation studies. Applications were studied that allow people with diabetes to manage their condition by logging daily information such as blood glucose level, carbohydrate intake and insulin dose. This domain was selected because it represents the broader category of productivity applications with a narrow focus and a limited range of primary functionality. This is a common style of mobile application, and some other examples of such applications include those for note taking, calorie counting and scheduling.

Type 1 diabetes occurs when the insulin producing cells of the pancreas are destroyed leaving the body unable to control its blood glucose levels. People with type 1 diabetes have to take insulin regularly to try to stop their glucose levels from becoming too high, but if they take too much insulin, their glucose levels may also drop too low, causing a number of symptoms including dizziness and palpitations.

The vast majority of patients with type 1 diabetes administer their insulin through multiple daily injections, and the remaining proportion use insulin pumps. Most people are offered an education programme to help them self manage their condition. This teaches them how to calculate the amount of insulin to administer at each meal according to the current blood glucose level, number of carbohydrates consumed and various other factors such as time of day, exercise and illness. The daily glucose levels are usually stored in a hand-written diary, which can be shared with a healthcare team at regular intervals. It is surprisingly difficult for patients to keep their blood glucose levels within the target range, and yet failure to do so can lead to serious complications, which can be life-threatening and are a huge burden on the health service.

Most insulin pumps have dose calculators to help patients determine how much insulin to administer, but some people on multiple daily injections do not have this support, and tend to do the calculations themselves. This trend is beginning to change, with the advent of glucose monitors such as Accu-Chek Expert (Roche, 2012) and Insulinx (Abbott, 2011) which do have dose calculators, but their prohibitive cost means that these have not yet become widely used. Consequently, a growing number of mobile phone applications have been developed both to log healthcare data and to offer help with calculations.

Business: Spreadsheets

We chose spreadsheets for our second validation study. These allow users to complete a wide variety of tasks from financial planning to statistical analysis. The spreadsheet domain was selected to represent general productivity applications in which the specific tasks can vary hugely from user to user. Other examples of such applications include those for text creation, photo editing and music creation.

Spreadsheets are ubiquitous software tools used for a variety of tasks from financial planning to statistical analysis. The mobile nature of business has increased the need for users to access spreadsheets while on the move. Consequently mobile spreadsheet applications are becoming everyday tools and the requirements of users are expanding to include advanced functionality such as specialist functions and features.

Flood et al. (2011) conducted an online survey of experienced spreadsheet users to determine how much mobile spreadsheet applications are used and for what purpose. The results showed that 79% of respondents required access to a spreadsheet while away from a traditional computing device and that most respondents only use mobile spreadsheet applications for examining existing spreadsheets, such as those they receive electronically, or for editing spreadsheets. The survey also revealed that none of the respondents needed to create a spreadsheet while away from a traditional computing device. One of the biggest problems for mobile spreadsheet applications is the small screen size associated with portable devices. However, mobile device are still used despite this problem in order to multitask.

Results

The results of the validation in the healthcare and business domains are discussed below.

Step 1: Identify all Potentially Relevant Applications

The first step of the protocol requires selection of appropriate key words for each domain, which are used to search the app stores of each of the chosen platforms. In the case of diabetes management the keyword *diabetes* was used while the word *spreadsheets* was used to search for spreadsheet applications.

Table 2 shows the number of applications that were returned for each of the keywords searched. The table shows that a large number of applications were available for both domains on the iOS and Android platforms but a much smaller number was available for the Blackberry.

Table 2. Number of apps returned by App store search

	iOS	Android	Blackberry
Diabetes	231	168	28
Spreadsheets	105	179	58

Step 2: Remove Light or Old Versions of Each Application

After compiling the results from each of the searches, any light or old versions of other applications included within the search results were removed because these versions often contain less functionality than the full or newer version. Table 3 shows that there are only a small number of such versions available in each domain.

Step 3: Identify the Essential Functional Requirements and Exclude all Applications that do not offer this Functionality.

The next step of the protocol involves the identification of applications that offer the essential functional requirements that are required for a given domain. For the diabetes management, these requirements were as follows:

1. Set Measurement Units.
2. Log Blood Glucose Level.
3. Log Carbohydrate Intake.
4. Log Insulin dose.
5. Display Data graphically.
6. Export data via email or similar.

Table 3. Number of light or old versions

	iOS	Android	Blackberry
Diabetes Management	9	6	1
Spreadsheets	16	14	10

The reason for choosing the first requirement is that different units are used for measuring insulin in the United States and Europe and many applications only offer the units that are used in the country where the application was developed. The next three requirements are for logging blood glucose level, carbohydrate intake and insulin dose respectively as recommended by medical professionals. Viewing this data graphically can help patients identify trends which can help to avoid health problems in the future. The final functional requirement was to export the data so that users can back-up their entries in case their mobile device is lost or damaged. This is a crucial part of the health management process.

For the spreadsheet application the essential functionality identified was the ability to view and alter existing Microsoft© Excel spreadsheets.

The search for applications with the required functionality was conducted by examining the descriptions of each application on the online app store. Table 4 shows that only a small number of applications offered the essential functionality. The remaining applications were designed for a wide variety of purposes. In the case of diabetes management, these applications included general health information, cookbooks with recipes specific for diabetes sufferers and Body Mass Index (BMI) calculators. In the case of spreadsheets, the search returned a number of training books and video applications, along with those for specific purposes such as calculating tips. In addition to this, a number of applications were returned which exported data to spreadsheets, such as *Contacts Export*, which allows users to export some or all of their phone contacts to a spreadsheet. These

Table 4. Number of applications offering essential functional requirements

	iOS	Android	Blackberry
Diabetes Management	8	6	1
Spreadsheets	11	7	1

results highlight the inefficiency of existing search algorithms on mobile app stores. The source code for the algorithms is not freely available however, so it is difficult to determine why they are so ineffective.

Step 4: Identify all Desirable Requirements

The fourth step of the protocol consisted of identifying all of the desirable functionality of the evaluated applications. This is carried out in order to allow developers to analyse the full range of functionality offered by competing applications and to determine opportunities for additional features, which can help to differentiate new applications from existing ones.

Within the diabetes management domain a wide range of desirable requirements were identified. It is beyond the scope of this chapter to include a full list of these requirements but the interested reader is referred to Garcia et al. (2011) for further details. Some of the most common desirable requirements are listed below:

- **Log physical activities and other medications:** These factors can impact the appropriate dose of insulin that needs to be taken and therefore a number of applications allow users to store this type of information.
- **Allow personal settings:** Most applications allowed users to input personal information such as email address, target glucose level, and weight.
- **Carbohydrate database:** Some of the applications contained a carbohydrate database, which allows users to look up particular food items to determine the amount of carbohydrates contained. This is important in the diabetes domain, as users are required to calculate the amount of insulin required from the amount of carbohydrates consumed.
- **Insulin dose calculator:** Some applications feature the ability to calculate an appropriate dose of insulin to inject based on the carbohydrates consumed and blood glucose level.

A range of desirable functionality for spreadsheet applications was also identified, most of which is already available in desktop spreadsheet applications. A summary of the key desirable functionality is presented here.

- **Freeze Panes:** these allow a user to keep a number of rows and/or columns locked on screen while they move through the rest of the spreadsheet. This feature is particularly useful when a user needs to keep headings available on-screen while searching through a large spreadsheet.
- **Sorting:** A number of applications allow users to sort tables of data on a given column. The small screen size associated with mobile applications makes it difficult to view all of the required data at one time, therefore making it necessary for users to re-order the information to see what they are most interested in.
- **Go to Cell:** When looking at large spreadsheets navigation can be time consuming. A number of applications allow users to type a cell address and the application will move to this cell, saving time.
- **Formatting:** A number of spreadsheet applications allow users to format cells on the spreadsheet. They allow users to set the foreground and background colour of the cell, change the font and size of the text in a cell as well as making it bold and italic.

One notable omission in the functionality of applications in both domains is the ability to filter data. When examining a large quantity of data, such as that associated with patient records, users often need to limit the data in order to focus on a particular subset. The limited screen size of mobile devices can make it particularly important to limit datasets so that users can examine them more easily.

Step 5: Construct Tasks to Test the Essential Functional Requirements using Each of the Methods Below

The final step in the protocol was designed to examine the usability of the existing applications both in terms of efficiency and through adherence to our heuristics (Table 1).

Each of the applications from Step 3 was downloaded from the relevant app store in order to conduct the efficiency testing using a form of keystroke level modelling. This was carried out by a single evaluator since it is purely a quantitative measure. The four most efficient applications from each domain on each platform were then subjected to further heuristic evaluation 3, 4 or 5 expert evaluators. The other applications were excluded partly because of a lack of resources and partly because of duplication among the various applications. By restricting attention to a maximum of four applications in each domain, we reduced the number of applications to 24 and the number of evaluations to 120. The following sections outline the essential findings for each of the two stages of the usability evaluation.

Tasks

Each of the evaluations was based upon a series of tasks that were devised to test the essential functionality associated with each domain in Step 3. For diabetes management the tasks were as follows:

1. Change the insulin units to "mmol/L".
2. Add a new entry with the following values.
    a. Blood glucose level of 6.7 mmol/L.
    b. Carbohydrate intake of 50 grams or 5 portions.
    c. Insulin dose of 5.5 units.
3. Display a visual representation of the data.
4. Export logged data via email.

For spreadsheets, the tasks were as follows. First, a simple spreadsheet was created containing five numbers in the first column together with a sum formula to total the five numbers. This allowed the examination of the following tasks.

1. Enter a single digit value in a cell.
2. Enter a formula in a cell.

KLM Evaluation

As mentioned earlier some interaction methods have not yet been incorporated into the traditional KLM model. Therefore, it was not possible to predict the time it would take to complete the tasks according to the operators used but instead the number of interactions to complete each task was used as an indirect measure of its efficiency.

The results for the diabetes applications (Garcia, et al., 2011) showed large variations in the efficiency of mobile applications. For example on the iOS platform it was found that it took three interactions to set the units to the desired value on one application compared with ten interactions on another application. This is because the latter application required users to exit the application and enter the device settings to change the units while other applications contained their own settings tab.

Similar results were found for the spreadsheet applications: it took between four and thirteen keystrokes to enter the sum formula. This variation occurred because the most efficient applications allowed users to select the sum formula from a menu. In contrast the least efficient application required the user manually to enter entire formulae, including cell references.

One similarity between all of the spreadsheet applications was the choice of input method, which was typically the keyboard. In contrast, applications for the diabetes domain offered a wide range of input methods for entering the insulin, glucose and carbohydrate data including the keyboard, sliders, pickers, and other bespoke techniques.

Heuristic Evaluation

The heuristic evaluation was conducted by giving each evaluator a mobile device with the relevant applications installed, and an evaluation sheet upon which to record results. The evaluators were asked to carry out the selected tasks independently at least once. The evaluation sheet contained a section for each mobile heuristic (see Table 1), which was then broken down into domain specific sub-divisions. For example the subdivisions of Heuristic E for the diabetes domain were as follows:

- E1 It is easy to input the numbers.
- E2 It is easy to see what the information on each screen means.
- E3 You can easily navigate around the app.
- E4 The screens have a 'back' button.
- E5 The user can get crucial information 'at a glance.'

Details of the other subdivisions are given in (Martin, et al., 2011). Each attribute was given a ranking between zero and four using Nielsen's Severity Ranking Scale (Nielsen, 2012), which was corroborated by evidence and justification for that ranking.

The heuristic evaluation exposed a different set of problems from the KLM, because the mobile heuristics that guided it are not only related to efficiency. It is beyond the scope of this chapter to give a full qualitative summary of the evaluation (the interested reader may refer to Martin et al. [2011]), but the most commonly breached heuristics for the diabetes applications were A, C, D, G, and H (Table 1). This led us to the following associated guidelines:

- The battery status and time should be visible while using the application.
- The network status should be visible while sending data.
- Options should be provided for backing up and restoring data.
- Unnecessary options and irrelevant information should be avoided. A screen should not be overloaded with too many elements.
- All data transmission should be encrypted.
- Unrecoverable errors should be prevented.

Heuristics were also breached in the spreadsheet domain. The most frequently breached heuristics were B, E, F, G, and H. One of the most common problems uncovered during the evaluation was the data input method. Most applications presented the user with the alphabetic keyboard rather than a numerical one each time they tried to input data, despite the most common form of input in a spreadsheet being numerical. The following guidelines resulted from the heuristic evaluation:

- It should be possible to encrypt data and files.
- The input method should allow quick entry of numerical data.
- Symbols should be distinct and not easily confused.
- Users should be able to undo previous actions.
- The menu structure should be easily discovered.

CRITICAL EVALUATION OF THE PROTOCOL

This section contains a summary of some of the advantages and disadvantages of this protocol, together with an outline of some of the difficulties that were encountered when using it. The evaluation protocol was devised in order to gain a broad overview of the status and trends of all of the mobile applications currently available for a given task domain. Our results show that the method provides a useful way to reduce the search for applications by offering key functionality in each domain, and gives a systematic evaluation of those that remain. Even though the style of keystroke level modelling used here was not sophisticated, it provided a very quick way to compare the efficiency of the applications before embarking on a more in-depth analysis of the better performing applications using heuristics. Some of the differences in efficiency highlighted by the KLM were caused by the devices themselves, and others were due to the design choices of each implementation. For example, the hard keyboard of the Blackberry proved to be less efficient than the touch screens of the other two devices. Examples of implementation decisions that affected efficiency include the choice of data entry method and the ability to set defaults. The heuristic evaluation uncovered a different set of problems from the KLM since most of the heuristics used were not related to efficiency. Instead, some fundamental problems were exposed concerning security, error recovery, minimalist design, visibility of system status and backup and restore options. The analysis of desirable functional requirements generated many recommendations in each domain, which could then be further distilled using additional requirements analysis techniques.

Advantages

The protocol has a number of strengths. For example, it is highly adaptable in the sense that it is platform independent and indeed may not even be restricted to a mobile platform. The first step relies on the existence of an app store, but such distribution mechanisms are now becoming commonplace for desktop and Web applications because of the inception of the Mac App Store, Chrome Web Store, and Windows 8 App Store. The protocol also offers flexibility in terms of application domains as the two contrasting validation studies described here have shown. As well as being flexible, the protocol establishes the state of the art within a specific task domain in a cost effective manner. It focuses on applications that provide a core set of functionality of interest to developers thus eliminating a large number of applications returned by a preliminary search.

The protocol is economical because it can be applied using a small number of people, indeed the majority of the work can be conducted by a single individual. The only exception to this is the heuristic evaluation, for which multiple usability experts are required (Nielsen, et al., 1990). However, since this is the final step of the protocol, only a small number of applications are evaluated and therefore the time commitment required by the experts is relatively small.

The protocol outlined above has proved to be cost effective not only in terms of money but also in terms of time because many applications are eliminated at the start (Step 3). Consequently, developers do not waste time looking at irrelevant applications that are returned by the initial search of the app stores. The results presented in Section 4 show that the identification of essential functionality is a particularly useful mechanism for excluding a large number of applications from subsequent evaluations.

The time needed to learn how to apply the protocol is low. During the evaluation presented above two postgraduate students were introduced to the protocol and quickly became efficient with its application. By the time they applied it to the second platform were able to use the protocol without supervision.

The protocol produces a comprehensive set of the desirable functionality provided by existing applications. Thus, developers of new applications can see which functionality is most common, and which therefore should be considered for inclusion within a new application, and which functionality is missing. This information provides developers with opportunities to find gaps in the market. Since the protocol gathers functionality from multiple platforms, it may uncover features that do not yet exist on the platform for which the new application is to be developed.

The application of the protocol described in Section 4 showed that the use of a combination of two different analytical usability techniques generated an extensive list of pitfalls to avoid. The KLM evaluation (Card, et al., 1980) exposed some fundamental efficiency issues whereas the heuristic evaluation (Nielsen, 1990; Bertini, et al., 2009) showed that there were problems concerning security and error recovery. The two techniques therefore complement each other in generating a wide range of issues associated with each task domain. Although further issues can be uncovered through the use of user studies, such studies can be expensive and time consuming to perform. The approaches adopted here can be performed in-house using limited resources, which is an important consideration for small software development companies, which are typical within the mobile application domain.

Disadvantages

The protocol has some weaknesses. For example, some of the steps are time consuming, and may be tedious to perform. In addition, it is not always possible to determine the full functionality of each application from its description on the online store, but this is an important step in reducing the number of applications in order to keep expenditure low. It can be difficult to identify all of the relevant keywords to use in the initial search, yet a poor choice can return a high number of irrelevant applications, or mean that the search is not exhaustive. There are also problems with measuring the number of keystrokes used to carry out a task, since it may depend on the user's dexterity when using certain kinds of interaction objects such as pickers and sliders. Since the heuristic evaluation is qualitative rather than quantitative, some results are necessarily subjective, according to the personal taste of the evaluator.

The problem with the keyword search could be reduced by conducting multiple searches to ensure that the broadest range of applications is returned. However, in this case the developer will encounter the same application multiple times. One way to circumvent this problem is to automate the search procedure. Through automated multi-objective search algorithms, multiple searches could be done together, presenting the user with a complete set of unique applications returned by any search term.

Although the protocol is designed to be efficient, the volume of results returned by the search algorithms may mean it takes time to evaluate all of the applications returned. It would be possible to use app store ratings to further refine the results to eliminate some of the lower quality applications but this may result in essential functionality being missed.

Mobile applications are released on a frequent basis (Holden, 2010). This means that the results obtained will be time sensitive. Through automated search, a system could poll for new applications on a regular basis during the development lifecycle. Developers can then quickly evaluate these newly released apps.

The protocol tends to highlight the problems with applications and ignore their positive features and so information may be lost. Such functionality could be elicited by involving users in the evaluation.

Threats to Validity

This protocol is not intended as a complete method for eliciting requirements or designing an interface because it has some known and deliberate limitations. Most obviously, it ignores any kind of user evaluation, including the star rating of the app store. It also excludes price as a factor. As a method of gathering requirements, it is restricted in the sense that it only examines the functionality of existing applications. The results could be used to form the basis of subsequent requirements gathering by informing the design of questionnaires and structured interviews. With regard to interface design, it can be useful to show which interaction objects work best for each task, and which designs to avoid, but it is no substitute for traditional design techniques such as iterative prototyping.

Another limitation of the validation is that the results were restricted to applications running on the iOS, Android, and Blackberry smart phone platforms. These were chosen because they have the highest market share in the UK (Online Marketing Trends, 2011), and jointly account for 75.5% of the total market. It was also restricted to native applications, since such applications can take advantage of device-specific features, which are particularly interesting to developers.

There was also a difficulty when searching the Android app store from the mobile device: a lower number of applications were returned than with the same search using the app store website on a desktop computer. This may have occurred because the app store limited the results to only those applications that would function with the appropriate version of Android.

FUTURE RESEARCH DIRECTIONS

This evaluation protocol may evolve as it is a prototype, which will need refining and improvement through further empirical studies. For example, the version of KLM used here was quick and effective but unlike traditional KLM, it did not use predictive timings. Instead, it used a simple count for each operator, so more research is needed to determine reliable timings for the commonly used interaction methods.

The heuristic evaluation was relatively fast to perform, but it is only one of many other evaluation techniques that could have been chosen including analytics, cognitive walkthroughs, and heuristic walkthroughs (Billi, et al., 2010; Sears, 1997). A future iteration of this methodology could include additional or modified steps involving alternative techniques.

The potential for future research in the domain of usability evaluation of pervasive applications is enormous. A number of suggestions are given in Zhang et al. (2005) including study into the best ways to design presentation methods, menu structures, database facilities, data entry methods, and connectivity methods. In addition, some of the problems of precision using pickers and sliders noted in Section 4 suggest that fundamental experiments are still needed to determine the optimal granularity of such objects in the same way that much research took place in the early days of GUIs to decide how many buttons to include on a mouse.

CONCLUSION

When developing mobile applications it may be too costly and time consuming to evaluate all existing applications within a given domain. Consequently, we developed a systematic evaluation protocol for determining the state of the art in a given domain. This protocol has been validated in two different domains: healthcare and business. This exercise gave us a detailed snapshot of the status and trends of the applications in a subset of each domain. The protocol was found to be both cost effective and efficient for evaluating the subsets of applications within these two domains. The protocol has been applied to three of the most popular mobile platforms in the UK and it was found that on each platform the protocol enabled an efficient examination of the applications under study.

The protocol highlighted a number of issues that should be avoided when developing new applications within the healthcare and business domains and helped to produce guidelines for the design of new mobile applications within these domains.

ACKNOWLEDGMENT

The authors would like to thank Oxford Brookes University for generous support towards this research and the anonymous referees for their detailed and pertinent comments.

REFERENCES

Abbott Insulinx. (2011). *Website*. Retrieved from http://www.abbottdiabetescare.co.uk/your-products/freestyle-insulinx

Android. (2011). *Developers user interface guidelines*. Retrieved from http://developer.android.com/guide/practices/ui_guidelines/index.html

Apple. (2011). *iOS human interface guidelines*. Retrieved from http://developer.apple.com

Bertini, E., Catarci, T., Dix, A., Gabrielli, S., Kimani, S., & Santucci, G. (2009). Appropriating heuristic evaluation methods for mobile. *International Journal of Mobile Human Computer Interaction*, *1*(1), 20–41. doi:10.4018/jmhci.2009010102

Billi, M., Burzagli, L., Catarci, T., Santucci, G., Bertini, E., Gabbanini, F., & Palchetti, E. (2010). A unified methodology for the evaluation of accessibility and usability of mobile applications. *Universal Access in the Information Society*, *9*(4), 337–356. doi:10.1007/s10209-009-0180-1

Card, S., Moran, T. P., & Newell, A. (1983). *The psychology of human computer interaction*. Mahwah, NJ: Lawrence Erlbaum Associates.

Card, S. K., & Moran, T. P. (1980). The keystroke-level model for user performance: Time with interactive systems. *Communications of the ACM*, *23*(7), 396–410. doi:10.1145/358886.358895

Dunlop, M. D., & Crossan, A. (2000). Predictive text entry methods for mobile phones. *Personal Technologies*, *4*(2-3), 134–143. doi:10.1007/BF01324120

Finkelstein, A. (1993). *Requirements engineering: An overview*. Paper presented in the 2nd Asia-Pacific Software Engineering Conference (APSEC 1993). Tokyo, Japan.

Fling, B. (2009). *Mobile design and development*. San Francisco, CA: O' Reilly Media.

Flood, D., Harrison, R., & Duce, D. (2011). *Using mobile apps: Investigating the usability of mobile apps from the users' perspective*. Unpublished.

Flood, D., Harrison, R., & McDaid, K. (2011). *Spreadsheets on the move: An evaluation of mobile spreadsheets*. Paper presented in the European Spreadsheet Risk Interest Group (EuSpRIG) Annual Conference. London, UK.

Garcia, E., Martin, C., Garcia, A., Harrison, R., & Flood, D. (2011). Systematic analysis of mobile diabetes management applications on different platforms. In *Proceedings of the Information Quality in e-Health 7th Conference of the Workgroup Human-Computer Interaction and Usability Engineering of the Austrian Computer Society*, (pp. 379-396). Graz, Austria: Springer.

Holden, W. (2010). *A world of apps*. New York, NY: Juniper Research Limited.

Holleis, P., Otto, F., Hussmann, H., & Schmidt, A. (2007). Keystroke-level model for advanced mobile phone interaction. In *Proceedings of the SIGCHI Conference on Human Factors in Computing Systems (CHI 2007)*, (pp. 1505-1514). New York, NY: ACM.

Holleis, P., Scherr, M., & Broll, G. (2011). A revised mobile KLM for interaction with multiple NFC-tags. In *Proceedings of the 13th IFIP TC 13 International Conference on Human-Computer Interaction*, (pp. 204-221). Berlin, Germany: Springer-Verlag.

ITU. (2011). [*ICT facts and figures*. Retrieved from http://www.itu.int]. *WORLD (Oakland, Calif.), 2011*.

Li, H., Liu, Y., Liu, J., Wang, X., Li, Y., & Rau, P. (2010). Extended KLM for mobile phone interaction: A user study result. In *Proceedings of the 28th International Conference Extended Abstracts on Human Factors in Computing Systems (CHI EA 2010)*, (pp. 3517-3522). New York, NY: ACM.

Lou, L., & John, B. E. (2005). Predicting task execution time on handheld devices using keystroke level modelling. In *Proceedings of CHI 2005 Extended Abstracts on Human Factors in Computing Systems (CHI EA 2005)*, (pp. 1605-1608). New York, NY: ACM.

Martin, C., Flood, D., Sutton, D., Aldea, A., Waite, M., Garcia, E., ... Harrison, R. (2011). *A systematic evaluation of mobile applications for diabetes management using heuristics*. Unpublished.

Nielsen, J. (2011). *Website*. Retrieved from http://www.useit.com/papers/heuristic/heuristic_evaluation.html

Nielsen, J., & Molich, R. (1990). Heuristic evaluation of user interfaces. In *Proceedings of the ACM CHI 1990 Conference,* (pp. 249-256). Seattle, WA: ACM.

Nielsen Norman Group. (2012). *Report usability of mobile websites & applications*. Retrieved from http://www.nngroup.com/reports/mobile/

Online Marketing Trends. (2011). *Website*. Retrieved from http://www.onlinemarketing-trends.com/2011/04/smartphone-marketshare-2011-italyus-aus.html

Roche Accu-chek Expert. (2012). *Website*. Retrieved from http://www.accu-chek.co.uk/gb/products/metersystems/avivaexpert.html

Rogers, Y., Sharp, H., & Preece, J. (2011). *Interaction design* (3rd ed.). Mahwah, NJ: John Wiley & Sons Ltd.

Schulz, T. (2008). Using the keystroke-level model to evaluate mobile phones. In *Proceedings of IRIS 31*. IRIS.

Sears, A. (1997). Heuristic walkthroughs: Finding the problems without the noise. *International Journal of Human-Computer Interaction, 9*, 213–234. doi:10.1207/s15327590ijhc0903_2

Streefkerk, J. W., van Esch-Bussemakers, M. P., Neerincx, M. A., & Looije, R. (2008). Evaluating context-aware mobile interfaces for professionals. In Lumsden, J. (Ed.), *Handbook of Research on User Interface Design and Evaluation for Mobile Technology* (pp. 759–779). Hershey, PA: IGI Global Publishing. doi:10.4018/978-1-59904-871-0.ch045

User Focus. (2011). *Red route usability: The key user journeys with your web site*. Retrieved from http://userfocus.co.uk/articles/redroutes.html

Wroblewski, L. (2010). *Touch gesture reference guide*. Retrieved from http://www.lukew.com/ff/entry.asp?1071

Zhang, D., & Adipat, B. (2005). Challenges, methodologies, and issues in the usability testing of mobile applications. *International Journal of Human-Computer Interaction, 18*(3), 293–308. doi:10.1207/s15327590ijhc1803_3

KEY TERMS AND DEFINITIONS

GOMS: The Goals, Operators, Methods, and Selection model is a description of the way that a user carries out tasks on a device or system in terms of each of these four components. Each method is a series of steps that the user must undertake in order to carry out the goal using the given operators. If there is more than one method then the user can select between them.

Heuristic Evaluation: A technique for usability evaluation where experts judge whether each component of a user interface adheres to a list of established usability heuristics.

Keystroke Level Model: A method for predicting the efficiency of an interface for an expert user by estimating the total time required to complete a given task by analyzing the steps required in the process.

Mobile Device: A small, hand-held computing device with a display screen and input via touch, stylus or a miniature keyboard.

Mobile Platform: The hardware and software environment for smartphones, tablets, personal digital assistants and other portable devices.

Productivity Application: A software program that is primarily designed to facilitate work.

Usability Evaluation: A technique for evaluating the design of a user interface.

Compilation of References

Abbott Insulinx. (2011). *Website.* Retrieved from http://www.abbottdiabetescare.co.uk/your-products/freestyle-insulinx

Abra, J. (1989). Changes in creativity with age: Data, explanations, and further predictions. *International Journal of Aging & Human Development*, *28*(2), 105–126. doi:10.2190/E0YT-K1YQ-3T2T-Y3EQ

Agar, M. H. (1980). *The professional stranger: An informal introduction to ethnography*. New York, NY: Academic Press.

Ågerfalk, P. J. (2010). Editorial: Getting pragmatic. *European Journal of Information Systems*, *19*, 251–256. doi:10.1057/ejis.2010.22

AGESIC. (2008). *Memoria anual 2008*. Retrieved from http://www.agesic.gub.uy/innovaportal/file/267/1/Memoria_Anual2008.pdf

AGESIC. (2008). *Modelo de madurez de gobierno electrónico*. AGESIC.

AGESIC. (2009). *Memoria anual 2009*. Retrieved from http://www.agesic.gub.uy/innovaportal/file/267/1/Memoria2009.pdf

AGESIC. (2011). *Plan estratégico 2011 - 2015*. Retrieved from http://www.agesic.gub.uy/innovaportal/file/265/1/planestrategico2011-2015.pdf

Ahlström, D. (2005). Modeling and improving selection in cascading pull-down menus using fitts' law, the steering law and force fields. In *Proceedings of the SIGCHI Conference on Human Factors in Computing Systems, CHI 2005*, (pp. 61-70). ACM.

Ahn, Y.-Y., Han, S., Kwak, H., Moon, S., & Jeong, H. (2007). Analysis of topological characteristics of huge online social networking services. In *Proceedings of the 16th International Conference on World Wide Web - WWW 2007*, (pp. 835-844). New York, NY: ACM Press. Retrieved August 25, 2011, from http://www2007.org/papers/paper676.pdf

Ajzen, I. (1991). The theory of planned behaviour. *Organizational Behavior and Human Decision Processes*, *50*, 179–211. doi:10.1016/0749-5978(91)90020-T

Ajzen, I., & Fishbein, M. (1980). *Understanding attitude and predicting social behavior*. Englewood Cliffs, NJ: Prentice-Hall.

Aleixo, C., Nunes, M., & Isaias, P. (2012). Usability and digital inclusion: Standard and guidelines. *International Journal of Public Administration*, *35*, 221–239. doi:10.1080/01900692.2011.646568

Allan, G. (2007). The use of grounded theory methodology in investigating practitioners' integration of COTS components in information systems. In *Proceedings of the 28th International Conference on Information Systems*, (pp. 1-10). Montreal, Canada: IEEE.

Allan, B. (2007). *Blended learning: Tools for teaching and training*. London, UK: Facet.

Alvarez, R. (2005). Taking a critical linguistic turn. In Howcroft, D., & Trauth, E. (Eds.), *Handbook of Critical Information Systems Research* (pp. 104–122). Cheltenham, UK: Edward Elgar Publishing Inc.

American Association of School Librarians. (2010). *Best websites for teaching and learning: Top 25 websites for teaching and learning*. Retrieved April 4, 2011 from: http://ala.org/ala/mgrps/divs/aasl/guidelinesandstandards/bestlist/bestwebsitestop25.cfm

Ammenwerth, E., Graber, S., Herrman, G., Burke, T., & Konig, J. (2003). Evaluation of health information systems-problems and challenges. *International Journal of Medical Informatics*, *71*, 125–135. doi:10.1016/S1386-5056(03)00131-X

Anders, G. (2001). Business fights back: eBay learns to trust again. *Fast Company*. Retrieved 10th September, 2011 from http://www.fastcompany.com/magazine/53/ebay.html

Andrews, R., Freeman, A., Hou, D., McGuinn, N., Robinson, A., & Zhu, J. (2007). The effectiveness of information and communication technology on the learning of written English for 5 To 16-year-olds. *British Journal of Educational Technology*, *38*(2), 325–336. doi:10.1111/j.1467-8535.2006.00628.x

Android. (2011). *Developers user interface guidelines*. Retrieved from http://developer.android.com/guide/practices/ui_guidelines/index.html

Annansingh, F., & Baptista Nunes, J. M. (2005). Validating interpretivist research: Using a cross-sectional survey to validate case study elicitation of knowledge leakage risks associated with the use of virtual reality models. In *Proceedings of the 4th European Conference on Research Methodology for Business and Management Studies*. Paris, France: Université Paris Dauphine.

APC. (2011). *American power conversion corporation*. Retrieved from http://www.apcc.com/

Apple. (2011). *iOS human interface guidelines*. Retrieved from http://developer.apple.com

Arber, S. (1993). Designing samples. In Gilbert, N. (Ed.), *Researching Social Life* (pp. 68–93). London, UK: Sage.

Arnowitz, J., Arent, M., & Berger, N. (2006). *Effective prototyping for software makers*. San Francisco, CA: Morgan Kaufmann.

Atkinson, R., & Flint, J. (2001). Accessing hidden and hard-to-reach populations: Snowball research strategies. *Social Research Update, 33*. Retrieved August 28, 2011 from http://sru.soc.surrey.ac.uk/SRU33.pdf

Atterer, R., Wnuk, M., & Schmidt, A. (2006). Knowing the user's every move - User activity tracking for website usability evaluation and implicit interaction. In *Proceedings of the 15th International Conference on World Wide Web (WWW 2006)*, (pp. 203-212). IEEE.

ATutor. (2011). *AChecker: Web accessibility checker*. Retrieved February 15, 2011, from http://achecker.ca/checker/index.php

Augustine, S. (2005). *Managing agile projects*. Upper Saddle River, NJ: Prentice Hall PTR.

Avgerou, C., & Madon, S. (2004). Framing IS studies: Understanding the social context of IS innovation. In Avgerou, C., Ciborra, C., & Land, F. (Eds.), *The Social Study of Information and Communication Technology* (pp. 162–182). Oxford, UK: Oxford University Press.

Avison, D. E., Lau, F., Myers, M. D., & Nielsen, P. A. (1999). Action research. *Communications of the ACM*, *42*(1). doi:10.1145/291469.291479

Avison, D., & Fitzgerald, G. (1993). *Information systems development: Methodologies, techniques and tools*. New York, NY: Alfred Waller Ltd, Publishers.

Avison, D., Fitzgerald, G., & Powell, P. (2004). Editorial. *Information Systems Journal*, *14*(1), 1–2. doi:10.1111/j.1365-2575.2004.00163.x

Babaian, T., Lucas, W., & Topi, H. (2006). Making memories: Apply user input logs to interface design and evaluation. In *Proceedings of CHI 2006 Extended Abstracts on Human Factors in Computing Systems*, (pp. 496-501). ACM.

Bagozzi, R. P. (2007). The legacy of the technology acceptance model and a proposal for a paradigm shift. *Journal of the Association for Information Systems*, *8*(4), 243–254.

Bailey, J., & Burd, E. (2005). Tree-map visualisation for web accessibility. In *Proceedings of the Computer Software and Applications Conference*, (pp. 275-280). IEEE Press.

Baldwin, E., & Curley, M. (2007). *Managing IT innovation for business value: Practical strategies for IT and business managers*. New York, NY: Intel Press.

Barber, B. (1983). *The logic and limits of trust*. New Brunswick, NJ: Rutgers University Press.

Barnum, C. M. (2001). *Usability testing and research*. London, UK: Longman.

Baskerville, R. L. (1999). Investigating information systems with action research. *Communications of the Association for Information Systems*, *2*(19), 2–31.

Baskerville, R., & Myers, M. D. (2004). Special issue on action research in information systems: Making IS research relevant to practice—Foreword. *Management Information Systems Quarterly*, *28*(3), 329–335.

Bauer, R. A. (1960). Consumer behavior as risk-taking. In Hancock, R. S. (Ed.), *Dynamic Marketing for a Changing World* (pp. 389–398). Chicago, IL: American Marketing Association.

Baumbach, D. (2009, March/April). Web 2.0 and you. *Knowledge Quest.*

Baxter, P., & Jack, S. (2008). Qualitative case study methodology: Study design and implementation for novice researchers. *Qualitative Report*, *11*(4), 544–559.

Beaudouin-Lafon, M., & Mackay, W. (2003). Prototyping tools and techniques. In *The Human-Computer Interaction Handbook: Fundamentals, Evolving Technologies and Emerging Applications* (2nd ed., pp. 1006–1031). Hillsdale, NJ: Lawrence Erlbaum Associates.

Beaulieu, A., & Estalella, A. (2009). Rethinking research ethics for mediated settings. In *Proceedings of 5th International Conference on e-Social Science,* (pp. 1-15). Cologne, Germany: IEEE.

Beaulieu, A. (2004). Mediating ethnography: Objectivity and the making of ethnographies in the internet. *Social Epistemology*, *18*(2-3), 139–163. doi:10.1080/0269172042000249264

Beck, K. (2003). *Test-driven development by example*. Reading, MA: Addison Wesley.

Behr, A. L. (1983). *Empirical research methods for the human sciences: An introductory text for students of education, psychology and the social sciences*. Pretoria, South Africa: Butterworths.

Belanger, F., Hiller, J. S., & Smith, W. J. (2002). Trustworthiness in electronic commerce: The role of privacy, security and site attributes. *The Journal of Strategic Information Systems*, *11*, 245–270. doi:10.1016/S0963-8687(02)00018-5

Bell, M. (2008). Introduction to service-oriented modeling. In *Service-Oriented Modeling: Service Analysis, Design, and Architecture*. New York, NY: Wiley & Sons.

Benbasat, I., & Dexter, A. S. (1986). An investigation of the effectiveness of color and graphical presentation under varying time constraints. *Management Information Systems Quarterly*, *10*(1), 59–84. doi:10.2307/248881

Benbasat, I., Goldstein, D. K., & Mead, M. (1987). The case research strategy in studies of information systems. *Management Information Systems Quarterly*, *11*(3), 369–386. doi:10.2307/248684

Benbasat, I., & Zmud, R. W. (1999). Empirical research in information systems: The practice of relevance. *Management Information Systems Quarterly*, *23*(1), 3–16. doi:10.2307/249403

Bennett, J. (2003). *Evaluation method in research*. New York, NY: Continuum.

Benson, R., & Brack, C. (2009). Developing the scholarship of teaching: What is the role of e-teaching and learning? *Teaching in Higher Education*, *14*(1), 71–80. doi:10.1080/13562510802602590

Berg, B. L. (1995). *Qualitative research methods for the social scientist*. London, UK: Allyn and Bacon.

Berger, P., & Stripling, B. (2010). Student inquiry and web 2.0. *School Library Monthly*, *26*(5). Retrieved April 4, 2011 from http://www.schoollibrarymonthly.com/articles/Berger2010-v26n5p14.html

Berners-Lee, T., Hendler, J., & Lassila, O. (2001, May 17). The semantic web. *Scientific American*, 35–43.

Bertini, E., Catarci, T., Dix, A., Gabrielli, S., Kimani, S., & Santucci, G. (2009). Appropriating heuristic evaluation methods for mobile. *International Journal of Mobile Human Computer Interaction*, *1*(1), 20–41. doi:10.4018/jmhci.2009010102

Bertot, J. C., Snead, J. T., Jaeger, P. T., & McClure, C. R. (2006). Functionality, usability, and accessibility: Iterative user-centered evaluation strategies for digital libraries. *Performance Measurement and Metrics*, *7*(1), 17–28. doi:10.1108/14678040610654828

Beyer, H. (2010). *User-centered agile methods*. New York, NY: Morgan & Claypool Publishers.

Beyer, H., & Holtzblatt, K. (1997). *Contextual design: Defining customer-centered systems*. San Fransisco, CA: Morgan Kaufmann Publishers Inc.

Bhattacherjee, A. (2002). Individual trust in online firms: Scale development and initial test. *Journal of Management Information Systems*, *19*, 211–241.

Bhattacherjee, A., & Premkumar, G. (2004). Understanding changes in belief and attitude toward information technology usage: a theoretical model and longitudinal test. *Management Information Systems Quarterly*, *28*(2), 229–254.

Billi, M., Burzagli, L., Catarci, T., Santucci, G., Bertini, E., Gabbanini, F., & Palchetti, E. (2010). A unified methodology for the evaluation of accessibility and usability of mobile applications. *Universal Access in the Information Society*, *9*(4), 337–356. doi:10.1007/s10209-009-0180-1

Blackmon, M. H., Polson, P. G., Kitajima, M., & Lewis, C. (2002). Cognitive walkthrough for the web. In *Proceedings of the CHI 2002 Conference on Human Factors in Computing Systems,* (pp. 463-470). ACM Press.

Black, R., & Wood, A. (2003). Utilising information communication technology to assist the education of individuals with Down syndrome. In *Down Syndrome Issues and Information*. Portsmouth, UK: The Sarah Duffen Center.

Blaikie, N. W. H. (2003). *Analyzing quantitative data: From description to explanation*. London, UK: Sage Publications Ltd.

Blank, R., Heizer, W., & Von Voss, H. (1999). Externally guided control of static grip forces by visual feedback age and task effects in 3-6 year old children and in adults. *Neuroscience Letters*, *271*, 41–44. doi:10.1016/S0304-3940(99)00517-0

Blomberg, J., Giacomi, J., Mosher, A., & Swenton-Wall, P. (1933). Ethnographic field methods and their relation to design. In *Participatory Design: Principles and Practices* (pp. 123–155). Hillsdale, NJ: Lawrence Erlbaum Associates.

Blommaert, J. (2005). *Discourse: A critical introduction*. Cambridge, UK: Cambridge University Press. doi:10.1017/CBO9780511610295

Blum, F. H. (1955). Action research: A scientific approach? *Philosophy of Science*, *22*(1), 1–7. doi:10.1086/287381

Bluteau, J., Coquillart, S., Payan, Y., & Gentaz, E. (2008). Haptic guidance improves the visuo-manual tracking of trajectories. *PLoS ONE, 3*(3), e1775.

Boden, M. A. (2003). *The creative mind: Myths and mechanisms* (2nd ed.). London, UK: Routledge.

Bodker, S., Nielsen, C., & Graves Petersen, M. (2000). Creativity, coopeartion and interactive design. In *Proceedings of the 3rd Conference on Designing Interactive Systems: Processes, Practices, Methods, and Techniques,* (pp. 252-261). ACM.

Boehm, B. W. (1988). A spiral model of software development and enhancement. *ACM SIGSOFT Software Engineering Notes*, *11*(4), 14–24.

Bonnardel, N., Piolat, A., & Le Bigot, L. (2011). The impact of colour on website appeal and users' cognitive processes. *Displays*, *32*(2), 69–80. doi:10.1016/j.displa.2010.12.002

Bonoma, T. V. (1985). Case research in marketing: Opportunities, problems, and a process. *JMR, Journal of Marketing Research*, *22*(2), 199–208. doi:10.2307/3151365

Botero, A., Kommonen, K., Koskijoki, M., & Ollinki, I. (2003). *Codesigning visions, uses and applications*. Retrieved from http://arki.uiah.fi/arkipapers/codesigning_ead.pdf

Boudreau, M. C., Gefen, D., & Straub, D. W. (2001). Validation in information systems research: A state-of-the-art assessment. *Management Information Systems Quarterly*, *25*(1), 1–16. doi:10.2307/3250956

Bowen, G. A. (2006). Grounded theory and sensitizing concepts. *International Journal of Qualitative Methods*, *5*(3), 1–9.

Bowker, G. (1994). Information mythology and infrastructure. In Bud-Frierman, L. (Ed.), *Information Acumen: The Understanding and Use of Knowledge in Modern Business* (pp. 231–247). London, UK: Routledge.

Boyer, K., Olson, J., Calantone, R., & Jackson, E. (2002). Print verus electronic surveys: A comparison of two data collection methodologies. *Journal of Operations Management*, *20*, 357–373. doi:10.1016/S0272-6963(02)00004-9

Brackertz, N. (2007). Who is hard to reach and why? *Institute for Social Research*. Retrieved August 30, 2011 from http://www.sisr.net/publications/0701brackertz.pdf

Brajnik, G. (2006). *Web accessibility testing with barriers walkthrough*. Retrieved February 10, 2011, from http://www.dimi.uniud.it/giorgio/projects/bw Brajnik, G. (2008). Beyond conformance: The role of accessibility evaluation methods. In S. Hartmann, et al. (Eds.), *2nd International Workshop on Web Usability and Accessibility*, (pp. 63-80). Auckland, New Zealand: Springer.

Brajnik, G. (2008). Measuring web accessibility by estimating severity of barriers. In S. Hartmann, X. Zhou, & M. Kirchberg (Eds.), *2nd International Workshop on Web Usability and Accessibility*, (pp. 112-121). Auckland, New Zealand: Springer.

Brajnik, G., & Lomuscio, R. (2007). SAMBA: A semi automatic method for measuring barriers of accessibility. In *Proceedings of the ACM SIGACCESS Conference on Computers and Accessibility*, (pp. 43-49). ACM Press.

Braun, V., & Clarke, V. (2006). Using thematic analysis in psychology. *Qualitative Research in Psychology*, *3*, 77–101. doi:10.1191/1478088706qp063oa

Brealey, R. A., Marcus, A. J., & Myers, S. C. (1996). *Principios de dirección financiera*. Madrid, Spain: McGraw-Hill.

Breslin, J. G., Passant, A., & Decker, S. (2010). *The social semantic web*. Berlin, Germany: Springer-Verlag.

Briggs, P., Simpson, B., & De Angeli, A. (2004). Trust and personalisation: A reciprocal relationship? In Karat, C.-M., Blom, J., & Karat, J. (Eds.), *Designing Personalized User Experiences for eCommerce*. Dordrecht, The Netherlands: Kluwer Academic Publishers. doi:10.1007/1-4020-2148-8_4

Broekman, I., Enslin, P., & Pendlebury, S. (2002). Distributive justice and information communication technologies in higher education in South Africa. *South African Journal of Higher Education*, *16*(1), 29–35. doi:10.4314/sajhe.v16i1.25269

Bronfenbrenner, U. (2005). Lewinian space and ecological substance. In Bronfenbrenner, U. (Ed.), *Making Human Beings Human: Bioecological Perspectives on Human Development* (pp. 41–49). Thousand Oaks, CA: Sage Publications.

Brooks-Young, S. (2008, May). Web tools: The second generation. *District Administration*.

Brown, C. (2011). *Excavating the meaning of information and communication technology use amongst South African university students: A critical discourse analysis*. (PhD Thesis). = University of Cape Town. Cape Town, South Africa.

Brown, C. (2011). The influence of the recontextualisation of globalization: Discourses on higher education students' technological identities. In *Proceedings of the IADIS International Conference: ICT, Society and Human Beings 2011*, (pp. 73-80). IADIS Press.

Brown, C., & Czerniewicz, L. (2008). Trends in student use of ICTs in higher education in South Africa. In *Proceedings of the 10th Annual Conference of WWW Applications*. Cape Town, South Africa: Cape Peninsula University of Technology.

Brown, C., Thomas, H., van der Merwe, A., & van Dyk, L. (2008). The impact of South Africa's ICT infrastructure on higher education. In *Proceedings of the 3rd International Conference of E-Learning*, (pp. 69-76). Cape Town, South Africa: Academic Publishing Limited.

Bruner, G. C., & Kumar, A. (2005). Explaining consumer acceptance of handheld Internet devices. *Journal of Business Research*, *58*(5), 553–558. doi:10.1016/j.jbusres.2003.08.002

Brut, M., & Buraga, S. C. (2008). An ontology-based approach for modelling grid services in the context of e-learning. *International Journal of Web and Grid Services, 4*(4), 379–394. doi:10.1504/IJWGS.2008.022543

Bryant, A. (2002). Grounding systems research: Re-establishing grounded theory. *The Journal of Information Technology. Theory and Application, 4*(1), 25–42.

Bryant, A., & Charmaz, K. (2010). Grounded theory in historical perspective: An epistemological account. In Byant, A., & Charmaz, K. (Eds.), *The Sage Handbook of Grounded Theory* (pp. 31–57). London, UK: Sage.

Bryman, A. (2004). *Social research methods* (2nd ed.). Oxford, UK: Oxford University Press.

Bryman, A. (2006). Paradigm peace and the implications for quality. *International Journal of Social Research Methodology, 9*(2), 111–126. doi:10.1080/13645570600595280

Bryman, A. (2007). Barriers to integrating quantitative and qualitative research. *Journal of Mixed Methods Research, 1*(8), 8–22. doi:10.1177/2345678906290531

Bryman, A., Becker, S., & Sempik, J. (2008). Quality criteria for quantitative, qualitative and mixed methods research: A view from social policy. *International Journal of Social Research Methodology, 11*(4), 261–276. doi:10.1080/13645570701401644

Bryman, A., & Cramer, D. (2005). *Quantitative data analysis with SPSS 12 and 13: A guide for social scientists.* East Sussex, UK: Routledge.

Brynjolfsson, E. (1993). The productivity paradox of information technology: Review and assessment. *ACM, 36*(12), 66–77. doi:10.1145/163298.163309

Brynjolfsson, E., & Hitt, L. M. (1998). Beyond the productivity paradox. *Communications of the ACM, 41*(8), 49–55. doi:10.1145/280324.280332

Brynjolfsson, E., & Saunders, A. (2009). *Wired for innovation: How information technology is reshaping the economy.* Cambridge, MA: MIT Press.

Bühler, C., Heck, H., Perlick, O., Nietzio, A., & Ulltveit-Moe, N. (2006). Interpreting results from large scale automatic evaluation of web accessibility. In Miesenberger, K., Klaus, J., Zagler, W., & Karshmer, A. (Eds.), *Computers Helping People with Special Needs* (pp. 184–191). Berlin, Germany: Springer. doi:10.1007/11788713_28

Bullinger, H.-J., Müller-Spahn, F., & Rössler, A. (1996). *Encouraging creativity - Support of mental processes by virtual experience. Virtual Reality World.* IDG.

Burdea, G. C., & Coiffet, P. (2003). *Virtual reality technology* (2nd ed.). Mahwah, NJ: John Wiley & Sons Inc.

Campbell, D. T., & Fiske, D. W. (1959). Convergent and discriminate validation by the multitrait-multimethod matrix. *Psychological Bulletin, 56*, 81–105. doi:10.1037/h0046016

Cañibano Calvo, L., Sánchez Medina, A. J., García-Ayuso, M., & Chaminade, C. (Eds.). (2002). *Directrices para la gestión y difusión de información sobre intangibles.* Madrid, Spain: Fundación Airtel Móvil.

Cañibano, L. (2001). *La relevancia de los intangibles en el análisis de la situación financiera de la empresa.* Paper presented at the IVIE Universidad Autónoma de Madrid. Madrid, Spain.

Cañibano, L., Covarsí, M. G.-A., & Sánchez, M. P. (1999). *The value relevance and managerial implications of intangibles: A literature review.* Retrieved from http://www.oecd.org/industry/industryandglobalisation/1947974.pdf

Cardenas-Claros, M. S., & Gruba, P. A. (2010). *Bridging CALL & HCI: Input from participatory design.* Retrieved from http://www.postgradolinguistica.ucv.cl/pr_curriculum_pub_doc.php?pid=430

Card, S. K., & Moran, T. P. (1980). The keystroke-level model for user performance: Time with interactive systems. *Communications of the ACM, 23*(7), 396–410. doi:10.1145/358886.358895

Card, S., Moran, T. P., & Newell, A. (1983). *The psychology of human computer interaction.* Mahwah, NJ: Lawrence Erlbaum Associates.

Carlbring, P., Brunt, S., Bohman, S., Austin, D., Richards, J., & Ost, L. (2007). Internet vs. paper and pencil administration of questionnaires commonly used in panic/agoraphobia research. *Computers in Human Behavior, 23*, 1421–1434. doi:10.1016/j.chb.2005.05.002

Carliner, S. (2004). *An overview of online learning*. Amherst, MA: HRD Press.

Carlsson, C., & Walden, P. (2002). Mobile commerce: Some extensions of core concepts and key issues. In *Proceedings of the SSGRR 2002s Conference*. L'Aquila, Italy: SSGRR.

Carter, M., Thatcher, J. B., Applefield, C., & Mcalpine, J. (2011). What cell phones mean in young people's daily lives and social interactions. *SAIS 2011*. Retrieved from http://aisel.aisnet.org/sais2011/29

Castillo, J. J. (2009). *Snowball sampling*. Retrieved July 30, 2011, from http://www.experiment-resources.com/snowball-sampling.html

Cavana, R. Y., Delahaye, B. L., & Sekaran, U. (2001). *Applied business research: Qualitative and quantitative methods*. Canberra, Australia: John Wiley & Sons.

Cecez-Kecmanovic, D. (2005). Basic assumptions of the critical research perspectives in information systems. In Howcroft, D., & Trauth, E. (Eds.), *Handbook of Critical Information Systems Research* (pp. 19–46). Cheltenham, UK: Edward Elgar Publishing Inc.

Cellular-News. (2009, April 23). Vodafone sees loss of UK market share and lower ARPUs. *Cellular-News*. Retrieved from http://www.cellular-news.com/story/37159.php

Chamaz, K. (2006). *Constructing grounded theory: A practical guide through qualitative analysis*. Thousand Oaks, CA: SAGE Publication Ltd.

Charmaz, K. (2004). Grounded theory. In Hesse-Biber, S. N., & Leavy, P. (Eds.), *Approaches to Qualitative Research: A Reader on Theory and Practice. Oxford, UK*. Oxford: Oxford University Press.

Charmaz, K. (2006). Coding in grounded theory practice. In Charmaz, K. (Ed.), *Constructing Grounded Theory: A Practical Guide through Qualitative Analysis* (pp. 42–71). Thousand Oaks, CA: SAGE Publication Ltd.

Checkland, P. B. (1981). *Systems thinking, systems practice*. Chichester, UK: John Wiley.

Chen, H., Nunes, M. B., Zhou, L., & Peng, G. (2011). Expanding the concept of requirements traceability: The role of electronic records management in gathering evidence of crucial communications and negotiations. *Aslib Proceedings, 63*(2/3), 168–187. doi:10.1108/00012531111135646

Chen, S. C., & Dhillon, G. S. (2003). Interpreting dimensions of consumer trust in e-commerce. *Information Technology Management, 4*, 303–313. doi:10.1023/A:1022962631249

Chen, S. Y., & Macredie, R. (2010). Web-based interaction: A review of three important human factors. *International Journal of Information Management, 30*(5), 379–387. doi:10.1016/j.ijinfomgt.2010.02.009

Chesbrough, H. W. (2006). Open innovation: A new paradigm for understanding industrial innovation In H. Chesbrough, W. Vanhaverbeke, & J. West (Eds.), *Open Innovation: Researching a New Paradigm*, (pp. 1-14). Oxford, UK: Oxford University Press.

Cheung, C., & Lee, M. (2000). Trust in internet shopping: A proposed model and measurement instrument. In *Proceedings of the 2000 Americas Conference on Information Systems (AMCIS)*, (pp. 681-689). Los Angeles, CA: AMCIS.

Chigona, W., Mjali, P., & Denzl, N. (2007). *Role of ICT in national development: A critical discourse analysis of South Africa's government statements*. Paper presented at QualIT 2007 Qualitative Research in IT. Wellington, New Zealand.

Chigona, A., & Chigona, W. (2008). MXit up it up in the media: Media discourse analysis on a mobile instant messaging system. *Southern Africa Journal of Information and Communication, 9*, 42–57.

Cho, H., & LaRose, R. (1999). Privacy issues in internet surveys. *Social Science Computer Review, 17*(4), 1–19. doi:10.1177/089443939901700402

Chopra, K., & Wallace, W. A. (2003). Trust in electronic environments. In *Proceedings of the 36th Hawaii International Conference on System Sciences*. IEEE.

Cho, Y., Park, J., & Han, S. K. S. (2011). Development of a web-based survey system for evaluating affective satisfaction. *International Journal of Industrial Ergonomics*, *41*, 247–254. doi:10.1016/j.ergon.2011.01.009

Chun, M., Sohn, K., Arling, P., & Granados, N. (2009). Applying systems thinking to knowledge management systems: The case of Pratt-Whitney Rocketdyne. *Journal of Information Technology Case and Application Research*, *11*(3), 43–67.

CIC-IADE. (2003). *Modelo intellectus. CIC-IADE. CMMi_Product_Teem. (2002). Capability maturity model® integration (CMMISM), version 1.1*. Pittsburgh, PA: Carnegie Mellon University.

Clarke, A. (1999). *Evaluation research: An introduction to principles, methods and practice*. Thousand Oaks, CA: Sage Publications.

Coghlan, D., & Brannick, T. (2005). *Doing action research in your own organization* (2nd ed.). London, UK: Sage Publications.

Cohen, L. (2007, October). Information literacy in the age of social scholarship. *Library 2.0: An Academic's Perspective.*

Coleman, G., & O'Connor, R. (2008). Investigating software process in practice: A grounded theory perspective. *Journal of Systems and Software*, *81*(5), 772–784. doi:10.1016/j.jss.2007.07.027

Colwell, C., Petrie, H., & Kornbrot, D. (1998). *Haptic virtual reality for blind computer users*. Paper presented at the Assets 1998 Conference. Los Angeles, CA.

Condos, C., James, A., Every, P., & Simpson, T. (2002). Ten usability principles for the development of effective WAP and m-commerce services. *Aslib Proceedings*, *54*(6), 345–355. doi:10.1108/00012530210452546

Cooper, A. (2007). *About face 3.0: The essentials of interaction design*. New York, NY: John Wiley & Sons, Inc.

Cooper, D., & Subotzky, G. (2001). *The skewed revolution: Trends in South African higher education: 1988-1998*. Cape Town, South Africa: University of Western Cape.

Corbin, J., & Strauss, A. (1990). Grounded theory research: Procedures, canons, and evaluative criteria. *Qualitative Sociology*, *13*(1), 3–21. doi:10.1007/BF00988593

Cornicelli, L., & Grund, M. (2011). Assessing deer hunter attitudes toward regulatory change using self-selected respondents. *Human Dimensions of Wildlife*, *16*(3), 174–182. doi:10.1080/10871209.2011.559529

Council on Higher Education. (2010). Access and throughput in South African higher education: Three case studies. *Higher Education Monitor*, *9*. Retrieved from http://www.che.ac.za/documents/d000206/

Couper, M. P., Traugott, M. W., & Lamias, M. J. (2001). Web survey design and administration. *Public Opinion Quarterly*, *65*, 230–253. doi:10.1086/322199

Courage, C., & Baxter, K. (2004). *Understanding your users: A practical guide to user requirements methods, tools, and techniques*. San Francisco, CA: Morgan Kaufmann Publishers Inc.

Coyle, D., & Doherty, G. (2008). Towards ontologies for technology in mental health interventions. In *Proceedings of the 2008 First International Workshop on Ontologies in Interactive Systems*. Washington, DC: IEEE Computer Society.

Crabtree, B. F., & Miller, W. L. (1992). Primary care research: A multi-method typology and qualitative road map. In Crabtree, B. F., & Miller, W. L. (Eds.), *Doing Qualitative Research*. Thousand Oaks, CA: Sage.

Cranefield, J., & Yoong, P. (2007). To whom should information systems research be relevant: The case for an ecological perspective. In *Proceedings of the 15th European Conference on Information Systems (ECIS 2007)*. ECIS.

Creswell, J. W. (2003). *Research design qualitative, quantitative, and mixed methods approaches* (2nd ed.). Thousand Oaks, CA: SAGE Publications.

Creswell, J. W. (2003). *Research design: Qualitative, quantitative, and mixed methods approaches* (2nd ed.). Thousand Oaks, CA: SAGE Publications.

Creswell, J. W. (2012). *Educational research: Planning, concucting, and evaluating quantitiative and qualitative research* (4th ed.). Boston, MA: Pearson.

Cronholm, S., & Goldkuhl, G. (2003). Strategies for information systems evaluation: Six generic types. In *Proceedings of the Tenth European Conference on Information Technology Evaluation*, (pp. 65-74). Madrid, Spain: IEEE.

Cronholm, S., & Goldkuhl, G. (2003). Strategies for information systems evaluation: Six generic types. *Electronic Journal of Information Systems Evaluation*, *6*(2), 65–74.

Cross, M., & Adams, F. (2007). ICT policies and strategies in higher education in South Africa: National and institutional pathways. *Higher Education Policy*, *20*, 73–95. doi:10.1057/palgrave.hep.8300144

Crump, B., & Logan, K. (2008). A framework for mixed stakeholders and mixed methods. *The Electronic Journal of Business Research Methods*, *6*(1), 21–28.

Csikszentmihalyi, M. (1997). *Creativity: Flow and the psychology of discovery and invention*. New York, NY: Harper Perennial.

Cukier, W., Bauer, R., & Middleton, C. (2004). Applying Habermas' validity claims as a stanadard for critical discourse analysis. In *Information Systems Research: Relevant Theory and Informed Practice* (pp. 233–258). London, UK: Kluwer Academic Publishers. doi:10.1007/1-4020-8095-6_14

Cukier, W., Ngwenyama, O., Bauer, R., & Middleton, C. (2009). A critical analysis of media discourse on information technology: Preliminary results of a proposed method for critical discourse analysis. *Information Systems Journal*, *19*, 175–196. doi:10.1111/j.1365-2575.2008.00296.x

Cunningham, B. (1997). Case study principles for different types of cases. *Quality & Quantity*, *31*, 401–423. doi:10.1023/A:1004254420302

Curley, M. (2007). *Introducing an IT capability maturity framework*. Paper presented at the International Conference for Enterprise Information Systems, ICEIS. Madeira, Portugal.

Curley, M. (2004). *Managing information technology for business value*. New York, NY: Intel Press.

Cybulski, J., Nguyen, L., Thanasankit, T., & Lichtenstein, S. (2003). Understanding problem solving in requirements engineering: Debating creativity with IS practitioners. *PACIS 2003 Proceedings*. Retrieved from http://aisel.aisnet.org/pacis2003/32

Cyr, D., Bonanni, C., Bowes, J., & Ilsever, J. (2005). Beyond trust: Website design preferences across cultures. *Journal of Global Information Management*, *13*(4), 24–52. doi:10.4018/jgim.2005100102

Cyr, D., Head, M., & Ivanov, A. (2006). Design aesthetics leading to m-loyalty in mobile commerce. *Information & Management*, *43*(8), 950–963. doi:10.1016/j.im.2006.08.009

Cyr, D., Head, M., & Larios, H. (2010). Colour appeal in website design within and across cultures: A multi-method evaluation. *International Journal of Human-Computer Studies*, *68*, 1–21. doi:10.1016/j.ijhcs.2009.08.005

Czerniewicz, L., & Brown, C. (2007). Disciplinary differences in the use of educational technology. In *Proceedings of ICEL 2007: 2nd International Conference on e-Learning*, (pp. 117-130). New York, NY: ICEL.

Czerniewicz, L., & Brown, C. (2009). A virtual wheel of fortune? Enablers and constraints of ICTs in higher education in South Africa. In Marshall, S., Kinuthia, W., & Taylor, W. (Eds.), *Bridging the Knowledge Divide: Educational Technology for Development*. Denver, CO: Information Age Publishing.

Dabholkar, P. A. (1996). Consumer evaluations of new technology-based self-service options: An investigation of alternative models of service quality. *International Journal of Research in Marketing*, *13*(1), 29–51. doi:10.1016/0167-8116(95)00027-5

Daft, R., & Wiginton, J. (1979). Language and organization. *Academy of Management Review*, *4*(2), 179–191.

Dai, H., & Palvi, P. C. (2009). Mobile commerce adoption in China and the United States: A cross-cultural study. *ACM SIGMIS Database*, *40*(4), 43–61. doi:10.1145/1644953.1644958

Dalsgaard, P., & Halskov, K. (2010). Innovation in participatory desing. In *Proceedings of the 11th Biennial Participatory Design Conference,* (pp. 281-282). ACM.

Daly, J., Kellehear, A., & Gliksman, M. (1997). *The public health researcher: A methodological approach*. Oxford, UK: Oxford University Press.

Daniel, E. (1999). Provision of electronic banking in the UK and the Republic of Ireland. *International Journal of Bank Marketing*, *17*(2), 72–82. doi:10.1108/02652329910258934

Daniels, A., Rosenberg, R., Anderson, C., Law, J., Marvin, A., & Law, P. (2011). Verification of parent-report of child autism spectrum disorder diagnosis to a web-based autism registry. *Journal of Autism and Developmental Disorders*, *42*(2), 257–265. doi:10.1007/s10803-011-1236-7

Darken, R. P., & Peterson, B. (2002). Spatial orientation, wayfinding, and representation. In Stanney, K. M. (Ed.), *Handbook of Virtual Environments Design, Implementation and Applications* (pp. 493–518). Mahwah, NJ: Lawrence Erlbaum Associates.

Dart, J., Petheram, R. J., & Straw, W. (1998). *Review of evaluation in agricultural extension*. Retrieved from https://rirdc.infoservices.com.au/downloads/98-136.pdf

Daskalantonakis, M. K. (1992). A practical view of software measurement and implementation experiences within Motorola. *IEEE Transactions on Software Engineering*, *18*(11), 998–1010. doi:10.1109/32.177369

Davidson, E. J. (1999). Joint application design (JAD) in practice. *Journal of Systems and Software*, *45*, 215–223. doi:10.1016/S0164-1212(98)10080-8

Davis, D. N., & Vijayakumar, M. V. (2010). A "society of mind" cognitive architecture based on the principles of artificial economics. *International Journal of Artificial Life Research*, *1*(1), 51–71. doi:10.4018/jalr.2010102104

Davis, F. (1989). Perceived usefulness, perceived ease of use, and user acceptance of information technology. *Management Information Systems Quarterly*, *13*(3), 319–340. doi:10.2307/249008

Davis, F. D., Bagozzi, R. P., & Warshaw, P. R. (1989). User acceptance of computer technology: A comparison of two theoretical models. *Management Science*, *35*(8), 982–1003. doi:10.1287/mnsc.35.8.982

Davis, F. D., Bagozzi, R. P., & Warshaw, P. R. (1992). Extrinsic and intrinsic motivation to use computers in the workplace. *Journal of Applied Social Psychology*, *22*(14), 1111–1132. doi:10.1111/j.1559-1816.1992.tb00945.x

Davis, F., Bagozzi, R. P., & Warshaw, P. R. (1989). User acceptance of computer technology: A comparison of two theoretical models. *Management Science*, *35*(8), 982–1002. doi:10.1287/mnsc.35.8.982

Davison, R. M., Martinsons, M. G., & Kock, N. (2004). Principles of canonical action research. *Information Systems Journal*, *14*(1), 43–63. doi:10.1111/j.1365-2575.2004.00162.x

De Bra, P., et al. (1999). AHAM: A dexter-based reference model for adaptive hypermedia. In *Proceedings of the ACM Conference on Hypertext and Hypermedia*. ACM Press.

Denzin, J., Norman, K., & Lincoln, Y. S. (2005). *The Sage handbook of qualitative research* (3rd ed.). Thousand Oaks, CA: Sage.

Denzin, N. K. (1978). *Sociological methods*. New York, NY: McGraw-Hill.

Denzin, N. K., & Lincoln, Y. S. (2003). Introduction: The discipline and practice of qualitative research. In Denzin, N. K., & Lincoln, Y. S. (Eds.), *The Landscape of Qualitative Research* (pp. 1–45). New Delhi, India: SAGE Publications.

Denzin, N. K., & Lincoln, Y. S. (2008). *The landscape of qualitative research*. Thousand Oaks, CA: Sage Publications.

Department of Education and Department of Communication. (2001). *A strategy for information and communication technology in education*. Retrieved from http://www.info.gov.za/otherdocs/2001/ict_doe.pdf

Department of Education. (2001). *The national plan on higher education*. Pretoria, South Africa: Government Printers. Retrieved 12 June 2006, from http://www.polity.org.za/html/govdocs/misc/higheredu1.htm?rebookmark=1

Department of Education. (2003). *White paper on e-education: Transforming learning and teaching through ICT*. Pretoria, South Africa: Government Printers. Retrieved 1 June 2006, from http://www.info.gov.za/whitepapers/2003/e-education.pdf

Department of Training and Industry. (2000). *South African ICT sector development framework*. Retrieved from http://www.dti.gov.za/saitis/docs/html/chap06.html

DeSanctis, G., & Poole, M. S. (1994). Capturing the complexity in advanced technology use: Adaptive structuration theory. *Organization Science*, *5*(2), 121–147. doi:10.1287/orsc.5.2.121

Deutsch, M. (1962). Cooperation and trust: Some theoretical notes. In M. R. Jones (Ed.), *Nebraska Symposium on Motivation*, (pp. 275-319). Lincoln, NE: University of Nebraska Press.

Devaraj, S., & Kohli, R. (2003). Performance impacts of information technology: Is actual usage the missing link? *Management Science*, *49*(3), 273–289. doi:10.1287/mnsc.49.3.273.12736

Dexter, L. A. (1970). *Elite and specialized interviewing*. London, UK: Colchester.

Dey, A. (2004). *Accessibility evaluation practices–survey, results*. Retrieved February 15, 2011, from http://deyalexander.com/publications/accessibility-evaluation-practices.html

Díaz, B., & Cachero, C. (2009). Fiabilidad de herramientas automáticas para el aseguramiento de la accesibilidad en Web. In *Proceedings of IADIS Ibero-Americana WWW/Internet 2009*. Madrid, Spain.: IEEE.

Dick, B. (1993). *You want to do an action research thesis?* Retrieved from http://www.scu.edu.au/schools/gcm/ar/art/arthesis.html

Dijkstra, E. W. (1972). The humble programmer. *Communications of the ACM*, *15*(10), 859–866. doi:10.1145/355604.361591

Dillman, D. (2007). *Mail and internet surveys "the tailored design method"* (2nd ed.). New York, NY: John Wiley & Sons, Inc.

Dillman, D., Glenn, P., Tortora, R., Swift, K., Kohrell, J., & Berck, J. (2009). Response rate and measurement differences in mixed-mode surveys using mail, telephone, interactive voice reponse (IVR) and the internet. *Social Science Research*, *38*, 1–18. doi:10.1016/j.ssresearch.2008.03.007

Dillman, D., Reipus, U., & Matzat, U. (2010). Advice in surveying the general public over the internet. *International Journal of Internet Science*, *5*(1), 1–4.

Dirksen, V., Huizing, A., & Smit, B. (2010). Piling on layers of understanding: The use of connective ethnography for the study of (online) work practices. *New Media & Society*, *12*(7), 1045–1063. doi:10.1177/1461444809341437

Disability Rights Commission. (2006). *Guide to good practice in commissioning accessible websites*. London, UK: British Standards Institution.

Dockter, J., Haug, D., & Lewis, C. (2010, February). Redefining rigor: Critical engagement, digital media, and the new English/language arts. *Journal of Adolescent & Adult Literacy*. doi:10.1598/JAAL.53.5.7

Dodgson, M., Gann, D., & Salter, A. (2008). *Management of technological innovation: Strategy and practice*. Oxford, UK: Oxford University Press.

Doney, P. M., Cannon, J. P., & Mullen, M. R. (1998). Understanding the influence of national culture on the development of trust. *Academy of Management Review*, *23*(3), 601–620.

Doolin, B. (2009). Information systems and power: A Foucauldian perspective. In Brooke, C. (Ed.), *Critical Management Perspectives on Information Systems* (pp. 211–230). Oxford, UK: Butterworth Heineman.

Dubé, L., & Paré, G. (2003). Rigor in information systems positivist case research: Current practices, trends, and recommendations. *Management Information Systems Quarterly*, *27*(4), 597–635.

Duncan, D. F., White, J. B., & Nicholson, T. (2003). Using internet-based surveys to reach hidden populations: Case of nonabusive illicit drug users. *American Journal of Health Behavior, 27*(3), 208-218. Retrieved September 10, 2011 from http://www.duncan-associates.com/hiddenpop.pdf

Dunlop, M. D., & Crossan, A. (2000). Predictive text entry methods for mobile phones. *Personal Technologies*, *4*(2-3), 134–143. doi:10.1007/BF01324120

Edvinsson, L. (1997). Developing intellectual capital at Skandia. *Long Range Planning*, *30*(3), 366–373. doi:10.1016/S0024-6301(97)90248-X

Egger, F. N., & de Groot, B. (2000). *Developing a model of trust for electronic commerce: An application to a permissive marketing web site*. Paper presented at 9th International World Wide Web Conference. Amsterdam, The Netherlands.

Eisenberg, M. (2008). The parallel information universe. *Library Journal*, *133*(8).

Ekeh, P. P. (1974). *Social exchange theory: The two traditions*. London, UK: Heinemann Educational.

Erickson, G. S., Komaromi, K., & Unsal, F. (2010). Social networks and trust in e-commerce. *International Journal of Dependable and Trustworthy Information Systems*, *1*(1), 45–59. doi:10.4018/jdtis.2010010103

Esteves, J., Ramos, I., & Carvalho, J. (2002). Use of grounded theory in information systems area: An exploratory analysis. In *Proceedings of the 1st European Conference on Research Methodology for Business and Management Studies*, (pp. 129-136). Reading, UK: Business and Management Studies.

Euroforum. (1998). *Medición del capital intelectual: Modelo intelect*. Madrid, Spain: IUEE.

European Union. (2002). *European parliament resolution on the commission communication eEurope 2002: Accessibility of public web sites and their content (COM(2001) 529 - C5-0074/2002 - 2002/2032(COS))*. Retrieved February 5, 2011, from http://www.europarl.europa.eu/omk/omnsapir.so/pv2?PRG=DOCPV&APP=PV2&LANGUE=EN&SDOCTA=18&TXTLST=1&POS=1&Type_Doc=RESOL&TPV=DEF&DATE=130602&PrgPrev=PRG@TITRE%7cAPP@PV2%7cTYPEF@TITRE%7cYEAR@02%7cFind@%2577%2565%2562%2520%7cFILE@BIBLIO02%7cPLAGE@1&TYPEF=TITRE&NUMB=1&DATEF=020613

European Union. (2010). *Digital agenda 2010*. Retrieved February 14, 2011, from http://ec.europa.eu/information_society/digital-agenda/index_en.htm

eXaminator. (2005). *eXaminator: Evaluación automática de la accesibilidad*. Retrieved February 12, 2011, from http://examinator.ws/

Eysenback, G., & Till, J. E. (2001). Ethical issues in qualitative research on internet communities. *British Medical Journal*, *323*, 1103–1105. doi:10.1136/bmj.323.7321.1103

Fairclough, N. (2001). *Language and power* (2nd ed.). London, UK: Longman.

Fairclough, N. (2006). *Language and globalisation*. Oxon, UK: Routledge.

Fallman, D. (2003). Design-oriented human-computer interaction. In *Proceedings of the SIGCHI Conference on Human Factors in Computing Systems*, (pp. 225-232). ACM.

Fang, X., Chan, S., Brzezinski, J., & Xu, S. (2005). Moderating effects of task type on wireless technology acceptance. *Journal of Management Information Systems*, *22*(3), 123–157. doi:10.2753/MIS0742-1222220305

Fan, W., & Yan, Z. (2010). Factors affecting response rates of the web survey: A systematic review. *Computers in Human Behavior*, *26*, 132–139. doi:10.1016/j.chb.2009.10.015

Feagin, J., Orum, A., & Sjoberg, G. (Eds.). (1991). *A case for case study*. Chapel Hill, NC: University of North Carolina Press.

Feilzer, M. Y. (2010). Doing mixed methods research pragmatically: Implications for the rediscovery of pragmatism as a research paradigm. *Journal of Mixed Methods Research*, *4*(1), 6–16. doi:10.1177/1558689809349691

Felder, R. M. (2009). *Index of learning styles*. Retrieved September, 19, 2009, from http://www4.ncsu.edu/unity/lockers/users/f/felder/public/ILSpage.html

Felder, R. M., & Silverman, L. K. (1988). Learning and teaching styles in engineering education. *Journal of Engineering Education*, *7*(7), 674–681.

Fendt, J., & Sachs, W. (2007). Grounded theory method in management research: Users' perspectives. *Organizational Research Methods*, *11*(3), 430–455. doi:10.1177/1094428106297812

Fenton, N. E., & Pfleeger, S. L. (1998). *Software metrics: A rigorous and practical approach*. Boston, MA: PWS Publishing Co.

Fereday, J., & Muir-Cochrane, E. (2006). Demonstrating rigor using thematic analysis: A hybrid approach of inductive and deductive coding and theme development. *International Journal of Qualitative Methods*, *5*(1), 1–11.

Fernandez, W., Lehmann, H., & Underwood, A. (2002). Rigour and relevance in studies of information systems innovation: A grounded theory methodology approach. In *Proceedings of the 10th European Conference on Information Systems*, (pp. 110-119). Gdansk, Poland: IEEE.

Fernandez, W. (2004). Using the Glaserian approach in grounded studies of emerging business practices. *Electronic Journal of Business Research Methods, 2*(2), 83–94.

Fernandez, W. (2005). The grounded theory method and case study data in IS research: Issues and design. In Hart, D., & Gregor, S. (Eds.), *Information Systems Foundations: Constructing and Criticising* (pp. 43–60). Canberra, Australia: The Australian National University.

Ferreira, J., Sharp, H., & Robinson, H. (2011). User experience design and agile development: Managing cooperation through articulation work. *Software, Practice & Experience, 41*(9), 963–974. doi:10.1002/spe.1012

Feygin, D., Keehner, M., & Tendick, F. (2002). Haptic guidance: Experimental evaluation of a haptic training method for a perceptual motor skill. In *Proceedings of the Sympathetic Haptic Interfaces for Virtual Environment and Teleoperator Systems*, (pp. 40-47). IEEE.

Fidel, R. (2008). Are we there yet? Mixed methods research in library and information science. *Library & Information Science Research, 30*, 265–272. doi:10.1016/j.lisr.2008.04.001

Fidler, D. J., Hodapp, R. M., & Dykens, M. E. (2002). Behavioral phenotypes and special education: Parent report of educational issues for children with Down syndrome, Prader-Willi syndrome, and Williams syndrome. *The Journal of Special Education, 36*, 80–88. doi:10.1177/00224669020360020301

Field, A. (2005). *Discovering statistics using SPSS: And sex, drugs and rock'n'roll* (2nd ed.). London, UK: SAGE Publications.

Fields, D., & Kafai, Y. (2008). Knowing and throwing mudballs, hearts, pies, and flowers: A connective ethnography of gaming practices. In V. Jonker et al. (Eds.), *8th International Conference on Learning Sciences*. Utrecht, The Netherlands: University of Utrecht.

Fields, D., & Kafai, Y. (2009). A connective ethnography of peer knowledge sharing and diffusion in a tween virtual world. *Computer-Supported Learning, 4*(1), 47–68. doi:10.1007/s11412-008-9057-1

Findlay, C. S., & Lumsden, C. J. (1988). The creative mind: Toward an evolutionary theory of discovery and innovation. *Journal of Social and Biological Systems, 11*(1), 3–55. doi:10.1016/0140-1750(88)90025-5

Fink, A. (2010). Survey research methods. *Education Research Methodology: Quantitative Methods and Research*, 152 – 160.

Finkelstein, A. (1993). *Requirements engineering: An overview*. Paper presented in the 2nd Asia-Pacific Software Engineering Conference (APSEC 1993). Tokyo, Japan.

Fiora, T. W. A., Baker, S. W., & Dobbie, G. (2008). Automated usability testing framework. In *Proceedings of the Ninth Conference on Australasian User Interface (AUIC 2008)*, (pp. 55-64). AUIC.

Fleming, C., & Bowden, M. (2009). The most commonly cited disadvantages of web-based surveys are sample frame and non-response bias. *Journal of Environmental Management, 90*, 284–292. doi:10.1016/j.jenvman.2007.09.011

Fleming, C., & Bowden, M. (2009). Web-based surveys as an alternative to traditional mail methods. *Journal of Environmental Management, 90*(1), 284–292. doi:10.1016/j.jenvman.2007.09.011

Fling, B. (2009). *Mobile design and development*. San Francisco, CA: O' Reilly Media.

Flood, D., Harrison, R., & Duce, D. (2011). *Using mobile apps: Investigating the usability of mobile apps from the users' perspective*. Unpublished.

Flood, D., Harrison, R., & McDaid, K. (2011). *Spreadsheets on the move: An evaluation of mobile spreadsheets*. Paper presented in the European Spreadsheet Risk Interest Group (EuSpRIG) Annual Conference. London, UK.

Flood, R. L., & Jackson, M. C. (1991). *Creative problem solving*. New York, NY: John Wiley.

Fodor, J. A. (1983). *The modularity of the mind*. Cambridge, MA: MIT Press.

Forsythe, S., & Shi, B. (2003). Consumer patronage and risk perceptions in internet shopping. *Journal of Business Research, 56*(11), 867–875. doi:10.1016/S0148-2963(01)00273-9

Foucault, M. (1994). The subject and power. In Faubion, J. (Ed.), *Power* (pp. 326–348). New York, NY: The New Press.

Foucault, M. (1994). Truth and power. In Faubion, J. (Ed.), *Power* (pp. 111–133). New York, NY: The New Press.

Fowler, M. (2006, July 28). *Customer affinity*. Retrieved from http://martinfowler.com/bliki/CustomerAffinity.html

Fredrick, K. (2010). In the driver's seat: Learning and library 2.0 tools. *School Library Monthly, 26*(6).

Freire, A. P., Fortes, R., Turine, M., & Paiva, D. (2008). An evaluation of web accessibility metrics based on their attributes. In *Proceedings of the 26th Annual ACM International Conference on Design of Communication,* (pp. 73-80). Lisbon, Portugal: ACM Press.

Fricker, R. (2008). Sampling methods for web and e-mail surveys. *The Sage Handbook of Online Research Methods, 1*, 195–217. Retrieved August 26, 2011 from http://faculty.nps.edu/rdfricke/docs/5123-Fielding-Ch11.pdf

Fukuyama, F. (1995). *Trust: The social virtues and the creation of prosperity*. New York, NY: Free Press.

Fukuzumi, S., Ikegami, T., & Okada, H. (2009). Development of quantitative usability evaluation method. In *Proceedings of the 13th International Conference on Human-Computer Interaction (HCI International 2009)*, (pp. 252-258). ACM.

Fundación, C. T. I. C. (2011). *TAW*. Retrieved February 10, 2011, from http://www.tawdis.net/

Gable, G. (1994). Integrating case study and survey research methods: An example in information systems. *European Journal of Information Systems, 3*(2), 112–126. doi:10.1057/ejis.1994.12

Galanouli, D., Murphy, C., & Gardner, J. (2004). Teachers' perceptions of the effectiveness of ICT-competence training. *Computers & Education, 43*(1-2), 63–79. doi:10.1016/j.compedu.2003.12.005

Galliers, R. D., Mylonopoudos, N. A., Morris, C., & Meadows, M. (1997). IS research agendas and practices in the UK. In D. E. Avison (Ed.), *Key Issues in Information Systems: Proceedings of the 2nd UKAIS Conference*, (pp. 143-172). New York, NY: McGraw-Hill.

Galliers, R. (1991). Choosing appropriate information systems research approaches: A revised taxonomy. In Galliers, R. (Ed.), *Information Systems Research: Issues, Methods and Practical Guidelines* (pp. 144–162). Oxford, UK: Blackwell.

Galliers, R. D. (1995). A manifesto for information management research. *British Journal of Management, 6*, 45–52. doi:10.1111/j.1467-8551.1995.tb00137.x

Gall, J., Gall, M., & Borg, W. (1999). *Applying educational research: A practical guide* (4th ed.). New York, NY: Addison Wesley Longman, Inc.

Gambetta, D. G. (Ed.). (1988). *Trust: Making and breaking cooperative relations*. New York, NY: Basil Blackwell.

GAO. (1979). *Contracting for computer software development--Serious problems require management attention to avoid wasting additional millions*. Washington, DC: General Accounting Office.

Garbarino, H., & Delgado, B. (2011). *Taxonomía de indicadores de activos intangibles de TI para la administración pública: Caso república oriental del Uruguay*. Paper presented at the CISTI 2011. Madrid, Spain.

Garbarino, H., Delgado, B., & Carrillo Verdún, J. (2011). *Taxonomy of IT intangibles based on the electronic government maturity model in Uruguay*. Paper presented at the MCCSIS 2011. Madrid, Spain.

García de Castro, M. A., Merino Moreno, C., Plaz Landaeta, R., & Villar Mártil, L. (2004). *La gestión de activos intangibles en la administración pública*.

Garcia, E., Martin, C., Garcia, A., Harrison, R., & Flood, D. (2011). Systematic analysis of mobile diabetes management applications on different platforms. In *Proceedings of the Information Quality in e-Health 7th Conference of the Workgroup Human-Computer Interaction and Usability Engineering of the Austrian Computer Society*, (pp. 379-396). Graz, Austria: Springer.

Garrison, D. R., & Kanuka, H. (2004). Blended learning: Uncovering its transformative potential in higher education. *The Internet and Higher Education, 7*, 95–105. doi:10.1016/j.iheduc.2004.02.001

Gause, D. C., & Weinberg, G. M. (1989). *Exploring requirements: Quality before design*. New York, NY: Dorset House Pub.

Gaver, B., Dunne, T., & Pacenti, E. (1999). Design: Cultural probes. *Interaction, 6*, 21–29. doi:10.1145/291224.291235

Gee, J. (2005). *An introduction to discourse analysis*. New York, NY: Routledge.

Geertz, C. (1973). *The interpretation of cultures: Selected essays*. New York, NY: Basic Books.

Gefen, D., & Heart, T. (2006). On the need to include national culture as a central issue in e- commerce trust beliefs. *Journal of Global Information Management*, *14*(4), 1–30. doi:10.4018/jgim.2006100101

Gefen, D., Karahanna, E., & Straub, D. W. (2003). Trust and TAM in online shopping: An integrated model. *Management Information Systems Quarterly*, *27*(1), 51–90.

Gefen, D., Rose, M., Warkentin, M., & Pavlou, P. (2004). Cultural diversity and trust in IT adoption: A comparison of USA and South African e-voters. *Journal of Global Information Systems*, *13*(1), 54–78. doi:10.4018/jgim.2005010103

Gerring, J. (2007). *Case study research: Principles and practice*. Cambridge, UK: Cambridge University Press.

Gershenfeld, N., Krikorian, R., & Cohen, D. (2004, October). The internet of things. *Scientific American*. doi:10.1038/scientificamerican1004-76

Geyman, J. P., Deyo, R. A., & Ramsey, S. D. (Eds.). (2000). *Evidence-based clinical practice: Concepts and approaches*. Boston, MA: Butterworth-Heinemann.

Giddens, A. (Ed.). (1974). *Positivism and sociology*. London, UK: Heinemann.

Giess, C., Evers, H., & Meinzer, H. P. (1998). *Haptic volume rendering in different scenarios of surgical planning*. Paper presented at the Third PHANTOM Users Group Workshop. Cambridge, MA.

Gilbert, T. (2006). Mixed methods and mixed methodologies: The practical, the technical and the political. *Journal of Research in Nursing*, *11*(3), 205–217. doi:10.1177/1744987106064634

Gile, K. J., & Handcock, M. S. (2010). Respondent-driven sampling: An assessment of current methodology. *Sociological Methodology*, *40*, 285–327. Retrieved August 25, 2011, from http://arxiv.org/PS_cache/arxiv/pdf/0904/0904.1855v1.pdf

Gillham, B. (2000). *Case study research methods*. London, UK: Continuum.

Glaser, B. (1978). *Theoretical sensitivity*. Mill Valley, CA: Sociology Press.

Glaser, B. (1992). *Emergent vs. forcing: Basics of grounded theory analysis*. Mill Valley, CA: Sociology Press.

Glaser, B. G. (1978). *Theoretical sensitivity: Advances in the methodology of grounded theory*. Mill Valley, CA: Sociology Press.

Glaser, B. G., & Strauss, A. L. (1967). *The discovery of grounded theory: Strategies for qualitative research*. New York, NY: Aldine de Gruyter. doi:10.1097/00006199-196807000-00014

Glaser, B. G., & Strauss, A. L. (1999). *The discovery of grounded theory: Strategies for qualitative research*. New York, NY: Aldine de Gruyter. doi:10.1097/00006199-196807000-00014

Glaser, B., & Strauss, A. (1967). *The discovery of grounded theory*. London, UK: Aldine Transaction.

Golan, O. (2010). Trust over the net: The case of Israeli youth. *International Journal of Dependable and Trustworthy Information Systems*, *1*(2), 70–85. doi:10.4018/jdtis.2010040104

Goldberg, A., & Robson, D. (1983). *Smalltalk-80: The language and its implementation*. Reading, MA: Addison-Wesley.

Gordon, J., & McNew, R. (2008). Developing the online survey. *The Nursing Clinics of North America*, *43*, 605–619. doi:10.1016/j.cnur.2008.06.011

Gordon, S. (2008). A case for case-based research: Editorial preface. *Journal of Information Technology Case and Application Research*, *10*(1), 1–6.

Gorman, P. J., Lieser, J. D., Murray, W. B., Haluck, R. S., & Krummel, T. M. (1998). *Assessment and validation of force feedback virtual reality based surgical simulator*. Paper presented at the Third PHANToM Users Group Workshop. Cambridge, MA.

Goulding, C. (1999). *Grounded theory: Some reflections on paradigm, procedures and misconceptions*. Technical Working Paper. Wolverhampton, UK: University of Wolverhampton.

Goulding, C. (2002). *Grounded theory – A practical guide for management, business, and market researchers*. London, UK: Sage.

Gower, J. (1975). Generalized procrustes analysis. *Psychometrika*, *40*, 33–51. doi:10.1007/BF02291478

Grabner-Krauter, S., & Kaluscha, E. A. (2003). Empirical research in on-line trust: A review and critical assessment. *International Journal of Human-Computer Studies*, *58*, 783–812. doi:10.1016/S1071-5819(03)00043-0

Graefe, A., Mowen, A., Covelli, E., & Trauntvein, N. (2011). Recreation participation and conservation attitudes: Differences between mail and online respondents in a mixed mode survey. *Human Dimensions of Wildlife*, *16*(3), 183–199. doi:10.1080/10871209.2011.571750

Granic, A., & Cukusic, M. (2011). Usability testing and expert inspections complemented by educational evaluation: A case study of an e-learning platform. *Journal of Educational Technology & Society*, *14*(2), 107–123.

Gras-Velazquez, A., Joyce, A., & Derby, M. (2009). *Women and ICT: Why are girls still not attracted to ICT studies and careers?* Brussels, Belgium: European Schoolnet (EUN Partnership AISBL).

Gray, B. (2004). Informal learning in an online community of practice. *Journal of Distance Education*, *19*(1), 20–35.

Green, R. (2000). The internet unplugged. *eAI Journal*, 82-86.

Greenhalgh, T. (2006). Computer assisted learning in undergraduate medical education. *British Medical Journal*, *322*, 40–44. doi:10.1136/bmj.322.7277.40

Greenhow, C. (2009). *Social scholarship: Applying social networking technologies to research practices*. Knowledge Quest.

Green, N. (1999). Disrupting the field: Virtual reality technologies and "multisited" ethnographic methods. *The American Behavioral Scientist*, *43*(3), 409–421. doi:10.1177/00027649921955344

Grudin, J., & Grinter, E. R. (1995). Ethnography and design. *Computer Supported Cooperative Work Journal*, *3*(1), 55–59. doi:10.1007/BF01305846

Guba, E. G., & Lincoln, Y. S. (1981). *Effective evaluation*. San Francisco, CA: Jossey-Bass.

Guba, E. G., & Lincoln, Y. S. (1989). *Fourth generation evaluation*. Thousand Oaks, CA: Sage.

Guenther, T., & Möllering, G. (2010). A framework for studying the problem of trust in online settings. *International Journal of Dependable and Trustworthy Information Systems*, *1*(3), 14–31. doi:10.4018/jdtis.2010070102

Guindon, R. (1998). Designing the design process: Exploiting opportunistic thoughts. *Human-Computer Interaction*, *5*(2), 305–344. doi:10.1207/s15327051hci0502&3_6

Gu, J. C., Lee, S. C., & Suh, Y. H. (2009). Determinants of behavioral intention to mobile banking. *Expert Systems with Applications*, *36*(9), 11605–11616. doi:10.1016/j.eswa.2009.03.024

Gummesson, E. (2000). *Qualitative methods in management research* (2nd ed.). Thousand Oaks, CA: Sage.

Gurtman, M. B. (1992). Trust, distrust, and interpersonal problems: A circumplex analysis. *Journal of Personality and Social Psychology*, *62*, 989–1002. doi:10.1037/0022-3514.62.6.989

Hacker, S. K., Willard, M. A., & Couturier, L. (2002). *The trust imperative*. New York, NY: American Society of Quality.

Hackos, J. A., & Redish, J. C. (1998). *User and task analysis for interface design*. New Your, NY: John Wiley & Sons, Inc.

Håkansson, A., Nguyen, N. T., Hartung, R. L., Howlett, R. J., & Jain, L. C. (Eds.). (2009). Agent and multi-agent systems: Technologies and applications. *Lecture Notes in Artificial Intelligence, 5559*.

Hamilton, S., & Chervany, N. L. (1981). Evaluating information system effectiveness-Part 1: Comparing evaluation approaches. *Management Information Systems Quarterly*, *5*(3), 55–69. doi:10.2307/249291

Handcock, M. S., & Gile, K. J. (2011). On the concept of snowball sampling. *Sociological Methodology*, *1554*, 1–5.

Hansen, B., & Kautz, K. (2005). Grounded theory applied – Studying information systems development methodologies in practice. In *Proceedings of the 38th Hawaii International Conference on System Science,* (pp. 1-10). Waikoloa, HI: IEEE.

Hartung, R. L., & Håkansson, A. (2010). Meta agents, ontologies and search, a proposed synthesis. *Lecture Notes in Computer Science*, *6277*, 273–281. doi:10.1007/978-3-642-15390-7_28

Harvey, L. (1990). *Critical social research*. London, UK: Unwin Hyman.

Hazdic, M., Wongthongtham, P., Dillon, T., & Chang, E. (2009). Ontology-based multi-agent systems. In *Studies in Computational Intelligence*. London, UK: Springer.

Heckathorn, D. (2002). Respondent-driven sampling II: Deriving valid population estimates from chain-referral samples of hidden populations. *Social Problems*. Retrieved September 08, 2011 from http://www.jstor.org/stable/10.1525/sp.2002.49.1.11

Heckathorn, D., Semaan, S., Broadhead, R., & Hughes, J. (2002). Extensions of respondent-driven sampling: A new approach to the study of injection drug users aged 18 – 25. *AIDS and Behavior, 6*(1), 18-25. Retrieved September 02, 2011 from http://www.springerlink.com/index/00G4W675B7EVVH4T.pdf

Hedberg, J. G., & Brudvik, O. C. (2008). Supporting dialogic literacy through mashing and moddling of places and spaces. *Theory into Practice*, *47*, 138–149. doi:10.1080/00405840801992363

Heiervang, E., & Goodman, R. (2011). Advantages and limitations of web-based surveys: Evidence from a child mental health survey. *Social Psychiatry and Psychiatric Epidemiology*, *46*(1), 69–76. doi:10.1007/s00127-009-0171-9

Heikkinen, H., Huttunen, R., & Syrjälä, L. (2007). Action research as narrative: Five principles for validation. *Educational Action Research*, *15*(1), 5–19. doi:10.1080/09650790601150709

Hemmi, A., Bayne, S., & Landt, R. (2009). The appropriation and repurposing of social technologies in higher education. *Journal of Computer Assisted Learning*, *25*, 19–30. doi:10.1111/j.1365-2729.2008.00306.x

Hendler, J. (2001). Agents and the semantic web. *IEEE Intelligent Systems*, *16*(2), 30–37. doi:10.1109/5254.920597

Henze, N., & Herrlich, M. (2004). The personal reader: A framework for enabling personalization services on the semantic web. In *Proceedings of the Twelfth GI Workshop on Adaptation and User Modeling in Interactive Systems (ABIS 2004)*. Berlin, Germany: ABIS.

Henze, N. (2004). Reasoning and ontologies for personalized e-learning in the semantic web. *Journal of Educational Technology & Society*, *7*(4), 82–97.

Herman, J. L., Morris, L. L., & Fitz-Gibbon, C. T. (1987). *Evaluator's handbook*. Newbury Park, CA: Sage.

HESA. (2009). Students and qualifiers data tables. *Higher Education Statistics Agency*. Retrieved from http://www.hesa.ac.uk/index.php?option=com_datatables&Itemid=121&task=show_category&catdex=3

Hesse-Biber, S. (2010). Emerging methodologies and methods practices in the field of mixed methods research. *Qualitative Inquiry*, *16*(6), 415–418. doi:10.1177/1077800410364607

Hevner, A. R., March, S. T., Park, J., & Ram, S. (2004). Design science in information systems research. *Management Information Systems Quarterly*, *28*(1), 75–105.

Highsmith, J., & Fowler, M. (2001). The agile manifesto. *Software Development Magazine*, *9*(8), 29–30.

Hill, S. R., & Troshani, I. (2009). Adoption of personalisation mobile services: Evidence from young australians. *BLED 2009 Proceedings*. Retrieved from http://aisel.aisnet.org/bled2009/35

Hindess, B. (1996). *Discourses of power*. Oxford, UK: Blackwell Publishers Ltd.

Hine, C. (2000). *Virtual ethnography*. London, UK: Sage.

Hine, C. (2007). Connective ethnography for the exploration of e-science. *Journal of Computer-Mediated Communication*, *12*(2). doi:10.1111/j.1083-6101.2007.00341.x

Hinton, L., Kurinczuk, J., & Ziebland, S. (2010). Infertility, isolation and the internet: A qualitative interview study. *Patient Education and Counseling*, *81*, 436–441. doi:10.1016/j.pec.2010.09.023

Hirscheim, R., & Lyytinen, K. (1996). Exploring the intellectual foundations of information systems. *Accounting. Management and Information Technology*, *2*(1-2), 1–64.

Hirschheim, R., Klein, H., & Lyytinen, K. (1995). *Information systems development and data modeling: Conceptual and philosophical foundations*. Cambridge, UK: Cambridge University Press. doi:10.1017/CBO9780511895425

Hirschheim, R., & Smithson, S. (1988). *A critical analysis of information systems evaluation*. Amsterdam, The Netherlands: North-Holland.

HiSoftware. (2003). *Cynthia says*. Retrieved February 15, 2011, from http://www.cynthiasays.com/

Hitt, L. M., & Brynjolfsson, E. (1996). Productivity, business profitability, and consumer surplus: Three different measures of information technology value. *Management Information Systems Quarterly*, *20*(2), 121–142. doi:10.2307/249475

Hoehle, H., & Huff, S. (2009). Electronic banking channels and task-channel fit. *ICIS 2009 Proceedings*. Retrieved from http://aisel.aisnet.org/icis2009/98

Hoffman, D. L., Novak, T. P., & Peralta, M. (1999). Building consumer trust on-line. *Communications of the ACM*, *42*(4), 80–85. doi:10.1145/299157.299175

Hofstede, G. (1980). Motivation, leadership, and organization: Do American theories apply abroad? *Organizational Dynamics*, *9*(1), 42–63. doi:10.1016/0090-2616(80)90013-3

Holden, W. (2010). *A world of apps*. New York, NY: Juniper Research Limited.

Holleis, P., Otto, F., Hussmann, H., & Schmidt, A. (2007). Keystroke-level model for advanced mobile phone interaction. In *Proceedings of the SIGCHI Conference on Human Factors in Computing Systems (CHI 2007)*, (pp. 1505-1514). New York, NY: ACM.

Holleis, P., Scherr, M., & Broll, G. (2011). A revised mobile KLM for interaction with multiple NFC-tags. In *Proceedings of the 13th IFIP TC 13 International Conference on Human-Computer Interaction*, (pp. 204-221). Berlin, Germany: Springer-Verlag.

Howard, P. N. (2002). Network ethnography and the hypermedia organization: New media, new organizations, new methods. *New Media & Society*, *4*(4), 550–574. doi:10.1177/146144402321466813

Howcroft, B., Hamilton, R., & Hewer, P. (2002). Consumer attitude and the usage and adoption of home-based banking in the United Kingdom. *International Journal of Bank Marketing*, *20*(3), 111–121. doi:10.1108/02652320210424205

Howcroft, D., & Trauth, E. (2005). Choosing critical IS research. In Howcroft, D., & Trauth, E. (Eds.), *Handbook of Critical Information Systems Research*. Chelteham, UK: Edward Elgar Publishing, Inc.

Hubbard, D. W. (2007). *How to measure anything: Finding the value of intangibles in business* (2nd ed.). Hoboken, NJ: John Wiley & Sons, Inc.

Hughes, J., & Jones, S. (2003). Reflections on the use of grounded theory in interpretive information systems research. *Electronic Journal of Information Systems Evaluation*, *6*(1).

Hughes, J. A. (1990). *The philosophy of social research* (2nd ed.). London, UK: Longman.

Hughes, J., & Howcroft, D. (2000). Grounded theory: Never knowingly understood. *Information Systems Research*, *4*(1), 181–197.

Hyatt, J., & Craig, A. (2009, October). Adapt for outreach: Taking technology on the road. *Computers in Libraries*.

IASC. (2009). *NIC 38 activos intangibles*. Retrieved from http://www.iasb.org/NR/rdonlyres/8C28675D-FE12-468C-A7B7-66F8C1628673/0/IAS38.pdf

Iivari, J., Isomäki, H., & Pekkola, S. (2010). The user – The great unknown of systems development: Reasons, forms, challenges, experiences and intellectual contributions of user involvement. *Information Systems Journal*, *20*, 109–117. doi:10.1111/j.1365-2575.2009.00336.x

Im, I., Kim, Y., & Han, H. J. (2008). The effects of perceived risk and technology type on users' acceptance of technologies. *Information & Management*, *45*(1), 1–9. doi:10.1016/j.im.2007.03.005

International, Q. S. R. (2011). *What is qualitative research*. Retrieved from http://www.qsrinternational.com/what-is-qualitative-research.aspx

Internet Archive. (2001). *The wayback machine*. Retrieved February 5, 2011, from http://www.archive.org/web/web.php

Irani, Z., & Love, P. (Eds.). (2008). *Evaluating information systems: Public and private sector*. Oxford, UK: Butterworth-Heinemann.

Isomäki, H. (2009). The human modes of being in investigating user experience. In Saariluoma, P., & Isomäki, H. (Eds.), *Future Interaction Design II* (pp. 191–207). London, UK: Springer-Verlag. doi:10.1007/978-1-84800-385-9_10

Issa, T. (2008). *Development and evaluation of a methodology for developing websites.* (PhD Thesis). Curtin University. Perth, Australia. Retrieved from http://espace.library.curtin.edu.au:1802/view/action/nmets.do?DOCCHOICE=17908.xml&dvs=1235702350272~864&locale=en_US&search_terms=17908&usePid1=true&usePid2=true

Issa, T., & Turk, A. (2012). Applying usability and HCI principles in developing marketing websites. *International Journal of Computer Information Systems and Industrial Management Applications*, *4*, 76–82.

Issa, T., Turk, A., & West, M. (2010). Development and evaluation of a methodology for developing marketing websites. In Martako, D., Kouroupetroglou, G., & Papadopoulou, P. (Eds.), *Integrating Usability Engineering for Designing the Web Experience: Methodologies and Principles*. Hershey, PA: IGI Global. doi:10.4018/978-1-60566-896-3.ch006

ITU. (2008). ICT statistics news log - Global mobile phone subscribers to reach 4.5 billion by 2012. *ITU*. Retrieved from http://www.itu.int/ITUD/ict/newslog/Global+Mobile+Phone+Subscribers+To+Reach+45+Billion+By+2012.aspx

ITU. (2011). [*ICT facts and figures*. Retrieved from http://www.itu.int]. *WORLD (Oakland, Calif.)*, *2011*.

Ivankova, N. V., Creswell, J. W., & Stick, S. L. (2006). Using mixed-methods sequential explanatory design: From theory to practice. *Field Methods*, *18*(1), 3–20. doi:10.1177/1525822X05282260

Jackson, M. A. (1983). *Systems development I*. Englewood Cliffs, NJ: Prentice Hall.

Jacobson, R. D., Kitchin, R., Garling, T., Golledge, R., & Blades, M. (1998). *Learning a complex urban route without sight: Comparing naturalistic versus laboratory measures*. Paper presented at the International Conference of the Cognitive Science Society of Ireland. Dublin, Ireland.

Jaeger, P. T. (2006). Assessing Section 508 compliance on federal e-government web sites: A multi-method, user-centered evaluation of accessibility for persons with disabilities. *Government Information Quarterly*, *23*(2), 169–190. doi:10.1016/j.giq.2006.03.002

Jans, R., & Dittrich, K. (2008). A review of case studies in business research. In Dul, J., & Hak, T. (Eds.), *Case Study Methodology in Business Research*. Oxford, UK: Butterworth-Heinemann.

Jarvenpaa, S. L., & Grazioli, S. (1999). Surfing among sharks: How to gain trust in cyberspace. *Financial Times, Mastering Information Management, 15,* 2-3.

Jayaratna, N. (1994). *Understanding and evaluating methodologies -NIMSAD- A systemic framework*. London, UK: McGraw-Hill International.

Jefatura del Estado de España. (2002). *LEY 34/2002, de 11 de julio, de servicios de la sociedad de la información y de comercio electrónico (BOE nº 166)*. Retrieved February 5, 2011, from http://www.boe.es/boe/dias/2002/07/12/pdfs/A25388-25403.pdf

Jensen, M. (2007). The new metrics of scholarly authority. *The Chronicle*, *53*(41).

Jiang, J. J., & Klein, G. (1996). User perceptions of evaluation criteria for three system types. *The Data Base for Advances in Information Systems*, *27*(3), 63–69. doi:10.1145/264417.264435

Jick, T. D. (1979). Mixing qualitative and quantitative methods: Triangulation in action. *Administrative Science Quarterly*, *24*(4), 602–611. doi:10.2307/2392366

Johnson, R. B., & Onwuegbuzie, A. J. (2004). Mixed method research: A research paradigm whose time has come. *Educational Researcher*, *33*(7), 14–26. doi:10.3102/0013189X033007014

Johnson, S., & Aragon, S. (2003). An instructional strategy framework for online learning environments. *Bureau of Educational Research*, *100*, 31–43.

Jones, A., Issroff, K., & Scanlon, E. (2007). Affective factors in learning with mobile devices. In *Big Issues in Mobile Learning* (pp. 17–22). Nottingham, UK: University of Nottingham.

Jones, L. A., & Lederman, S. J. (2006). *Human hand function*. Oxford, UK: Oxford University Press. doi:10.1093/acprof:oso/9780195173154.001.0001

Kanaracus, C. (2008). Gartner: Global IT spending growth stable. *InfoWorld*. Retrieved from http://www.infoworld.com/t/business/gartner-global-it-spending-growth-stable-523

Kaplan, A., & Haenlein, M. (2010). Users of the world unite! The challenges and opportunities of social media. *Business Horizons*, *53*(1), 59–68. doi:10.1016/j.bushor.2009.09.003

Kaplan, B., & Duchon, D. (1988). Combining qualitative and quantitative methods in information systems research: A case study. *Management Information Systems Quarterly*, *12*(4), 571–586. doi:10.2307/249133

Kaplan, R. S., & Norton, D. P. (Eds.). (1996). *The balanced scorecard*. Boston, MA: Harvard Business School Press.

Kaplan, R. W., & Saccuzzo, D. P. (2008). *Psycho- logical testing: Principles, applications, and issues*. Monterey, CA: Brooks/Cole.

Kaplan, R., & Norton, D. (2004). La disponibilidad estratégica de los activos intangibles. *Harvard Business Review*, *122*, 38–51.

Karasti, H. (2001). *Increasing sensitivity towards everyday work practice in system design*. (Unpublished Doctoral Dissertation). University of Oulu. Oulu, Finland.

Karoulis, A., Demetriadis, S., & Pombortsis, A. (2006). Comparison of expert-based and empirical evaluation methodologies in the case of a CBL environment: The "Orestis" experience. *Computers & Education*, *47*(2), 172–185. doi:10.1016/j.compedu.2004.09.002

Katz, J. (1983). A theory of qualitative methodology: The social system of analytical fieldwork. In Emerson, R. (Ed.), *Contemporary Field Research: A Collection of Readings* (pp. 127–148). Boston, MA: Little Brown Company.

Kaufman, D. (2006). Tools keep web content flowing. *Television Week*, *25*(30), 60.

Kawalek, J. P. (2008). *Rethinking information systems in organizations: Integrating organizational problem solving*. New York, NY: Routledge.

Kay, A. (1996). The early history of smalltalk. In Bergin, T. J., & Gibson, R. G. (Eds.), *History of Programming Languages II* (p. 511). Reading, MA: Addison-Wesley. doi:10.1145/234286.1057828

Kayyali, R., Shirmohammadi, S., & El Saddik, A. (2008). Measurement of progress for haptic motor rehabilitation patients, In *Proceedings of the IEEE International Workshop on Medical Measurements and Applications*, (pp. 108-113). IEEE.

Keats, D. M. (2000). *Interviewing: A practical guide for students and professionals*. Buckingham, UK: Open University Press.

Keet, C. M., et al. (2008). Enhancing web portals with ontology-based data access: The case study of South Africa's accessibility portal for people with disabilities. In *Proceedings of the Fifth International Workshop OWL: Experiences and Directions (OWLED 2008)*. Karlsruhe, Germany: OWLED.

Kelly, S. (2004). *ICT and social/organisational change: A praxiological perspective on groupware innovation*. (Ph.D. Thesis). University of Cambridge. Cambridge, UK.

Kendall, C., Kerr, L. R. F. S., Gondim, R. C., Werneck, G. L., Macena, R. H. M., & Pontes, M. K. (2008). An empirical comparison of respondent-driven sampling, time location sampling, and snowball sampling for behavioral surveillance in men who have sex with men, Fortaleza, Brazil. *AIDS and Behavior*, *12*(4), 97–104. doi:10.1007/s10461-008-9390-4

Kenny, A. (2010). *A new history of western philosophy*. Oxford, UK: Oxford University Press.

Kerres, M., & De Witt, C. (2003). A didactical framework for the design of blended learning arrangements. *Journal of Educational Media*, *28*(2/3), 101–113. doi:10.1080/1358165032000165653

Khalifa, M., & Shen, N. K. (2008). Explaining the adoption of transactional B2C mobile commerce. *Journal of Enterprise Information Management*, *21*(2), 110–124. doi:10.1108/17410390810851372

Kim, S.-C., Kim, C.-H., Yang, G.-H., Yang, T.-H., Han, B.-K., Kang, S.-C., & Kwon, D.-S. (2009). Small and lightweight tactile display (salt) and its application. In *Proceedings of the World Haptics Conference*, (pp. 69–74). World Haptics.

Kim, C., Mirusmonov, M., & Lee, I. (2010). An empirical examination of factors influencing the intention to use mobile payment. *Computers in Human Behavior*, *26*(3), 310–322. doi:10.1016/j.chb.2009.10.013

Kim, H. W., Chan, H. C., & Gupta, S. (2007). Value-based adoption of mobile internet: An empirical investigation. *Decision Support Systems*, *43*(1), 111–126. doi:10.1016/j.dss.2005.05.009

Kim, J., & Forsythe, S. (2007). Hedonic usage of product virtualization technologies in online apparel shopping. *International Journal of Retail and Distribution Management*, *35*(6), 502–514. doi:10.1108/09590550710750368

Kim, K. H. (2006). Can we trust creativity tests? A review of the Torrance tests of creative thinking. *Creativity Research Journal*, *18*(1), 3–14. doi:10.1207/s15326934crj1801_2

Kim, K. K., & Prabhakar, B. (2004). Initial trust and the adoption of B2C e-commerce: The case of internet banking. *The Data Base for Advances in Information Systems*, *35*(2), 50–65. doi:10.1145/1007965.1007970

King, A. (2011). *WebbIE*. Retrieved February 5, 2011, from http://www.webbie.org.uk/

Kini, A., & Choobineh, J. (1998). *Trust in electronic commerce: Definition and theoretical considerations*. Paper presented at the 31st Hawaii International Conference on System Sciences. Hawaii, HI.

Klecun, E. (2008). Bringing lost sheep into the fold: Questioning the discourse of the digital divide. *Information Technology & People*, *21*(3), 267–282. doi:10.1108/09593840810896028

Klein, H. K., & Myers, M. D. (1999). A set of principles for conducting and evaluating interpretive field studies in information systems. *Management Information Systems Quarterly*, *23*(1), 67–94. doi:10.2307/249410

Klein, H., & Myers, M. (1999). A set of principles for conducting and evaluating interpretive field studies in information systems. *Management Information Systems Quarterly*, *23*(1), 67–97. doi:10.2307/249410

Knowbility. (2011). *Accessibility internet rally (AIR)*. Retrieved January 16, 2012, from http://www.knowbility.org/v/air/

Koch, H. (2010). Developing dynamic capabilities in electronic marketplaces: A cross-case study. *The Journal of Strategic Information Systems*, *19*(1), 28–38. doi:10.1016/j.jsis.2010.02.001

Ko, H., Jung, J., Kim, J. Y., & Shim, S. W. (2004). Cross-cultural differences in perceived risk of online shopping. *Journal of Interactive Advertising*, *4*(2). Retrieved from http://jiad.org/article46

Koivumaki, T., Ristola, A., & Kesti, M. (2008). The perceptions towards mobile services: An empirical analysis of the role of use facilitators. *Personal and Ubiquitous Computing*, *12*(1), 67–75. doi:10.1007/s00779-006-0128-x

Koufaris, M., & Hampton-Sosa, W. (2004). The development of initial trust in an online company by new customers. *Information & Management*, *41*(3), 377–397. doi:10.1016/j.im.2003.08.004

Kozinets, R. V. (2010). *Netnography – Doing ethnographic research online*. London, UK: Sage.

KPMG. (2009). UK consumers prefer to pay for digital content in time, not cash. *KPMG*. Retrieved from http://rd.kpmg.co.uk/mediareleases/15612.htm

Kramer, M. A. M. (2009). *The case for MobileHCI and mobile design research methods in mobile and informal learning contexts. Researching Mobile Learning: Frameworks*. Methods, and Research Designs.

Kroenke, D. (1992). *Management information systems*. New York, NY: McGraw-Hill.

Kueng, P., & Kawalek, P. (1996). Goal-based business process models: Creation and evaluation. *Business Process Management Journal*, *3*(1), 17–28. doi:10.1108/14637159710161567

Kuhn, T. S. (1970). *The structure of scientific revolutions* (2nd ed.). Chicago, IL: University of Chicago Press.

Kuhn, T. S. (1977). *The essential tension: Selected studies in scientific tradition and change*. Chicago, IL: University of Chicago Press.

Kumar, K. (1990). Post implementation evaluation of computer-based information systems: Current practices, computing practices. *Communications of the ACM*, *33*(2), 203–212. doi:10.1145/75577.75585

Kumar, M., & Sareen, M. (2009). Impact of technology-related environment issues on trust in B2B e-commerce. *International Journal of Information Communication Technologies and Human Development*, *3*(1), 21–40. doi:10.4018/jicthd.2011010102

Kung, L., Picard, R., & Towse, R. (2008). *The internet and the mass media*. Thousand Oaks, CA: SAGE Publications Ltd.

Kurillo, G., Gregorič, M., Goljar, N., & Bajd, T. (2005). Grip force tracking system for assessment and rehabilitation of hand function. *Technology and Health Care*, *13*, 137–149.

Kurosu, M., Urokohara, H., & Sato, D. (2002). A new data collection method for usability testing. In Leventhalm, L., & Barnes, J. (Eds.), *Usability Engineering: Process, Products and Examples*. Upper Saddle River, NJ: Pearson Prentice Hall.

Kushiniruk, A. (2002). Evaluation in the design of health information systems: application of approaches emerging from usability engineering. *Computers in Biology and Medicine*, *32*, 141–149. doi:10.1016/S0010-4825(02)00011-2

Kushniruk, A. W., Patel, V. L., & Cimiho, J. J. (1997). Usability testing in medical informatics: Cognitive approaches to evaluation of information systems and user interfaces. In *Proceedings of AMIA Annual Fall Symposium*. Montreal, Canada: AMIA.

Kushniruk, A. W., & Patel, V. L. (2004). Cognitive and usability engineering methods for evaluation of clinical information systems. *Journal of Biomedical Informatics*, *37*, 56–76. doi:10.1016/j.jbi.2004.01.003

Kvasny, L., & Trauth, E. (2002). The digital divide at work and at home: Discourses about power and underrepresented groups in the information society. In Whitley, E., Wynn, E., & DeGross, J. (Eds.), *Global and Organizational Discourse About Information Technology* (pp. 273–294). New York, NY: Kluwer Academic Publishers.

Laforet, S., & Li, X. (2005). Consumers' attitudes towards online and mobile banking in China. *International Journal of Bank Marketing*, *23*(5), 362–380. doi:10.1108/02652320510629250

Lagsten, J., & Goldkuhl, G. (2008). Interpretative IS evaluation: Results and uses. *The Electronic Journal Information Systems Evaluation*, *11*(2), 97–108.

Laham, D. (1997). Latent semantic analysis approaches to categorization. In *Proceedings of the 19th Annual Conference of the Cognitive Science Society*, (pp. 412-417). Mahwah, NJ: Lawrence Erlbaum.

Lam, W., & Chua, A. (2005). Knowledge management project abandonment: An exploratory examination of root causes. *Communications of the AIS*, *16*(23), 723–743.

Landauer, T. K., & Dumais, S. T. (1997). A solution to Plato's problem: The latent semantic analysis theory of acquisition, induction, and representation of knowledge. *Psychological Review*, *104*, 211–240. doi:10.1037/0033-295X.104.2.211

Landauer, T. K., Foltz, P. W., & Laham, D. (1998). Introduction to latent semantic analysis. *Discourse Processes*, *25*, 259–284. doi:10.1080/01638539809545028

Langendoerfer, P. (2002). M-commerce: Why it does not fly (yet?). In *Proceedings of the SSGRR 2002s Conference*. L'Aquila, Italy: SSGRR.

Larsen, K. R., & Monarchi, D. E. (2004). A mathematical approach to categorization and labeling of qualitative data: The latent categorization method. *Sociological Methodology*, *34*(1), 349–392. doi:10.1111/j.0081-1750.2004.00156.x

Larsen, K. R., Monarchi, D. E., Hovorka, D., & Bailey, C. (2008). Analyzing unstructured text data: Using latent categorization to identify intellectual communities in information systems. *Decision Support Systems*, *45*(4), 884–896. doi:10.1016/j.dss.2008.02.009

Laukkanen, T. (2005). Comparing consumer value creation in Internet and mobile banking. In *Proceedings of the International Conference on Mobile Business (ICMB 2005)*. ICMB.

Laurel, B., & Lunenfeld, P. (2003). *Design research: Methods and perspectives*. Cambridge, MA: The MIT Press.

Lave, J., & Wenger, E. (1991). *Situated learning: Legitimate peripheral participation*. Cambridge, UK: Cambridge University Press. doi:10.1017/CBO9780511815355

Layder, D. (1993). *New strategies in social research: An introduction and guide*. Cambridge, UK: Polity Press.

Lee, S., Kim, P.-J., & Jeong, H. (2006). Statistical properties of sampled networks. *Physical Review E, 73*(1), 1-7. Retrieved August 24, 2011, from http://stat.kaist.ac.kr/~pj/sampling06.pdf

Lee, A. (1989, March). A scientific methodology for MIS case studies. *Management Information Systems Quarterly*, 33–50. doi:10.2307/248698

Lee, A. (1991). Integrating positivist and interpretivist approaches to organizational research. *Organization Science, 2*, 342–365. doi:10.1287/orsc.2.4.342

Lee, A. (2001). Challenges to qualitative researchers in information systems. In Trauth, E. M. (Ed.), *Qualitative Research in IS: Issues and Trends* (p. 240). Hershey, PA: IGI Global. doi:10.4018/978-1-930708-06-8.ch010

Lee, A. S. (1999). Researching MIS. In Currie, W. L., & Galliers, R. (Eds.), *Rethinking Management Information Systems* (pp. 7–27). Oxford, UK: Oxford University Press.

Leech, N. L., & Onwuegbuzie, A. J. (2009). Atypology of mixed-methods research designs. *Quality & Quantity, 43*(2), 265–275. doi:10.1007/s11135-007-9105-3

Lee, J. C., & Myers, M. (2004). Dominant actors, political agendas, and strategic shifts over time: A critical ethnography of an enterprise systems implementation. *The Journal of Strategic Information Systems, 13*, 355–374. doi:10.1016/j.jsis.2004.11.005

Lee, M., & Turban, E. (2001). A trust model for consumer internet shopping. *International Journal of Electronic Commerce, 6*(1), 75–91.

Lee, Y., Kozar, K. A., & Larsen, K. R. T. (2003). The technology acceptance model: Past, present, and future. *Communications of the Association for Information Systems, 12*, Retrieved from http://aisel.aisnet.org/cais/vol12/iss1/50

Legris, P., Ingham, J., & Collerette, P. (2003). Why do people use information technology? A critical review of the technology acceptance model. *Information & Management, 40*(3), 191–204. doi:10.1016/S0378-7206(01)00143-4

Lehmann, H., & Fernandez, W. (2007). Adapting the grounded theory method for information systems research. In *Proceedings of the 4th QUALIT Conference Qualitative Research in IT & IT in Qualitative Research*, (pp. 1-8). Wellington, New Zealand: QUALIT.

Lehmann, H. (2010). *The dynamics of international information systems: Anatomy of a grounded theory*. New York, NY: Springer. doi:10.1007/978-1-4419-5750-4

Leicester, M. (2001). A moral education in an ethical system. *Journal of Moral Education, 30*(2), 251–260. doi:10.1080/03057240120077255

Leimbach, L. (2010). Bloom's mashup: Bloom's taxonomy tools and tutorials. *Teacher Tech: Changing Education One Byte at a Time.* Retrieved April 4, 2011 from http://rsu2teachertech.wordpress.com/2010/12/04/blooms-mashup/

Lesser, V. M., Yang, D. K., & Newton, L. D. (2011). Assessing hunters' opinions based on a mail and a mixed-mode survey. *Human Dimensions of Wildlife, 16*(3), 164–173. doi:10.1080/10871209.2011.542554

Lester, R. K., & Piore, M. J. (2004). *Innovation-The missing dimension*. Boston, MA: Harvard University Press.

Lev, B. (Ed.). (2001). *Intangibles: Management, measurement, and reporting*. New York, NY: Brookings Institution Press.

Levy, Y., & Ellis, T. J. (2006). A systems approach to conduct an effective literature review in support of information systems research. *Informing Science Journal, 9*, 181–212.

Lewin, K. (1946). Action research and minority problems. *The Journal of Social Issues*, *2*, 34–46. doi:10.1111/j.1540-4560.1946.tb02295.x

Lewin, K. (1947). Frontiers in group dynamics. *Human Relations*, *1*(1), 5–41. doi:10.1177/001872674700100103

Li, H., Liu, Y., Liu, J., Wang, X., Li, Y., & Rau, P. (2010). Extended KLM for mobile phone interaction: A user study result. In *Proceedings of the 28th International Conference Extended Abstracts on Human Factors in Computing Systems (CHI EA 2010)*, (pp. 3517-3522). New York, NY: ACM.

Liao, C. H., Tsou, C. W., & Huang, M. F. (2007). Factors influencing the usage of 3G mobile services in Taiwan. *Online Information Review*, *31*(6), 759–774. doi:10.1108/14684520710841757

Likert, R. (1932). A technique for the measurement of attitudes. *Archives de Psychologie*, *22*(140), 55.

Lim, Y.-K., Stolterman, E., & Tenenberg, J. (2008). The anatomy of prototypes: Prototypes as filters, prototypes as manifestations of design ideas. *ACM Transactions on Computer-Human Interaction*, *15*(7), 1–7, 27. doi:10.1145/1375761.1375762

Lin, H.-H., & Wang, Y.-S. (2006). An examination of the determinants of customer loyalty in mobile commerce contexts. *Information & Management*, *43*(3), 271–282. doi:10.1016/j.im.2005.08.001

Littlejohn, A., & Pegler, C. (2007). *Preparing for blended e-learning: Connection with e-learning*. Florence, KY: Routledge.

Liu, Z., Min, Q., & Ji, S. (2009). An empirical study on mobile banking adoption: The role of trust. In *Proceedings of the 2nd International Symposium on Electronic Commerce and Security, ISECS*, (Vol. 2, pp. 7-13). Washington, DC: IEEE Computer Society.

Liu, A., Tendick, F., Cleary, K., & Kaufmann, C. (2003). A survey of surgical simulation: Applications, technology, and education. *Presence: Teleoperating a Virtual Environment*, *12*, 599–614. doi:10.1162/105474603322955905

Liu, C. (2010). Human-machine trust interaction: A technical overview. *International Journal of Dependable and Trustworthy Information Systems*, *1*(4), 61–74. doi:10.4018/jdtis.2010100104

Liu, J., Cramer, S. C., & Reinkensmeyer, D. J. (2006). Learning to perform a new movement with robotic assistance: Comparison of haptic guidance and visual demonstration. *Journal of Neuroengineering and Rehabilitation*, *3*(1), 20. doi:10.1186/1743-0003-3-20

Liu, X., Carroll, C. B., Wang, S., Zajicek, J., & Bain, P. G. (2005). Quantifying drug-induced dyskinesias in the arms using digitised spiral-drawing tasks. *Journal of Neuroscience Methods*, *144*, 47–52. doi:10.1016/j.jneumeth.2004.10.005

Locke, K. D. (2000). *Grounded theory in management research*. London, UK: SAGE.

Longpradit, P. (2008). An inquiry-led personalised navigation system (IPNS) using multi-dimensional linkbases. *New Review of Hypermedia and Multimedia*, *14*(1), 33–55. doi:10.1080/13614560802316095

López, L. (2010). *AWA: Marco metodológico especifico en el dominio de la accesibilidad web para el desarrollo de aplicaciones web*. (Unpublished Doctoral Dissertation). Universidad Carlos III. Madrid, Spain.

Lorenz, M., & Horstmann, M. (2004). Semantic access to graphical web resources for blind users. In *Proceedings of the 3rd International Semantic Web Conference (ISWC 2004)*. Hiroshima, Japan: ISWC.

Lou, L., & John, B. E. (2005). Predicting task execution time on handheld devices using keystroke level modelling. In *Proceedings of CHI 2005 Extended Abstracts on Human Factors in Computing Systems (CHI EA 2005)*, (pp. 1605-1608). New York, NY: ACM.

Luarn, P., & Lin, H. H. (2005). Toward an understanding of the behavioral intention to use mobile banking. *Computers in Human Behavior*, *21*(6), 873–891. doi:10.1016/j.chb.2004.03.003

Lubbe, S. (2003). The development of a case study methodology in the information technology field: A step by step approach. *Ubiquity: A Web-Based Publication of the ACM*, *4*(27). Retrieved October 5, 2003 from http://www.acm.org/ubiquity/views/v4i27_lubbe.pdf

Lucero, A., & Arrasvuori, J. (2010). PLEX cards: A source of inspiration when designing for playfulness. In *Proceedings of the 3rd International Conference on Fun and Games*, (pp. 28-37). New York, NY: ACM.

Lugo-Villeda, L. I., Frisoli, A., Sandoval-Gonzalez, O., Padilla, M. A., Parra-Vega, V., & Avizzano, C. A. … Bergamasco, M. (2009). Haptic guidance of light-exoskeleton for arm-rehabilitation tasks, In *Proceedings of the 18th IEEE International Symposium on Robot and Human Interactive Communication,* (pp. 903-908). IEEE.

Lu, J., Yu, C. S., Liu, C., & Yao, J. E. (2003). Technology acceptance model for wireless internet. *Internet Research, 13*(3), 206–222. doi:10.1108/10662240310478222

Luján-Mora, S. (2011). *Análisis de la accesibilidad del sitio web del Senado de España.* Retrieved February 12, 2011, from http://accesibilidadweb.dlsi.ua.es/?menu=ej-analisis-senado-parte-1

Luke, S., Spector, L., Rager, D., & Hendler, J. (1997). Ontology-based web agents. In *Proceedings of the First International Conference on Autonomous Agents.* New York, NY: ACM.

Lycett, M., & Gialis, G. M. (2000). Component-based information systems: Toward a framework for evaluation. In *Proceedings of the 33rd Hawaii International Conference on System Sciences.* Hawaii, HI: IEEE.

Mahoney, C. (1997). Overview of qualitative methods and analytic techniques. In J. Frechtling, L. Sharp, & Westat (Eds.), *User-Friendly Handbook for Mixed Method Evaluations.* Washington, DC: NSF Program Officer Conrad Katzenmeyer.

Maiden, N., & Gizikis, A. (2001). Where do requirements come from? *IEEE Software, 18*(5), 10–12. doi:10.1109/52.951486

Mankoff, J., Fait, H., & Tran, T. (2005). Is your web page accessible? A comparative study of methods for assessing web page accessibility for the blind. In *Proceedings of the SIGCHI Conference on Human Factors in Computing Systems,* (pp. 41-50). Portland, OR: SIGCHI.

Marcus, G. E. (1995). Ethnography in/of the world system: The emergence of multi-sited ethnography. *Annual Review of Anthropology, 24,* 95–117. doi:10.1146/annurev.an.24.100195.000523

Mårtensson, P., & Lee, A. S. (2004). Dialogical action research at omega corporation. *Management Information Systems Quarterly, 28*(3), 507–536.

Martin, C., Flood, D., Sutton, D., Aldea, A., Waite, M., Garcia, E., … Harrison, R. (2011). *A systematic evaluation of mobile applications for diabetes management using heuristics.* Unpublished.

Martin, A., Biddle, R., & Noble, J. (2009). *XP customer practices: A grounded theory.* New York, NY: IEEE.

Martin, J. (1991). *Rapid application development.* New York, NY: Macmillan Publishing Company.

Martins, J. T., & Nunes, M. B. (2009). Methodological constituents of faculty technology perception and appropriation: Does from follow function? In *Proceedings of the IADIS International Conference on e-Learning.* Algarve, Portugal: IADIS.

Martins, J., & Nunes, M. (2009). Methodological constituents of faculty technology perception and appropriation: Does form follow function? In *Proceedings of the IADIS International Conference eLearning 2009,* (pp. 124-131). Algarve, Portugal: IADIS Press.

Martins, J., & Nunes, M. (2009). Translucent proclivity: Cognitive catalysts of faculty's preference for adaptable e-learning institutional planning. In *Proceedings of the World Conference on E-Learning in Corporate, Government, Healthcare & Higher Education ELEARN 2009,* (pp. 1359-1366). Vancouver, Canada: ELEARN.

Martins, J., & Nunes, M. (2010). Grounded theory-based trajectories of Portuguese faculty effort-reward imbalance in e-learning development. In *Proceedings of the 9th European Conference on Research Methodology for Business and Management Studies,* (pp. 310-119). Madrid, Spain: IEEE.

Martins, J., & Nunes, M. (2010). Polychronicity and multi-presence: A grounded theory of e-learning time-awareness as expressed by Portuguese academics. In *Proceedings of the 9th WSEAS International Conference on Advances in e-Activities, Information Security and Privacy,* (pp. 32-40). Merida, Venezuela: WSEAS.

Masri, F., & Luján-Mora, S. (2010). *Análisis de los métodos de evaluación de la accesibilidad web.* Paper presented at Universidad 2010. Havana, Cuba.

Masri, F., & Luján-Mora, S. (2010). *Test de usuario: Un método empírico imprescindible para la evaluación de la accesibilidad web*. Paper presented at Actas de la 5ª Conferencia Ibérica de Sistemas y Tecnologías de Información. Santiago de Compostela, Spain.

Massimi, M., Baecker, R. M., & Wu, M. (2007). Using participatory activities with seniors to critique, build, and evaluate mobile phones. In *Proceedings of the 9th International ACM SIGACCESS Conference on Computers and Accessibility*, (pp. 155–162). New York, NY: ACM.

Masten, D., & Plowman, T. (2003). Digital ethnography: The next wave in understanding the consumer experience. *Design Management Journal*, *14*(2), 74–81.

Matavire, R., & Brown, I. (2008). Investigating the use of grounded theory in information systems research. In *Proceedings of the 2008 Annual Research Conference of the South African Institute of Computer Scientists and Information Technologists in IT research in developing countries: riding the Wave of Technology*, (pp.139-147). ACM Press.

Mathiassen, L., & Nielsen, P. A. (2008). Engaged scholarship in IS research. *Scandinavian Journal of Information Systems*, *20*(2).

Maudsley, G. (2011). Mixing it but not mixed-up: Mixed methods research in medical education: A critical narrative review. *Medical Teacher*, *33*, 92–104. doi:10.3109/0142159X.2011.542523

Maurer, M., & Githens, R. P. (2010). Toward a reframing of action research for human resource and organization development: Moving beyond problem solving and toward dialogue. *Action Research*, *8*(3), 267–292. doi:10.1177/1476750309351361

Mayer, R. C., Davis, J. D., & Schoorman, F. D. (1995). An integrative model of organisational trust. *Academy of Management Review*, *20*(3), 709–734.

May, T. (2003). *Social research: Issues, methods and process*. Maidenhead, UK: Open University Press.

Mbeki, T. (2005). *Address of the President of South Africa, Thabo Mbeki, at the world summit on the information society, Tunis, Tunisia*. Retrieved from http://www.thepresidency.gov.za/pebble.asp?relid=3168

McBurney, D. H., & White, T. L. (2007). *Research methods* (7th ed.). New York, NY: Thomson Learning.

McGrath, K. (2005). Doing critical research in information systems: A case of theory and practice not informing each other. *Information Systems*, *15*, 85–101. doi:10.1111/j.1365-2575.2005.00187.x

McKechnie, S., Winklhofer, H., & Ennew, C. (2006). Applying the technology acceptance model to the online retailing of financial services. *International Journal of Retail & Distribution Management*, *34*(4/5), 388–410. doi:10.1108/09590550610660297

McKnight, D. H., & Chervany, N. L. (2001). What trust means in e-commerce customer relationships: An interdisciplinary conceptual typology. *International Journal of Electronic Commerce*, *6*(2), 35–59.

McKnight, D. H., Choudhury, V., & Kacmar, C. (2002). Developing and validating trust measures for e-commerce: An integrative typology. *Information Systems Research*, *13*(3), 334–359. doi:10.1287/isre.13.3.334.81

McPherson, M. A., & Nunes, J. M. (2008). Critical issues for e-learning delivery: What may seem obvious is not always put into practice. *Journal of Computer Assisted Learning*, *24*(5), 433–445. doi:10.1111/j.1365-2729.2008.00281.x

Medina, A. J. S., González, A. M., & Falcón, J. M. G. (2003). *El capital intelectual: Concepto y dimensiones*. Retrieved from http://www.aedem-virtual.com/articulos/iedee/v13/132097.pdf

Medina, A. S. (2003). *Modelo para la medición del capital intelectual de territorios insulares: Una aplicación al caso de Gran Canaria*. Gran Canaria, Spain: Universidad de las Palmas de Gran Canaria.

Merino Rodríguez, B., Merino Moreno, C., Plaz Landaeta, R., & Villar Mártil, L. (2003). *Capital intelectual en la administración pública: El caso del instituto de estudios fiscales*. Madrid, Spain.

Midgley, G. (2000). *Systemic intervention: Philosophy, methodology and practice*. New York, NY: Kluwer Academic/Plenum.

Miles, M. B., & Huberman, M. A. (1994). *Qualitative data analysis: An expanded sourcebook*. Thousand Oaks, CA: Sage.

Miles, R. (2005). *Aspectj cookbook*. San Francisco, CA: O'Reilly & Associates Inc.

Millard, D., et al. (2000). FOHM: A fundamental open hypertext model for investigating interoperability between hypertext domains. In *Proceedings of the Eleventh ACM Conference on Hypertext and Hypermedia HT 2000*, (pp. 93-102). ACM Press.

Miller, J., & Glassner, B. (2006). The "inside" and the "outside": Finding realities in interviews. In Silverman, D. (Ed.), *Qualitative Research Theory, Method and Practice* (2nd ed., pp. 125–139). London, UK: SAGE Publications.

Mills, C. W. (1959). *The sociological imagination*. London, UK: Oxford University Press.

Mingers, J. (2001). Combining IS research methods: Towards a pluralist methodology. *Information Systems Research, 12*(3), 240–259. doi:10.1287/isre.12.3.240.9709

Mingers, J. (2003). The paucity of multimethod research: A review of the information systems literature. *Information Systems Journal, 13*, 233–249. doi:10.1046/j.1365-2575.2003.00143.x

Ministro per l'Innovazione e le Tecnologie de la Repubblica Italiana. (2005). *Requisiti tecnici e i diversi livelli per l'accessibilit`a agli strumenti informatici, (G.U.n.1838/8/2005)*. Retrieved February 10, 2011, from http://www.pubbliaccesso.it/normative/DM080705.htm

Ministry of Justice. (1995). *Disability discrimination act*. Retrieved February 20, 2011, from http://www.legislation.gov.uk/ukpga/1995/50/contents

Minsky, M. (1986). *The society of mind*. New York, NY: Simon and Schuster.

Mintzberg, H. (1979). An emerging strategy of "direct" research. *Administrative Science Quarterly, 24*(4), 582–589. doi:10.2307/2392364

Mintzberg, H. (1987a, July/August). Crafting strategy. *Harvard Business Review*.

Mintzberg, H. (1987). The strategy concept I: Five Ps for strategy. *California Management Review*, 11–24.

Mitchell, M., Lebow, J., Uribe, R., Grathouse, H., & Shoger, W. (2011). Internet use, happiness, social support and introversion: A more fine grained analysis of person variables and internet activity. *Computers in Human Behavior, 27*(6), 1857–1861. doi:10.1016/j.chb.2011.04.008

Moghaddam, A. (2006). Coding issues in grounded theory. *Issues in Educational Research, 16*(1), 52–66.

Mohamad, A., Eid, M. M., & Abdulmotaleb, H. El Saddik, & Iglesias, R. (2007). A haptic multimedia handwriting learning system. In *Proceedings of the International Workshop on Educational Multimedia and Multimedia Education (Emme 2007)*, (pp. 103-108). New York, NY: ACM.

Monk, A., Wright, P., Haber, J., & Davenport, L. (1993). *Improving your human-computer interface*. Englewood Cliffs, NJ: Prentice Hall Publishing.

Montero Perez, M., Senecaut, M., Clarebout, G., & Desmet, P. (2010). Designing for motivation: the case of mobile language learning. In *Proceedings of the XIVth International CALL Conference*. CALL.

Montoni, A., & Rocha, A. (2010). Applying grounded theory to understand software improvement implementation. In *Proceedings of the 7th International Conference on the Quality of Information and Communication Technology*, (pp. 25-34). Porto, Portugal: IEEE.

Moon, J. W., & Kim, Y. G. (2001). Extending the TAM for a world-wide-web context. *Information & Management, 38*(4), 217–230. doi:10.1016/S0378-7206(00)00061-6

Morris, M. G., & Venkatesh, V. (2000). Age differences in technology adoption decisions: Implications for a changing workforce. *Personnel Psychology, 53*(2), 375–403. doi:10.1111/j.1744-6570.2000.tb00206.x

Muller, M. J. (2003). Participatory design: The third space in HCI. In *The Human-Computer Interaction Handbook: Fundamentals, Evolving Technologies and Emerging Applications* (2nd ed., pp. 1051–1068). Hillsdale, NJ: Lawrence Erlbaum Associates.

Mulpuru, S. (2006). US ecommerce outlook for Q4 2006: A recap of Q3 2006 online retail sales and overview of the upcoming holiday. *Forrester Research Business View Research Document*. Retrieved December 14th 2011 from http://www.forrester.com/rb/Research/us_ecommerce_outlook_for_q4_2006/q/id/40586/t/2

Murthy, D. (2008). Digital ethnography: An examination of the use of new technologies for social research. *Sociology*, *42*(5), 837–855. doi:10.1177/0038038508094565

Myers, M. (1999). Investigating information systems with ethnographic research. *Communications of AIS, 2*.

Myers, M. (2003). *Qualitative research in information systems*. Retrieved from http://www.qual.auckland.ac.nz

Myers, M. D. (1997). Qualitative research in information systems. *Management Information Systems Quarterly*, *21*(2), 241–242. doi:10.2307/249422

Myers, M. D., & Newman, M. (2007). The qualitative interview in IS research: Examining the craft. *Information and Organization*, *17*, 2–26. doi:10.1016/j.infoandorg.2006.11.001

Neumann, L., & Star, S. L. (1996). Making infrastructure: The dream of a common language. In Blomberg, J., Kensing, F., & Dykstra-Erickson, E. (Eds.), *Computer Professionals for Social Responsibility* (pp. 231–240). Palo Alto, CA: ACM.

Neuman, W. L. (2000). *Social research methods: "Qualitative and quantitative approaches* (4th ed.). Reading, MA: Allyn & Bacon.

Ng'ambi, D. (2008). A critical discourse analysis of students anonymous online postings. *International Journal of Information and Communication Technology Education*, *4*(3), 31–39. doi:10.4018/jicte.2008070104

Nganji, J. T., Brayshaw, M., & Tompsett, B. (2011). Ontology-based e-learning personalisation for disabled students in higher education. *ITALICS*, *10*(1), 1–11.

Nganji, J. T., & Nggada, S. (2011). Disability-aware software engineering for improved system accessibility and usability. *International Journal of Software Engineering and Its Applications*, *5*(3), 47–62.

Ng, B. D., & Wiemer-Hastings, P. (2005). Addiction to the internet and online gaming. *Cyberpsychology & Behavior*, *8*(2), 100–113. doi:10.1089/cpb.2005.8.110

Niaz, M. (2008). A rationale for mixed method (integrative) research programmes in education. *Journal of Philosophy of Education*, *42*(2), 287–305. doi:10.1111/j.1467-9752.2008.00625.x

Nielsen Norman Group. (2012). *Report usability of mobile websites & applications*. Retrieved from http://www.nngroup.com/reports/mobile/

Nielsen, J. (1994). *Usability inspection methods*. Paper presented at CHI 1994 Conference Companion on Human Factors in Computing Systems. Boston, MA.

Nielsen, J. (2000). *Why you only need to test with 5 users*. Retrieved March 14, 2011, from http://www.useit.com/alertbox/20000319.html

Nielsen, J. (2011). *Website*. Retrieved from http://www.useit.com/papers/heuristic/heuristic_evaluation.html

Nielsen, J. (2012). *Usability 101: Introduction to usability*. Retrieved from http://www.useit.com/alertbox/20030825.html

Nielsen, J., & Molich, R. (1990). Heuristic evaluation of user interfaces. In *Proceedings of the ACM CHI 1990 Conference*, (pp. 249-256). Seattle, WA: ACM.

Nielsen. (2010). *Mobile youth around the world*. Retrieved from http://no.nielsen.com/site/documents/Nielsen-Mobile-Youth-Around-The-World-Dec-2010.pdf

Nielsen, J. (1993). *Usability engineering*. New York, NY: Academic Press.

Nielsen, J. (1994). Heuristic evaluation. In Nielsen, J., & Mack, R. L. (Eds.), *Usability Inspection Methods* (pp. 25–64). New York, NY: John Wiley.

Nielsen, J. (1994). *Usability engineering*. Boston, MA: Morgan Kaufmann.

Nijhoff, M., & Slappendel, C. (1996). Perspectives on innovation in organizations. *Organization Studies*, *17*(1), 107–129. doi:10.1177/017084069601700105

Nodine, M., Fowler, J., Ksiezyk, T., Perry, B., Taylor, M., & Unruh, A. (2000). Active information gathering in InfoSleuth. *International Journal of Cooperative Information Systems*, *9*(1/2), 3–28. doi:10.1142/S021884300000003X

Norman, D. A. (1988). *The psychology of everyday things*. New York, NY: Basic Books.

Novak, J. D., & Cañas, A. J. (2006). *The theory underlying concept maps and how to construct them*. Technical Report IHMC CmapTools. Tallahassee, FL: Florida Institute for Human and Machine Cognition. Retrieved from http://cmap.ihmc. us/Publications/ResearchPapers/TheoryUnderlyingConcept Maps.pdf

Noy, C. (2008). Sampling knowledge: The hermeneutics of snowball sampling in qualitative research. *International Journal of Social Research Methodology*, *11*(4), 327–344. doi:10.1080/13645570701401305

Nunes, M., Alajamy, M., Al-Mamari, S., Martins, J., & Zhou, L. (2010). The role of pilot studies in grounded theory: Understanding the context in which research is done! In *Proceedings of the 9th European Conference on Research Methodology for Business and Management Studies*, (pp. 34-43). Madrid, Spain: IEEE.

O'Brien, D., & Scharber, C. (2010). Teaching old dogs new tricks: The luxury of digital abundance. *Journal of Adolescent & Adult Literacy*, *53*(7).

Oates, B. J. (2006). *Researching information systems and computing*. London, UK: SAGE Publications.

Oblak, J., Cikajlo, I., & Matjačić, Z. (2010). Universal haptic drive: A robot for arm and wrist rehabilitation. *IEEE Transactions on Neural Systems and Rehabilitation Engineering*, *18*(3), 293–302. doi:10.1109/TNSRE.2009.2034162

O'Brien, H., & Toms, E. (2010). The development and evaluation of a survey to measure user engagement. *Journal of the American Society for Information Science and Technology*, *61*(1), 50–69. doi:10.1002/asi.21229

Office for National Statistics. (2007). Use of ICT at home. *Office for National statistics*. Retrieved from http://www.statistics.gov.uk/cci/nugget.asp?id=1710

Okoli, C., & Schabram, K. (2010). A guide to conducting a systematic literature review of information systems research. *Sprouts: Working Papers on Information Systems*, *10*(26). Retrieved February 13, 2011, from http://sprouts.aisnet.org/10-26

Olle, T. W., Hagelstein, J., Macdonald, I. G., Rolland, C., Sol, H. G., & Assche, F. J. M. V. (1988). *Information systems methodologies: "A framework for understanding*. Reading, MA: Addison-Wesley Publishing Company.

O'Neill, R. (2006). *The advantages and disadvantages of qualitative and quantitative research methods*. Retrieved from http://www.roboneill.co.uk/papers/research_methods.htm

Online Marketing Trends. (2011). *Website*. Retrieved from http://www.onlinemarketing-trends.com/2011/04/smartphone-marketshare-2011-italyus-aus.html

O'Reilly, T. (2005). *What is web 2.0? Design patterns and business models for the next generation of software*. Retrieved July 25, 2011, from http://www.oreillynet.com/pub/a/oreilly/tim/news/2005/09/30/what-is-web-20.html

Orlikowski, W. (1993). Case tools as organizational change: Investigating incremental and radical changes in systems development. *Management Information Systems Quarterly*, *17*(3), 309–340. doi:10.2307/249774

Orlikowski, W. J. (1991). Integrated information environment of matrix of control? *Accounting. Management and Information Technologies*, *1*(1), 9–42. doi:10.1016/0959-8022(91)90011-3

Orlikowski, W., & Baroudi, J. (1991). Studying information technology in organizations: Research approaches and assumptions. *Information Systems Research*, *2*(1), 1–28. doi:10.1287/isre.2.1.1

Pagani, M. (2004). Determinants of adoption of third generation mobile multimedia services. *Journal of Interactive Marketing*, *18*(3), 46–59. doi:10.1002/dir.20011

Pandit, N. (1996). The creation of theory: A recent application of the grounded theory method. *Qualitative Report*, *2*(4), 1–14.

Pan, G. (2005). Information system project abandonment: A stakeholder analysis. *International Journal of Information Management*, *25*(2), 173–184. doi:10.1016/j.ijinfomgt.2004.12.003

Pan, K., Nunes, J. M. B., & Peng, G. C. (2011). Risks affecting ERP post-implementation: Insights from a large Chinese manufacturing group. *Journal of Manufacturing Technology Management*, *22*(1), 107–130. doi:10.1108/17410381111099833

Parker, M., & Benson, R. (1988). *Information economics: Linking business performance to information technology*. Englewood Cliffs, NJ: Prentice Hall.

Park, N., Roman, R., Lee, S., & Chung, J. E. (2009). User acceptance of a digital library system in developing countries: An application of the technology acceptance model. *International Journal of Information Management*, *29*(3), 196–209. doi:10.1016/j.ijinfomgt.2008.07.001

Parmanto, B., & Zeng, X. M. (2005). Metric for web accessibility evaluation. *Journal of the American Society for Information Science and Technology*, *56*(13), 1394–1404. doi:10.1002/asi.20233

Parnas, D. L. (1972). On the criteria to be used in decomposing systems into modules. *Communications of the ACM*, *15*(12), 1053–1058. doi:10.1145/361598.361623

Parsons, D., Hokyoung, R., & Cranshaw, M. (2006). A study of design requirements for mobile learning environments. In *Proceedings of the Sixth International Conference on Advanced Learning Technologies*, (pp. 96-100). IEEE.

Patrizia, M., et al. (2009). A robotic toy for children with special needs: From requirements to design. In *Proceedings of the 11th International Conference on Rehabilitation Robotics*, (pp. 1070-1075). Kyoto, Japan: IEEE.

Patton, J. (2002). Hitting the target: Adding interaction design to agile software development. *OOPSLA 2002 Practitioners Reports*, (pp. 1–ff). New York, NY: ACM.

Patton, M. Q. (2005). Goal-based vs. goal-free evaluation. *Encyclopedia of Social Measurement*, *2*, 141-144.

Patton, M. Q. (1987). *Utilization-focused evaluation*. Newbury Park, CA: Sage Publisher.

Patton, M. Q. (1990). *Qualitative evaluation and research methods*. Newbury Park, CA: Sage Publisher.

Patton, M. Q. (1999). Enhancing the quality and credibility of qualitative analysis. *Health Services Research*, *34*(5), 1189–1208.

Patton, M. Q. (2002). *Qualitative research and evaluation methods* (3rd ed.). Thousand Oaks, CA: Sage.

Pavitt, K. (2005). Innovation process. In Fagerberg, J., Mowery, D., & Nelson, R. R. (Eds.), *The Oxford Handbook of Innovation*. Oxford, UK: Oxford University Press.

Pavlou, P. (2003). Consumer acceptance of electronic commerce – Integrating trust and risk in the technology acceptance model. *International Journal of Electronic Commerce*, *7*(3), 69–103.

Pederick, C. (2003). *Web developer for Firefox toolbar*. Retrieved February 12, 2011, from http://chrispederick.com/work/web-developer/

Peng, G. C., & Nunes, J. M. B. (2009). Surfacing ERP exploitation risks through a risk ontology. *Industrial Management & Data Systems*, *109*(7), 926–942. doi:10.1108/02635570910982283

Peng, G. C., & Nunes, J. M. B. (2009). Identification and assessment of risks associated with ERP post-implementation in China. *Journal of Enterprise Information Management*, *22*(5), 587–614. doi:10.1108/17410390910993554

Peng, G. C., & Nunes, J. M. B. (2012). Establishing and verifying a risk ontology for ERP post-implementation. In Ahmad, M., Colomb, R. M., & Abdullah, M. S. (Eds.), *Ontology-Based Applications for Enterprise Systems and Knowledge Management*. Hershey, PA: IGI Global. doi:10.4018/978-1-4666-1993-7.ch003

Peng, G. C., & Nunes, M. B. (2009). Identification and assessment of risks associated with ERP post-implementation in China. *Journal of Enterprise Information Management*, *22*(5), 587–614. doi:10.1108/17410390910993554

Penney, S. (2010). *Bloom's taxonomy of web 2.0 tools*. Retrieved April 4, 2011 from http://www.usi.edu/distance/bdt.htm

Pernice, K., & Nielsen, J. (2001). *Beyond alt text: Making the web easy to use for users with disabilities*. Retrieved Feb 5, 2011, from http://www.nngroup.com/reports/accessibility/testing/

Petrie, H., & Kheir, O. (2007). The relationship between accessibility and usability of websites. In *Proceedings of the SIGCHI Conference on Human Factors in Computing Systems*, (pp. 397–406). San Jose, CA: SIGCHI.

Petter, S. C., & Gallivan, M. J. (2004). Toward a framework for classifying and guiding mixed method research in information systems. In *Proceedings of the 37 Hawaii International Conference on Systems Sciences*. Hawaii, HI: IEEE.

PHANToM device. (2011). *SenSable technologies inc.* Retrieved from http://www.sensable.com

Pickard, A. (2007). *Research methods in information.* London, UK: Facet Publishing.

Pikkarainen, T., Pikkarainen, K., Karjaluoto, H., & Pahnila, S. (2004). Consumer acceptance of online banking: An extension of the technology acceptance model. *Internet Research, 14*(3), 224–235. doi:10.1108/10662240410542652

Pinsonneault, A., & Kraemer, K. L. (1993). Survey research methods in management information systems: An assessment. *Journal of Management Information Systems, 10*(2), 75–105.

Pogue, D. (2008). Art you taking advantage of web 2.0? *New York Times: Pogue's Posts: The latest in Technology from David Pogue*. Retrieved January 19, 2012 from http://pogue.blogs.nytimes.com/2008/03/27/are-you-taking-advantage-of-web-20/?scp=1&sq=Are%20you%20taking%20advantage%20of%20Web%202.0&st=cse

Polson, P., Lewis, C., Rieman, J., & Wharton, C. (1992). Cognitive walkthroughs: a method for theory-based evaluation of user interfaces. *International Journal of Man-Machine Studies, 6*, 741–773. doi:10.1016/0020-7373(92)90039-N

Pomson, A. (2008). Look who's talking: Emergent evidence for discriminating between differences in listserv participation. *Education and Information Technologies, 13*(2), 147–163. doi:10.1007/s10639-008-9056-x

Ponec, P. (2009). *jWorkSheet*. Retrieved from http://jworksheet.ponec.net/

Porter, M. F. (1980). An algorithm for suffix stripping. *Program, 14*, 130–137. doi:10.1108/eb046814

Porter, S. (2004). Pros and cons of paper and electronic surveys. *New Directions for Institutional Research, 121*, 91–97. doi:10.1002/ir.103

Pozzebon, M. (2004). Conducting and evaluating critical interpretive research: Examining criteria as a key component in building a research tradition. *International Federation for Information Processing, 143*, 275–292. doi:10.1007/1-4020-8095-6_16

Prahalad, C. K., & Rangaswami, M. R. (2009, September). Why sustainability now the key driver of innovation. *Harvard Business Review*, 56–64.

Prasad, A., Saha, S., Misra, P., Hooli, B., & Murakami, M. (2010). Back to green. *Journal of Green Engineering, 1*(1), 89–110.

Preece, J., Rogers, Y., Benyon, D., Holland, S., & Carey, T. (1994). *Human computer interaction.* Reading, MA: Addison-Wesley.

Preecem, J., Abras, C., & Maloney-Krichmar, D. (2004). Designing and evaluating online communities: Research speaks to emerging practice. *International Journal of Web Based Communities, 1*(1), 2–18. doi:10.1504/IJWBC.2004.004795

Prensky, M. (2001). Digital natives, digital immigrants, part 2: Do they really think differently? *Horizon, 9*(6), 1–9. doi:10.1108/10748120110424843

Pressman, R. S. (2000). *Software engineering: A practitioner's approach.* Berkshire, UK: McGraw-Hill.

Pressman, R. S. (2005). *Software engineering: A practitioner's approach* (6th ed.). Boston, MA: McGraw-Hill.

Prisco, G. M., Avizzano, C. A., Calcara, M., Ciancio, S., Pinna, S., & Bergamasco, M. (1998). A virtual environment with haptic feedback for the treatment of motor dexterity disabilities. In *Proceedings of the International Conference on Robotics and Automation, 1998,* (vol. 4, pp. 3721-3726). IEEE Press.

Public Accounts Committee Report. (2000). *Public accounts committee report, 1st report.* London, UK: House of Commons.

Pueschel, S. M., Gallagher, P. L., Zartler, A. S., & Pezzullo, J. C. (1987). Cognitive and learning processes in children with Down syndrome. *Research in Developmental Disabilities, 8*(1), 21–37. doi:10.1016/0891-4222(87)90038-2

Pura, M. (2005). Linking perceived value and loyalty in location-based mobile services. *Managing Service Quality, 15*(6), 509–553. doi:10.1108/09604520510634005

Puri, A. (2007). The web of insights – The art and practice of webnography. *International Journal of Market Research, 49*(3), 387–408.

Quinn, C. N. (1996). Pragmatic evaluation: Lessons from usability. In A. Christie & B. Vaughan (Eds.), *Proceedings of the 13th Annual Conference of the Australasian Society for Computers in Learning in Tertiary Education (ASCILITE 1996).* Adelaide, Australia: Australia Society for Computers in Learning in Tertiary Education.

Rabin, S. (2002). *AI game programming wisdom*. Boston, MA: Charles River Media Inc.

RAE. (2010). *Diccionario de la lengua española*. Retrieved 28/09/2010, from http://www.rae.es/RAE/Noticias.nsf/Home?ReadForm

Razmerita, L., & Lytras, M. D. (2008). Ontology-based user modeling personalization: Analyzing the requirements of a semantic learning portal. In *Proceedings of the 1st World Summit on the Knowledge Society (WSKS 2008),* (vol 5288, pp. 354-363). Rhodes, Greece: WSKS.

Reason, P., & Bradbury, H. (2001). Introduction: Inquiry and participation in search of a world worthy of human aspiration. In Reason, P., & Bradbury, H. (Eds.), *Handbook of Action Research: Participative Inquiry and Practice*. Thousand Oaks, CA: Sage.

Rego, H., Moreira, T., & Grarcia-Penalvo, F. (2010). Web-based learning information system for web 3.0. [Berlin, Germany: Springer-Verlag.]. *Proceedings of WSKS, 2010*, 196–201.

Reichheld, F. F., & Schefter, P. (2000). E-loyalty: Your secret weapon on the web. *Harvard Business Review, 78*, 105–113.

Reitsma, R. (2006). Europe's 2006 online shopping landscape. *Forrester Research, Business View Research Document.* Retrieved Dec 14th 2011 from http://www.forrester.com/rb/Research/europes_2006_online_shopping_landscape/q/id/40479/t/2

Remenyi, D., Williams, B., Money, A., & Swartz, E. (1998). *Doing research in business management: An introduction to process and method.* London, UK: Sage Publications.

Results, A. P. C. C. (2005). *American power conversion reports record revenue for the fourth quarter and full year 2005*. Retrieved from http://www.apcc.com/

Results, A. P. C. C. (2006). *American power conversion reports first quarter 2006 financial results*. Retrieved from http://www.apcc.com/

Rheingold, H. (1993). *The virtual community: Homesteading on the electronic frontier*. Reading, MA: Addison-Wesley.

Rice, A. J. (2007). *Mathematical statistics and data analysis* (3rd ed.). New York, NY: Cengage Learning, Inc.

Rice, P., & Ezzy, D. (1999). *Qualitative research methods: A health focus*. Oxford, UK: Oxford University Press.

Rieman, J., Franzke, M., & Redmiles, D. (1995). *Usability evaluation with the cognitive walkthrough.* Paper presented at CHI 1995 Mosaic of Creativity. Denver, CO.

Ries, E. (2011). *The lean startup: How today's entrepreneurs use continuous innovation to create radically successful businesses*. New York, NY: Crown Business.

Ritter, L. A., & Sue, V. M. (2007). *The survey questionnaire. Using Online Surveys in Evaluation: New Directions for Evaluation* (pp. 37–45). New York, NY: Wiley.

Ritter, L. A., & Valarie, M. S. (2007). Managing online survey data. In Ritter, L. A., & Valarie, M. S. (Eds.), *Using Online Surveys in Evaluation: New Directions for Evaluation* (pp. 51–55). New York, NY: Wiley.

Robertson, J. (2005). Requirements analysts must also be inventors. *IEEE Software, 22*(1), 48, 50.

Robertson, J., & Heitmeyer, C. (2005). Point/counterpoint. *IEEE Software, 22*(1), 48–51.

Robey, D. (1996). Diversity in information systems research: Threat, promise and responsibility. *Information Systems Research, 7*, 400–408. doi:10.1287/isre.7.4.400

Robey, D. R. (1996). Research commentary: Diversity in information systems research. *Information Systems Research, 7*(4), 400–408. doi:10.1287/isre.7.4.400

Robinson, J., & Martin, S. (2010). IT use and declining social capital? More cold water from the general social survey (GSS) and the American time-use survey (ATUS). *Social Science Computer Review*, *28*(1), 45–63. doi:10.1177/0894439309335230

Roblyer, M. D. (2003). *Integrating technology: Educational technology into teaching*. Upper Saddle River, NJ: Merrill Prentice-Hall.

Robrecht, L. (1995). Grounded theory: Evolving methods. *Qualitative Health Research*, *5*(2), 169–177. doi:10.1177/104973239500500203

Robson, C. (2002). *Real world research: A resource for social scientists and practitioner-researchers* (2nd ed.). Oxford, UK: Blackwell.

Robson, C. (2002). *Real world research: A resource for social scientists and practitioner-researchers*. Oxford, UK: Blackwell Publishing.

Rocco, T. S., Bliss, L. A., Gallagher, S., & Perez-Prado, A. (2003). Taking the next step: Mixed methods research in organisational systems. *Information Technology, Learning and Performance Journal*, *21*(1), 19–29.

Roche Accu-chek Expert. (2012). *Website*. Retrieved from http://www.accu-chek.co.uk/gb/products/metersystems/avivaexpert.html

Rodon, J., & Pastor, A. (2007). Applying grounded theory to study the implementation of an inter-organizational information system. *The Electronic Journal of Business Research Methods*, *5*(2), 71–82.

Roethlisberger, F. J. (1977). *The elusive phenomena*. Boston, MA: Harvard University Press.

Rofiq, A., & Mula, J. M. (2010). *Impact of cyber fraud and trust on e-commerce use: A proposed model by adopting theory of planned behaviour*. Paper presented at 21st Australasian Conference on Information Systems. Brisbane, Australia.

Rogers, R. (2004). Literate identities across contexts. In Rogers, R. (Ed.), *An Introduction to Critical Discourse Analysis* (pp. 51–78). Mahwah, NJ: Lawrence Erlbaum.

Rogers, Y., Sharp, H., & Preece, J. (2011). *Interaction design* (3rd ed.). Mahwah, NJ: John Wiley & Sons Ltd.

Romen, D., & Svanaes, D. (2008). Evaluating web site accessibility: Validating the WAI guidelines through usability testing with disabled users. In *Proceedings of the 5th Nordic Conference on Human-Computer Interaction: Building Bridges*, (pp. 535-538). Lund, Sweden: IEEE.

Roode, D., Speight, H., Pollock, M., & Webber, R. (2004). It's not the digital divide – It's the socio-techno divide. In *Proceedings of the 12th European Conference on Information Systems Turka*. IEEE.

Ropers, S. (2001). New business models for the mobile revolution. *eAI Journal*, 53-57.

Rose, J., & Fogarty, G. (2006). Determinants of perceived usefulness and perceived ease of use in the technology acceptance model: Senior consumers' adoption of self-service banking technologies. in *Proceedings of the Academy of World Business, Marketing & Management Development Conference*, (vol 2, pp. 122-129). Paris, France: World Business, Marketing, & Management Development.

Ross, I. (1975). Perceived risk and consumer behavior: a critical review. In Schlinger, M. J. (Ed.), *Advances in Consumer Research* (*Vol. 2*, pp. 1–20). New York, NY: Association for Consumer Research.

Rossi, P. H., & Williams, W. (1972). *Evaluating social programs: Theory, practice, and politics*. New York, NY: Seminar Press.

ROU. (2009). *Constitución de la república: Constitucion 1967 con las modificaciones plebiscitadas el 26 de noviembre de 1989, el 26 de noviembre de 1994, el 8 de diciembre de 1996 y el 31 de octubre de 2004*. ROU.

Rowe, S. (2004). Discourse in activity and activity as discourse. In Rogers, R. (Ed.), *An Introduction to Critical Discourse Analysis in Education*. Mahwah, NJ: Lawrence Erlbaum Associates.

Royce, W. W. (1987). Managing the development of large software systems. In *Proceedings of the 9th International Conference on Software Engineering*, (pp. 328-338). IEEE Computer Society.

Ruhleder, F., & Jordan, B. (1997). Capturing complex, distributed activities – Video-based interaction analysis as a component of workplace ethnography. In A. S. Lee, J. Liebenau, & J. DeGross (Eds.), *The IS and Qualitative Research Conference,* (pp. 246-275). London, UK: Chapman & Hall.

Rybas, N., & Gajjala, R. (2007). Developing cyberethnographic research methods for understanding digitally mediated identities. *Forum Qualitative Sozial Forschung, 8*(3), 1–33.

Ryle, G. (1968). Thinking and reflecting. *Royal Institute of Philosophy Lectures, 1,* 210–226. doi:10.1017/S0080443600011511

Saaty, T. L. (1980). *Analytic hierarchy process.* New York, NY: McGraw-Hill.

Sambamurthy, V., & Zmud, R. W. (2000). Research commentary: The organizing logic for an enterprise's IT activities in the digital era – A prognosis of practice and a call for research. *Information Systems Research, 11*(2). doi:10.1287/isre.11.2.105.11780

Sánchez Medina, A. J., Melián González, A., & García Falcón, J. M. (2004). *El capital intelectual: Concepto y dimensiones.* Gran Canaria, Spain: Campus Universitario de Tafira Las Palmas de Gran Canaria.

Sanders, L. (2008). On modeling: An evolving map of design practice and design research. *Interaction, 15,* 13–17. doi:10.1145/1409040.1409043

Saprikis, V., Chouliara, A., & Vlachopoulou, M. (2010). Perceptions towards online shopping: Analyzing the Greek university students' attitude. In *Proceedings of the Communications of the IBIMA.* IBIMA.

Saunders, M., Lewis, P., & Thornhill, A. (2003). *Research methods for business students* (3rd ed.). Essex, UK: Pearson Education.

Sax, G. (1997). *Principles of educational psychological measurement and evaluation* (4th ed.). New York, NY: Wadsworth Publishing Company.

Schell, C. (1992). *The value of case-study as a research methodology. Research Report.* Manchester, UK: Manchester Business School.

Schleckler, M., & Hirsch, E. (2001). Incomplete knowledge: Ethnography and the crisis of context in studies of media, science and technology. *History of the Human Sciences, 14*(1), 69–87. doi:10.1177/09526951010101400104

Schuh, P. (2004). *Integrating agile development in the real world.* New York, NY: Cengage Learning.

Schuler, D., & Namioka, A. (1993). *Participatory design: Principles and practices.* Mahwah, NJ: Lawrence Erlbaum Associates.

Schultze, U. (2000). A confessional account of an ethnography about knowledge work. *Management Information Systems Quarterly, 24*(1), 3–41. doi:10.2307/3250978

Schulz, T. (2008). Using the keystroke-level model to evaluate mobile phones. In *Proceedings of IRIS 31.* IRIS.

Schutz, A. (1962). *Collected papers I: The problem of social reality.* The Hague, The Netherlands: Martinus Nijhoff.

Schutz, A. (1962). Concept and theory formation in the social sciences. In Nijhoff, M. (Ed.), *Collected Papers* (*Vol. I*, pp. 48–66). The Hague, The Netherlands.

Scriven, M. (1967). The methodology of evaluation. In Tyler, R. W., Gagne, R. M., & Scriven, M. (Eds.), *Perspectives of Curriculum Evaluation* (pp. 39–83). Chicago, IL: Rand MacNally.

Scriven, M. (1972). Objectivity and subjectivity in educational research. In Thomas, L. G. (Ed.), *Philosophical Redirection of Educational Research: The Seventy-First Year Book of the National Security for the Study of Education.* Chicago, IL: University of Chicago Press.

Scriven, M. (1991). Prose and cons about goal-free evaluation. *The American Journal of Evaluation, 12*(1), 55–62. doi:10.1177/109821409101200108

Sears, A. (1997). Heuristic walkthroughs: Finding the problems without the noise. *International Journal of Human-Computer Interaction, 9,* 213–234. doi:10.1207/s15327590ijhc0903_2

Seddon, P. B. (2002). Measuring organizational IS effectiveness: An overview and update of senior management perpectives. *Advances in Information Systems, 33*(2), 11–28. doi:10.1145/513264.513270

Seidek, S., & Recker, J. (2009). Using grounded theory for studying business process management phenomena. In *Proceedings of the 17th European Conference on Information Systems,* (pp. 1-13). Verona, Italy: IEEE.

Sekaran, U. (2003). *Research methods for business "a skill building approach"* (4th ed.). New York, NY: John Wiley & Sons.

Selwyn, N. (2006). The use of computer technology in university teaching and learning: A critical perspective. *Journal of Computer Assisted Learning, 23,* 83–94. doi:10.1111/j.1365-2729.2006.00204.x

Senatore, F., et al. (2008). Development of a generic assistive platform to aid patients with motor disabilities. In *Proceedings of the 14th Nordic-Baltic on Biomedical Engineering and Medical Physics,* (pp. 168-171). Riga, Latvia: IEEE.

Serbanati, A., Medaglia, C. M., & Ceipidor, U. B. (2012). *Building blocks of the internet of things: State of the art and beyond.* Retrieved from http://cdn.intechopen.com/pdfs/17872/InTech-Building_blocks_of_the_internet_of_things_state_of_the_art_and_beyond.pdf

Sexton, N., Miller, H., & Dietsch, A. (2011). Appropriate uses and considerations for online surveying in human dimensions research. *Human Dimensions of Wildlife, 16*(3), 154–163. doi:10.1080/10871209.2011.572142

Seymour, L., & Fourie, R. (2010). *ICT literacy in higher education: Influences and inequality.* Paper presented at the Southern African Computer Lecturers' Association Conference Pretoria. Pretoria, South Africa.

Shaffer, T. R., & O'Hara, B. S. (1995). The effects of country of origin on trust and ethical perceptions. *The Service Industries Journal, 15,* 162–179. doi:10.1080/02642069500000019

Shannak, R., & Aldhmour, F. (2009). Grounded theory as methodology for theory generation in information system research. *European Journal of Economics. Finance and Administrative Sciences, 15,* 33–50.

Sharples, M. (2009). Methods for evaluating mobile learning. In *Researching Mobile Learning: Frameworks, Tools and Research Designs.* Retrieved from http://oro.open.ac.uk/31417/

Sharrock, W., & Randall, D. (2004). Ethnography, ethnomethodology and the problem of generalisation in design. *European Journal of Information Systems, 13,* 186–194. doi:10.1057/palgrave.ejis.3000502

Shavelson, R. (1981). *Statistical reasoning for the behavioral sciences* (3rd ed.). Needham Heights, MA: Allyn and Bacon.

Shneiderman, B. (2000). Creating creativity: User interfaces for supporting innovation. *ACM Transactions on Computer-Human Interaction, 7*(1), 114–138. doi:10.1145/344949.345077

Shneiderman, B. (2007). Creativity support tools: Accelerating discovery and innovation. *Communications of the ACM, 50*(12), 20–32. doi:10.1145/1323688.1323689

Sichel, D. E. (1997). *The computer revolution: An economic perspective.* Washington, DC: Brookings Institution Press.

Sidorova, A., Evangelopoulos, N., Valacich, J. S., & Ramakrishnan, T. (2008). Uncovering the intellectual core of the information systems discipline. *Management Information Systems Quarterly, 32,* 467–482.

Sillence, E., Briggs, P., & Fishwick, L. (2004). Trust and mistrust of online health sites. In *Proceedings of Computer Human Interaction CHI 2004.* Vienna, Austria: ACM. doi:10.1145/985692.985776

Silva, L., & Hirschheim, R. (2007). Fighting against windmills: Strategic information systems and organizational deep structures. *Management Information Systems Quarterly, 31*(2), 327–354.

Sirithumgul, P., Suchato, A., & Punyabukkana, P. (2009). Quantitative evaluation for web accessibility with respect to disabled groups. In *Proceedings of the 2009 International Cross-Disciplinary Conference on Web Accessibililty,* (pp. 136-141). Madrid, Spain: IEEE.

Sjostrom, C. (2001). Designing haptic computer interfaces for blind people. *Proceedings of ISSPA, 1,* 68–71.

Slatin, J., & Rush, S. (2003). *Maximum accessibility: Making your web site more usable for everyone.* Boston, MA: Addison-Wesley.

Smith, D. (2006). *Exploring innovation.* Maidenhead, UK: McGraw-Hill Education.

Smithson, S., & Hirschheim, R. (1998). Analysing information systems evaluation: Another look at an old problem. *European Journal of Information Systems*, *7*(3), 158–174. doi:10.1057/palgrave.ejis.3000304

Smyth, J., Dillman, D., Christian, L., & O'Neill, A. (2010). Using the internet to survey small towns and communities: Limitations and possibilities in the early 21st century. *The American Behavioral Scientist*, *53*, 1423–1448. doi:10.1177/0002764210361695

Snider, C., Dekoninck, E., & Culley, S. (2011). Studying the appearance and effect of creativity within the latter stages of the product development process. In *Proceedings of the DESIRE 2011 Conference on Creativity and Innovation in Design*, (pp. 317-328). ACM.

Somekh, B., & Lewin, C. (2005). *Research methods in the social sciences*. London, UK: Sage Publications.

Sommerville, I. (2007). *Software engineering* (8th ed.). Harlow, UK: Pearson Education. Retrieved from http://www.pearsoned.co.uk/HigherEducation/Booksby/Sommerville/

Song, L., Singleton, E. S., Hill, J. R., & Koh, M. H. (2004). Improving online learning: Student perceptions of useful and challenging characteristics. *The Internet and Higher Education*, *7*, 59–70. doi:10.1016/j.iheduc.2003.11.003

Spikol, D. (2009). Exploring novel learning practices through co-designing mobile games. *Researching Mobile Learning: Frameworks, Methods, and Research Designs*. Retrieved from http://www.ungkommunikation.se/documents/ungkommunikation/documents/ungkommunikation/rapporter/spikol,%20david.pdf

Squires, D., & Preece, J. (1999). Predicting quality in educational software: Evaluating for learning, usability and the synergy between them. *Interacting with Computers*, *11*, 467–483. doi:10.1016/S0953-5438(98)00063-0

Stacey, E., Smith, P. J., & Barty, K. (2004). Adult learners in the workplace: Online learning and communities of practice. *Distance Education*, *25*(1), 107–123. doi:10.1080/0158791042000212486

Stahl, B. (2004). Whose discourse? A comparison of the Foucauldian and Habermasian concepts of discourse in critical IS research. In *Proceedings of the Tenth America's Conference on Information Systems*, (pp. 4329-4336). New York, NY: IEEE.

Stake, R. E. (1995). *The art of case study research*. Thousand Oaks, CA: Sage Publications.

Standish Group Inc. (2001). *Extreme chaos*. Retrieved from http://standishgroup.com/sample_research/extreme_chaos.pdf

Standish Group. (2009). *Chaos report*. Boston, MA: Standish Group. Retrieved from http://www. standishgroup.com/newsroom/chaos_2009.php

Star, S. L. (1999). The ethnography of infrastructure. *The American Behavioral Scientist*, *43*(3), 377–391. doi:10.1177/00027649921955326

Star, S. L. (2002). Infrastructure and the ethnographic practice: Working on the fringes. *Scandinavian Journal of Information Systems*, *14*(2).

Star, S. L., & Ruhleder, K. (1996). Steps toward an ecology of infrastructure: Design and access for large information spaces. *Information Systems Research*, *7*(1), 111–134. doi:10.1287/isre.7.1.111

Steen, M. (2001). Cooperation, curiosity and creativity as virtues in participatory design. In *Proceedings of the DESIRE 2011 Conference on Creativity and Innovation in Design*, (pp. 171-174). ACM.

Steyaert, J. (2005). Web-based higher education: The inclusion/exclusion paradox. *Journal of Technology in Human Services*, *23*(1), 67–78. doi:10.1300/J017v23n01_05

Steyn, A. G. W., & de Villiers, A. P. (2007). Public funding of higher education in South Africa by means of formula. *Review of Higher Education in South Africa*. Retrieved from http://www.che.ac.za/documents/d000146/5-Review_HE_SA_2007.pdf

Straub, D. W., Boudreau, M., & Gefen, D. (2004). Validation guidelines for IS positivist research. *Communications of the Association for Information Systems*, *13*, 380–427.

Straub, D., Limayem, M., & Karahanna-Evaristo, E. (1995). Measuring system usage implications for IS theory testing. *Management Science*, *41*(8), 1328–1342. doi:10.1287/mnsc.41.8.1328

Strauss, A., & Corbin, J. (1990). *Basics of qualitative research: Grounded theory procedures and techniques*. Newburry Park, CA: Sage.

Strauss, A., & Corbin, J. (1998). *Basics of qualitative research*. New Delhi, India: SAGE Publications.

Strauss, A., & Corbin, J. (1998). *Basics of qualitative research: Grounded theory procedures and techniques*. London, UK: Sage.

Strauss, A., Fagerhaugh, S., Suczek, B., & Wiener, C. (1985). *Social organization of medical work*. Chicago, IL: University of Chicago Press.

Streefkerk, J. W., van Esch-Bussemakers, M. P., Neerincx, M. A., & Looije, R. (2008). Evaluating context-aware mobile interfaces for professionals. In Lumsden, J. (Ed.), *Handbook of Research on User Interface Design and Evaluation for Mobile Technology* (pp. 759–779). Hershey, PA: IGI Global Publishing. doi:10.4018/978-1-59904-871-0.ch045

Subrahmanyam, K., & Smahel, D. (2011). Internet use and well-being: Physical and psychological effects. *Advancing Responsible Adolescent Development*, 123-142.

Sullivan, O. (2011). An end to gender display through the performance of housework? A review and reassessment of the quantitative literature using insights from the qualitative literature. *Journal of Family Theory & Review*, *3*(1), 1–13. doi:10.1111/j.1756-2589.2010.00074.x

Sun, S. (2011). The internet effects on students communication at Zhengzhou Institute of Aeronautical Industry Management. *Advances in Computer Science, Environment. Ecoinformatics and Education*, *218*, 418–422. doi:10.1007/978-3-642-23357-9_74

Susman, G. I., & Evered, R. D. (1978). An assessment of the scientific merits of action research. *Administrative Science Quarterly*, *23*(4), 582. doi:10.2307/2392581

Sveiby, K. E. (1997). The intangible assets monitor. *Journal of Human Resource Costing and Accounting*, *2*(1), 73–97. doi:10.1108/eb029036

Swanson, E. B. (1988). *Information system implementation: Bridging the gap between design and utilization*. Homewood, IL: Irwin.

Symonds, J. E., & Gorard, S. (2010). Death of mixed methods? Or the rebirth of research as a craft. *Evaluation and Research in Education*, *23*(2), 121–136. doi:10.1080/09500790.2010.483514

Szyperski, C. (2002). *Component software: Beyond object-oriented programming* (2nd ed.). Boston, MA: Addison-Wesley Professional.

Tansey, O. (2007). Process tracing and elite interviewing: A case for non-probability sampling. *PS: Political Science & Politics*, *40*(4), 765-772. Retrieved August 30, 2011 from http://www.springerlink.com/index/Q6Q17272M32757X6.pdf

Tan, Y.-H., & Thoen, W. (2000). Toward a generic model of trust for electronic commerce. *International Journal of Electronic Commerce*, *5*(2), 61–74.

Tashakkori, A., & Creswell, J. W. (2007). Editorial: The new era of mixed methods. *Journal of Mixed Methods Research*, *1*, 3–7. doi:10.1177/2345678906293042

Tashakkori, A., & Teddlie, C. (1998). *Mixed methdology: Combining qualitative and quantitative approaches*. Thousand Oaks, CA: SAGE.

Taylor, S., & Todd, P. A. (1995). Understanding information technology usage: A test of competing models. *Information Systems Research*, *6*(2), 144–176. doi:10.1287/isre.6.2.144

Teddlie, C. B., & Tashakkori, A. (2009). *Foundations of mixed methods research: Integrating quantitative and qualitative approaches in the social and behavioral sciences*. Thousand Oaks, CA: SAGE Publications.

Teddlie, C., & Tashakkori, A. (2003). Major issues and controversies in the use of mixed methods in the social and behavioral sciences. In Tashakkori, A., & Teddlie, C. (Eds.), *Handbook on Mixed Methods in the Behavioural and Social Sciences* (pp. 3–50). Thousand Oaks, CA: Sage.

Teddlie, C., & Tashakkori, A. (2009). *Foundations of mixed methods research - Integrating quantitative and qualitative approaches in the social and behavioral sciences*. Thousand Oaks, CA: SAGE Publisher.

Teo, C., Burdet, E., & Lim, H. (2002). A robotic teacher of Chinese handwriting. In *Proceedings of the 10th Symposium on Haptic Interfaces for Virtual Environment and Teleoperator Systems, HAPTICS 2002*. IEEE Computer Society.

Teo, T. S. H., & Pok, S. H. (2003). Adoption of WAP-enabled mobile phones among internet users. *Omega: The International Journal of Management Science*, *31*(6), 483–498. doi:10.1016/j.omega.2003.08.005

Thatcher, J., Burks, M., Heilmann, C., Henry, S., Kirkpatrick, A., & Lauke, P. ... Waddell, C. (2006). *Web accessibility: Web standards and regulatory compliance*. Berkeley, CA: Apress.

Thomas, G., & James, D. (2006). Re-inventing grounded theory: Some questions about theory, ground and discovery. *British Educational Research Journal*, *32*(6), 767–795. doi:10.1080/01411920600989412

Thompson, M. (2004). ICT, power, and developmental discourse: A critical view. *Electronic Journal on Information Systems in Developing Countries*, *20*(4), 1–25.

Tidd, J., Bessant, J., & Pavitt, K. (2005). *Managing innovation: Integrating technological, market and organizational change*. Chichester, UK: John Wiley & Sons.

Tompsett, B. C. (2008). Experiencias de enseñanza a estudiantes de informática con discapacidad en universidades del reino unido. *Technologías de la Información Y las Comunicaciones en la Autonomía Personal, Dependencia Y Accesibilidad*. Fundación Alfredo Brañas. *Colección Informática Número*, *16*, 371–398.

Toraskar, K. (1991). How managerial users evaluate their decision-support: A grounded theory approach. *Journal of Computer Information Systems*, *7*, 195–225.

Trauth, E. M. (2001). *Qualitative research in IS: Issues and trends*. Hershey, PA: IGI Global.

Trevino, L. K., & Webster, J. (1992). Flow in computer-mediated communication: Electronic mail and voice evaluation. *Communication Research*, *19*(2), 539–573. doi:10.1177/009365092019005001

Trillo, M. A., & Sánchez, S. M. (2006). Influencia de la cultura organizativa en el concepto de capital intelectual. *Intangible Capital*, *11*(2), 164–180.

Tsai, H. F., Cheng, S. H., Yeh, T. L., Shih, C.-C., Chen, K. C., & Yang, Y. C. (2009). The risk factors of internet addiction--A survey of university freshmen. *Psychiatry Research*, *167*(3), 294–299. doi:10.1016/j.psychres.2008.01.015

Tu, L., & Kvasny, L. (2006). American discourses of the digital divide and economic development: A sisyphean order to catch up? In D. Howcroft, E. Trauth, & DeGross (Eds.), Social Inclusion: Societal & Organizational Implications for Information Systems, (pp. 51-66). London, UK: Kluwer Academic Publishers.

Turner, M., Kitchenham, B., Brereton, P., Charters, S., & Budgen, D. (2010). Does the technology acceptance model predict actual use? A systematic literature review. *Information and Software Technology*, *52*(5), 463–479. doi:10.1016/j.infsof.2009.11.005

Tyler, R. W. (1942). General statement on evaluation. *The Journal of Educational Research*, *35*, 492–501.

UCA & MEF. (2011). *Unidad centralizada de adquisiciones*. Retrieved from http://uca.mef.gub.uy/portal/web/guest/portada

Umbach, P. (2004). Web surveys: Best practices. *New Directions for Institutional Research*, *121*, 23–38. doi:10.1002/ir.98

Universidad del País Vasco. (2011). *EvalAccess 2.0: Web service tool for evaluating web accessibility*. Retrieved February 10, 2011, from http://sipt07.si.ehu.es/evalaccess2/index.html

University of Toronto. (2011). *A-prompt: Web accessibility verifier*. Retrieved February 10, 2011, from http://aprompt.snow.utoronto.ca/download.html

Urban, G. L., Amyx, C., & Lorenzon, A. (2009). Online trust: State of the art, new frontiers, and research potential. *Journal of Interactive Marketing*, *23*, 179–190. doi:10.1016/j.intmar.2009.03.001

Urquhart, C., & Fernandez, W. (2006). Grounded theory method: The researcher as a blank slate and other myths. In *Proceedings of the ICIS 2006 International Conference on Information Systems*. Milwaukee, WI: ICIS.

Urquhart, C. (1997). Exploring analyst-client communication: Using grounded theory techniques to investigate interaction in informal requirements gathering. In Lee, A., Liebenau, J., & DeGross, J. (Eds.), *Information Systems and Qualitative Research* (pp. 149–181). London, UK: Chapman and Hall.

Urquhart, C. (2001). An encounter with grounded theory: Tackling the practical and philosophical issues. In Trauth, E. (Ed.), *Qualitative Research in IS: Issues and Trends* (pp. 104–140). Hershey, PA: IGI Global. doi:10.4018/978-1-930708-06-8.ch005

Urquhart, C. (2007). The evolving nature of grounded theory method: The case of the information systems discipline. In Charmaz, K., & Bryant, T. (Eds.), *The Handbook of Grounded Theory* (pp. 311–331). Thousand Oakes, CA: Sage Publications.

Urquhart, C., Lehmann, H., & Myers, M. (2010). Putting the 'theory' back into grounded theory: Guidelines for grounded theory studies in information systems. *Information Systems Journal, 20*(4), 357–381. doi:10.1111/j.1365-2575.2009.00328.x

US Government. (1998). *Section 508 standards guide: 1194.22 web-based intranet and internet information and applications.* Retrieved February 10, 2011, from http://www.section508.gov/index.cfm?fuseAction=stdsdoc#Web

User Focus. (2011). *Red route usability: The key user journeys with your web site.* Retrieved from http://userfocus.co.uk/articles/redroutes.html

Vaishnavi, V., & Kuechler, W. (2004). *Design science research in information systems.* Retrieved from http://www.desrist.org/desrist

Vaishnavi, V., & Kuechler, W. (2007). *Design science research methods and patterns: Innovating information and communication technology.* New York, NY: Auerbach Publications. doi:10.1201/9781420059335

Valenza, J. K., & Johnson, D. (2009, October). Things that keep us up at night. *School Library Journal.*

Van Aken, J. E. (2005). Management research as a design science: Articulating the research products of mode 2 knowledge production in management. *British Journal of Management, 16,* 19–36. doi:10.1111/j.1467-8551.2005.00437.x

Van de Ven, A. H. (2010). *Reflections on engaged scholarship.* Paper delivered to Carlson School of Management, University of Minnesota. Minneapolis, MN.

Van de Ven, A. H. (2007). *Engaged scholarship: A guide for organizational and social research.* Oxford, UK: Oxford University Press.

Van der Heijden, H., Verhagen, T., & Creemers, M. (2003). Understanding online purchase intentions: Contributions from technology and trust perspectives. *European Journal of Information Systems, 12*(1), 41–48. doi:10.1057/palgrave.ejis.3000445

Van Deursen, A. J. A. M., & Van Dijk, J. A. G. M. (2009). Using the internet: Skill related problems in users' online behavior. *Interacting with Computers, 21*(5-6), 393–402. doi:10.1016/j.intcom.2009.06.005

Van Dijk, T. (2001). Multidisciplinary CDA: A plea for diversity. In Wodak, R., & Myers, M. (Eds.), *Methods of Critical Discourse Analysis.* London, UK: Sage Publications. doi:10.4135/9780857028020.d7

Van Niekerk, J., & Roode, J. (2009). Glaserian and Straussian grounded theory: Similar or completely different? In *Proceedings of the 2009 Annual Research Conference of the South African Institute of Computer Scientists and Information Technologists,* (pp. 96-103). Johannesburg, South Africa: SAICSIT.

Vanhaverbeke, W., & Cloodt, M. (2006). Open innovation in value networks. In H. Chesbrough, W. Vanhaverbeke, & J. West (Eds.), *Open Innovation: Researching a New Paradigm,* (pp. 258-284). Oxford, UK: Oxford University Press.

Vasconcelos, A. (2007). The use of grounded theory and of arenas/social worlds theory in discourse studies: A case study on the discursive adaptation of information systems. *The Electronic Journal of Business Research. Methods (San Diego, Calif.), 5*(2), 125–136.

Vaske, J. (2011). Advantages and disadvantages of internet surveys: Introduction to the special issue. *Human Dimensions of Wildlife: An International Journal*, *16*, 149–153. doi:10.1080/10871209.2011.572143

Vass, M., Carroll, J. M., & Clifford, A. (2002). Supporting creativity in problem solving environments. In *Proceedings of the 4th Conference on Creativity & Cognition*, (pp. 31-37). New York, NY: ACM.

Velleman, E., Meerbeld, C., Strobbe, C., Koch, J., Velasco, C. A., Snaprud, M., & Nietzio, A. (2007). *D-WAB4, unified web evaluation methodology (UWEM 1.2 Core)*. Retrieved January 14, 2012, from http://www.wabcluster.org/uwem1_2/

Venkatesh, V., & Davis, F. D. (2000). A theoretical extension of the technology acceptance model: Four longitudinal field. *Management Science*, *46*(2), 186–204. doi:10.1287/mnsc.46.2.186.11926

Venkatesh, V., Morris, M. G., Davis, G. B., & Davis, F. D. (2003). User acceptance of information technology: Toward a unified view. *Management Information Systems Quarterly*, *27*(3), 425–478.

Verhagen, T., Meents, S., & Tan, Y. (2006). *Perceived risk and trust associated with purchasing at electronic marketplaces*. Retrieved 28th October, 2011 from http://ideas.repec.org/s/dgr/vuarem.html

Verhagen, T., Tan, Y., & Meents, S. (2004). *An empirical exploration of trust and risk associated with purchasing at electronic marketplaces*. Paper presented at 17th Bled eCommerce Conference. Bled, Slovenia.

Videria Lopes, C., Maeda, C., & Mendhekar, A. (1996, December). Aspect-oriented programming. *ACM Computing Surveys*.

Vigo, M., Arrue, M., Brajnik, G., Lomuscio, R., & Abascal, J. (2007). Quantitative metrics for measuring web accessibility. In *Proceedings of the 2007 International Cross-Disciplinary Conference on Web Accessibility*, (pp. 99-107). Banff, Canada:IEEE.

Vigo, M., & Brajnik, G. (2011). Automatic web accessibility metrics: Where we are and where we can go. *Interacting with Computers*, *23*(2), 137-155. doi:10.1016/j.intcom.2011.01.001

Villegas, E., Pifarré, M., & Fonseca, D. (2010). *Diseño metodológico de experiencia de usuario aplicado al campo de la accesibilidad*. Paper presented at Actas de la 5ª Conferencia Ibérica de Sistemas y Tecnologías de Información. Santiago de Compostela, Spain.

Vittet-Philippe, P., & Navarro, J. M. (2000). Mobile e-business (m-commerce): State of play and implications for European enterprise policy. *European Commission Enterprise Directorate-General E-Business Report*. Retrieved from http://www.ncits.org/tc_home/v3htm/v301008.pdf

von Hippel, E. (2005). *Democratizing innovation*. Boston, MA: The MIT Press.

Vrechopoulos, A. P., Constantiou, I. D., Mylonopoulos, N., & Sideris, I. (2002). Critical success factors for accelerating mobile commerce diffusion in Europe. In *Proceedings of the 15th Bled Electronic Commerce Conference*. Bled, Slovenia. IEEE.

Vygotsky, L. S. (1978). *Mind in society: The development of higher psychological process* (Cole, M., Lopez-Morillas, M., Luria, A. R., & Wertsch, J., Trans.). Cambridge, MA: Harvard University Press.

W3C. (1994). *Evaluation, repair, and transformation tools for web content accessibility*. Retrieved February 10, 2011, from http://www.w3.org/WAI/ER/existingtools.html

W3C. (1995). *The W3C markup validation service*. Retrieved February 15, 2011, from http://validator.w3.org/

W3C. (1995). *The W3C CSS validation service*. Retrieved February 15, 2011, from http://jigsaw.w3.org/css-validator/

W3C. (1999). *Web content accessibility guidelines 1.0*. Retrieved February 10, 2011, from http://www.w3.org/TR/WCAG10/

W3C. (1999). *Checklist of checkpoints for web content accessibility guidelines 1.0*. Retrieved February 10, 2011, from http://www.w3.org/TR/WCAG10/full-checklist.html

W3C. (2000). *Authoring tool accessibility guidelines 1.0*. Retrieved February 10, 2011, from http://www.w3.org/TR/ATAG10/

W3C. (2002). *User agent accessibility guidelines 1.0*. Retrieved February 10, 2011, from http://www.w3.org/TR/UAAG10/

W3C. (2002). *Template for accessibility evaluation reports*. Retrieved February 10, 2011, from http://www.w3.org/WAI/eval/template.html

W3C. (2006). *Web accessibility evaluation tools: Overview*. Retrieved February 20, 2011, from http://www.w3.org/WAI/ER/tools/

W3C. (2008). *Web content accessibility guidelines 2.0*. Retrieved February 10, 2011, from http://www.w3.org/TR/WCAG20/

W3C. (2011). Web *accessibility initiative (WAI)*. Retrieved February 15, 2011, from http://www.w3.org/WAI/

Wagner, C. (2004). Enterprise strategy management systems: Current and next generation. *The Journal of Strategic Information Systems*, *13*, 105–128. doi:10.1016/j.jsis.2004.02.005

Walling, D. R. (2009). Tech-savvy teaching and student-produced media: Idea networking and creative sharing. *Tech Trends: Linking Research and Practice to Improve Learning.* Retrieved April 4, 2011 from http://webcast.oii.ox.ac.uk/?view=Webcast&ID=20051130_109

Walsh, A. (2010). Mobile phone services and UK higher education students, what do they want from the library? *Library and Information Research*, *34*(106), 22–36.

Walsham, G. (1995). The emergence of interpretivism in IS research. *Information Systems Research*, *6*(4), 376–394. doi:10.1287/isre.6.4.376

Walz, S. (2005). Constituents of hybrid reality: Cultural anthropological reflections and a serious game design experiment merging mobility, media and computing. In Buurman, G. (Ed.), *Total Interaction: Theory and Practice of a New Paradign for the Design Disciplines* (pp. 122–141). Boston, MA: Birkhauser. doi:10.1007/3-7643-7677-5_9

Wang, H., & Liu, X. P. (2011). Haptic interaction for mobile assistive robots. *IEEE Transactions on Instrumentation and Measurement*, *60*(11), 3501–3509. doi:10.1109/TIM.2011.2161141

Wang, Y. S., Wang, Y. M., Lin, H. H., & Tang, T. I. (2003). Determinants of user acceptance of internet banking: An empirical study. *International Journal of Service Industry Management*, *14*(5), 501–519. doi:10.1108/09564230310500192

Ward, K. J. (1999). The cyber-ethnographic (re)construction of two feminist online communities. *Sociological Research Online*, *4*(1). doi:10.5153/sro.222

Warner, T. (1996). *Communication skills for information systems*. London, UK: Pitman Publishing.

Warren, L., Hitchin, L., & Brayshaw, M. (1997). IS: The challenge of neo-disciplinary research. In D. E. Avison (Ed.), *Key Issues in Information Systems: Proceedings of the 2nd UKAIS Conference*, (pp. 187-194). New York, NY: McGraw-Hill.

Warwick, D. (1984). *Interviews and interviewing*. London, UK: Education for Industrial Society-Management in Schools.

Watt, S., Zdrahal, Z., & Brayshaw, M. (1995). A multi-agent approach to configuration and design tasks. In *Proceedings of Articial Intelligence and Simulation of Behaviour Conference*. IEEE.

Web Accessibility Benchmarking Cluster. (2007). *Unified web evaluation methodology 1.2*. Retrieved February 5, 2011, from http://www.wabcluster.org/uwem1_2/

Web Accessibility in Mind. (2011). *WebAIM: World laws*. Retrieved January 14, 2012, from http://www.webaim.org/articles/laws/world/

Web Accessibility in Mind. (2011). *Web accessibility evaluation tool (WAVE) 4.0*. Retrieved January 14, 2012, from http://wave.webaim.org/

Webster, J., & Watson, R. T. (2002). Analyzing the past to prepare for the future: Writing a literature review. *Management Information Systems Quarterly*, *26*(2), 13–23.

Weick, K. (1996). Drop your tools: An allegory for organizational studies. *Administrative Science Quarterly*, *41*(2), 301–313. doi:10.2307/2393722

Weinreich, N. K. (1996). *Integrating quantitative and qualitative methods in social marketing research.* Retrieved from http://www.social-marketing.com/research.html

Weisberg, R. W. (1986). *Creativity: Genius and other myths*. New York, NY: W H Freeman & Co.

Weiss, C. H. (1972). *Evaluation research: Methods for assessing program effectiveness*. Englewood Cliffs, NJ: Prentice Hall. doi:10.1016/S0886-1633(96)90023-9

Weiss, G., & Wodak, R. (2003). Introduction: Theory, interdisciplinarity and critical discourse analysis. In Weiss, G., & Wodak, R. (Eds.), *Critical Discourse Analysis: Theory and Interdisciplinarity* (pp. 1–34). New York, NY: Palgrave Macmillian.

Wei, T. T., Marthandan, G., Chong, A. Y. L., Ooi, K. B., & Arumugam, S. (2009). What drives Malaysian m-commerce adoption? An empirical analysis. *Industrial Management & Data Systems*, *109*(3), 370–388. doi:10.1108/02635570910939399

Wenger, E. (1999). *Communities of practice: Learning, meaning and identity*. Cambridge, UK: Cambridge University Press.

Wesch, M. (2009). YouTube and you –Experiences of self-awareness in the context collapse of the recording webcam. *Explorations in Media Technology*, *8*(2), 19–34.

Wharton, C., Rieman, J., Lewis, C., & Poison, P. (1994). The cognitive walkthrough method: A practitioner's guide. In Nielsen, J., & Mack, R. (Eds.), *Usability Inspection Method* (pp. 105–141). New York, NY: John Wiley & Sons.

Wiggins, B. (2011). Confronting the dilemma of mixed methods. *Journal of Theoretical and Philosophical Psychology*, *31*(1), 44–60. doi:10.1037/a0022612

Willcocks, L. P., & Lester, S. (1996). *The evaluation and management of information systems investments: Feasibility to routine operations, investing in information systems*. London, UK: Chapman and Hall.

Williamson, J. B., Karp, D. A., & Dalphin, J. R. (1982). *The research craft*. Boston, MA: Little Brown.

Witmer, B. G., Bailey, J. H., Knerr, B. W., & Parsons, K. C. (1996). Virtual spaces and real world places: Transfer of route knowledge. *International Journal of Human-Computer Studies*, *45*, 413–428. doi:10.1006/ijhc.1996.0060

Wittel, A. (2000). Ethnography on the move: From field to internet. *Forum Qualitative Sozial Forschung*, *1*(1).

Wong, C. C., & Hiew, P. L. (2005). Diffusion of mobile entertainment in Malaysia: Drivers and barriers. *Enformatika*, *5*, 263–266.

Wright, K. B. (2005). Researching internet-based populations: Advantages and disadvantages of online survey research, online questionnaire authoring software packages, and web survey services. *Journal of Computer-Mediated Communication*, *10*(3. Retrieved July 20, 2011, from http:// jcmc.indiana.edu/vol10/issue3/wright.html

Wroblewski, L. (2010). *Touch gesture reference guide*. Retrieved from http://www.lukew.com/ff/entry.asp?1071

Wu, P. F. (2009). Opening the black boxes of TAM: Towards a mixed methods approach. *ICIS* 2009. Retrieved from http:// aisle.aisnet.org/icis2009/101

Wu, J. H., & Wang, S. C. (2005). What drives mobile commerce? An empirical evaluation of the revised technology acceptance model. *Information & Management*, *42*(5), 719–729. doi:10.1016/j.im.2004.07.001

Xu, Y. (2010). *The study of online shopping perceived risk under the China's e-business circumstance*. Paper presented at the Information Management, Innovation Management and Industrial Engineering (ICIII), 2010 International Conference. Kunming, China.

Yan, J. (2008, Winter). Social technology as a new medium in the classroom. *New England Journal of Higher Education*.

Yates, J., & Orlikowski, W. J. (1992). Genres of organizational communication: A structurational approach to studying communication and media. *Academy of Management Review*, *17*(1), 299–326.

Yeates, D., & Cadle, J. (2001). *Project management for information systems* (3rd ed.). Harlow, UK: Prentice Hall.

Yin, R. (2003). *Case study research: Design and methods* (3rd ed.). Thousand Oaks, CA: Sage.

Yin, R. K. (1994). *Case study research: Design and methods*. London, UK: Sage Publications.

Yin, R. K. (2003). *Case study research design and method* (3rd ed.). Thousand Oaks, CA: Sage Publisher.

Young, J. (2004). Cultural models and discourses of masculinity: Being a boy in a literacy classroom. In Rogers, R. (Ed.), *An Introduction to Critical Discourse Analysis in Education* (pp. 147–172). Mahwah, NJ: Lawrence Erlbaum Associates.

Young, T. R. (1984). The lonely micro. *Datamation*, *30*(4), 100–114.

Zadrozny, W. (2005). *Making the intangibles visible: How emerging technologies will redefine enterprise dashboards. On-Demand Innovation Services*. New York, NY: IBM Research.

Zaphiris, P., & Constantinou, P. (2007). Using participatory design in the development of a language learning tool. *Interactive Technology and Smart Education*, *4*(2), 79–90. doi:10.1108/17415650780000305

Zhang, D., & Adipat, B. (2005). Challenges, methodologies, and issues in the usability testing of mobile applications. *International Journal of Human-Computer Interaction*, *18*(3), 293–308. doi:10.1207/s15327590ijhc1803_3

Zhong, N., Ma, J. H., Huang, R. H., Liu, J. M., Yao, Y. Y., Zhang, Y. X., & Chen, J. H. (2010). Research challenges and perspectives on wisdom web of things (W2T). *Journal of Supercomputing*. Retrieved from http://kis-lab.com/zhong/open_publish/Research_challenges_and_perspectives_on_Wisdom_Web_of_Things.pdf

Zmud, R. (1996). Editor's comments: On rigor and relevancy. *Management Information Systems Quarterly*, *20*(3).

Zucker, L. G. (1986). Production of trust: Institutional sources of economic structure, 1840 – 1920. *Research in Organizational Behavior*, *8*, 53–111.

About the Contributors

Pedro Isaías is a Professor at the Universidade Aberta (Portuguese Open University) in Lisbon, Portugal, responsible for several courses and Director of the Master degree program in Electronic Commerce and Internet since its start in 2003. He holds a PhD in Information Management (in the speciality of information and decision systems) from the New University of Lisbon. Author of several books, book chapters, papers, and research reports, all in the information systems area, he has headed several conferences and workshops within the mentioned area. He has also been responsible for the scientific coordination of several EU funded research projects. He is also member of the editorial board of several journals and program committee member of several conferences and workshops. At the moment, he conducts research activity related to information systems in general, e-learning, e-commerce, and WWW related areas.

Miguel Baptista Nunes, BSc, MSc, PhD, MBCS, FHEA, FIMIS, is a Senior Lecturer in Information Management at the Information School, University of Sheffield. In terms of research, Miguel is actively involved both in information systems and educational informatics research. He is currently the Head of the Information Systems Research Group in the Department. In terms of information systems, he has done research in the areas of information systems modelling, design and development, project management, and risk management. In terms of educational informatics, Miguel is particularly interested in social constructivist approaches to curriculum and course design, instructional systems design, and Web-based learning environments for active and distance learning. He has published more than 150 refereed articles in both academic conferences and academic journals, published a book on action research for e-learning, and served as programme chair for a number of international conferences.

* * *

Maram Alajamy, BSc, MSc, is a Doctoral Researcher at the Information School, University of Sheffield, UK. Previously, she worked as a Teaching Assistant at the University of Damascus, Syria, for 4 years. Her main research interest is the role of information professionals during information systems strategic planning processes, particularly within higher education institutions.

Fenio Annansingh is a Lecturer in Information and Knowledge Management at Plymouth Business School at the University of Plymouth (UK). Fenio joined Plymouth Business School in 2007. Her main area of research is information systems in two collateral areas, namely information management and knowledge management. Her research interest includes knowledge leakage, information systems risk management, and security concerns in social networking. She received her Masters degree in Information Systems and her PhD in Information Science from the University of Sheffield. Fenio has several publications and professional papers with a number of affiliates and has conducted peer review for a number of leading IS journals.

Roger Blake is an Assistant Professor in the Management Science and Information Systems Department of the College of Management at the University of Massachusetts Boston. His primary areas of interest are in data mining, data and information quality, text analysis, and business intelligence, and his research has appeared journals including *Behaviour and Information Technology*, *Computers in Human Behavior*, and the *ACM Journal of Data and Information Quality*. He holds an S.M. from the Massachusetts Institute of Technology and a Ph.D. from the University of Rhode Island.

Monica Bordegoni is Full Professor at the Faculty of Industrial Design of Politecnico di Milano since 2004. She has published a total of about 200 scientific papers. More than 30 articles have been published in scientific journals and more than 25 in proceedings of Scopus-indexed conferences (including ASME, SPIE, IEEE conferences) on the major research subjects. Her research activities focus on methods and tools for virtual prototyping of products, interaction techniques and multimodal technologies, haptic technologies and interaction, and their application in the engineering and industrial design sectors, emotional engineering. Some of her major scientific contributions of the last twenty years are summarized in the following: development of a method and tool for the semi-automatic generation of graphical user interfaces; development of an algorithm for dynamic gestures recognition based on pattern recognition; development of a methodology based on interactive virtual prototyping for the test and the validation of conceptual product design; development and use of novel haptic interfaces for product design. Specifically, two prototypes have been developed in the context of the EU-IST T'nD and SATIN projects, and the results have been included in CAD tools, developed and distributed by think3 Inc.

Mike Brayshaw is interested in new ways of living, learning, and interacting. His main areas of research interest are currently artificial intelligence and education, artificial intelligence tools and environments, and knowledge, society, and new ways of working. Recently this has focused on the role of the Semantic Web in Internet-based education. Specifically of interest is the learning computer of skills—often learning to program. Historically, the use of such techniques as software visualisation was investigated, but more recently the use of semantic technology has been the chosen vehicle to achieving these goals. He has also worked on large software visualisation systems for tracing and debugging large programs. He holds a PhD in Artificial Intelligence and has been a UK University Lecturer since 1994.

Cheryl Brown is a Lecturer in the Centre for Educational Technology at the University of Cape Town. She recently completed her PhD in Information Systems focusing on what technology means to students and how this influences the way they use technology at university. Her research focuses are digital literacy and identity, particularly amongst first year university students. She also teaches in the center's postgraduate program in Information and Communication Technologies in Education.

José Carrillo is a Professor of the Faculty of Computer Science, Academic Director of Master of Computer Auditing and Computer Security and Director of Master in Business Management Consulting at the Polytechnic University of Madrid. PhD in Computer Science and Forestry Engineer from the Polytechnic University of Madrid, he obtained a Master in Business Administration at IESE CEPADE.

ISACA Academic Advocate since 2001, he is President of the Spanish Association for Governance, Management, and Measurement Systems (AEMES) since 1997 and Member of the Board of ISBSG of Australia since 1999. Member of the Spanish Association of Planning (AESPLAN) and the Spanish Association of Accounting and Business Administration (AECA), he was also Director of Institutional Relations of the Foundation COTEC from 1997 to 2000 and an expert on technological innovation. He has held management positions at IBM, Seinca, Banco Hispano, Confederation Spanish Savings Banks, and Caja Madrid.

Si Chen holds a BSc in Information Engineering at the University of Southeast in China, and a MSc in Information Systems Management at the University of Sheffield, UK. She is currently doing her PhD at the Information School at Sheffield. Her PhD research focuses on the evaluation of Information Systems (IS) in Chinese large group enterprises. Her research interests include IS evaluation, IS integration, and strategic alignment of IS.

Regina Connolly is a Senior Lecturer in Information Systems at Dublin City University Business School, Dublin, Ireland, and is Programme Director of the MSC in Electronic Commerce. In her undergraduate degree she received the Kellogg Award for outstanding dissertation and her MSc degree was awarded with distinction. She was conferred with a PhD in Information Systems from Trinity College Dublin. Her research interests include eCommerce trust and privacy issues, website service quality evaluation, eGovernment, IT value and evaluation in the public sector, and strategic information systems. She has published extensively in leading academic journals and is editor of the *Journal of Internet Commerce*. Dr. Connolly has served on the expert eCommerce advisory group for Dublin Chamber of Commerce, which has advised national government on eCommerce strategic planning.

Gabriel J. Costello is a Lecturer in Engineering at the Galway-Mayo Institute of Technology (www.gmit.ie). Prior to this, he worked for twenty years in the telecommunications industry where he held engineering, new product introduction, and product line management positions. He completed a PhD in Management Information Systems at the J.E. Cairnes School of Business and Economics, National University of Ireland, Galway, in the area of Information Systems Innovation. Research interests include: innovative enterprises and systems, speech-enabled self-service IT systems, and using IS to support RUE (Rational Use of Energy) in organisations. His publications include: *Journal of Information Technology, International Conference on Information Systems (ICIS), International Federation for Information Processing (IFIP), European Conference on Information Systems (ECIS),* and *European Academy of Management Conference (EURAM).*

Mario Covarrubias graduated in Mechanical Engineering at Instituto Tecnologico de Pachuca (México) in 1996, and received a Master degree in Manufacturing Systems from ITESM-MTY (México) in 2002. He obtained the status of Full Professor in the Mechanical Department at ITESM-CCM on 2003. From 2004 to 2006, he was the Coordinator of the CECCC (Centro de Entrenamiento en Sistemas CAD-CAE-CAM). From 2007 to 2010, he was a PhD student at the Virtual Prototypes and Real Product Course of Politecnico di Milano. Since June 2010, he has been an Assistant Professor at the Department of Mechanical Engineering of Politecnico di Milano. His research interests are in virtual prototyping, visualization systems, and haptic interfaces.

Umberto Cugini is Full Professor at the Faculty of Industrial Engineering of Politecnico di Milano. In 1979, Umberto Cugini founded the KAEMaRT research group, which carries out research activities in several areas related to product development methodologies and tools. He has published more than 250 journal papers and refereed international conference papers. He has also acted as reviewer for several international journals and conferences. Over the years, his primary research interests have been: computer graphics, geometric and feature-based modelling, physics-based modelling and simulation, robotics and automation systems, knowledge-based systems for product development, engineering knowledge management, business process reengineering, human computer interaction, multimodal and haptic-based interaction, virtual/augmented reality, virtual prototyping, emotional engineering, topological optimization, and systematic innovation.

Bruno Delgado is a Professor of Software Engineering and Project Management in the Department of Software Engineering and Professor of Corporate Governance and IT Governance in the Department of Information Systems at Faculty Engineering Universidad ORT Uruguay. Also he teaches Software Production Process in the Department of Innovation and Technology Management at the Faculty of Social Sciences and Administration. He is Systems Engineer and obtained a Master in Banking and Finance (MBF) at I.E.B. – Complutense University of Madrid (UCM) in 2007. Bruno is PhD (c) in Computer Science Engineering at the Polytechnic University of Madrid (UPM), where he is developing his doctoral thesis. He has worked to date in banking and as IT consultant in the Ministry of Economy and Finance and Ministry of Foreign Affairs, participating in various projects in consumer protection, development of international commerce, and free trade zones.

Jan Derboven graduated as a Master in Germanic Languages and Cultural Studies. Before joining the Centre for User Experience Research (CUO) at KU Leuven in 2007, he worked as a usability consultant and as a technical writer. As a Senior Researcher at CUO, he is currently pursuing his PhD and working on a range of projects, mainly about innovative interaction design and serious gaming. His research interests include the design of innovative interactions and the integration of insights from critical theory in HCI practices.

Brian Donnellan is Professor of Information Systems Innovation at the National University of Ireland Maynooth (www.nuim.ie) and Academic Director of the Innovation Value Institute (www.ivi.ie). Prior to joining NUI Maynooth, Prof. Donnellan was a faculty member in the National University of Ireland, Galway. He has spent 20 years working in the ICT industry where he was responsible for the provision of IS to support computer-aided design and new product development. He is an Expert Evaluator for the European Commission and has been guest and associate editor of several leading IS journals including *Journal of IT*, *Journal of Strategic Information Systems*, and *MIS Quarterly*.

Derek Flood is a Post Doctoral Research Fellow with the Regulated Software Research Group at Dundalk Institute of Technology (DkIT). After receiving his PhD in the area of Natural Language Processing for Spreadsheet Information Retrieval, he took up a Post Doctoral Research Assistant position at Oxford Brookes University where he worked on the PACMAD project. This project examined the usability issues associated with mobile applications. Through this work, Derek has published at a number of international conferences including the IADIS International Conference on Interfaces and Human Computer Interaction, the Psychology of Programming Interest Groups Annual Conference, and the European Spreadsheet Risk Interest Groups' Annual Conference, where he has been a regular contributor for the last 5 years. His current research interests include software process improvement for medical devices, spreadsheet risk, usability of mobile applications, and the psychological impact of mobile applications on users.

Yoshiaki Fukazawa received the B.E., M.E., and D.E. degrees in Electrical Engineering from Waseda University, Tokyo, Japan, in 1976, 1978, and 1986, respectively. He joined Department of Computer Science of Sagami Institute of Technology as a Lecturer in 1983 and Department of Electrical Engineering of Waseda University as an Associate Professor in 1987. He is now a Professor of Department of Information and Computer Science, Waseda University. His research interests are various aspects of software engineering, especially software reuse, user interface, and embedded software. He is a member of IPSJ, IEICE Japan, JSSST, ACM, and IEEE.

Helena Garbarino is a Professor at the Universidad ORT Uruguay. She is PhD (c) in Computer Science Engineering at the Polytechnic University of Madrid (UPM) and MsC from Universidad ORT Uruguay in Information Systems and Software Engineering. She has a degree in Information Systems from Oxford University. She is the Director of Studies of a Management Information degree at the Faculty of Engineering in Universidad ORT Uruguay. She teaches at the Computer Theory Institute, and her research interests include IT governance, knowledge management, and IT value management. Helena has several scholarly publications and participates as a reviewer too. She is a consultant in IT government implementation processes, especially in SMEs.

Elia Gatti graduated cum laude in Cognitive Neuroscience at Vita- Salute San Raffaele University with a thesis on "Force Perception at Rest and during Movements," in 2010. Currently, he is a PhD student working within the KAEMaRT Research Unit at Politecnico di Milano and collaborates with the LAPCO laboratory (action, perception, and cognition laboratory) at Vita-Salute San-Raffaele University. His research interests focus on the study of human-multimodal integration and cognition in real and VR environments, through psychophysical and statistical methods.

Roelien Goede is a Senior Lecturer at the Vaal Triangle campus of the North-West University. She obtained a PhD in Information Technology in 2005 from the University of Pretoria. Her research focus is in systems thinking, data warehousing, and education. She has an interest in philosophy of science and often speaks on research methodology.

Steven Gordon is Professor of Information Technology Management at Babson College. His research focuses on how information systems and technology can improve the process of corporate innovation. He is the former Editor-in-Chief of the *Journal of Information Technology Cases and Applications Research* and serves on the Advisory Board of the *International Journal of e-Politics*. He has been published widely in the academic press and is the editor of three text books and two research anthologies. Before coming to Babson, Dr. Gordon consulted to the airline industry at Simat, Helliesen, and Eichner, Inc. He also founded and served as President of Beta Principles, Inc., a developer and marketer of accounting software and reseller of computer hardware. He received his PhD degree from the Massachusetts Institute of Technology.

Rachel Harrison is Professor of Computer Science and Director of Research in the Department of Computing and Communication Technologies at Oxford Brookes University. Previously, she was Professor and Head of Computer Science at the University of Reading. Prof. Harrison's research includes work on software metrics, systems evolution, requirements engineering, and search-based software engineering. She has published over 120 refereed publications and has consulted widely with industry, working with organizations such as IBM, the DERA, Philips Research Labs, Praxis Critical Systems, and The Open Group. She is Editor-in-Chief of the *Software Quality Journal*, published by Springer.

Mike Hart is an Emeritus Professor in the Department of Information Systems at the University of Cape Town. He is a National Research Foundation-rated researcher and his major research themes have involved the adoption and strategic use of information technologies and systems in organizations; management and decision-support use of I.S. He is currently researching "Customer Service Delivery in Contact Centres," with an emphasis on effective use of customer information systems and integration of information across channels.

Janet Holland completed a Ph.D. in Teaching and Leadership, Instructional Design and Technology, with a minor in Communications from the University of Kansas in Lawrence, Kansas. Dr. Holland currently serves as an Associate Professor at Emporia State University in Emporia, Kansas, teaching pre-service teachers and master degree students in Instructional Design and Technology. Research interests include improving curriculum pedagogy issues including instructional design, online learning, mobile learning, service learning, affective learning communities, peer mentoring, and the globalization of instruction. As an instructional designer, new technologies are continually examined in an effort to inspire innovative teaching and learning practices.

Dusti Howell, Ph.D., is a Professor of Instructional Design and Technology at Emporia State University in Kansas. He earned a Ph.D. major in Educational Communications and Technology and a Ph.D. minor in Educational Psychology from the University of Wisconsin. He has written over a dozen books including *Using PowerPoint in the Classroom* by Corwin Press and *Digital Storytelling: Creating an e-Story* by Linworth Publishing. His expertise includes innovations and research in digital storytelling, learning strategies, and interventions for the negative impacts technology can have on children. Dr. Howell has taught each grade level from first grade through graduate school. He is a frequent speaker at state, national, and international conferences, and conducts workshops and professional development seminars.

Hannakaisa Isomäki received her PhD in Information Systems at the University of Tampere, Finland, in 2002. She has worked as Academy of Finland's Project Researcher at the University of Jyväskylä, Finland (1996 – 2001), Assistant Professor in the Department of Computer Science and Information Systems at the University of Jyväskylä (2001 – 2003), Professor of Applied Information Technology in the Department of Methodology Sciences at the University of Lapland, Finland (2003 – 2006), and Research Director in the Faculty of Information Technology at the University of Jyväskylä (2006 –). Her field of expertise includes various topics of human-centered information systems. Her current research interests concern information security culture, user-centered software development, and privacy in human-computer interaction. She has over 100 scientific publications and has been involved in many national and international scientific evaluation tasks.

Tomayess Issa is a Senior Lecturer, Postgraduate Course Leader, and Postgraduate Online Coordinator at the School of Information Systems at Curtin University, Australia. She completed her Doctoral research in Web Development and Human Factors, and she is a member of an international conference program committee. Currently, she conducts research locally and globally in information systems, HCI, usability, Internet, sustainability and green IT, social network, teaching, and learning.

Hajime Iwata received the B.E., M.E., and D.E. degrees in Information and Computer Science from Waseda University, Tokyo, Japan, in 2002, 2004, and 2008, respectively. He joined Media Network Center of Waseda University as a Research Assistant in 2005. He is now an Assistant Professor of Department of Network and Communication of Kanagawa Institute of Technology from 2008. His research interests include support for learning operating method of software. His main research theme is automatic generation of learning support system of operation method. He is a member of IPSJ Japan and ACM.

Sergio Lujan Mora is Lecturer of the Department of Languages and Systems, University of Alicante. He earned his Ph.D. in Computer Engineering at the University of Alicante (Spain) in 2005. His main research interests include Web accessibility and usability, Web applications and business intelligence, data warehouse design, analysis, and object-oriented design with UML, MDA, etc. He has published over 80 refereed papers in international conferences and journals. Among them, the Congress ER, UML, and DOLAP. Among the international journals indexed in the JCR include DKE, JCS, and JDBM. He has also published several books related to programming and Web design. He has participated in several research projects funded by both national and international organizations, such as the Spanish Ministry of Science and Technology and the European Commission.

Clare Martin is a Principal Lecturer in the Department of Computing and Communication Technologies at Oxford Brookes University. Previously, she was a Lecturer at the University of Buckingham, and prior to that a Research Student at Oxford University. Her background is in formal methods, and she has a particular interest in the use of mathematics to reason about the semantics and general algebraic properties of programming and specification languages. Her other research activity is in the area of human computer interaction, especially in relation to mobile applications for use in healthcare.

Jorge Tiago Martins, BSc, PgC LTHE, FHEA, is a Doctoral Researcher at the Information School, University of Sheffield, UK. He joined the Information Systems Research Group in 2008 and is the recipient of a research grant by the National Portuguese Foundation for Science and Technology (FCT). He is currently involved in a research project that looks at academics' perceptions and attitudinal alignment regarding the embedding and provision of e-learning in higher education institutions. He is the author of around 10 refereed articles published in academic conferences and academic journals. His research interests include e-learning, organisational trust, change managemen,t and research methods in information systems research.

Firas Masri received his Advanced Studies Diploma from the University of Alicante. He is now continuing PhD studies at the Department of Languages and Information Systems at the University of Alicante, researching about Web accessibility evaluation methods. Firas Masri has contributed significantly to the good understanding of combined Web accessibility evaluations methods which frequently need to be applied in the certification duties of websites compliance with accessibility guidelines. He has published several articles about Web accessibility evaluation methods at some international congresses. His publications reflect his research interests in this field and his social commitment with disabled people. Currently, he is also working as a Researcher at Sistrade Software Consulting Company, applying the scientific experience and knowledge to the field of business and encouraging private companies to implement Web content accessibility guidelines.

Paula Miranda is an Assistant Professor in the Department of Informatics and Systems at School of Technology at the Polytechnic Institute of Setubal, Portugal. She holds a Master degree from University of Coimbra with the dissertation in the area of Decision Support Systems (DSSs). Currently, she is working in new research area related with the Web 3.0 and learning. At the present, she is a PhD student in Information Technology and Science at the ISCTE Lisbon University Institute. She has authored several research papers in the aforementioned area of DSSs. Her areas of interest include information systems in general, more specifically database systems, systems analysis and design, decision support systems, and programming environments. Over the last years her research interests have been concentrated in the information society.

Reza Mojtahed, BSc, MSc, is PhD Researcher in the Information School at the University of Sheffield. He was educated at London School of Economics where he obtained a BSc (Hons) and then he accomplished his MSc (Hons) in the field of Information System and Management at the University of Sheffield. He is a member of the Information Systems Research Group at the Sheffield iSchool since 2010. He applied the TAM model in a number of his previous research projects (e.g. on electronic and mobile banking areas). His research interests include e-government, provision of new services, G2C, IT investment, IT improvement, and decision making.

Maribel Montero Perez is a Ph.D. candidate in the Interdisciplinary Research Group ITEC-IBBT, KU Leuven Kulak. Her research focuses on the use and effectiveness of input enhancement for foreign language listening comprehension and vocabulary learning. She has been working on different CALL projects and is currently involved in the iRead+ project (funded by IBBT as an ICON-project) which consists of creating an intelligent reading companion for language learners and struggling readers.

Julius T. Nganji is currently in the final phase of his Ph.D. in Computer Science at the University of Hull where his thesis focuses on personalising learning resources for disabled students. He also works part-time during term time as a Mentor with the Disability Services at the University of Hull and holds an M.Sc in Website Design and Development from the same university. His research interests include e-learning on the Semantic Web for disabled students, ICT accessibility and usability for disabled people, adoption and diffusion of ICTs by developing countries, and telemedicine. Since 2002, he has worked in Cameroon and the UK in various roles in the information and communication technology sector.

Nor Mardziah Osman, BA, MSc, is a PhD researcher at the Information School, The University of Sheffield, United Kingdom. She had been working with Royal Malaysian Customs Department (RMCD) for almost 20 years. She is the recipient of scholarship by the Public Services Department of Malaysia, and she joined the Information Systems Research Group in October 2010. Her current PhD research focuses on the evaluation of Custom Information Systems by taking the RMCD system as a case study.

Guo Chao Peng, BSc, CiLT, PhD, is a Lecturer in Information Systems in the Information School at the University of Sheffield, UK. He holds a BSc in Information Management (1st Class Honours) and a PhD in Information Systems. Dr. Peng has published regularly in the IS field in high-quality international journals, such as *Industrial Management and Data Systems* (IMDS), *Aslib Proceedings*, *Journal of Manufacturing and Technology Management* (JMTM), and *Journal of Enterprise Information Management* (JEIM), as well as in books and referred international conferences. He also conducted peer review of submissions to more than 15 leading IS journals and international conferences. He is the co-founder and co-chair of the IADIS Information Systems Post-Implementation and Change Management (ISPCM) Conference.

Sara Pífano is a PhD candidate devoted to the study of Web 2.0's Critical Success Factors at the Universidade Aberta (Portuguese Open University) in Portugal. She holds a degree in International Relations from the University of Minho (Braga, Portugal) and a MA in Refugee Studies from City University (London, United Kingdom). She's held the position of Research Assistant at the Information Society Research Lab, in Lisbon, Portugal, and is currently working as a freelance researcher and writer. Besides her focus on the broad field of the information society, her main research interests and recent publications place an emphasis in Web 2.0 and its repercussions in the several domains of society, particularly education, business, and politics.

G. Shankaranarayanan received the PhD degree in Management Information Systems from the University of Arizona Eller School of Management (1998). His research interests include data modeling and design, database schema evolution, metadata modeling and management, data quality management, and the economics of data management. His research has appeared in several journals including the *Journal of Database Management, Decision Support Systems, Communications of the ACM, DATABASE for Advances in Information Systems, IEEE Transactions on Data and Knowledge Engineering*, and the *ACM Journal for Data and Information Quality*. He serves as the Area Editor (NA) for the *International Journal of Information Quality* and as an Associate Editor for the *ACM Journal for Data and Information Quality*.

Johanna Silvennoinen received her Master's degree in Art at Aalto University's School of Art, Design, and Architecture in Helsinki, Finland, in 2011. Currently, she is a PhD Candidate of Cognitive Science in the University of Jyväskylä. Her research experience concerns the study of online communities applying online ethnography. At present she works as a Research Assistant in the Department of Mathematical Information Technology at the University of Jyväskylä, Finland. Her main research interests are online ethnography and visual user interface design. The topic of her PhD thesis is "Visual Variables in User Interfaces: Enhancing Design for Positive User Experience."

Junko Shirogane received the B.E., M.E., and D.E. degrees in Information and Computer Science from Waseda University, Tokyo, Japan, in 1997, 1999, and 2002, respectively. She joined Media Network Center of Waseda University as a Research Assistant in 2000 and Department of Communication of Tokyo Woman's Christian University as a Lecturer in 2003. She is now an Associate Professor of School of Arts and Sciences, Tokyo Woman's Christian University. Her research interests include support tools for development of software with graphical user interfaces. She is a member of IPSJ, IEICE Japan, JSSST, HIS, IEEE, and ACM.

Nicky Sulmon is a Human-Computer Interaction Researcher at the Centre for User Experience Research (CUO) at the University of Leuven and IBBT Future Health Department. Nicky obtained a Bachelor's degree in Applied Informatics before graduating as a Master in Visual Arts at the Media Arts Design Faculty in Genk. He also holds a Postgraduate degree in Usability Design. Nicky is currently involved in different research projects on media architecture, e-health, and serious games at school.

Estelle Taylor is a Senior Lecturer at the Potchefstroom Campus of the North-West University. She obtained a PhD in Computer Science in 2008 from the North-West University. Her research focus is in Education of Computer Science and Information Systems.

Christoffel van Aardt is a Technical Specialist at Vaal University of Technology. He obtained a Master's degree in Computer Science in 2011 from the North-West University. His research focus is on the use of technology in education.

Yuichiro Yashita received the B.E. and M.E. degrees in Computer Science from Waseda University, Tokyo, Japan, in 2009 and 2011, respectively. His research interests include support for usability evaluations of software with graphical user interfaces. His main research theme is development of method to automatically evaluate graphical user interfaces of software using operation histories by end users.

Bieke Zaman is a Senior Researcher at the Centre for User Experience Research (CUO) and IBBT Future Health Department. She coordinates research on games, user experience, and user-centred design with children. Bieke holds a Doctoral degree in Social Sciences at the K.U. Leuven, a Master degree in Communication Science, as well as a Postgraduate degree in Usability Design and Web Development. Bieke has organized several workshops and tutorials at international conferences, and teaches the courses Human-Computer Interaction, Usability Design, and Multimedia Production for Master students of the Faculties of Social Sciences, Economy, Information Management, and Computer Science. Bieke is involved in various nationally funded as well as EU-funded projects. Bieke is a reviewer for several conferences. She is part of the Interaction Design and Children program committee, the Fun and Games steering committee, and was Program Chair of the Fun and Games 2010 edition.

Lihong Zhou, BSc, MSc, PhD, is an Associate Professor at the School of Information Management in Wuhan University, China. He obtained his MSc and PhD in Information Studies from the Information School, University of Sheffield, UK. His research interests are in healthcare knowledge sharing, risk mitigation and management, and information systems project management.

Index

E

F

G

H

I

J

K

L

M

N

O

P

Q

R

S

T

U

V

W